T0206405

CAMBRIDGE LIBRARY COLLECTION

Books of enduring scholarly value

Darwin

Two hundred years after his birth and 150 years after the publication of 'On the Origin of Species', Charles Darwin and his theories are still the focus of worldwide attention. This series offers not only works by Darwin, but also the writings of his mentors in Cambridge and elsewhere, and a survey of the impassioned scientific, philosophical and theological debates sparked by his 'dangerous idea'.

Principles of Mental Physiology

William Carpenter (1813–85) was trained as a doctor; he was apprenticed to an eye surgeon, and later attended University College London and the University of Edinburgh, obtaining his M.D. in 1839. Rather than practising medicine, he became a teacher, specialising in neurology, and it was his work as a zoologist on marine invertebrates that brought him wide scientific recognition. His Principles of Mental Physiology, published in 1874, developed the ideas he had first expounded in the 1850s, and expounds the arguments for and against the two models of psychology then current – automatism, which assumed that the mind operates under the control of the physiology of the body for all human activity, and free will, 'an independent Power, controlling and directing that activity'. Drawing on animal as well as human examples, his arguments, especially on the acquisition of mental traits in the individual, are much influenced by Darwin.

Cambridge University Press has long been a pioneer in the reissuing of out-of-print titles from its own backlist, producing digital reprints of books that are still sought after by scholars and students but could not be reprinted economically using traditional technology. The Cambridge Library Collection extends this activity to a wider range of books which are still of importance to researchers and professionals, either for the source material they contain, or as landmarks in the history of their academic discipline.

Drawing from the world-renowned collections in the Cambridge University Library, and guided by the advice of experts in each subject area, Cambridge University Press is using state-of-the-art scanning machines in its own Printing House to capture the content of each book selected for inclusion. The files are processed to give a consistently clear, crisp image, and the books finished to the high quality standard for which the Press is recognised around the world. The latest print-on-demand technology ensures that the books will remain available indefinitely, and that orders for single or multiple copies can quickly be supplied.

The Cambridge Library Collection will bring back to life books of enduring scholarly value across a wide range of disciplines in the humanities and social sciences and in science and technology.

Principles of Mental Physiology

With their Applications to the Training and Discipline of the Mind, and the Study of its Morbid Conditions

William Benjamin Carpenter

CAMBRIDGE UNIVERSITY PRESS

Cambridge New York Melbourne Madrid Cape Town Singapore São Paolo Delhi

Published in the United States of America by Cambridge University Press, New York

www.cambridge.org
Information on this title: www.cambridge.org/9781108005289

© in this compilation Cambridge University Press 2009

This edition first published 1874
This digitally printed version 2009

ISBN 978-1-108-00528-9

This book reproduces the text of the original edition. The content and language reflect the beliefs, practices and terminology of their time, and have not been updated.

PRINCIPLES OF

MENTAL PHYSIOLOGY.

PRINCIPLES

OF

MENTAL PHYSIOLOGY,

WITH THEIR APPLICATIONS TO

THE TRAINING AND DISCIPLINE OF THE MIND,

AND

THE STUDY OF ITS MORBID CONDITIONS.

BY

WILLIAM B. CARPENTER, M.D., LL.D.,
F.R.S., F.L.S., F.G.S.,

REGISTRAR OF THE UNIVERSITY OF LONDON;
CORRESPONDING MEMBER OF THE INSTITUTE OF FRANCE,
AND OF THE AMERICAN PHILOSOPHICAL SOCIETY;
ETC. ETC.

LONDON:
HENRY S. KING & CO.,
65, CORNHILL, AND 12, PATERNOSTER ROW.
1874.

TO THE MEMORY OF

THE LATE

SIR HENRY HOLLAND, BART., D.C.L., F.R.S.,

PRESIDENT OF THE ROYAL INSTITUTION,

ETC. ETC.

This Volume is Dedicated,

AS AN EXPRESSION OF THE AUTHOR'S GRATITUDE

FOR BENEFITS DERIVED FROM

HIS SCIENTIFIC WRITINGS,

HIS WISE COUNSELS,

AND HIS CORDIAL ENCOURAGEMENT,

DURING A FRIENDSHIP OF THIRTY YEARS

PREFACE.

The following Treatise is an expansion of the Outline of Psychology contained in the Fourth and Fifth Editions of my "Principles of Human Physiology" (1852 and 1855), but omitted from the later editions of that work, to make room for new matter more strictly Physiological. The appreciation of that Outline expressed at the time by several friends to whose opinions I attached great value, made me contemplate the separate reproduction of it, at some future date, in an enlarged form: but the fulfilment of that intention has been delayed, in the first instance, by the pressure of Official duties; and, since this has been lightened, by the diversion of all the time and thought I could spare into an entirely different line of Scientific investigation. That investigation, however, having been taken in hand by Her Majesty's Government for systematic prosecution by the "Challenger" Expedition, I found myself free to entertain a proposal made to me by the projectors of the "International Scientific Series," to republish my Outline in an enlarged form, as one of their Popular Treatises.

Not having seen reason to make any important change in

my own Psychological views since I first put them forward, but, on the contrary, having found them confirmed and extended by the experience and reflection of twenty years, I set myself to revise my former exposition of them, with the idea of simply introducing such illustrations as might lead to the more ready apprehension of the principles I aimed to enforce, and of filling-up such deficiencies as it might seem most desirable to supply. But, as I proceeded with the work, I found it grow under my hands; and it has at last so far exceeded the limits originally contemplated, as to become unsuitable to the Series for which it had been designed, from which it has been accordingly withdrawn, for issue as an independent Treatise.

I now send it forth as a contribution to that Science of Human Nature, which has yet (as it seems to me) to be built-up on a much broader basis than any Philosopher has hitherto taken as his foundation. To the character of a System of Psychology, this treatise makes no pretension whatever; being simply designed to supplement existing Systems of Physiology and Metaphysics, by dealing with a group of subjects, which, occupying the border-ground between the two, have been almost entirely neglected in both. Hence, in treating of Sensation, I have not entered into those details on the Physiology of the Senses which are readily accessible elsewhere; but have especially applied myself to the elucidation of the share which the Mind has, not only in the *interpretation* of Sense-impressions, but in the *production* of Sensorial states not less real to the Ego who experiences them than are those called-forth by external objects,

—a topic of the greatest importance in reference to the value of all Testimony given under a Mental preconception. And, in like manner, I have done no more than enumerate a large proportion of those principal modes of Mental activity, which are commonly designated as Intellectual Faculties, Propensities, and Emotions; in order that I might have space to bring into clear view that distinction between their *automatic* and their *volitional* operation, which has long appeared to me the only sound basis, on the one hand, for Education and Self-discipline, and, on the other, for that Scientific study of the various forms of *abnormal* Mental activity, which, rightly cultivated, is probably the most promising field of Psychological inquiry.

Although the doctrine of *automatic* Mental activity is here presented in a Physiological form, it may be fully accepted, and turned to practical account, by such as repudiate *in toto* the idea that Thought and Feeling can be regarded as the expression of Brain-change. This is evident from the entire accordance which exists between Sir William Hamilton and Mr. John S. Mill as to the phenomena which I have grouped together under the term "Unconscious Cerebration;" though the former states them in terms of Metaphysics, and the latter in terms of Physiology.

It will, I doubt not, be considered by many, that there is a palpable inconsistency between the two fundamental doctrines which are here upheld;—that of the dependence of the Automatic activity of the Mind upon conditions which bring it within the *nexus* of Physical Causation; and that of the existence of an independent Power, controlling and directing that activity, which we call Will*. I can only say that both are equally true to my own consciousness; as I believe they are to the common consciousness of Mankind. I cannot regard myself, either Intellectually or Morally, as a mere puppet, pulled by suggesting-strings; any more than I can *dis*regard that vast body of Physiological evidence, which proves the direct and immediate relation between Mental and Corporeal agency. The same difficulty (if it be a difficulty) is experienced by some of the greatest thinkers of the day. Thus even John S. Mill, the most powerful advocate of Automatism, found himself brought by his own Mental experiences to what is virtually an acceptance of the independence of the Will:—"I saw," he says (*Autobiography*, p. 169), "that though our "character is formed by circumstances, our own desires can "do much to shape those circumstances; and that what is "really inspiriting and ennobling in the doctrine of free "will, is the conviction that *we have real power over the* "*formation of our own character;* that our will, by influencing

* The two doctrines seem to me to have been more clearly presented by Hartley, than by any other Writer:—"By the Mechanism of human Actions, I mean that each action results from the previous circumstances of Body and Mind, in the same manner, and with the same certainty, as other effects do from their mechanical causes; so that a person cannot do indifferently either the action A, or its contrary *a*, while the previous circumstances are the same; but is under an absolute Necessity of doing one of them, and that only. Agreeably to this I suppose, that by Free-will is meant a Power of doing either the action A, or its contrary *a*, while the previous circumstances remain the same."—(*On Man*, Conclusion to Part I.) This simple and definite issue has been greatly obscured by the attempts which have been made on both sides to evade it; but it is one which had much better be faced openly and directly. Are we, or are we not, altogether what "circumstances" make us?

"some of our circumstances, can modify our future habits "or capacities of willing." On the other hand, Archbishop Manning and Mr. Martineau, who may be considered as typical Metaphysicians, and who hold the Freedom of the Will as a fundamental article of their Religious and Ethical creeds, seem not less satisfied than myself, that our succession of Thoughts and Feelings is in great degree determined by antecedent conditions, which are intimately related to those of our Physiological Mechanism.

That these two Agencies must both be accepted as fundamental facts of Man's composite nature, which can only be viewed aright in their mutual relation, is, I believe, a conclusion towards which there is now a general convergence amongst intelligent thinkers, whose minds are not trammelled by System, or obscured by the dust that has been so continually raised in philosophical discussion. A better type of such thinkers could be scarcely found than the late Charles Buxton; who, looking at the subject from a point of view very different from mine, expressed himself in language almost identical with that which I had used (pp. 2, 3), as to the necessity of no longer attempting to keep apart in our Scientific investigations that which Nature has so inseparably united:—"Irresistible, undeniable facts demonstrate "that man is not a den wherein two enemies are chained "together; but *one being*—that soul and body are one—one "and indivisible. We had better face this great fact. 'Tis "no good to blink it. Our knowledge of Physiology has "come to a point where the old idea of Man's consti-"tution must be thrown aside. To struggle against the

"overwhelming force of science, under the notion of shield-
"ing religion, is mere folly."—(*Notes of Thought*, p. 266).

These well-considered conclusions of a deeply-religious mind may be specially commended to the consideration of those, who may be disposed to condemn without examination anything that savours of a "Materialism" which they have been accustomed to regard as philosophically absurd and morally detestable. And those who assume that a Physiological Psychology strikes at the root of Morals and Religion, may be fearlessly asked to show in what a system which leaves the Will of Man free to make the best use he can of the Intellectual and Moral capacities with which his Bodily Organism has been endowed by his Creator, and which gives him the strongest and noblest motives both for Self-discipline and for Philanthropic exertion, is unworthy of the nature and destiny of the being whose creation "in the image of God" can have no higher meaning than his capacity for *infinite progress*.

It would be ungrateful were I not to take this opportunity of renewing the expression of the special obligations I owe, in the original construction of my "fabric of thought" on this great subject, to the writings and conversation of my valued friends the late Sir Henry Holland, Professor Laycock, Dr. J. D. Morell, and Dr. Noble.—To the first of these it had been my intention to dedicate this Treatise, the title of which he had kindly permitted me to borrow from one of his own; and I have gladly, therefore, complied with the wish of his Family, that I should dedicate it to his

Memory. No one can know the wonderful *suggestiveness* of Sir H. Holland's scientific writings, who has not had the occasion (which has often occurred to myself) to trace back to them some of the best of the thoughts which he had honestly believed to be his own. This I have found particularly the case in regard to the subject of Attention; the fundamental importance of which in relation to the Will, I first learned from him to appreciate.

Other obligations to later writers on Psychology are noted in their proper places: but all the general doctrines of importance herein set forth, will be found, I believe, either explicitly stated or clearly indicated in my original Outline; and in their fuller development I have preferred to draw either upon my own mental experience and that of others, or upon that very large group of *abnormal* phenomena, which has not yet (so far as I am aware) been discussed by any professed Psychologist, but of which the careful study seems to me absolutely essential to a due understanding of the relation of the Will to the Automatic activity of the Mind, and of both to the Physiological Mechanism. Some apology may be thought due for the introduction of so many old and familiar illustrations; and especially for such numerous citations from the well-known work of Dr. Abercrombie on "The Intellectual Powers." But I have not hesitated to bring in old stories whenever they were specially to the point; and I believe that in many instances I have been able to give them an entirely new application.

In conclusion I venture to ask for a fair measure of indulgence for such errors and shortcomings (especially on the Psychological side) as will doubtless be discovered in this Treatise; on the ground that it has been impossible for me to devote to it that *continuous* thought, which is especially required for the systematic prosecution of any inquiry of this kind, and for the exposition of its results. Had I kept the work back longer in the hope of a more favourable opportunity for its production, I might have altogether lost, with the advance of years, the power of producing it. Such as it is, I offer it, on the one hand, to those who are interested in the progress of Psychological Science, and are disposed to widen its area of investigation; and, on the other, to those who desire a definite basis and aim in the Intellectual and Moral training either of others or themselves;—with the hope that I may at any rate stimulate some other investigator to follow-out the path I have tried to open, who shall bring to the Scientific interpretation of Physiological phenomena a knowledge of Metaphysics to which I can lay no claim, and a Mind better trained in abstract thought.

CONTENTS.

BOOK I.

GENERAL PHYSIOLOGY.

CHAPTER I.

OF THE GENERAL RELATIONS BETWEEN MIND AND BODY.

CHAPTER II.

OF THE NERVOUS SYSTEM AND ITS FUNCTIONS.

CHAPTER III.

Of Attention.

CHAPTER IV.

Of Sensation.

CHAPTER V.

Of Perception and Instinct.

CHAPTER VI.

Of Ideation and Ideo-Motor Action.

CHAPTER VII.

Of the Emotions.

CHAPTER VIII.

Of Habit.

CHAPTER IX.

Of the Will.

BOOK II.

SPECIAL PHYSIOLOGY.

CHAPTER X.

OF MEMORY.

CHAPTER XI.

OF COMMON SENSE.

CHAPTER XII.

OF IMAGINATION.

CHAPTER XIII.

Of Unconscious Cerebration.

CHAPTER XIV.

Of Reverie and Abstraction :—Electro-Biology.

CHAPTER XV.

Of Sleep, Dreaming, and Somnambulism.

CHAPTER XVI.

Of Mesmerism and Spiritualism.

CHAPTER XVII.

Of Intoxication and Delirium.

CHAPTER XVIII.

Of Insanity.

CHAPTER XIX.

Influence of Mental States on the Organic Functions.

CHAPTER XX.

Of Mind and Will in Nature.

APPENDIX.

Dr. Ferrier's Experimental Researches on the Brain.

BOOK I.

GENERAL PHYSIOLOGY.

CHAPTER I.

OF THE GENERAL RELATIONS BETWEEN MIND AND BODY.

1. THE Conscious Life of every individual Man essentially consists in an action and reaction between his Mind and all that is outside it,—the *Ego* and the *Non-Ego*. But this action and re-action cannot take place, in his present stage of existence, without the intervention of a Material Instrument; whose function it is to bridge over the *hiatus* between the individual Consciousness and the External World, and thus to bring them into mutual communication. And it is the object of this Treatise to take up and extend the inquiry into the action of Body upon Mind, as well as of Mind upon Body, on the basis of our existing knowledge; so as to elucidate, as far as may be at present possible, the working of that Physiological Mechanism which takes a most important share in our Psychical operations; and thus to distinguish what may be called the *automatic* activity of the Mind, from that which is under *volitional* direction and control.—This inquiry has been started more than once, but has not until recently been systematically prosecuted. "There is one view of the connection between Mind and Matter," says Prof. Dugald Stewart, "which is perfectly agreeable to the just rules of philosophy. The object of this is, to ascertain the laws which regulate their union, without attempt-

ing to explain in what manner they are united. Lord Bacon was, I believe, the first who gave a distinct idea of this kind of speculation; and I do not know that much progress has yet been made in it." Considering his own province, however, to be purely Metaphysical, the eminent Professor just quoted gave no further attention to the subject; and those who have more recently taken it up, having for the most part been Physiologists and Physicians, rather than professed Psychologists, have been too often looked upon by the latter as opponents rather than as allies. But so long as either the Mental or the Bodily part of Man's nature is studied *to the exclusion* of the other, it seems to the Writer that no real progress can be made in Psychological Science; for that which "God hath joined together," it must be vain for Man to try to "put asunder."

2. To the prevalent neglect of the study of the *mutual relations* of Mind and Body, may be traced many of the fallacies discernible in the arguments adduced on each side, in the oft-repeated controversies between the advocates of the *Materialistic* and the *Spiritualistic* hypotheses;—controversies in themselves almost as absurd as that mortal contest, which (as fable tells us) was once carried on by two knights respecting the material of a shield seen by them from opposite sides, the one maintaining it to be made of gold, the other of silver, and each proving to be in the right as regarded the half seen by himself. Now the Moral of this fable, as respects our present inquiry, is, that as the entire shield was really made-up of a gold-half and a silver-half *which joined each other midway*, so the Mind and the Brain, notwithstanding those differences in *properties* which place them in different philosophical categories, are so intimately blended in their *actions*, that more valuable information is to be gained by seeking for it at the points of contact, than can be obtained by the prosecution of those older methods of research, in which the Mind has been studied by Metaphysicians altogether without reference to its material instrument,

whilst the Brain has been dissected by Anatomists and analyzed by Chemists, as if they expected to map-out the course of Thought, or to weigh or measure the intensity of Emotion. The Psychologist who looks at his subject in the light of that more advanced Philosophy of the present day, which regards Matter merely as the vehicle of Force, has no difficulty in seeing where both sets of disputants were right and both wrong; and, laying the foundations of his Science broad and deep in the *whole* constitution of the individual Man and his relations to the Universe external to him, aims to build it up with the materials furnished by experience of every kind, Mental and Bodily, normal and abnormal,—ignoring no fact, however strange, that can be attested by valid evidence, and accepting none, however authoritatively sanctioned, that will not stand the test of thorough scrutiny.

3. Although few (if any) Philosophers would be disposed to question that the Brain is the instrument of our higher Psychical powers, the ideas which are entertained of the nature of this instrumentality have been seldom clearly or consistently defined. Some, who have attended exclusively to the close relationship which indubitably exists between Corporeal and Mental states, have thought that *all* the operations of the Mind are but manifestations or expressions of material changes in the Brain:—that thus Man is but a *thinking machine*, his conduct being entirely determined by his original constitution, modified by subsequent conditions over which he has no control, and his fancied power of self-direction being altogether a delusion;—and hence that notions of *duty* or *responsibility* have no real foundation, Man's character being formed *for* him, and not *by* him, and his mode of action in each individual case being simply the consequence of the reaction of his Brain upon the impressions which called it into play. On this creed, what is commonly termed Criminality is but one form of Insanity, and ought to be treated as such; Insanity itself is nothing else than a disordered action of the Brain; and the

B 2

highest elevation of Man's *psychical* nature is to be attained by due attention to all the conditions which favour his *physical* development.

The most thorough-going expression of this doctrine will be found in the "Letters on the Laws of Man's Nature and Development," by Henry G. Atkinson and Harriet Martineau. A few extracts will suffice to show the character of this system of Philosophy. "Instinct, passion, thought, &c., are effects of organized substances." "All causes are material causes." "In material conditions I find the origin of all religions, all philosophies, all opinions, all virtues, all 'spiritual conditions and influences,' in the same manner that I find the origin of all diseases and of all insanities in material conditions and causes." "I am what I am; a creature of necessity; I claim neither merit nor demerit." "I feel that I am as completely the result of my nature, and impelled to do what I do, as the needle to point to the north, or the puppet to move according as the string is pulled." "I cannot alter my will, or be other than what I am, and cannot deserve either reward or punishment."

It seems to the Writer that every system of Philosophy which regards the succession of Mental Phenomena as determined *solely* by the ordinary laws of Physical Causation, and which rejects the *self-determining* power of the Will (or, which is the same thing, regards the Will as only another expression for the *preponderance of Motives*, or as the *general resultant* of the action of the Physiological Mechanism), virtually leads to the same conclusion.

4. Now this honestly-expressed *Materialistic* doctrine recognises certain great facts, as to which the unprejudiced and observant Physiologist can entertain no doubt; notwithstanding that their validity may be denied by those who have had comparatively little opportunity of studying them, or who have so made up their minds to a foregone conclusion, as to be ready to admit nothing which is not in accordance with it. The whole series of phenomena which so plainly mark the influence of the Body on the Mind, of *physical* upon *psychical* states,—the obvious dependence of the

normal activity of the Mind upon the healthful nutrition of the Brain, and upon its due supply of Oxygenated Blood,—the effect of Intoxicating agents and of Morbid Poisons in perverting that activity, and especially in withdrawing the "Mechanism of Thought and Feeling" from Volitional control,—the remarkable influence of local affections of the Brain, traceable in some cases to defective supply of blood, in others to blows on the head, in producing strange disturbances of Memory,—the large share which certain states of bodily disorder on the part of Parents, or conditions tending to induce defective nutrition during the periods of Infancy and Childhood, have been proved to possess in the induction of Idiocy and Cretinism,—the distinct Hereditary Transmission of *acquired habits*, which, modifying the Bodily constitution of the Parent, repeat themselves in that which he communicates to his Offspring,—these and numerous other phenomena (hereafter to be considered) might be cited in support of the Materialistic doctrine, and *must* be taken account of by any one who would seek the solution of this mystery.

5. But these phenomena are not to be looked-at to the exclusion of the facts of our own internal Consciousness. In reducing the Thinking Man to the level of "a puppet that moves according as its strings are pulled," the Materialistic Philosopher places himself in complete antagonism to the positive conviction, which—like that of the existence of an External World—is felt by every right-minded Man who does not trouble himself by speculating upon the matter, that *he really does possess a self-determining power*, which can rise above all the promptings of Suggestion, and can, *within certain limits* (§ 25), mould external circumstances to its own requirements, instead of being completely subjugated by them.

The Writer entirely agrees with Archbishop Manning, in maintaining that we have exactly the same evidence of the existence of this *self-determining power within ourselves*, that we have of the existence of *a material world outside ourselves.* For however intimate may be the functional correlation between Mind and Brain (§§ 11, 12),—and Archbishop Manning seems disposed to go as far as the Writer in recognizing this intimacy — "there is still another faculty, and more than this, another Agent, distinct from the thinking brain." * * * "That we are conscious of Thought and Will, is a fact of our internal experience. It is a fact also of the universal experience of all men; this is an immediate and intuitive truth of absolute certainty. Dr. Carpenter lays down as an axiomatic truth 'that the Common-sense decision of Mankind, in regard to the existence of an External World, is practically worth more than all the arguments of all the logicians who have discussed the basis of our belief in it.' What is true in this case of a judgment formed upon the report of Sense, by the interpretation of the Intellect, is still more evidently true of the decisions of our Consciousness on such interior facts as Thought or Will, and of the existence of an Internal World which is our living Personality, the Agent who thinks and wills. I may therefore lay it down as another axiom, side by side with that of Dr. Carpenter, that the decision of Mankind, derived from consciousness of the existence of our living self or personality, whereby we think, will, or act, is practically worth more than all the arguments of all the logicians who have discussed the basis of our belief in it." (*Contemporary Review*, Feb. 1871, p. 469.)

We can scarcely desire a better proof that our possession of this power is a reality and not a self-delusion, than is afforded by the comparison of the *normal* condition of the Mind with those variou *abnormal* conditions hereafter to be described (Chaps. XIV.—XVI.), in which the directing power of the Will is in abeyance. For the "subjects" of these conditions may *really* be considered (so long as they remain in them) as mere thinking Automata, puppets pulled by directing-strings; their whole course of thought and of action being determined by Suggestions conveyed from without, and their own Will having no power to modify or direct this, owing to the temporary suspension of its influence.—To whatever extent, then, we may be ready to admit the dependence of our Mental operations upon the organization and functional activity of our Nervous System, we must also admit that there is *something beyond and above* all this, to which, in the fully-developed and self-regulating Intellect, that activity is subordinated: whilst, in rudely trampling on the noblest conceptions of our Moral Nature as mere delusions, the purely Materialistic hypothesis is so thoroughly repugnant to the intuitive convictions of Mankind in general, that those who really experience these are made to *feel* its fallacy, with a certainty that renders logical proof unnecessary.

6. Let us turn now to the opposite doctrine held by *Spiritualists* * in regard to the nature and source of Mental phenomena; and consider this in its Physiological relations. To them the Mind appears in the light of a separate Immaterial existence, mysteriously connected, indeed, with a Bodily instrument, but not dependent upon this in any other way for the conditions of its operation, than as deriving its knowledge of external things through its Organs of Sense, and as making use of it to execute its determinations—so far as these are accomplished by Muscular effort. On this hypothesis, the operations of the Mind itself, having no dependence whatever on those of Matter, are never themselves affected by conditions of the Bodily organism; whose irregularities or defects of activity only pervert or obscure the outward manifestations of the Mind, just as the light of the brightest lamp may be dimmed or distorted by passing through a bad medium: while, further, as the Mind is thus independent of its Material tenement, and of the circumstances in which this may chance to be placed, but is endowed with a complete power of Self-government, it is responsible for *all* its actions, which must be judged-of by certain fixed standards.

7. Now this doctrine fully recognizes all that is ignored in the preceding; but, on the other hand, it ignores all that *it* recognized

* This term is here used in its older or Philosophical sense; not as designating *modern* "Spiritualists" (or rather "Spirit-rappers"), who have little in common with those whose name they have adopted.

and served to account for; and is not less opposed to facts of most familiar experience. For in placing the Mind altogether *outside* the Body, and in denying that its action is ever disordered by Bodily conditions, the Spiritualist puts us in the dilemma of either rejecting the plainest evidence, or of admitting that, after all we know nothing of the nature of the Mind itself; all that we *do* know, being that lower part of our Mental nature which operates on the Body, and is in its turn affected through it.—Those who would fully and consistently carry out this doctrine, are driven to maintain that even in the state of Intoxication there is no truly *mental* perversion; and that, in spite of appearances, the *mind* of the Lunatic (*divinæ particula auræ*) is perfectly sound, its bodily instrument being alone disordered. But it cannot be overlooked, that in the delirious ravings of Intoxication or of Fever, or in the conversation and actions of the Lunatic, we have precisely the same evidence of *mental* operation, that we have in the sayings and doings of the same individuals in a state of sanity; and ample testimony to this effect is borne by those who have observed their own mental state during the access of these conditions, and who have described the alteration which took place in the course of their Thoughts, when as yet neither the Sensorial nor the Motor apparatus was in the least perturbed (§ 537). Nothing can be more plain to the unprejudiced observer, than that the introduction of Intoxicating agents into the Blood-circulation really perverts the action of the *Mind;* disordering the usual sequence of phenomena most purely *psychical,* and occasioning new and strange results which are altogether at variance with those of its normal action. And when once the reality of this influence of Physical conditions upon purely-Mental states is forced upon the Physiologist, he cannot avoid recognizing it as a general fact of our nature; so that he comes to be impressed by the conviction, that whilst there is something in our Moral constitution beyond and above any agency which can be attributed to Matter, the operations of the Mind are *in a great degree* determined (in our present state of being) by the Material conditions with which they are so intimately associated.

8. This combination of two distinct agencies in the Mental constitution of each individual, is recognized in the whole theory and practice of Education. For whilst, in its earlier stages, the Educator aims to call-forth and train the Intellectual Faculties of his Pupil, and to form his Moral Character, by bringing appropriate *external* influences to bear upon him, every one who really understands his profession will make it his special object to foster the development, and to promote the right exercise, of that *internal* power, by the exertion of which each Individual becomes the director of his own conduct, and *so far* the arbiter of his own destinies. This power is exercised by the Will, in virtue of its domination over the *automatic* operations of the Mind, as over the *automatic* movements of the Body (§ 14); the real *self*-formation of the Ego commencing with his consciousness of the ability to determine *his own* course of thought and action. Until this self-directing power has been acquired, the Character *is* the resultant of the individual's original constitution, and of the circumstances in which he may have been placed; and so long as the circumstances are unfavourable to its development, and to the operation of those higher tendencies which should furnish the best motives to its exercise, so long the Character of the individual *is* formed *for* him rather than *by* him. A being entirely governed by the lower passions and instincts, whose higher Moral Sense has been repressed from its earliest dawn by the degrading influence of the conditions in which he is placed, who has never learned to exercise any kind of self-restraint (or, if he has learned it, has only been trained to use it for the lowest purposes), who has never heard of a God, of Immortality, or of the worth of his Soul,—such a being, one of those heathen outcasts of whom all our great towns are unhappily but too productive,—can surely be no more morally responsible for his

actions, than the Lunatic who has lost whatever self-control he once possessed, and whose moral sense has been altogether perverted by bodily disorder. But let the former be subjected to the training of one of those benevolent individuals who know how to find out "the holy spot in every child's heart;" let patient kindness, continually appealing to the highest motives which the child *can* understand, progressively raise his Moral standard, and awaken within him the dormant susceptibilities which enable him to feel that he has a Conscience and a Duty, that he has a power within himself of controlling and directing his thoughts and actions, and that the highest happiness is to be found in the determinate pursuit of the *true* and the *good*,—then, but not till then, can he be justly considered *responsible* for his actions, either morally or religiously,—then only does he rise above the level of the brute, and begin to show that he is indeed made in the image of his Creator.

9. Thus we see that the Materialistic and the Spiritualistic doctrines alike recognize, and alike ignore, certain great truths of Human Nature; and the question returns upon us, whether any general expression *can* be framed, which may be in harmony alike with the results of Scientific inquiry into the relation of Mental to Physiological action, and with those simple teachings of our own Consciousness, which must be recognized as affording the ultimate test of the truth of all Psychological doctrines. Towards such an expression we may make a step, as it appears to the Writer, in strict accordance with true Philosophy, by withdrawing ourselves entirely from the futile attempt to bring Matter and Mind into the same category, and by fixing our attention exclusively on the relation between *Mind* and *Force*. Although far from thinking that the views here offered express the *whole* truth, or solve *all* the difficulties of the subject (the *originating power* of the Human Will,—i. e. its independence of Physical Causation,—being the essential difficulty of *every* system

which recognizes it), he ventures to think that they deserve the attention of such as feel, with him, the importance of fearlessly pushing the inquiry to its utmost practicable limits, and of attaining such definite conceptions as the present state of Scientific knowledge may justify.

10. It is now generally admitted that we neither know, nor can know, anything of *Matter*, save through the medium of the impressions it makes on our Senses; and those impressions are only derived from the *Forces* of which Matter is the vehicle. Thus, of those most general Properties of Matter, *resistance* and *ponderosity*, our information is entirely derived through our own Tactile Sense (under which general head may be combined the Sense of Touch, the Sense of Muscular Exertion, and the Mental Sense of Effort), by which we recognize the Forces that attract its particles to each other and to the Earth; and what is ordinarily regarded as its distinctive characteristic, its "extension" or occupation of Space, we know only as an inference from our own Sense-perceptions. In fact, instead of Matter (as some affirm) being the object of our immediate cognizance, and the Laws of Matter our most certain form of knowledge, there seems valid ground for the assertion that our notion of *Matter* is a conception of the Intellect, *Force* being that externality of which we have the *most* direct—perhaps even the *only* direct—cognizance. And in this way, Force—of the existence of which we are rendered cognizant by the direct testimony of our own Consciousness, which is to us the most certain of realities—comes into immediate relation with Mind. Moreover, while *Matter* is essentially *passive*,—since, when left to itself, it always impresses our Consciousness in one and the same mode, any change in that impression being the consequence of an agency external to itself,—all its *Activities* are manifestations of the *Forces* of which it is the vehicle, and to the exercise of which all the phenomena of the Material Universe are due.

Water, for example, would continue unchanged so long as its Temperature remains the same, and no decomposing agency is brought to bear upon it: but Heat communicated to it occasions that repulsion between its particles, which transforms it from a non-elastic liquid into an elastic vapour exerting a proportionate Mechanical Force; and the same measure of Power is again given forth from it, either as Heat or as Motion, with the transformation of the aqueous vapour back to the liquid state.—In like manner, the transmission of a sufficiently strong Electric current through Water resolves it into its two component gases, which, when made to re-unite, give off the equivalent, in the form of Light and Heat, of the Elastic Force which kept their particles asunder, and which was itself more remotely derived from Electricity, developed by Chemical change.

But *Mind*, like *Force*, is essentially *active;* all its states are states of *change;* and of these changes we become directly or immediately conscious by our own experience of them. In fact, every term—as Sensation, Perception, Idea, Emotion,—which expresses a Mental state, is a designation of a phase of Mental existence that intervenes between other phases, in the *continual succession* of which our idea of Mind consists; and Consciousness itself is nothing else than the designation which we give to the condition which is common to all these forms of activity.

11. Now, nothing can be more certain, than that the primary form of Mental activity,—*Sensational* consciousness,—is excited through Physiological instrumentality. A certain Physical impression is made, for example, by the formation of a luminous image upon the Retina of the Eye; a change being thereby produced in that Nervous expansion, which is clearly analogous to that which a similar image would make upon a sensitive Photographic surface. But instead of recording itself by a permanent effect upon the Retinal surface, the effect of this Visual impression is to excite the activity of the Optic Nerve; through the instrumentality of which, again, an active condition is excited in the Optic Ganglion to which it proceeds,—just as, in the transmission of a Telegraphic

message, the movements of the signalling needle at one end of the wire repeat themselves in the movements of the magnetic needle at the other. So far, we are concerned with a Physiological mechanism alone; through which (probably by Chemical changes in the Nerve-substance) Light excites Nerve-force, and the transmission of this Nerve-force excites the activity of that part of the Brain which is the instrument of our Visual Consciousness. Now in what way the *physical* change thus excited in the Sensorium is translated (so to speak) into that *psychical* change which we call *seeing* the object whose image was formed upon our Retina, we know nothing whatever; but we are equally ignorant of the way in which Light produces Chemical change, and Chemical change excites Nerve-force. And all we can say is, that there is just as close a succession of sequences—as intimate a causal relation between antecedent and consequent—in the one case, as there is in the other. In other words, there is just the same evidence of what has been termed *Correlation*, between *Nerve-force* and that primary state of Mental activity which we call *Sensation*, that there is between *Light* and *Nerve-force*;—each antecedent, when the Physiological mechanism is in working order, being invariably followed by its corresponding consequent. And true Visual consciousness of an external object can no more be excited without an active condition of the Sensorium corresponding to it, than that active condition of the Sensorium can be called forth without the transmission of Nerve-force from the Retina; or than that active condition of the Retina which generates and transmits the Nerve-force, can be produced without Light or some other equivalent Force.*

12. The like Correlation may be shown to exist between Mental states and the form of Nerve-force which calls forth *Motion* through the Muscular apparatus. We shall hereafter see that

* The case of those "Subjective Sensations" which imitate the sensations called up by external objects, will be considered in its proper place (§§ 139—147).

each kind of Mental activity,—Sensational, Instinctive, Emotional, Ideational, and Volitional,—may express itself in Bodily movement; and it is clear that every such movement is called forth by an active state of a certain part of the Brain, which excites a corresponding activity in the Motor Nerves issuing from it, whereby particular Muscles are called into contraction. No Physiologist can doubt that the Mechanical force exerted by the Muscles is the expression of certain Chemical changes which take place between their own substance and the oxygenated Blood that circulates through them; or that the Nerve-force which calls forth those changes, is intimately related to Electricity and other Physical forces. But this Nervous activity has its source in molecular changes in the Nerve-centres; the transmission of Nerve-force along the motor nerve being just as dependent upon Chemical changes taking place between the substance of the Ganglionic centre from which it proceeds and the oxygenated Blood that circulates through it, as is the transmission of an Electric current along the Telegraph-wire upon the Chemical changes taking place between the metals and the exciting liquid of the Galvanic battery. But these changes are themselves capable of being brought about by the various forms of Mental activity just enumerated. Just as a perfectly constructed Galvanic battery is *inactive* while the circuit is "interrupted," but becomes *active* the instant that the circuit is "closed," so does a Sensation, an Instinctive tendency, an Emotion, an Idea, or a Volition, which attains an intensity adequate to "close" the circuit, liberate the Nerve-force with which a certain part of the Brain, while in a state of wakeful activity, is always "charged." That Mental antecedents can thus call forth Physical consequents, is just as certain as that Physical antecedents can call forth Mental consequents; and thus the Correlation between Mind-force and Nerve-force is shown to be complete *both ways*, each being able to excite the other.

13. Now using these facts as our basis, we seem justified in going further; and in asserting that the same kind of evidence justifies the belief, that a Physiological mechanism of the like nature furnishes the instrumentality through which all kinds of Mental operation take place. For no Scientific Psychologist has any doubt that there are "Laws of Thought" expressing sequences of Mental activity, which (if we could thoroughly acquaint ourselves with them) would be found as fixed and determinate as the "Laws of Matter;" the difficulty in ascertaining them arising solely from the difficulty in subjecting Mental phenomena to precise observation, and in analysing the complex conditions under which they occur. And whilst these laws comprehend that large part of our Mental activity which may be designated as *automatic*,—consisting in a succession of Mental states, of which each calls forth the next by Suggestion, without any interference from the Will,—it will be further shown that there are a great number of Mental phenomena which cannot be accounted for in any other way, than as resulting from the operation of a Physiological mechanism, which may go on not only *automatically*, but even *unconsciously* (Chap. XIII.). That we are not always conscious of the working of this Mechanism, is simply because our Sensorium is otherwise engaged: for just as we may not see things which are passing before our eyes, or be conscious of the movements of our legs in walking, if our Attention be wholly engrossed by our Cerebral "train of thought," so may we not be conscious of what is going on in our Cerebrum, whilst our Attention is wholly concentrated upon what is passing before our Eyes (§ 117). But the Physiological mechanism has this peculiarity,—that it *forms itself* according to the mode in which it is *habitually* exercised; and thus not only its *automatic* but even its *unconscious* action comes to be indirectly modified by the controlling power of the Will (§ 95).

14. It may serve to promote the right understanding of the general doctrine as to the relation of Will to Thought which it is the

chief object of this Treatise to set forth, if we briefly inquire into the relation of the Will to Bodily Movements. It has been customary to classify these as *voluntary* or *involuntary*, but it will be found preferable to distinguish them as *volitional* and *automatic:* the former being those which are called forth by a distinct effort of Will, and are directed to the execution of a definite *purpose;* whilst the latter are performed in respondence to an internal prompting of which we may or may not be conscious, and are not dependent on any preformed intention,—being executed, to use a common expression, "mechanically." Some of these are *primarily* or *originally* Automatic; whilst others, which were Volitional in the first instance, come by frequent repetition to be performed independently of the Will, and thus become *secondarily* Automatic.* Some of the Automatic movements, again, can be controlled by the Will; whilst others take place in opposition to the strongest Volitional effort. There is a large class of secondarily-automatic actions, which the Will can initiate, and which then go on of themselves in sequences established by previous Habit; but which the Will can stop, or of which it can change the direction, as easily as it set them going; and these it will be convenient to term *voluntary*, as being entirely under the control of the Will, although actually maintained Automatically.

15. Those movements of which the uninterrupted performance is essential to the maintenance of Life, are *primarily* automatic; and are not only independent of the Will, but entirely beyond its control. The "beating of the Heart," which is a typical example of such movements, though liable to be affected by *emotional* disturbance, cannot be altered either in force or frequency by any *volitional* effort. And only one degree removed from this is

* The sagacity of Hartley enabled him to anticipate on this point the discoveries of modern Physiology; for in designating as *secondarily automatic* the whole of the actions which come to be performed by Habit without Will or even Consciousness, though *originally* learned and practised with *conscious intent*, he showed a discernment of their true character which later researches have entirely justified.

the act of Respiration; which, though capable in Man of being so *regulated* by the Will as to be made subservient to the uses of Speech, cannot be *checked* by the strongest exertion of it for more than a few moments. If we try to "hold our breath," for such a period that the aëration of the blood is seriously interfered with, a feeling of distress is experienced, which every moment increases in intensity until it becomes absolutely unbearable; so that the automatic impulse which prompts its relief can no longer be resisted. So when a crumb of bread or a drop of water passes "the wrong way," the presence of an irritation in the windpipe automatically excites a combination of muscular movements, which tends to an expulsion of the offending particle by an explosive Cough. The strongest exertion of the Will is powerless to prevent this action; which is repeated in spite of every effort to repress it, until that result has been obtained. If the irritation be applied to the nasal entrance of the air-passages, as in snuff-taking, a peculiar valvular action at the back of the mouth automatically directs a part of the explosive blast through the nose; and this Sneeze, if the stimulus be applied in sufficient strength, is altogether beyond Volitional control.—It is worthy of note that whilst the act of *coughing* can be excited by a mandate of the Will, through the instrumentality to be hereafter explained (§ 47), we cannot thus execute a true *sneeze*, the stage-imitation of which is ludicrously unlike the reality.

16. There can be no doubt that in the lower tribes of Animals, a large part of the ordinary movements of Locomotion are of the same *primarily* automatic character; being executed in direct respondence to a stimulus that acts through the Nervous centres with which the locomotive members are directly connected, and being performed by the headless trunk with just the same perfect coordination as by the entire creature (§ 54). In Man, however, the power of performing these movements is *acquired* by a process of education; and no one can watch this process, without perceiving

how gradual is the acquirement of the co-ordinating power, especially in the *balancing* of the body during each successive step. As Paley says: "A child learning to walk is the greatest posture master in the world." Yet, when this co-ordination has been once established, the ordinary movements of Locomotion—though involving the combined action of almost every muscle in the body—are performed automatically; the Will being only concerned in starting, directing, or checking them.—Of this we have familiar experience in the continuance of the act of *walking*, whilst the attention is occupied by some "train of thought" which completely and continuously engrosses it. Though we set out with the intention of proceeding in a certain direction, after a few minutes we may lose all consciousness of where we are, or of whither our legs are carrying us; yet we continue to walk-on steadily, and may unexpectedly find ourselves at the end of our journey before we are aware of having done more than commence it (§ 71). Each individual movement here *suggests* the succeeding one, and the repetition continues, until, the Attention having been recalled, the automatic impulse is superseded by the control of the Will. Further, the direction of the movement is given by the sense of Sight, which so guides the motions of our legs that we do not jostle our fellow passengers or run up against lamp-posts; and the same sense directs also their general course along the line that *habit* has rendered most familiar, although at the commencement of our walk we may have intended to take some other.—Suppose our walk to be so prolonged, however, that the sense of *fatigue* comes-on before we have reached its appointed conclusion. This calls off our Attention from what is going on in the *mind*, to the condition of the *body*; and in order to sustain the movements of locomotion, a distinct exertion of the Will comes to be requisite for each. With the increasing sense of fatigue, an increased effort becomes necessary; and at last even the most determined Volition may find itself unable to evoke a respondent movement from the exhausted Muscles.

17. In this familiar experience we can clearly trace three distinct modes of action,—the Automatic, the Voluntary, and the Volitional. Whilst we are all unconscious of the movements which our legs are executing for us, those movements are purely *automatic*. When our attention is not so completely engrossed elsewhere, but that we know where we are and what we are doing, the movements of locomotion are not only *permitted* by the Will, but may be *guided* by it into some unusual direction; such movements are *voluntary*. But when the sense of fatigue attending each movement makes it necessary that a distinct effort of the Will shall be exerted for its repetition, the act comes to be *volitional*.—The explanation of these phenomena lies in the fact, that the Nervo-muscular mechanism immediately concerned in *executing* the movement (of which an account will be given hereafter, §§ 54, 71) is the same throughout, but that it is *started* by different means; the Will replacing the stimulus to action otherwise furnished by an external impression. Of this we have a typical example in the act of Coughing. When we *will* to cough (as for the purpose of giving a signal, or putting down a tedious speaker), we merely touch the spring, as it were, of a mechanism, which *automatically* combines the multitude of separate actions that are required to produce the result (§ 47); just as when we pull the trigger of a gun, or open the valve which admits steam into the steam-engine. And the only difference in kind between the act of Coughing and that of Walking consists in this,—that whilst the mechanism concerned in the *former* is ready for action from the first, that by which the *latter* is performed requires to have its various springs and levers adjusted to harmonious operation. But when this adjustment has been once made, it remains good for life; in virtue of that remarkable peculiarity of our Bodily constitution, which keeps up the Nutrition of each part in accordance with the use that is made of it (§ 276).

18. There may still be Metaphysicians who maintain that

actions which were originally prompted by the Will with a distinct intention, and which are still entirely under its control, can never cease to be Volitional; and that either an infinitesimally small amount of will is required to sustain them when they have been once set going, or that the will is in a sort of pendulum-like oscillation between the two actions,—the maintenance of the train of *thought*, and the maintenance of the train of *movement*. But if only an infinitesimally small amount of Will is necessary to sustain them, is not this tantamount to saying that they go on by a force of their own? And does not the experience of the *perfect continuity* of our trains of thought during the performance of movements that have become habitual, entirely negative the hypothesis of oscillation? Besides, if such an oscillation existed, there must be *intervals* in which each action goes on *of itself*; so that its essentially automatic character is virtually admitted. The Physiological explanation, that the Mechanism of Locomotion, as of other habitual movements, *grows to* the mode in which it is early exercised, and that it then works automatically under the general control and direction of the Will, can scarcely be put down by any assumption of a hypothetical necessity, which rests only on the basis of ignorance of one side of our composite nature.

19. But we may go a step further, and assert that it may now be regarded as a well-established Physiological fact, that even in the most purely Volitional movements—those which are prompted by *a distinct purposive effort*,—the Will does not *directly* produce the result; but plays, as it were, upon the Automatic apparatus by which the requisite Nervo-muscular combination is brought into action.

20. No better illustration of this doctrine could be adduced, than that which is furnished by the act of *Vocalization*; either in articulate Speech, or in the production of Musical tones. In each of these acts, the co-ordination of a large number of muscular movements is required; and so complex are their combinations, that the professed Anatomist would be unable, without careful

study, to determine what is the precise state of each of the muscles concerned in the production of a given musical note, or the enunciation of a particular syllable. Yet we simply *conceive* the tone or the syllable we wish to utter, and say to our automatic Self "Do this:" and the well-trained Automaton does it. The delicate gradations in the action of each individual muscle, and the harmonious combination of the whole, are effected under the guidance of the Ear, without (save in exceptional cases) the smallest knowledge on our own parts of the nature of the mechanism we are putting in action. In fact, the most perfect acquaintance with that mechanism would scarcely afford the least assistance in the acquirement of the power to use it. The "training" which develops the inarticulate Cry of the infant into articulate Speech or melodious Song, mainly consists in the fixation of the Attention on the *audible result*, the *selection* of that one of the imitative efforts to produce it which is most nearly successful, and the *repetition* of this until it has become habitual or *secondarily automatic.* The Will can thenceforwards reproduce any sound once acquired, by calling upon the Automatic apparatus for the particular combination of movements which it has *grown into* the power of executing in respondence to each preconception; provided, at least, that the apparatus has not been allowed to become rusty by disuse, or been stiffened by training into a different mode of action. Even the strongest Will, however, may fail to acquire complete control over the complex Automatic mechanism. The articulation of the Stammerer is disturbed by spasmodic impulses, which he vainly endeavours to keep under subjection:—the Vocalist's ear may tell him that he is singing out of tune, and yet he may be unable to correct his fault:—and even a Viardot or a Patti would feel unfit either for the performance of a new *rôle*, or for the repetition of an old one long laid by, however perfect might be her mental conception of it, until she had trained or re-trained her organ to execute that conception.

21. Another illustration, drawn from the movements of the Eyes, may place this doctrine in a still clearer light; inasmuch as the action of the living Automaton can be watched either by a bystander, or by the *Ego* that calls it forth. Let the reader *will* to fix his gaze on the face of a person directly opposite to him, and then *will* to move his head from side to side; his eyeballs will be seen to roll in their sockets *in the contrary direction*, and this not only without any volitional effort on his part, but even without his being in any way conscious of the act, except by a process of reasoning. Or, if he move his head upwards and downwards, his eyes (still fixed on the opposite face) will roll conversely *downwards and upwards*. And if, instead of looking at the face of another, he fix his gaze upon the reflection of his own eyes in a mirror, and then move his head as before, he will be able to satisfy himself that his Automaton is directing his eyes for him; every alteration in the position of his head being accompanied by a roll of his eyeballs *in the opposite direction*, so that their axes continue to be turned towards the reflected image, so long as he *wills* to keep them so.

22. The same may be shown to be true of all the so-called Voluntary movements. What we *will* is, not to throw this or that muscle into contraction, but *to produce a certain preconceived result*. That result may be within the capacity of our ordinary Mechanism; but, if it be not, we have to create a new mechanism by a course of training or practice; the effect of which (as already shown) is to make the Automatic apparatus *grow to* the mode in which it is habitually exercised.—That this is the true theory of these movements, is evident from several considerations, of which a few must here suffice. If the performance of a Voluntary movement required a transmission of Nervous power direct from the Brain (which may be assumed to be the instrument of the Will) to the Muscles concerned in its production, then we should need to know what those muscles are, and to select and combine

them intentionally; which is so far from being the fact, that the consummate anatomist is no better able than the completest ignoramus to execute a movement he has never practised. Again, if our Muscles were under the direct control of the Will, we could single out any one of them, and make it contract by itself; which we cannot really do, except in the few instances in which *willing* the result calls only a single muscle into action. So again, if an accomplished Musician should wish to play upon an instrument he has never practised, but of which he thoroughly understands the mechanism, it would be sufficient for him to *will* the movements he knows to be requisite for the production of the desired tones, instead of having to acquire the power of performing them by a laborious course of training; and the man who, on being asked whether he could play the fiddle, said that "he did not know till he had tried," *might* have shown himself a very Joachim when the instrument was put into his hands.

23. The doctrine that the Will, which carries into action the determinations of the Intellect, has no direct power over the muscles which execute its mandates, but operates through the automatic mechanism, is in entire harmony with the knowledge acquired of late years in regard to the relative functions of the *Cerebrum* and of the *Axial Cord* on which it is superimposed. For it will be shown (Chap. II.) that the latter, which receives all the nerves of Sense, and gives forth all the nerves of Motion, constitutes the fundamental and essential part of the Nervous System, and is alone concerned in the performance of all those movements which are *primarily* automatic or Instinctive: whilst the Cerebrum, the development of which seems to bear a pretty constant relation to the degree in which Intelligence supersedes Instinct as a source of action, is superadded to this Axial Cord; through which, on the one hand, it receives Sense-impressions, whilst, on the other, it calls the Muscles into action. And thus, when we *will* to cough, certain Cerebral fibres (§ 89)

convey the same stimulus to the centre of Respiratory movement, that is brought to it by the Sensory nerves when a crumb of bread or a drop of water "goes the wrong way," and calls forth the same respondent action.

24. Thus, then, the relation between the Automatic activity of the body, and the Volitional direction by which it is utilized and directed, may be compared to the independent locomotive power of a horse under the guidance and control of a skilful rider. It is not the rider's whip or spur that furnishes the *power*, but the nerves and muscles of the horse; and when these have been exhausted, no further action can be got out of them by the sharpest stimulation. But the rate and direction of the movement are determined by the Will of the rider, who impresses his mandates on the well-trained steed with as much readiness and certainty as if he were acting on his own limbs. Now and then, it is true, some unusual excitement calls forth the essential independence of the equine nature; the horse takes the bit between his teeth, and runs away with his master; and it is for the time uncertain whether the independent energy of the one, or the controlling power of the other, will obtain the mastery. This is just what we see in those Spasms and Convulsions which occur without loss of consciousness, and in which the muscles that we are accustomed to regard as "voluntary" are called into violent contraction, in spite of the strongest Volitional resistance. On the other hand, the horse will quietly find his way home, whilst his rider, wrapped in a profound reverie, entirely ceases to guide him; just as our own legs carry us along a course which habit has made familiar, while our Mind is engaged only upon its own operations, and our Will is altogether in abeyance. And, to complete the parallel, the process by which a Horse is taught any unusual performance—as when in "training" for the Circus or the Stage—entirely corresponds with that by which we "train" our own automatic mechanism to any novel action: the *result* desired by the master being indicated to the

learner, every effort that tends to produce it being encouraged and fixed by repetition, and every unsuitable action being repressed; until the entire sequence comes to be automatically executed at the first touch of the suggesting spring which expresses the directing Will.

25. Now all this will be found to be as true of the *Mind*, as it is of the body. Our Mental activity is, in the first instance, entirely *spontaneous* or *automatic;* being determined by our congenital nervous Organization, and by the conditions of its early development. It may be stated as a fundamental principle, that the Will can never *originate* any form of Mental activity. Thus, no one has ever *acquired* the creative power of Genius, or *made himself* a great Artist or a great Poet, or *gained by practice* that peculiar insight which characterises the original Discoverer; for these gifts are Mental Instincts or Intuitions (§ 408), which, though capable of being developed and strengthened by due cultivation, can never be generated *de novo.* But the power of the Will is exerted in the *purposive selection,* from among those objects of consciousness which Sensations from without and the working of the internal "Mechanism of Thought and Feeling" bring before the Ego (whether simultaneously or successively), of that which shall be determinately followed up; and in the *intensification of the force of its impression,* which seems the direct consequence of such limitation. This state is what is termed *Attention;* in regard to which it was well said by Sir William Hamilton, that its *intensity* is in a precisely inverse ratio to its *extensity.* And it will be the Writer's object to show, that it is solely by the Volitional *direction of the attention* that the Will exerts its domination; so that the acquirement of this power, which is within the reach of every one, should be the primary object of all Mental discipline. It is thus that each individual can pe ct and utilize his natural gifts; by rigorously training them in the first instance, and then by exercising them only in the manner most fitted to expand and elevate, while restraining

them from all that would limit or debase.—In regard to every kind of Mental activity that does *not* involve origination, the power of the Will, though limited to *selection*, is almost unbounded. For although it cannot directly bring objects before the consciousness which are not present to it (§ 371), yet, by concentrating the Mental gaze (so to speak) upon any object that may be within its reach, it can make use of this to bring in other objects by associative Suggestion. And, moreover, it can virtually determine what shall *not* be regarded by the Mind, through its power of keeping the Attention fixed *in some other direction;* and thus it can subdue the force of violent impulse, and give to the conflict of opposing motives a result quite different from that which would ensue without its interference (§ 332). This exercise of the Will, moreover, if habitually exerted in certain directions, will tend to form the Character, by establishing a set of *acquired habitudes;* which, no less than those dependent upon original constitution and circumstances, help to determine the working of the "Mechanism of Thought and Feeling." In so *utilising* it, the Will can also *improve* it by appropriate discipline; repressing its activities where too strong, fostering and developing them where originally feeble, directing all healthful energy into the most fitting channel for its exercise, and training the entire Mental as it does the Bodily organism to harmonious and effective working. And thus in proportion as our Will acquires domination over our Automatic tendencies, the spontaneous succession of our Ideas and the play of our Emotions show the influence of its habitual control; while our Character and Conduct in Life come to be the expression of our best Intellectual energies, directed by the Motives which we *determinately elect* as our guiding principles of action.

26. It is obvious that the view here taken does not in the least militate against the idea, that Mind may have an existence altogether independent of the Body which serves as its instrument. All which has been contended for is, that the connexion between Mind and

Body is such, that the actions of each have, in this present state of existence (which is all of which Science can legitimately take cognizance), a definite *causal relation* to those of the other; so that the actions of our Minds, *in so far as they are carried on without any interference from our Will*, may be considered as "Functions of the Brain."—On the other hand, in the control which the Will can exert over the *direction* of the thoughts, and over the *motive force* exerted by the feelings, we have the evidence of a new and independent Power, which may either oppose or concur with the automatic tendencies, and which, according as it is habitually exerted, tends to render the Ego *a free agent*. And, truly, in the existence of this Power, which is capable of thus regulating the very highest of those operations that are causally related to corporeal states, we find a better evidence than we gain from the study of any other part of our Psychical nature, that there *is* an entity wherein Man's nobility essentially consists, which does not depend for its existence on any play of Physical or Vital forces, but which makes these forces subservient to its determinations. It is, in fact, in virtue of the Will, that we are *not* mere thinking Automata, mere puppets to be pulled by suggesting-strings, capable of being played-upon by every one who shall have made himself master of our springs of action.

27. It may be freely admitted, however, that such thinking Automata *do* exist: for there are many individuals whose Will has never been called into due exercise, and who gradually or almost entirely lose the power of exerting it, becoming the mere creatures of habit and impulse; and there are others in whom (as we shall hereafter see) such Automatic states are of occasional occurrence, whilst in others, again, they may be artificially induced. And it is (1) by the study of those conditions in which the Will is completely in abeyance,—the course of thought being *entirely* determined by the influence of suggestions upon the Mind, whose mode of reaction upon them depends upon its original peculiarities and its sub-

sequently-acquired habits,—and (2) by the comparison of such abnormal states with that in which the Ego, in full possession of all his faculties, and accustomed to the habitual direction of his thoughts and control of his feelings, determinately applies his judgment to the formation of a decision between contending impulses, and carries that decision into action,—that we shall obtain the most satisfactory ideas of what share the Will really takes in the operations of our Minds and in the direction of our conduct, and of what must be set down to that automatic activity of our Psychical nature, which is correlated with Cerebral changes.

28. Thus, then, the Psychologist may fearlessly throw himself into the deepest waters of speculative inquiry in regard to the relation between his Mind and its Bodily instrument, provided that he trusts to the inherent buoyancy of that great fact of Consciousness, that *we have within us a self-determining Power which we call Will.* And he may even find in the evidence of the intimate relation between Mental activity and Physical changes in the Brain, the most satisfactory grounds which Science can afford, for his belief that the phenomena of the Material Universe are the expressions of an Infinite Mind and Will, of which Man's is the finite representative. (See Chap. XX.)

CHAPTER II.

OF THE NERVOUS SYSTEM AND ITS FUNCTIONS.

SECTION 1. *Relation of the Nervous System to the Body generally.*

29. THE Body of Man, or of any one of the higher Animals, may be regarded as made up of two portions which are essentially distinct, though intimately blended as well in their structure as in their actions,—viz. (1), the Apparatus of *Animal Life*, and (2) the Apparatus of *Vegetative* or *Organic Life.*

30. To the Apparatus of Animal Life belongs the whole Mechanism of those actions which essentially distinguish the Animal from the Plant; namely, Sensation, the higher Psychical changes which Sensation initiates, and the Movements which are consequent upon them. And thus the Apparatus of Animal Life may be said to consist of the Nervous System, the Organs of Sense, and the Organs of Motion,—these last including the Skeleton or jointed framework (composed of bones, cartilages, and ligaments), and the Muscles which give motion to its parts. It is in virtue of the *contractility* possessed by the Muscles, that all the sensible movements of the higher Animals are performed: the skeletal framework being merely *passive*, and furnishing a system of levers by which the contractile *power* of the muscles may be advantageously applied; and the muscles being either directly united to the bones, or being connected with them by means of the cords termed Tendons, which simply communicate the tension or "pull" produced by the contraction of the muscles. Thus, the closure of the fingers in grasping is for the most part produced by the contraction of

Muscles that form the fleshy part of the fore-arm, the strong tendons of which may be felt on the front of the wrist-joint; and in like manner, the propulsive movement of the foot in walking is effected by the large Muscles forming the calf of the leg,—these pulling upwards the heel by means of the great Tendo Achillis into which they are continued.

31. The Apparatus of *Organic* Life, on the other hand, serves in the first instance to construct or *build-up* the Apparatus of Animal Life, and then to *maintain* it in "working order." For all expenditure of Force involves not only a certain "wear and tear" of the apparatus which furnishes its instrumentality; but also a certain equivalent amount of Chemical change, either in the substance of the apparatus itself, or in the blood which circulates through it, or in both. Thus when a Muscle is called into contraction, there is a certain disintegration or "waste" of its tissue, which needs repair by Nutrition; but there is also an oxidation of Organic Compounds, by which Carbon and Hydrogen originally derived from the food are converted into Carbonic acid and Water; and what would elsewhere produce *Heat*, here takes the form of the mechanical equivalent of heat, namely *Motion*. How much of these Organic Compounds is supplied by the *muscle*, and how much by the *blood*, has not yet been satisfactorily determined: it may be regarded, however, as certain that the whole of the *motor* force generated in the contraction of a muscle is *not* derived (as Liebig maintained it to be) from the "waste" of the muscle itself, and the oxidation of its components; but that a large part of it is supplied by the oxidation of non-nitrogenous constituents of the blood.—The generation of Nerve-force involves a still more active change in that part of the Nervous system which is the instrument of its production (§ 41); and though we are not yet able to state precisely in what this change consists, yet we may affirm with certainty that it involves a reaction between Nerve-substance and oxygenated Blood, which requires a constant

supply of that fluid, and a no less constant removal of the products of the reaction to which it ministers.

32. Thus, then, the Apparatus of Organic Life may be said to consist of the organs by which Blood is *made*, those by which it is *kept in circulation*, and those by which it is *maintained in purity*; but the action of these has to be supplemented by that of the Apparatus of Animal Life. For, in the first place, the Animal must obtain its food by the exercise of its senses, of its psychical powers, and of its locomotive organs; and even in the Ingestion and Digestion of the food, when procured, the assistance of Muscles is required. So the Circulation of Blood is maintained by a muscular organ, the Heart, and is regulated locally by the muscularity of the walls of the Arteries; and both the rhythmical contraction of the Heart, and the calibre of the Arteries, are greatly influenced by the Nervous system. Again, the ordinary movements of Respiration, which constitute the most important of all the provisions by which the Blood is kept in the condition required for the development of the Nervous and Muscular forces, are dependent in the higher animals upon the Nervo-muscular apparatus; and although they are so completely *automatic* in their character, as to be performed not only *without* effort, but *in opposition* to effort, they are so far under the control of the higher Nervous centres, as to be subservient to the Vocal expression of Psychical states. So, again, although the action of the Excretory organs, by which the products of the "waste" are removed from the Blood, is essentially independent of the Nervo-muscular apparatus, this has a certain control over their outlets, which enables the excretions to be retained and discharged at suitable times.

33. We shall find, then, that in the higher Animals the Nervous system is the instrument, not only of those Psychical powers by which they are pre-eminently distinguished, but also of many operations which minister solely to the maintenance of the Organic Functions. But the portions of it which are directly concerned in

this latter duty, constitute an *automatic* apparatus, which is essentially independent of those higher centres that minister to the former. Thus not only does the Heart continue to beat, but the Respiratory movements are performed, as well in the sleeping as in the waking state; during the profoundest insensibility, as in the condition of fullest mental activity. It cannot be certainly affirmed how far the rhythmical contractions of the Heart are *dependent* upon Nervous agency; but there can be no doubt of this dependence in the case of the ordinary movements of Respiration; and they afford a typical example of what is known as "reflex" action (§ 47).

As neither the Physiological nor the Psychical action of the Nervous Mechanism can be properly understood, without some knowledge of its structure,—both as regards the Elementary parts of which it is composed, and the different modes in which these elements are combined and arranged in different Classes of Animals, —an account will now be given of what seems most essential to be known under each of these heads.

SECTION 2. *Elementary Structure of the Nervous System.*

34. Wherever a distinct Nervous system can be made out (which has not yet been found possible in the lowest Animals), it consists of two very different forms of structure, the presence of both of which, therefore, is essential to our idea of it as a whole. We observe, in the first place, that it is formed of *trunks*, which distribute branches to the different parts of the body, especially to the Muscles and to the Sensory surfaces; and of *ganglia*, which sometimes appear merely as knots or enlargements on these trunks, but which in other cases have rather the character of central masses from which the trunks proceed. Thus in Man, the "nervous system of animal life" consists of the Brain and Spinal Cord, which are aggregations of ganglia, and of the trunks and branches

that proceed from them (Fig. 1). In addition to this, he has also a "Nervous system of Organic life," the ganglionic centres of which are scattered through the body (§ 112). In both systems, the trunks are essentially composed of *nerve-fibres;* whilst the ganglionic centres are characterized by the presence of peculiar *cells* connected with these fibres.

35. It is easily established by experiment that the *active powers* of the Nervous system are concentrated in the *ganglia*, while the *trunks* serve as *conductors* of the influence which is to be propagated towards or from them. For, if a trunk be divided in any part of its course, all the parts to which the portion thus cut off from the ganglionic centre is distributed, are completely paralysed; that is, no impression made upon them is felt as a Sensation, and no Motion can be excited in them by any act of the mind. Or, if the substance of the ganglion be destroyed, all the parts which are exclusively supplied by nervous trunks proceeding from it, are in like manner paralysed. But if, when a trunk is divided, the portion still connected with the ganglionic centre which constitutes the Sensorium be pinched, or otherwise irritated, Sensations are felt, which are referred to the points supplied by the separated portion of the trunk; thus showing that the part remaining in connection with the centre is still capable of conveying impressions, and that the ganglion itself receives these impressions and makes them felt as sensations. On the other hand, if the separated portion of the trunk be irritated, Motions are excited in the muscles which it supplies; showing that it is still capable of conveying the motor influence, though cut off from the usual source of that influence.

36. Each *Nerve-fibre* in its most complete form (Fig. 2) consists of a membranous tube[1], lined by a peculiar material composed of a combination of fat and albumen, which is known as the "white substance of Schwann[2];" and this encloses an "axis cylinder[3]," composed of a protoplasmic substance, which seems

Fig. 1.

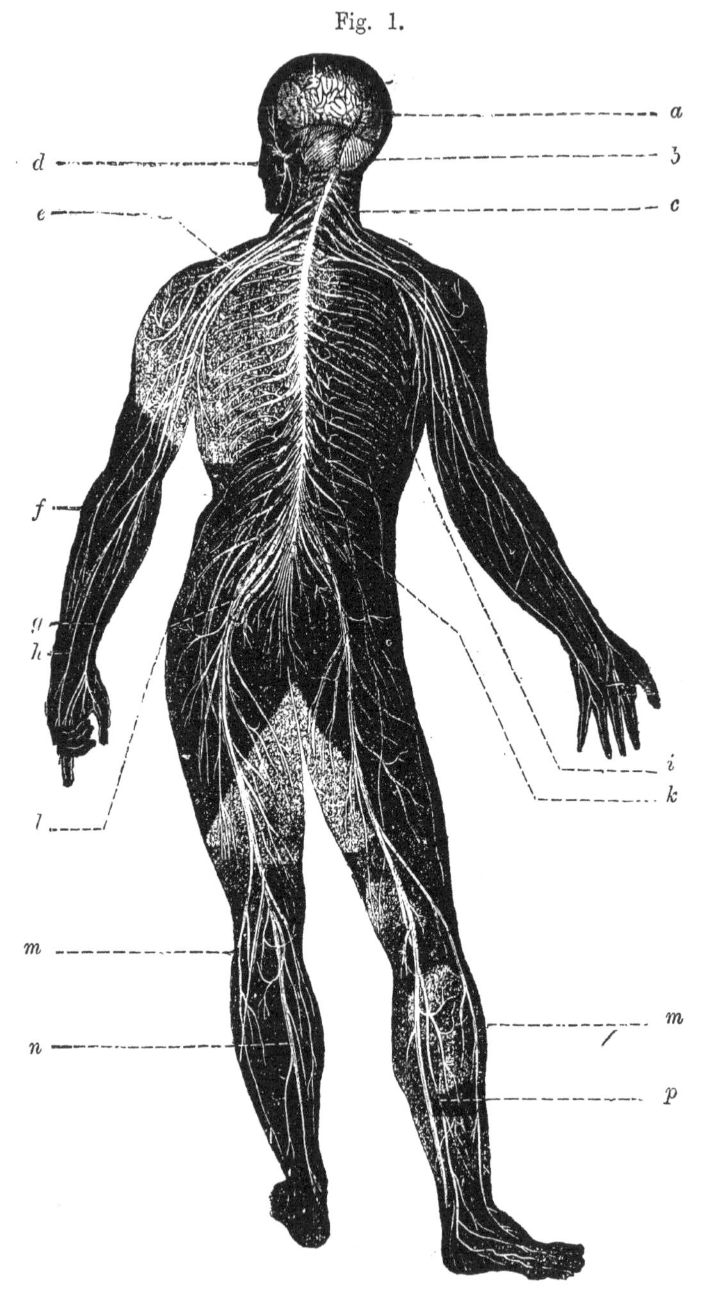

Nervous system of man.

to be the *essential* constituent of the Nerve-fibre. Each fibre appears to maintain its continuity uninterruptedly from its origin to its termination, without any union with other fibres, though bound up closely with them in the same nerve-trunk; and there is strong reason to believe that the "white substance of Schwann" serves as an *insulator*, whereby the axis-cylinders of the contiguous nerve-fibres are kept apart from one another, just as are the numerous wires, each having its own origin and termination, which are bound up together in the aërial cable of the District Telegraph. — The typical form of the *Nerve-cells* or "ganglion-globules" (Fig. 3) may be regarded as globular; but they generally, if not always, have two or more long extensions, which become continuous either with the axis-cylinders of nerve-fibres or with other cells. The nerve-cells, which do not seem to possess a definite cell-wall, are composed of a finely-granular substance, with which pigment-granules are mingled, especially in the warm-blooded Vertebrata; thus giving to their *ganglionic* nerve-substance that reddish-brown hue which causes it to be often designated *grey* or *cineritious* matter; the *tubular* nerve-substance, which contains no pigment-granules, being known as *white* matter. This difference of colour marks the distribution of the two substances in the Nervous centres of Man and the higher Animals

Fig. 2.

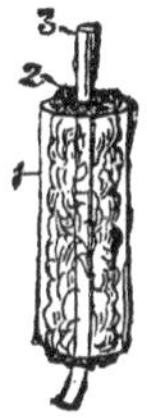

STRUCTURE OF NERVE-FIBRE.

Fig. 3.

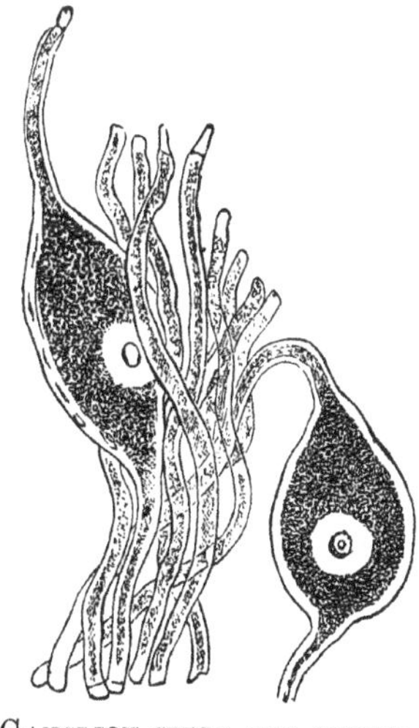

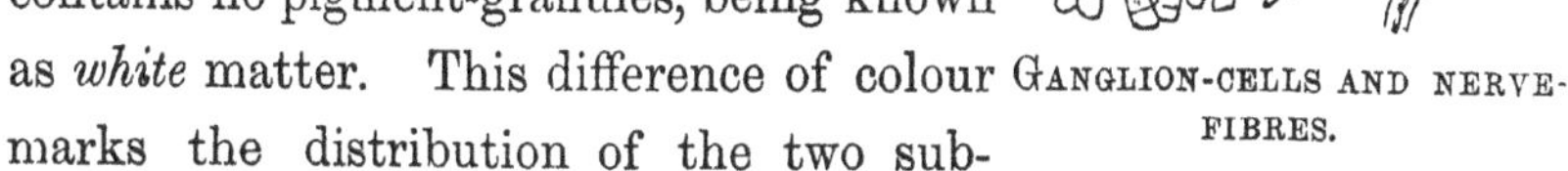

GANGLION-CELLS AND NERVE-FIBRES.

Fig. 1. Nervous System of Man:—*a*, Cerebrum; *b*, Cerebellum; *c*, Spinal Cord; *d*, facial nerve; *e*, brachial plexus, for supply of arm; *f*, radial nerve; *g*, median nerve; *h*, ulnar nerve; *i*, intercostal nerves; *k*, lumbar plexus, and *l*, sacral plexus, for supply of leg; *m m*, fibular nerve; *n*, tibial nerve; *p*, external saphenous nerve.

(Figs. 11—13); but as the pigment-cells are wanting in the lower Classes, the distinction between the two substances is not there recognizable by the eye, and is only to be discerned by the microscope.

37. Every Nerve-fibre, there is reason to believe, is connected at its ganglionic centre with a Nerve-cell, an extension of which forms its axis-cylinder; and through other extensions of the same nerve-cell, it may be brought into connection with other nerve-cells in the same ganglion. The axis-cylinder soon receives its insulating investments, and retains these through almost its whole length. But near its termination, where the fibre separates itself from others, and is proceeding to its ultimate destination, the axis-cylinder escapes (as it were) from its envelopes, and comes into immediate relation with the tissue to which it is distributed. Thus, when supplying a Muscle, the axis-cylinder breaks up into very minute fibrillæ, which seem to inosculate with each other, so as to form a network closely resembling that formed by the *pseudopodia* of *Rhizopods* (Fig. 5); and the like subdivision appears to take place in the axis-cylinders of the fibres which are distributed to the general substance of tissues that are to be endowed only with *ordinary* sensibility. But each of the *papillæ* which constitute the special organs of Touch has a nerve-fibre proceeding to it alone, of which the ultimate subdivisions are distributed upon a little cushion-like pad which it contains; and the ultimate distribution of the nerves in the papillæ of the tongue, which minister to the sense of Taste, seems to be of like character.

38. In the organs of Sight, Hearing, and Smell, however, there is a more special provision for the reception of the peculiar impressions to which they minister. For the Retina of the Eye may be said to be an expanded ganglion, consisting of layers of nerve-cells that seem to be the immediate recipients of the luminous impressions; and the first effect of those impressions appears to be to generate Nerve-force in the nerve-fibres constituting the

Optic nerve, which transmits them to its ganglionic centre forming part of the Sensorium. The like seems to be the case with regard to the sensitive surface which receives the vibrations that excite the sense of Sound; and also with respect to that which is affected by those odorous emanations which excite the sense of Smell. And it is common to these three organs, that neither the ganglionic expansions which receive these special impressions, nor the nerves proceeding from them, minister to *common* sensation; so that either the Optic, the Olfactive, or the Aúditory nerve may be pricked or pinched, without any sign of suffering being called forth. On the other hand, the Eye, the internal Ear, and the interior of the Nose, are endowed with *common* sensibility by other nerves distributed to those parts; so that if these nerves be paralysed, the surface to which they proceed may be touched without the contact being perceived, although neither Sight, Smell, nor Hearing may be impaired, save indirectly.

39. The Nerve-fibres which convey *from* the various parts of the body *to* the ganglionic centres those impressions which there excite Sensations, are called *afferent* or *excitor.** On the other hand, the Nerve-fibres which convey *from* the Ganglionic centres *to* the Muscles the impressions which call forth contractions in the latter, are called *efferent* or *motor.* It is probable that the nature of the Nerve-force excited in each is the same; so that the same fibre might serve either purpose, if its terminals enabled it to do so,—just as the same wire in an Electric Telegraph can convey an electric current in either direction, and can thus serve alike for the transmission of a message and for its reply. But as the terminals of the two sets of Nerve-fibres are essentially distinct, one set serves for the reception of impressions at the circum-

* They were formerly called *sensory;* but this term is inappropriate, since the impressions they convey only affect our Consciousness—*i. e.* excite sensations—when they reach the *Sensorium;* and often excite respondent motions without doing so.

ference, and for their transmission *to* the ganglionic centres; whilst the other serves for the transmission of the impressions that call forth Muscular contraction, *from* the ganglionic centres to the various parts of the circumference.—In most Nerve-trunks, *afferent* and *motor* fibres are bound up together; although, in the ordinary Spinal nerves of Vertebrata, these are connected by separate "roots" with the Spinal Cord which serves as their ganglionic centre (§ 62). But the nerves of special sense (the Olfactive, Optic, and Auditory), which proceed to those special ganglionic centres of which the aggregate constitutes the Sensorium, contain no motor fibres; and there are other nerves of the head in Vertebrata, which are either solely *afferent* or solely *motor* (Fig. 11).

40. The analogy just indicated between the two components of every Nervous System, and the two parts of an Electric Telegraph, —that in which change *originates*, and that which serves as the *conductor*,—holds good to this further extent; that as, for the origination of the Electric current, a certain Chemical reaction must take place between the exciting liquid and the galvanic combination of metals, so is it necessary, for the production of Nerve-force, that a reaction should take place between the Blood, on the one hand, and either the *central* nerve-cells, or the *peripheral* expansions of the nerve-fibres. We do not know, it is true, what is the precise nature of that reaction: but we have the evidence of it in the large supply of Blood which goes to all Organs of Sense,—*i.e.*, to organs which are adapted for receiving sensory impressions and transmitting them to the central Sensorium; and, yet more, in the extraordinary proportion that is transmitted to those central organs which receive those impressions, render the Mind cognizant of them as Sensations, and furnish the instrumental conditions of all Psychical operations, as well as of their action upon the Body. Thus, in the case of Man, although the Brain has not ordinarily more than about *one-fortieth* of the

weight of the body, yet it is estimated to receive from *one-sixth* to *one-fifth* of the whole circulating Blood.

41. The immediate dependence of the production of Nerve-force upon a reaction between the Nerve-substance and the Blood, is proved by the effects of suspension of the circulation, whether *local* or *general.* Thus, if the supply of blood to a limb be temporarily interrupted (as by pressure on its main artery), numbness, or diminution of Sensibility, is perceived in it, as well as loss of Muscular power (the hand or foot being "asleep"), until the circulation is re-established. The effect of complete interruption to the blood-supply of the Brain is extremely remarkable. That supply is conveyed into the cavity of the skull of Man and of the higher Vertebrata by *four* arterial trunks, which enter it at no great distance from one another, and then unite into the "circle of Willis;" from which are given off the various branches that distribute arterial blood to every part of the brain-substance. After traversing this, the blood returns by the veins, greatly altered in its chemical composition; especially as regards the loss of free Oxygen, and its replacement by various oxy-compounds of Carbon, Hydrogen, Phosphorus, &c., that have been formed by a process analogous to combustion. Now if *one, two,* or *three* of the arterial trunks be tied, the total quantity of blood supplied to the brain is diminished; but in virtue of the "circle of Willis," no part is entirely deprived of blood; and the functional activity of the brain, though enfeebled, is still maintained. If, however, the *fourth* artery be compressed so as entirely to prevent the passage of blood, there is an *immediate* and *complete* suspension of activity, the animal becoming as unconscious as if it had been stunned by a severe blow; whilst it recovers as soon as the blood is again allowed to flow through the artery. In fact, the "stunned" state produced by a blow on the head, is only secondarily dependent upon the effect of that blow on the Brain, which may have sustained no *perceptible* injury whatever; the state of insensibility being due to the paralysis

of the Heart and suspension of the Circulation, induced by the "shock." For the like insensibility may be the result of a blow on the "pit of the stomach" (acting on the great Solar plexus of nerves, § 112), or of the shock of some overpowering mental Emotion, either of which produces the like paralysis of the heart. Further, if the blood transmitted to the brain, though not deficient in quantity, be depraved in *quality* by the want of Oxygen and the accumulation of Carbonic acid (as happens in Asphyxia), there is a gradually increasing torpor of the mental faculties, ending in complete insensibility. (See also § 472, and Appendix.)

42. Thus, then, the dependence of Nervous power and of Mental activity upon the Physical changes kept up by the Circulation of oxygenated Blood through the brain, can be shown experimentally to be just as direct and immediate, as is the dependence of the Electric activity of a Galvanic battery upon the analogous changes taking place between its Metals and its exciting Liquid. And if we say that Electricity is the *expression* of Chemical change in the one case, how can we refuse to regard Thought as the *expression* of Chemical change in the other?—This view is not here advanced as *explaining* any Mental phenomenon. No Physicist would say that he can "explain" how it is that Electricity is generated by Chemical change: but he knows that such a relation of cause and effect exists between the two orders of phenomena, that every Chemical change is accompanied by a disturbance of Electricity; and thus, whenever he witnesses Electric disturbance, he is led to look for some Chemical change as its Physical cause. And in precisely the same sense, and no other, the Physiologist *must* regard some change in the substance of the Brain as the immediate Physical antecedent of all *automatic* Mental action.—It is the attribute of the Will to utilize this automatic power of the Brain, as it utilizes that of the Muscles; and thus to make the *Ego*, in proportion as he has acquired the mastery over it, a "free agent" (§§ 25—28).

SECTION 3. *Different Forms and Modes of Action of the Nervous Apparatus.*

43. The simplest type of an Animal consists of a minute mass of "protoplasm" or living jelly, which is not yet *differentiated* into "organs;" every part having the same endowments, and taking an equal share in every action which the creature performs. One of these "jelly-specks," the *Amœba* (Fig. 4), moves itself about by changing the form of its body, extemporising a foot (or *pseudopodium*) first in one direction and then in another; and then, when it has met with a nutritive particle, extemporises a stomach for its reception, by wrapping its soft body around it. Another, instead of

Fig. 4.

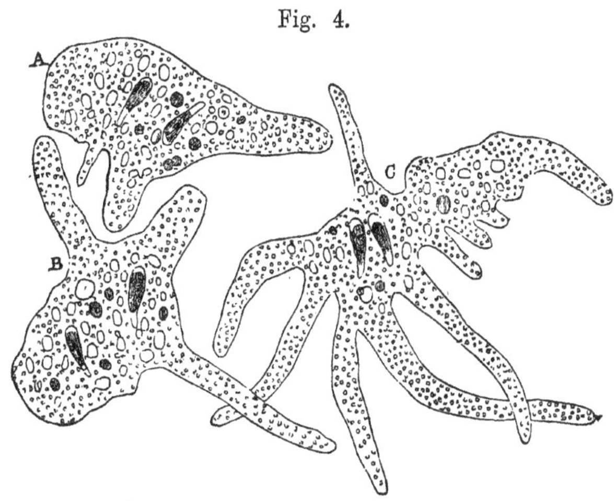

AMŒBA IN DIFFERENT FORMS, A, B, C,

going about in search of food, remains in one place, but projects its protoplasmic substance into long *pseudopodia* (Fig. 5), which entrap and draw-in very minute particles, or absorb nutrient material from the liquid through which they extend themselves, and are continually becoming fused (as it were) into the central body, which is itself continually giving off new pseudopodia.—Now we can scarcely conceive that a creature of such simplicity should possess any distinct *consciousness* of its needs, or that its actions should be directed by any *intention* of its own; and yet the Writer has lately found results of the most singular elaborateness to be wrought-out by the instrumentality of these minute "jelly-specks,"

Fig. 5.

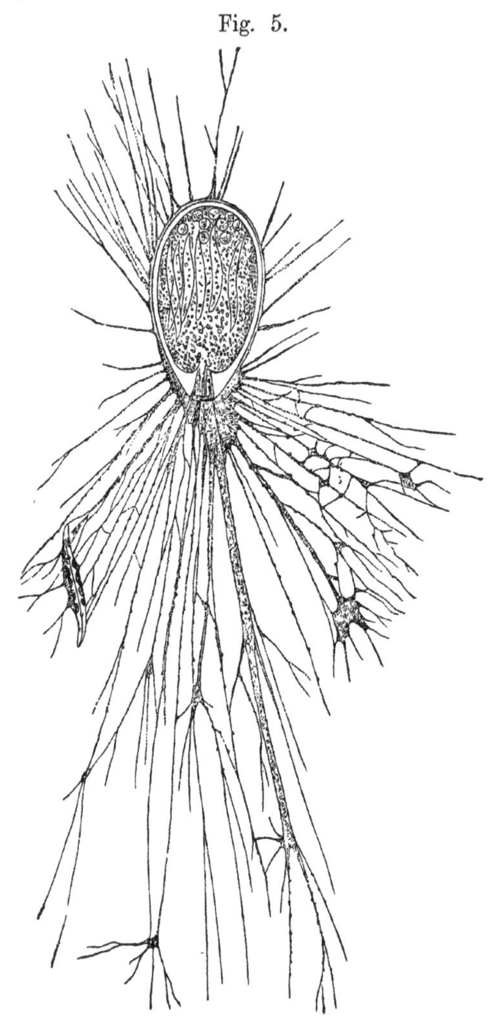

GROMIA, WITH EXTENDED PSEUDOPODIA.

which build-up "tests" or casings of the most regular geometrical symmetry of form, and of the most artificial construction.

a. Suppose a Human mason to be put down by the side of a pile of stones of various shapes and sizes, and to be told to build a dome of these, smooth on both surfaces, without using more than the least possible quantity of a very tenacious but very costly cement in holding the stones together. If he accomplished this well, he would receive credit for great intelligence and skill.—Yet this is exactly what these little "jelly-specks" do on a most minute scale; the "tests" they construct, when highly magnified, bearing comparison with the most skilful masonry of Man. From *the same sandy bottom*, one species picks up the *coarser* quartz-grains, cements them together with *phosphate of iron* secreted from its own substance, and thus constructs a flask-shaped "test" having a short neck and a single large orifice. Another picks up the *finest* grains, and puts them together with the same cement into perfectly spherical "tests" of the most extraordinary finish, perforated with numerous small pores, disposed at pretty regular intervals. Another selects the *minutest* sand-grains and the terminal portions of sponge-spicules, and works these up together,—apparently with no cement at all, by the mere "laying" of the spicules, — into perfect white spheres, like homœopathic globules, each having a single fissured orifice. And another, which makes a straight many-chambered "test," that resembles in form the chambered shell of an Orthoceratite—the conical mouth of each chamber projecting into the cavity of the next,—while forming the walls of its chambers of ordinary sand-grains rather loosely held together, shapes the conical mouths of the successive chambers by firmly cementing together grains of *ferruginous* quartz, which it must have picked out from the general mass.

To give these actions the vague designation "instinctive," does not in the least help us to account for them; since what we want, is to discover the *mechanism* by which they are worked out; and it is most difficult to conceive how so artificial a selection can be made by a creature so simple.

b. The Writer has often amused himself and others, when by the sea-side, with getting a *Terebella* (a marine Worm that cases its body in a sandy tube) out of its house, and then, putting it into a saucer of water with a supply of sand and comminuted shell, watching its appropriation of these materials in constructing a new tube. The extended tentacles soon spread themselves over the bottom of the saucer, and lay hold of whatever comes in their way, "all being fish that comes to their net;" and in half an hour or thereabouts the new house is finished, though on a very rude and inartificial type.—Now here the organization is far higher; the instrumentality obviously serves the needs of the animal, and suffices for them; and we characterize the action, on account of its uniformity and its apparent *un*-intelligence, as Instinctive.

44. We can only surmise that, in these humble *Rhizopods*, as the whole of each "jelly-speck" possesses the attribute of contractility elsewhere limited to Muscles, so may the attributes which are restricted in the higher types of Animal life to the Nervous apparatus, be there diffused through every particle,—the whole protoplasmic substance being endowed in a low degree with that power of receiving, conducting, and reacting upon external impressions, which is raised to a much more exalted degree when limited or *specialized* in the Nervous system. As we ascend the Animal series, and meet with a progressive differentiation of special structures, the general substance of the body loses the endowments which characterize it in the Rhizopod; and wherever we find a definite Muscular apparatus with Sensory organs, there is a strong presumption that there must also be a definite Nervous system, whose action may be purely *internuncial*,—that of calling forth Muscular movements in respondence to the impressions made by external agencies. The apparent absence of a Nervous system is doubtless to be attributed in many instances to the general softness of the tissues of the body, which prevents it from being clearly made-out among them. And we might justly expect to find it bearing a much smaller proportion to the entire structure, in these lowest Animals whose functions are chiefly Vegetative, than in the higher classes, in which the vegetative functions merely serve for the develop-

ment and subsequent maintenance of the Apparatus of Animal life (§ 30).

45. Perhaps the simplest form of a definite Nervous system is that presented by the *Ascidian Mollusks*: for, their bodies not possessing any repetition of similar parts,—either around a common centre as in the *Star-fish*, or longitudinally as in the *Centipede*,—their Nervous system is destitute of that multiplication of ganglia which we see in those animals; whilst the limited nature of their Animal powers involves a corresponding simplicity in their instrument. An Ascidian (Fig. 6) consists essentially of an external membranous bag or "mantle," within which is a Muscular envelope, and again within this a Respiratory sac, which may be considered as the dilated pharynx of the animal. At the bottom of this last is the entrance to the stomach, which, with the other viscera, lies at the lower end of the muscular sac. The external envelopes have two orifices; a mouth (*a*) to admit water into the pharyngeal sac; and a vent (*b*) for the expulsion of the water which has served for respiration, and of that which has passed through the alimentary canal, together with the fæcal matter, the ova, &c. A current of water is continually being drawn into the pharyngeal sac, by the vibration of the cilia that line it; and part of this is driven into the stomach, conveying to it the necessary supply of aliment in a very finely divided state; whilst a part is destined merely for the aëration of the circulating fluid, and is transmitted more directly to the vent after having served that purpose. These animals are for the most part fixed to one spot, during all save the earliest period of their existence; and they give but little external manifestation of life, beyond the continual entrance and

Fig. 6.

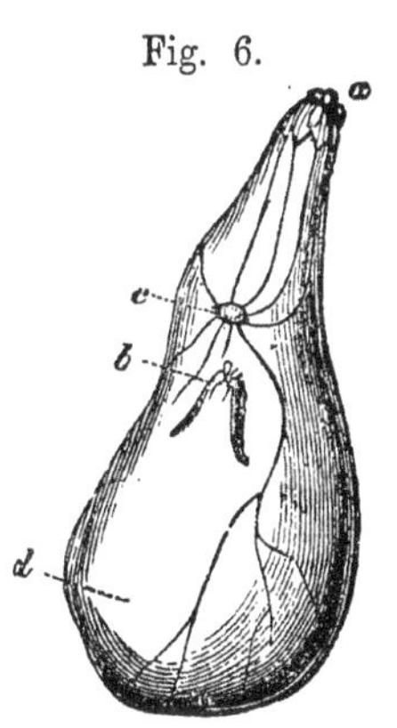

NERVOUS SYSTEM OF ASCIDIAN:—*a*, mouth; *b*, vent; *c*, ganglion; *d*, muscular sac.

exit of the currents just mentioned, which, being driven by ciliary action, are altogether independent of the Nervous system. When any substance, however, the entrance of which would be injurious, is drawn-in by the current, its presence excites a general contraction of the muscular envelope; and this causes a jet of water to issue from one or both orifices, which carries the offending body to a distance. And in the same manner, if the exterior of the body be touched, the muscular envelope suddenly and violently contracts, and expels the contents of the sac.

46. These are the only actions, so far as we know, to which the Nervous system of these animals is subservient. They scarcely exhibit a trace of eyes, or of other organs of special sense; and the only parts that appear peculiarly sensitive, are the small "tentacula" or feelers that guard the oral orifice. Between the two apertures in the mantle we find a solitary ganglion (*c*), which receives branches from both orifices, and sends others over the muscular sac (*d*). This simple apparatus seems to constitute the whole Nervous system of the animal; and it is fully sufficient to account for the movements which have been described. For the impression produced by the contact of any hard substance with the tentacula, or with the general surface of the mantle, being conveyed by the *afferent* fibres to this ganglion, will excite in it a motor impulse; which, being transmitted to the muscular fibres of the contractile sac, as well as to those circular bands that surround the orifices and act as *sphincters*, will call forth the movements in question.

47. We have here a characteristic example of what is designated as the *reflex* action of a Nerve-centre; being the response which it makes, through the *motor* fibres, to the impression that has been conveyed to it by the *afferent* or *excitor* fibres,—the whole constituting what has been termed the *nervous circle*. This response is purely *automatic* or involuntary; depending, like the contraction of a Muscle stimulated by electricity, upon the

inherent endowments of the Nervous apparatus. Whether such "reflex action" is or is not attended with Consciousness, depends on the other endowments of the ganglion which performs it; but it is certain that actions which *seem* to indicate a definite *purpose* and *will*, may be called forth by mere stimulation, under circumstances which forbid us to attribute them to anything else than the automatic and unconscious action of the Nerve-centre (§ 66).—Now the contraction of the muscular sac of the Ascidian, when called forth by the entrance of some irritating particle through the oral orifice, has its precise parallel in the act of *coughing* in ourselves. This is a combined succession of Respiratory movements, consisting of (1) a full inspiration; (2) a closure of the glottis (or aperture of the windpipe); and (3) the bursting open of the glottis by a violent expiratory blast, so that the offending body (such as a particle of food, or a drop of liquid, that has "gone the wrong way,"—or an irritating vapour that has been drawn in with the breath,—or a morbid secretion from the membrane of the air-passages) may be forcibly ejected. Now we are constantly made aware by our own experience, how completely *automatic* this action is; for not only is it performed *without* any will of our own, but even *against* the strongest volitional effort we may make to restrain it; and when we cough voluntarily, as to give a signal, or to put down a tedious speaker, we simply make use of the automatic apparatus. We could not ourselves devise or imagine anything better *adapted* than the above combination, to produce the required result. Yet that combination is assuredly made *for* us, not *by* us. An Infant coughs prior to all experience; and even in a state of entire insensibility, provided the patient can still swallow, coughing will be excited by the passage of any of the food or drink "the wrong way."

48. The act of *swallowing* affords another example of the same reflex action; for though we are accustomed to regard it as

altogether *voluntary*, inasmuch as we only swallow when we choose, yet it is not so in reality. For what the Will does, is to carry back the particle to be swallowed, by a movement of the tongue, so as to bring it into contact with the membrane lining the pharynx; and this contact serves to call the muscles of the pharynx into *automatic* action, whereby the particle is grasped and carried downwards into the gullet. It has several times happened that a feather, with which the back of the mouth was being tickled in order to excite vomiting (another form of reflex action), having been carried down a little too far, has been thus grasped by the pharyngeal muscles, and drawn out of the fingers of the operator.—In *sucking*, again, there is a combination of respiratory movements, producing the vacuum which draws forth the milk, with the movements by which it is swallowed; and the whole combination is a purely reflex action, performed by the instrumentality of a ganglionic centre which forms no part of the Brain proper, and called-forth by the contact, either of the nipple of the mother, or of something which produces the like impression, with the lips of the offspring (§ 69).—This last act is sometimes spoken of as *instinctive*, and has been even taken as a type of that class of operations; and in the broad sense of the term Instinct, it may doubtless be so regarded. But, in common with the ordinary and extraordinary movements of respiration, with swallowing, and, with many other actions that are *immediately* concerned in the maintenance of the Organic functions, it may be executed *unconsciously*; requiring nothing for its performance but an automatic Mechanism of nerves and muscles, which, in its normal state, responds as precisely to the stimulus made upon it, as the Locomotive steam-engine does to the directing actions of its driver.—The actions to which it seems preferable to limit the term *instinctive*, are those to which the prompting is given by *sensations*. These are not less "reflex" than the preceding in their essential nature, being the automatic responses given by the

Nervous mechanism to the impressions made upon it, in virtue of its original or acquired endowments; but the Nerve-centres concerned in them being of a higher order, their reflex activity cannot be called forth without affecting the *consciousness* of the Animal that executes them (§§ 57, 77, 78).

49. In ascending through the *Molluscous* series, we find the Nervous system increasing in complexity, in accordance with the increasing complexity of the general organization; the addition of new organs of special Sensation, and of new parts to be moved by Muscles, involving the addition of new ganglionic centres, whose functions are respectively adapted to these purposes. The possession of a distinct *head*, in which are located the organs of Vision, the rudimentary organs of Hearing, and the organs (if any such exist) of Smell and Taste, constitutes the distinction between the two primary divisions of the series,—the *cephalous* and the *acephalous;* the Snail and Whelk being typical examples of the former, the Oyster and Cockle of the latter. In the Cephalous Mollusks, we always find a pair of ganglia situated in the head; which pair, termed the *cephalic* ganglia, is really made up of several distinct ganglionic centres, and is connected by cords that pass round the œsophagus, with other ganglia disposed in various parts of the trunk. Still, generally speaking, the Nervous system bears but a small proportion to the whole mass of the body; and the ganglia which minister to its *general* movements, are often small in proportion to those which serve some *special* purpose, such as the actions of Respiration. This is what we should expect from the general inertness of the character of these animals (typified by the term *sluggish*), and from the small amount of Muscular structure which they possess.

50. Again, we find no other multiplication of *similar* centres, than a doubling on the two sides of the body; excepting in a few cases in which the organs they supply are correspondingly multiplied,— as in the arms of the *Cuttle-fish*, which are furnished with great

numbers of contractile suckers, every one possessing a ganglion of its own. Here we can trace very clearly the distinction between the *reflex* actions of each individual sucker, depending upon the powers of its own ganglion; and the actions prompted by Sensation, which are called forth through its connection with the Cephalic ganglia. For the Nerve-trunk which proceeds to each arm may be distinctly divided into two tracts; one containing the ganglia which appertain to the suckers and are connected with them by distinct filaments; whilst the other consists of fibres that form a direct communication between these and the Cephalic ganglia. Thus each sucker has a separate relation with a ganglion of its own, whilst all are alike connected with the Cephalic ganglia, and are placed under their control; and we see the results of this arrangement, in the mode in which the contractile power of the suckers may be called into operation. When the animal embraces any substance with its arm (being directed to this action by its Sight or some other sensation), it can bring all the suckers simultaneously to bear upon it; evidently by a determinate impulse transmitted along the connecting cords that proceed from the Cephalic ganglia to the ganglia of the suckers. On the other hand, any individual sucker may be made to contract and attach itself, by placing a substance in contact with it alone; and this action will take place equally well when the arm is separated from the body, or even in a small piece of the arm when recently severed from the rest,—thus proving that when it is directly excited by an impression made upon itself, it is a *reflex* act, quite independent of the Cephalic ganglia, not involving Sensation, and taking place through the medium of its own ganglion alone.*

51. In the *Articulated* series, on the other hand, in which the

* A very curious example of the *independent* activity of the gangliated cord in the arm of the Cuttle-fish, and of its similarity, both in structure and action, to the ventral cord of *Articulata*, is presented in the detached *Hectocotylus*-arm of the male of the *Argonaut* (Paper-Nautilus), which, when first discovered, was mistaken for a Worm.

Locomotive apparatus is highly developed, and its actions are of the most energetic kind, we find the Nervous system almost entirely subservient to this function. In its usual form, it consists of a chain of ganglia connected by a double cord; commencing in the head, and passing backwards through the body (Fig. 7). The ganglia, though they usually appear single, are really double; being composed of two equal halves closely united on the median line. In general we find a ganglion in each segment, giving-off nerves to the muscles of the legs, as in Insects, Centipedes, &c.; or to the muscles that move the rings of the body when no extremities are developed, as in the Leech, Worm, &c. In the lower Vermiform (or worm-like) tribes, especially in the marine species, the number of segments is frequently very great, amounting even to several hundreds; and the number of ganglia increases in the same proportion. But whatever be their degree of multiplication, they seem but repetitions of one another; the functions of each segment being the same with those of the rest. The *cephalic* ganglia, however, are always larger and more important; they are connected with the organs of special Sense; and they evidently possess a power of directing and controlling the movements of the entire body, whilst the power of each ganglion of the trunk is for the most part confined to its own segment.

Fig. 7.

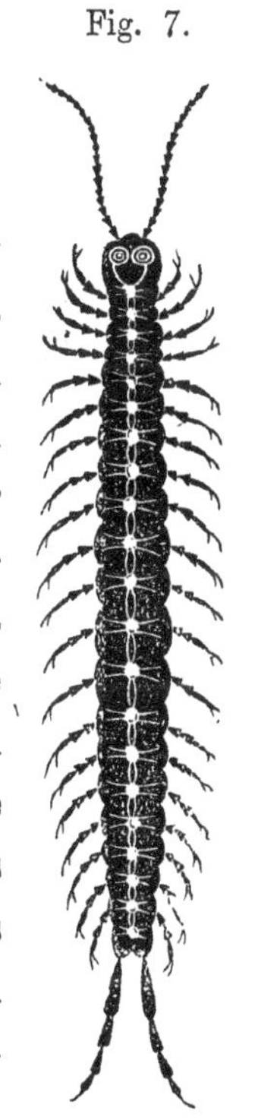

GANGLIATED NERVOUS CORD OF CENTIPEDE.

52. The Cephalic ganglia lie *above* the mouth, in the immediate neighbourhood of the eyes, with which they are connected by nerve-trunks. And from the constancy of the relation between the size of these ganglia and the development of the Visual organs, it cannot be doubted that they are to be regarded as essentially *optic* ganglia, though also containing the

ganglionic centres of the nerves of other Senses, altogether constituting the *Sensorium.*—These Cephalic ganglia are connected with the ganglion of the first segment of the trunk, by a band on either side; and the pair of bands, with the ganglia above and below, form a ring through which the œsophagus passes, so that the chain of ganglia comes to lie nearer the lower or *ventral* surface, *beneath* the alimentary canal, instead of just beneath the *dorsal* surface, *above* the alimentary canal, which is the position of the Spinal cord of Vertebrata. Hence the longitudinal gangliated chain of Articulated animals is often distinguished as the *ventral cord.*

53. A marked difference is observable in the arrangement of the ganglia of the Ventral cord, according as the act of Locomotion is performed by muscles uniformly repeated through the successive segments of the body, as in the crawling of the Maggot or Caterpillar; or by the muscles of special appendages, attached to particular segments, as in the perfect Insect. In the former case, the ganglionic chain is uniform throughout; whilst in the latter, the ganglia of the thorax, with which are connected the nerves that supply the legs and wings, are greatly increased in size, whilst those of the abdomen, the segments of which no longer take any share in the act of locomotion, are proportionally reduced. The change from one condition to the other takes place during the metamorphosis.—When the structure of the Ventral cord is more particularly inquired into, it is found to consist of two distinct *tracts;* one of which, composed of *nerve-fibres* only, passes backwards from the Cephalic ganglia over the surface of all the ganglia of the trunk: whilst the other includes the collections of nerve-cells which constitute *ganglia.* Hence every part of the body has two sets of nervous connections; one with the ganglion of its own segment, and another with the Cephalic ganglia. Each of the ganglia of the Ventral cord ministers to the reflex actions of its own segment, and, to a certain extent also, to those of other segments: for by

the peculiar arrangement of the fibres of the Cord, an impression conveyed by an *afferent* fibre to any one of these ganglia may excite contraction in the muscles of the *same* side of its own segment, or in those of the *opposite* side, or in those of segments at a greater or less distance, according to the point at which the *motor* fibres leave the cord. On the other hand, impressions made upon the afferent fibres which proceed from any part of the body to the Cephalic ganglia, give rise to *sensations* when conveyed to the latter; whilst, in response to these, the influence of the Sensations received through the Cephalic ganglia, being reflected through the motor fibres proceeding from them, harmonizes and directs the general movements of the body.

54. The general conformation of Articulated animals, and the arrangement of the parts of their Nervous system, render them peculiarly favourable subjects for the study of the *reflex* actions; some of the principal phenomena of which will now be described.—If the head of a Centipede be cut off whilst it is in motion, the body will continue to move onwards by the action of its legs; and the same will take place in the separate parts, if the body be divided into several distinct portions. After these actions have come to an end, they may be excited again by irritating any part of the Nerve-centres, or the cut extremity of the nervous cord. The body is moved forwards by the regular and successive action of the legs, as in the natural state; but its movements are always forwards, never backwards, and are only directed to one side when the forward movement is checked by an interposed obstacle. Hence, although they might *seem* to indicate Consciousness and a guiding Will, they do not do so in reality; for they are performed as it were "mechanically;" and show no direction of object, no avoidance of danger. If the body be opposed in its progress by an obstacle of not more than half of its own height, it mounts over it and moves directly onwards, as in its natural state; but if the obstacle be equal to its own height, its progress is arrested,

and the cut extremity of the body remains forced-up against the opposing substance, *the legs still continuing to move.*—If, again, the Ventral cord of a Centipede be divided in the middle of the trunk, so that the hinder legs are cut off from connection with the Cephalic ganglia, they will continue to move, but not in harmony with those of the fore-part of the body; being completely paralysed, as far as the animal's controlling power is concerned, though still capable of performing reflex movements by the influence of their own ganglia, which may thus continue to propel the body in opposition to the determinations of the animal itself. —The case is still more remarkable when the Ventral cord is not merely divided, but a portion of it is entirely removed from the middle of the trunk: for the anterior legs still remain obedient to the animal's control; the legs of the segments from which the nervous cord has been removed are altogether motionless; whilst those of the posterior segments continue to act through the reflex powers of their own ganglia, in a manner which shows that the animal has no power of checking or directing them.

55. Another curious phenomenon of this kind is presented by the *Mantis*, a large Insect allied to the Grasshoppers and Crickets, but of less active habits; its conformation fitting it to lie in wait for its prey, rather than to go in search of it. The first segment of its thorax is greatly prolonged, and is furnished with a pair of large and strong legs, ending in sharp claws; whilst the two posterior segments, and the legs attached to them, are of the ordinary type. From its resting on these last, and lifting up the first segment, with its legs stretched out as arms, in the attitude of prayer, though really in readiness for the capture of prey, the *Mantis* is regarded by the peasantry of Italy and the South of France, where it is common, with superstitious veneration, under the name of *Prie-Dieu*, and has hence acquired the specific name of *religiosa.* Now, if the head be cut off, the body still

retains its position, and resists attempts to overthrow it; while the arms close round anything that is introduced between them, and impress their claws upon it. But further, if the first segment of the thorax with its attached members be cut off, the posterior part of the body will still remain balanced upon the four legs that support it, not only resisting any attempts to overthrow it, but recovering its position when disturbed, and performing the same agitated movements of the wings and wing-covers as when the entire Insect is irritated; while the arms attached to the separated segment of the thorax will still act in the manner just described. Hence it is obvious that the ordinary movements of this Insect *immediately* depend on the reflex powers of the ganglia of the Ventral cord; and that while the prey is actually *captured* by their instrumentality, the control exercised over these movements by the Cephalic ganglia serves to *direct* them towards the prey,—just as our own movements in walking, which are themselves *acquired* reflex actions of the Spinal cord (§ 71), are still *directed* by the Sight, while maintained without either Volitional or even *conscious* effort.

56. The stimulus to the Reflex movements of the legs, in the foregoing cases, appears to be given by the contact of the extremities with the solid surface on which they rest. In other cases, the appropriate impression can only be made by the contact of liquid: thus a *Dytiscus* (a kind of water-beetle) from which the Cephalic ganglia had been removed, remained motionless so long as it rested upon a dry surface; but when cast into water, it executed the usual swimming motions with great energy and rapidity, striking all its comrades to one side by its violence, and persisting in these for more than half an hour.—Other movements, again, may be excited through the Respiratory surface. Thus, if the head of a *Centipede* be cut off, and, while the trunk remains at rest, some irritating vapour (such as that of ammonia or muriatic acid) be caused to enter the air-tubes on one side of it through the spiracles

or breathing-pores of that side, the body will be immediately bent in the opposite direction, so as to withdraw itself as much as possible from the influence of the vapour; if the same irritation be then applied on the other side, the reverse movement will take place; and the body may be caused to bend in two or three different curves, by bringing the irritating vapour into the neighbourhood of different parts of either side. This movement is evidently (like the acts of Coughing and Sneezing in the higher animals, § 47) a reflex one, and serves to withdraw the entrances of the air-tubes from the source of irritation.

57. From these and similar facts it appears that the *ordinary movements* of the legs and wings of Articulated animals are of a simply-reflex nature, being effected solely through the ganglia with which these organs are severally connected; whilst in the perfect creature they are harmonized, controlled, and directed by the guidance they receive from the Cephalic ganglia, which combines them into those *composite* movements which are distinguished as *instinctive*. This designation is now properly restricted to actions which, being performed without any guidance from experience, and executed in precisely the same manner (when the circumstances are similar) by all the individuals of a species, must be regarded as proceeding from an innate or constitutional tendency, corresponding with that which prompts our own primarily-automatic movements (§ 15). Instinctive actions, then, are as truly "reflex" in their character as are those we have been already considering, but differ from them only in their greater complexity; a combination of many separate impressions being needed to call them forth, and a combination of many distinct movements being concerned in their execution. The special *directing* power exerted by the Cephalic ganglia obviously depends upon their Sensorial attributes; for the directness of their connection with the organs of special Sense, and the constancy of the proportion which their size bears to the develop-

ment of the Eyes, places it beyond doubt that they furnish the instrumentality whereby (1) the Animal is rendered *conscious* of Sense-impressions, and (2) that Consciousness prompts and directs its actions. Thus the truly Instinctive actions of the lower Animals correspond in character with the *Sensori-motor* or *consensual* actions in Man (§§ 78, 79), but constitute a far larger proportion of their entire life-work. In fact, it would appear that *Instinct* culminates in the ARTICULATED series, and especially in the class of *Insects*; just as *Intelligence* does in the VERTEBRATED series, of which *Man* is the highest representative. In proportion as *Instinct* predominates, may we predict with certainty the actions of the individual, when we know the life-history of the species; its whole aim being to work out a design which is formed *for* it, not *by* it, and the tendency to which is embodied (as it were) in its organization. In proportion, on the other hand, as the lower animals possess any share of the Rational nature of Man, which enables them to profit by experience, the mental processes which determine their actions become more complex and seem more variable in their results, so that our power of accurate prediction proportionally diminishes. Of this we have a curious illustration in the contrast between the Architectural operations of Insects and those of Birds (§ 82).

58. The most remarkable examples of *instinctive* action that the entire Animal Kingdom can furnish, are presented in the operations of Bees, Wasps, Ants, and other Social Insects; which construct habitations for themselves upon a plan which the most enlightened Human intelligence, working according to the most refined geometrical principles, could not surpass; but which yet do so without education communicated by their parents, or progressive attempts of their own, and with no trace of hesitation, confusion, or interruption; the several individuals of a community all labouring effectively to one common end, because their Instinctive or Consensual impulses are the same.

—It *might*, indeed, be argued in the case of *Hive-Bees* (on whose life-history our notions of the range of Instinct are chiefly founded), that the extraordinary perfection of their workmanship, and the uniformity of the course they take under each of a great variety of contingencies, are to be accounted for by the experiential acquirement of *knowledge*, progressively improved, and transmitted from one generation to another ; but this cannot possibly be admitted in the case of certain of the *solitary Bees*. For with regard to these it may be positively affirmed that the offspring *can* know nothing of the construction of its nest, either from its own experience, or from instruction communicated by its parent ; so that when it makes a nest of the very same pattern, we cannot regard it as anything else than *a machine* acting in accordance with its Nervous organization,—unless we suppose its actions to be *directly* prompted by "an overruling mind or purpose" *outside* itself, which takes them out of the category of Scientific investigation.—Still, that even Insects *can* learn by experience, must be obvious to those who study the actions of Bees when they have been newly hived ; for if the hive be placed among several others having similar entrances, the bees are obviously undecided, for the first few days, which entrance to make for ; but soon come to recognize their own, as is shown by the straightness of their flight towards it. And Sir John Lubbock has succeeded in taming a Wasp to perform various actions that indicated a *purposive* direction guided by its individual experience.

59. In the change from the *larva* to the perfect or *imago* state of the Insect, besides the modifications already noted (§ 53), the Cephalic ganglia undergo a great increase in size. This evidently has reference to the increased development of the organs of special Sense in the latter ; the Eyes being much more perfectly formed, Antennæ and other appendages used for feeling being evolved, and organs of Hearing and Smell being added. In respondence to the new sensations which the animal must thus acquire, a great

number of new instinctive actions are manifested; indeed it may be said that the instincts of the perfect Insect have frequently nothing in common with those of the Larva. The former chiefly relate to the acts of reproduction, and to the provisions requisite for the deposit and protection of the eggs and for the early nutrition of the young; the latter have reference solely to the acquirement of food. The *larva*, indeed, may be regarded as a mere active embryo, which comes forth from the egg in an extremely immature condition, and then, having taken into itself an enormous amount of additional nutriment, goes back (as it were) into the quiescent state, in which this store of nutriment is applied to the development of the organs that characterize the perfect Insect. And there is evidence of an extremely curious kind, that the course of that development, and the nature of the instinctive tendencies which show themselves in the mature individual, are capable of being determined in certain cases by conditions purely Physical:—

a. The "workers" among Hive-Bees are not really "neuters," but are undeveloped females; every one of them being originally a *potential* Queen. They differ from the queen, or fertile female, however, not merely in the non-development of the reproductive organs (which shows itself in the inferior length of the abdomen), but also in the possession of the "pollen-baskets" on the thighs, which are used in the collection of pollen and propolis, and in the conformation of the jaws and antennæ. But they differ yet more in their instincts; for whilst the life-work of the Queen is to lay eggs, that of the Workers is to build cells for their reception, to collect and store up food, and to nurture the larvæ,—this nurturing process being continued as a sort of incubation during the pupa-state. The Worker-larvæ which come forth from the eggs that are laid in *ordinary* cells, are fed for three days upon a peculiar substance of jelly-like appearance, prepared in the stomachs of the workers; but afterwards upon "bee-bread" composed of a mixture of honey and pollen. The Queen-larvæ, on the other hand, are reared in larger *royal* cells of peculiar construction; and they are fed during the whole of the larva-period upon the substance prepared by the workers, which is hence known as "royal jelly." The length of time occupied in their development is different; the preliminary stages of the Queen being passed through in sixteen days, whilst those of the Worker require twenty-one.

b. Now it sometimes happens that, from some causes not understood, there is a failure in the production of young Queens, so that there are none forthcoming when wanted. The workers then select either worker-eggs or worker-larvæ not yet three days old; and around these they construct "royal cells," by throwing together several adjacent worker-cells, and destroying the larvæ they contain. The selected larvæ are fed with the "royal jelly," and are treated in every respect as Queen-larvæ; and in due time they come forth as *perfect Queens—thus having had not only their bodily organization, but their psychical nature, essentially altered by the nurture they have received.*

This last action is one which it is scarcely possible that either theory or experience could lead the Bees to perform: for not the most ingenious reasoning could have anticipated the fact, that by supplying a worker-larva with food of a different quality, and enlarging the cell around it, a change so remarkable should be produced in its structure, capacities, and instincts; and the circumstances of the case seem no less to forbid the notion that the Bees owe a knowledge of the process to experimental researches carried on either by themselves or by their ancestors, for the purpose of rocuring an artificial supply of queens when the natural supply fails. That recourse is uniformly had to it whenever the case requires, has been repeatedly shown by experiment; the removal of the parent-queen and of the royal larvæ from the hive, being always followed by the manufacture (so to speak) of worker-larvæ into new queens.—The *irrationality* of the impulse which prompts the Bees to this action, is evidenced by its occasional performance under circumstances which, if they could reason, would have shown them that it *must* be ineffective. A case has been recorded, in which a Queen, having only laid *drone* or male eggs, was stung to death by the workers, who cast her body out of the hive; but being thus left without a queen, and no royal larvæ being in process of development to replace her, the workers actually tried to obtain a queen by treating *drone*-larvæ in the usual manner,—of course without effect.

60. Thus, then, while the Human organism may be likened to a keyed instrument, from which any music it is capable of producing can be called-forth at the will of the performer, we may compare a Bee or any other Insect to a barrel-organ, which plays with the greatest exactness a certain number of tunes that are set upon it, but can do nothing else.—The following fact, mentioned by Pierre Huber, affords a curious example of the purely *automatic* nature of instinctive action :—

There is a Caterpillar that makes a very complicated hammock, the construction of which may be divided into six stages. One of these caterpillars which had completed its own hammock, having been transferred to another carried only to its third stage, completed this also by reperforming the fourth, fifth, and sixth stages. But another caterpillar taken out of a hammock which had been only carried to its third stage, and put into one already completed, appeared much embarrassed, and seemed forced to go back to the point at which it had itself left off, executing anew the fourth, fifth, and sixth stages which had been already wrought out.

61. While perfection in the Articulated series consists in the high development of that portion of the Nervous system which is immediately connected with the organs of Sense and of Motion, and which ministers to Instinct, perfection in the VERTEBRATED series shows itself in the high development of a superadded organ, the *Cerebrum* (Fig. 9), which is the instrument of Intelligence; of this scarcely any trace is found in the Invertebrated classes, whilst but a mere rudiment presents itself in the lowest class of Vertebrata.—Notwithstanding the marked difference in general plan of structure between an Insect and a FISH, the Physiologist recognizes a close correspondence in the essential characters of their Nervous systems. For the *Spinal cord* of the

latter is but a continuous series of ganglionic centres, directly connected with the Muscular apparatus of locomotion; whilst its *Brain* consists of several pairs of ganglia, which are for the most part, like the Cephalic ganglia of Insects, the immediate centres of the Sensory nerves. The *Spinal cord* (commonly termed the spinal marrow) is not, as was formerly supposed, a mere bundle of Nerves proceeding from the Brain; for, whilst serving to connect the Brain with the Nerve-trunks that supply the body generally, it is also an independent centre of reflex action. Although externally composed of longitudinal strands of *fibrous* substance, which, like that of the Nerve-trunks, acts mainly as a *conductor* of Nerve-force (§ 35), contains a sort of core of *ganglionic* substance, which enlarges in the parts of the Cord that give off the nerve-trunks supplying the locomotive members (§ 64). Although there is no actual division of this ganglionic matter into separate segments, as in the gangliated Ventral cord of Articulata (§ 51), yet their segmental division is marked in the regular succession of pairs of nerve-trunks (Fig. 1), which issue from it between the successive Vertebræ that make up the Spinal column. And these Nerve-trunks, like those of the gangliated cord of Articulata (§ 53), have two sets of connections with it: some of their fibres being traceable into its ganglionic substance, which is the centre of the reflex actions of each particular segment; whilst others are connected with its fibrous strands, and either pass into the ganglionic substance of the Cord at some distance above or below, or proceed continuously upwards towards the Brain. Thus, of their *afferent* fibres, some call forth reflex actions, either through their own segment of the Spinal cord, or through other segments above or below; whilst others convey those impressions to the Sensorium, which there call forth Sensations. And of their *motor* fibres, some are excited to action by the reflex power of the segment of the Cord from which they seem to issue, and others by that of segments above or below; whilst some execute the mandates of the Sensorial centres whose seat is in the head.

62. It is only in Vertebrate animals that a distinctness can be shown to exist between the *afferent* and the *motor* nerves: the proof of this distinctness being experimentally obtainable (1) through the separate origination of the two sets of fibres which are bound up in the *trunk* of each Spinal Nerve, by two bundles of *roots* (Fig. 11, 13, 14), of which the *posterior* are afferent, whilst the *anterior* are *motor;* and (2) through the distinct functions of some of the Nerves of the Head,—of which the Third, Fourth, and Sixth pairs, which supply the muscles of the Eye, the Seventh pair, which supplies the muscles of the Face generally, and the Ninth pair, which supplies the muscles of the Tongue, are *motor* only, whilst the Fifth pair is the general *sensory* nerve of the Face, having motor fibres only in its third division, which supplies the muscles of Mastication (Fig. 11). These nerves arise from that upward prolongation of the Spinal cord into the cavity of the skull, which is known as the *Medulla oblongata,* and which corresponds with the two lateral cords that diverge in Articulata to let the œsophagus go through (§ 52). As the whole Cerebro-spinal tract of Vertebrata lies between the Alimentary canal and the *dorsal* aspect of the body, there is no such divergence between the two lateral halves of their Medulla oblongata; but there is a *fissure* between them, which, obvious enough in Fishes (Fig. 8), is almost entirely closed in the higher Vertebrata, and is completely covered in by the Cerebellum. It is in the Medulla oblongata that the special ganglionic centre of the reflex movements of Respiration is lodged: the afferent or excitor nerves from the lungs (the Par vagum, Fig. 11), as well as from the face, proceeding to it; whilst other excitor fibres from the general surface, and the respondent motor fibres which call the respiratory muscles into action, are included in the ordinary nerve-trunks.

63. We should form a very erroneous notion of what essentially

constitutes the *brain* of a Vertebrated animal, and of the mutual relations of the aggregate of ganglionic centres of which it is composed, if we were only to study it in *Man.* For the great relative size and complexity of his *Cerebrum* tends to conceal the fundamental importance of those ganglionic centres on which it is superposed, and which constitute a no less essential part of *his* brain than they do of that of Fishes, although their proportional size is so much less, as to lead to their being commonly regarded as merely subordinate appendages to the Cerebrum. The Brain of a FISH is almost entirely composed of an aggregate of Ganglia of Sense, which may be regarded as collectively constituting its *Sensorium,*— that is, according to ordinary phraseology, the "seat of consciousness," but, more correctly, the Nerve-centre through the instrumentality of which the *Ego* becomes conscious of Sense-impressions. Putting aside the rudimentary *Cerebrum,* therefore, we may regard the *Axial cord* of the Fish (consisting of its Spinal cord with the Sensory ganglia) as the instrument, like the gangliated cord of the Insect, of its *automatic* movements; of which such as are executed through the Spinal centres do not involve Sensation, whilst in those of which the Sensory ganglia are the instruments, Sensation necessarily participates. When, on the other hand, in ascending the Vertebrate series from Fishes toward Man, we compare the different grades of development of the *Cerebrum* (Fig. 9) with the successively augmenting manifestations of *intelligence* (as exhibited in what we must regard as an *intentional* adaptation of means to ends under the direction of *experience*), we find so remarkable a correspondence, as scarcely to leave room for doubt that the Cerebrum is the instrument of those Psychical operations which we rank under the general designation *rational.* In proportion as the actions of an animal are directed by this endowment, the number of them that can be said to be *primarily* automatic, becomes not only *relatively* but *absolutely* limited; although many actions (especially in Man) which were in the first instance initiated by the Will, come after long habit to be as truly Automatic as if they had been so originally (§ 71).

64. In the curious little *Amphioxus* or *Lancelet*, which is the lowest known type of a Vertebrate animal, there is nothing that can be properly called a Brain; and we have here one of those "experiments prepared for us by Nature" (as Cuvier termed them), which show that the Áxial cord is the fundamental portion of the Nervous apparatus of the Vertebrate animal, as it is the first in order of development. The Amphioxus, having no eyes, has no Optic ganglia; and the Spinal cord has no ganglionic enlargement, indicative of any speciality of function, where it enters the head. But the mouth is furnished with a fringe of filaments, which are probably organs of Sense; and the ganglionic centre of their nerves may be considered as the Sensorium.—In others of the lowest Fishes having a Cartilaginous skeleton and a uniform worm-like body, such as the *Lamprey*, the Spinal cord has a like uniformity throughout; and the Brain consists merely of a cluster of ganglia within the skull, which scarcely bear a larger proportion to it, than do the Cephalic ganglia of Insects to the ganglia of their Ventral cord. But with the development of the Eyes and other organs of special Sense, we find the ganglionic centres of their nerves presenting a greatly increased size. The Brain of the *Cod*, viewed from above (Fig. 8, A), shows a series of three pairs of ganglia, lying in the same line with the Spinal cord: of which the first, *ol*, are the *olfactive* ganglia, or centres of the sense of Smell; while the third, *op*, which are the largest of all the ganglionic masses, are the *optic* ganglia, or centres of the sense of Sight. Between these is a pair of ganglionic masses, *ch*, which are usually designated as the rudiments of the *Cerebral Hemispheres*; but they may, perhaps, be more properly regarded as representing the bodies termed *corpora striata*, which, in the Brains of the higher Vertebrata, form part of that series of ganglionic masses

lying along the floor of the skull, on which the Cerebrum is superposed (Figs. 12, 13). Behind the Optic ganglia is a single ganglionic mass, *ce*, the *Cerebellum;* an organ which seems related rather to the regulation of the Movements of the Animal, than to its Psychical faculties, but of which the precise function has not been determined. The Spinal cord, *sp*, is seen to be divided at the top by a fissure, which is most wide and deep beneath the Cerebellum, where there is a complete separation between its two halves.—In the *Shark*, of which, though the skeleton is only cartilaginous, the general organization is very high, we find the *olfactive* ganglia, *ol*, lying at some distance in front of the Cerebral ganglia, *ch*, and connected with them by peduncles or footstalks; the *cerebral* Ganglia are not only relatively much larger, but contain a more distinct rudiment of true Hemispheres separated from the Corpora Striata by a "ventricle" or cavity; the Cerebellum, too, is relatively larger.—In some Fish, separate ganglionic centres of the nerves of Hearing and Taste are found on the under side of the Brain; whilst in others they are imbedded in the Medulla Oblongata, as is the case in Man. In the Vermiform Fishes, the Spinal cord is nearly uniform in size from one end to the other; but in those which have powerful pectoral and ventral fins (the representatives of the fore and hind limbs of land-animals) there is an enlargement of the Spinal cord in the segments which are connected with the nerves of each of these pairs of members.

Fig. 8.

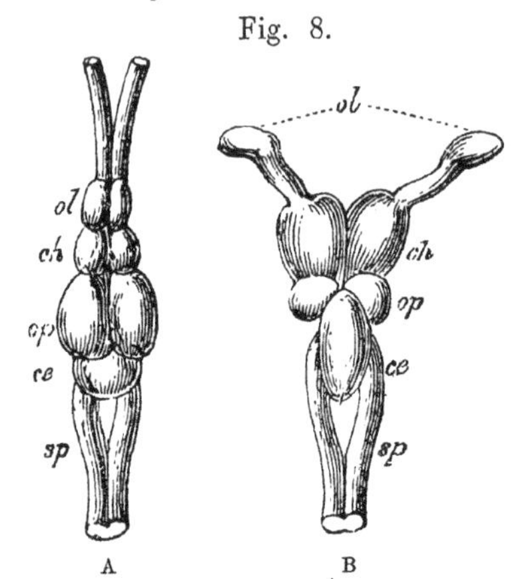

BRAINS OF FISH:—A, Cod; B, Shark; *ol*, Olfactory ganglia; *ch*, Cerebral ganglia; *op*, Optic ganglia; *ce*, Cerebellum; *sp*, Spinal Cord.

65. In REPTILES we do not find any considerable advance in the

development of the Brain, save that the Cerebral Hemispheres are somewhat larger, extending forwards so as to cover-in the Olfactive ganglia, and backwards so as partly to overlie the Optic ganglia (Fig. 9). The Cerebellum is almost invariably small, in conformity with the general inertness of these animals, and the want of variety in their movements. The Spinal cord is still very large in proportion to the Brain; and experiment proves (as will be presently shown) that the greater part of the ordinary movements of these animals are simply reflex, being excited through the afferent nerves proceeding to their ganglionic centre in the Spinal cord, which then reacts on the Muscles through the motor nerves. Where there is a uniformity of motor action through the whole series of Vertebral segments, as in the Serpent, we find a uniformity in the size of the Spinal cord, and in the amount of ganglionic matter it contains, throughout its whole length,—just as in the Ventral cord of the

Fig. 9.

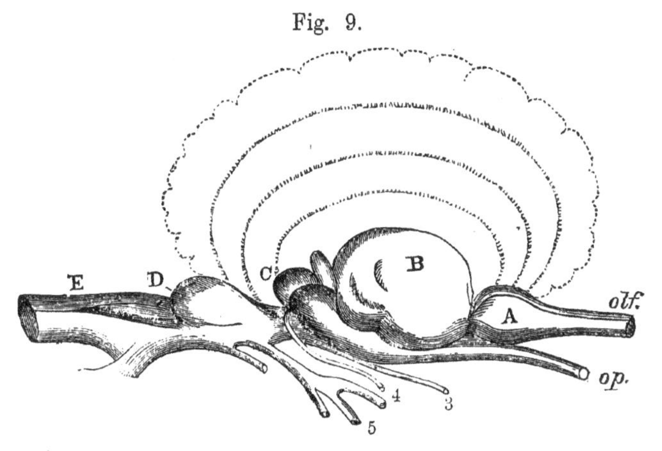

Brain of turtle, with diagrammatic representation of the increased development of the Cerebrum in higher Vertebrata:—A, Olfactory ganglia; B, Cerebral ganglia; C, Optic ganglia; D, Cerebellum; E, Spinal Cord; *olf*, Olfactory nerve; *op*, Optic nerve; 3, Third pair; 4, Fourth pair; 5, Fifth pair.

Centipede (§ 51). But where the locomotive power is delegated to limbs which are appendages of particular segments, we find special ganglionic enlargements of the Spinal cord in those segments,—

just as in the Ventral cord of the perfect Insect (§ 53). Thus in the Lizards and Turtles, whose two pairs of members are nearly similar in locomotive power, we find an anterior and a posterior enlargement of the Spinal cord at the origins of their nerves; whilst in the Frog, whose movements are chiefly effected by its hind legs, the posterior enlargement is the principal.

66. As it has been chiefly by experiments on Frogs and other Reptiles (in which the excitability of the Nervo-muscular apparatus is much longer retained after death than in warm-blooded animals) that the independent endowments of the *Spinal cord* as a centre of Reflex action, have been ascertained, it may be advantageous here to describe the results of these inquiries; and to compare them with what has been learned from observation of the results of disease or accident in Man.—When the Brain has been removed, or its functions have been suspended by a severe blow upon the head, a variety of motions may still be excited by appropriate stimuli. Thus, if the foot be pinched, or burned with a lighted taper, it is withdrawn; and (if the subject of the experiment be a Frog) the animal will leap away, as if to escape from the source of irritation. If the vent of a Frog be irritated with a probe, the hind-legs will endeavour to push it away. And if acetic acid be applied over the upper and inner part of the thigh, the foot of the same side will wipe it away; but if that foot be cut off, after some ineffectual efforts and a short period of inaction, the same movement will be made by the foot of the opposite side.

67. Now the performance of these as well as of many other movements that show a most remarkable adaptation to a purpose, might be supposed to indicate that *sensations* are called-up by the impressions; and that the animal can not only *feel*, but can *voluntarily direct* its movements, so as to get rid of the irritation which annoys it. But such an inference would be inconsistent with other facts.—In the first place, the motions performed by an animal under such circumstances are never spontaneous, but are

always excited by a *stimulus* of some kind. Thus, a decapitated Frog, after the first violent convulsive movements occasioned by the operation have passed away, remains at rest until it is touched; and then the leg or its whole body may be thrown into sudden action, which immediately subsides again. Again, we find that such movements may be performed, not only when the Brain has been removed, the Spinal cord remaining entire, but also when the Spinal cord has been itself cut across, so as to be divided into two or more portions, each of them completely isolated from each other, and from other parts of the nervous centres. Thus, if the head of a Frog be cut-off, and its Spinal cord be divided in the middle of the back, so that its fore-legs remain connected with the upper part, and its hind-legs with the lower, each pair of members may be excited to movement by a stimulus applied to itself; but the two pairs will not exhibit any consentaneous motions, as they will do when the Spinal cord is undivided. Or, if the Spinal cord be cut across, without the removal of the Brain, the lower limbs may be *excited* to movement by an appropriate stimulus, though the animal has clearly no power over them; whilst the upper remain under its control as completely as before. Now it is scarcely conceivable that, in this last case, Sensations should be felt and Volition exercised through the instrumentality of that portion of the Spinal cord which remains connected with the nerves of the posterior extremities, but which is cut-off from the Brain. For, if it were so, there must be two distinct centres of Sensation and Will in the same animal, the attributes of the Brain not being affected; and, by dividing the Spinal cord into two or more segments, we might thus create in the body of one animal two or more such independent centres, in addition to that which still holds its proper place within the head. To say that two or more distinct centres of Sensation and Will are present in such a case, would really be the same as saying that we have the power of constituting two or more distinct *Egos* in one body,—which is manifestly absurd.

68. But the best proofs of this limitation of the endowments of the Spinal Cord, are derived from the phenomena presented by the Human subject, in cases where that organ has suffered injury by disease or accident in the middle of the back. We find that when this injury has been severe enough to produce the effect of a complete division of the Cord, there is not only a total want of Volitional control over the lower extremities, but a complete absence of Sensibility also,—the individual not being in the least conscious of any impression made upon them. But when the lower segment of the Cord remains sound, and its nervous connexions with the limbs are unimpaired, distinct *reflex* movements may be excited in the limbs by stimuli directly applied to them; and this without the least Sensation on the part of the patient, either of the cause of the movement, or of the movement itself:—

a. Among the notes left by John Hunter, there was the record of a case of Paralysis of the lower extremities, in which it appeared that Hunter had witnessed reflex movements of the legs of the patient, occasioned by excitation which did not produce Sensation. When the patient was asked whether he *felt* the irritation by which the motions were excited, he significantly replied—glancing at his limbs,—"No, Sir, but you see *my legs* do." (Of this interesting fact, the Writer was informed by his friend Sir James Paget, to whom Hunter's notes furnished materials for the admirable Catalogue which he drew up of the Pathological portion of the Hunterian Museum.)

b. In a case of Paralysis, recorded by Dr. William Budd, in which injurious pressure on the Spinal cord in the back was produced by angular distortion of the spine, the sensibility of the legs was extremely feeble, and the power of voluntary motion was almost entirely lost. When, however, any part of the skin of the legs was pinched or pricked, the limb thus acted-on jumped with great vivacity; the toes were retracted towards the instep, the foot raised on the heel, and the knee so bent as to raise it off the bed; the limb was maintained in this state of tension for several seconds after the withdrawal of the stimulus, and then became suddenly relaxed. In general, while one leg was convulsed, its fellow remained quiet, unless stimulus was applied to both at once. In these instances, the pricking and pinching were perceived by the patient; but *much more violent* contractions were excited by a stimulus *of whose presence he was unconscious.* When a feather was passed lightly over the skin, in the hollow of the instep, as if to tickle, convulsions occurred in the corresponding limb, much more vigorous than those induced by pinching or pricking; they succeeded one another in a rapid series of jerks, and these were repeated as long as the stimulus was maintained. But when any other part of the limb was irritated in the same way, the convulsions which ensued were very feeble, and much less powerful than those induced by pricking or pinching.—This patient gradually regained both the sensibility of the lower extremities, and voluntary power over them; and as voluntary power increased, the susceptibility to involuntary movements diminished, as did also their extent and power.—This case, then, exhibits an increased tendency to perform reflex actions, when the control of the Brain was suspended; and it also shows that a *slight* impression upon the *surface*, of which the patient was *not conscious*, was more efficacious in exciting the Automatic movements, than were others of a more powerful nature which affected the Sensorium.

c. In another case recorded by Dr. W. Budd, the Paralysis was more extensive and complete, having been produced by an injury (resulting from a fall into the hold of a vessel) at the lower part of the neck. There was at first a *total loss* of Voluntary power over the lower extremities, trunk, and hands; slight voluntary power remained in the wrists, rather more in the elbows, and still more in the shoulders. The Sensibility of the hands and feet was greatly impaired. Recovery took place very gradually; and during its progress, several remarkable phenomena of reflex action were observed. At first, tickling one sole excited to movement that limb only which was acted upon; afterwards, tickling either sole excited both legs, and, on the 26th day, not only the lower extremities, but the trunk and upper extremities also. Irritating the soles, by tickling or otherwise, was at first the only method, and always the most efficient one, by which convulsions could be excited. On the 41st day, a hot plate of metal was applied to the soles, and was found to be a more powerful excitor of movement than any before tried. The movements continued as long as the hot plate was kept applied; but the same plate, at the common temperature, excited no movements after the first contact. Though the *contact* was distinctly felt by the patient, *no sensation of heat* was perceived by him, even when the plate was applied hot enough to cause blistering.—On the first return of Voluntary power, the patient was enabled to restrain in some measure the excited movements; but *this required a distinct effort of his Will;* and his first attempts to walk were curiously affected by the persistence of the susceptibility to reflex excitement. When he first attempted to stand, the knees immediately became forcibly bent under him; this action of the legs being excited by contact of the soles with the ground. On the 95th day this effect did not take place, until the patient had made a few steps; the legs then had a tendency to bend up, a movement which he counteracted by rubbing the surface of the abdomen; this rubbing excited the extensors to action, and the legs became extended with a jerk. A few more steps were then made, the manœuvre was repeated, and so on. This susceptibility to involuntary movements from impressions on the soles, gradually diminished; and on the 141st day, the patient was able to walk about, supporting himself on the back of a chair which he pushed before him; but his gait was unsteady. Sensation improved very slowly: it was on the 53rd day that he first slightly perceived the heat of the metal plate.—Now in this case, the abolition of Common Sensation was not so complete as in the former instance; but of the peculiar kind of impression which was found most efficacious in exciting reflex movements, *no consciousness whatever was experienced.* It is further interesting to remark, that the reflex actions were very feeble during the first seven days, in comparison with their subsequent energy; being limited to slight movements of the feet, which could not always be excited by tickling the soles. It is evident, then, that the Spinal Cord must have been in a state of concussion, which prevented the manifestation of its peculiar functions so long as this effect lasted; and it is easy, therefore, to perceive, that a still more severe shock might permanently destroy its power, so as to prevent the exhibition of any of the phenomena of reflex action.—(*Medico-Chirurgical Transactions*, vol. xxii., 1839.)

69. The dependence of the movements of Respiration and Deglutition (swallowing) upon the independent endowments of the ganglionic centres contained in the upper part of the Spinal cord, (§ 62) is equally well established.—It has occasionally happened that even Human infants have been born alive without any Brain; and have lived and breathed for some hours—crying and even sucking,—though they had no Nerve-centres above the Medulla Oblongata. And new-born puppies, reduced to the same condition by the removal of the whole contents of the skull except the Medulla Oblongata, have continued to perform the same actions. The *independence* of the Ganglionic centres, not only of the ordinary Spinal nerves, but of those which supply the muscles concerned in the acts of Breathing and Sucking, has been thus fully demonstrated. And the purely *reflex* character of the movement of Sucking in the new-born Mammal, is proved by the fact that it is immediately excited in a brainless puppy by introducing the finger moistened with milk between its lips. Now this act requires a combined contraction of a number of muscles,—those of grasping by the lips, those by which a vacuum is produced in the mouth, those of respiration, and those of swallowing,—all of which manifest the most perfect adaptation of means to ends; but it is clear that this adaptation is not made by any *intention* on the part of the Ego, but is the result of the working of its Nervous mechanism. If an Animal from which everything above the Medulla Oblongata has been removed has any Consciousness at all, it can be of no higher kind than that *sense of need,* which we ourselves experience when we hold our breath for a short time, and which directly prompts the movements that tend to its relief, without the least Idea, on our own parts, of the *purpose* which those movements will answer.

70. These facts, taken in connexion with the preceding experi-

ments both upon Vertebrated and Articulated animals, distinctly prove that Sensation is *not* a necessary link in the chain of reflex actions; but that all which is required is the "nervous circle" already described (§ 47). Thus these movements are all *necessarily* linked with the stimulus that excites them; that is, the same stimulus will always produce the same movement, when the condition of the body is the same. Hence it is evident that the Judgment and Will are *not* concerned in producing them, and that the *adaptiveness* of the movements is no proof of the existence of consciousness and discrimination in the being that executes them; such adaptation being made *for* the being—by the peculiar structure of its Nervous apparatus, which causes a certain movement to be executed in respondence to a given impression,—not *by* it. An animal thus circumstanced may be not unaptly compared to an Automaton, in which particular movements, each adapted to produce a given effect, are produced by touching certain springs.

71. It seems not improbable, however, that some of these reflex movements,—such as are performed by the legs of a Frog as if with the purpose of removing a source of irritation (§ 66),—were not *originally* automatic, but have *become* so by habit; these *secondarily* automatic actions (as Hartley well designated them) coming to be performed with the same absence of Will or Intention, as the *originally* or *primarily* automatic. Such is pretty certainly the character we are to assign to the ordinary Locomotive actions of Man. For though we are accustomed to regard these as Voluntary, and although they are so in the sense that we can commence and stop them at will, yet they continue *of themselves*, when—having been once set going—our Will has been entirely withdrawn from them; our whole attention being engrossed by some train of thought of our own, or by conversation with a companion. And it seems clear that in this case the succession of movements is purely reflex; being

sustained by the successive contacts of our feet with the ground, each exciting the next action (§ 16). For numerous instances are on record, in which Soldiers have continued to *march* in a sound sleep; *riding on horseback* (which requires a constant exercise of the balancing power) during sleep is a not unfrequent occurrence; and the Writer has been assured by an intelligent witness, that he has seen a very accomplished Pianist complete the performance of a piece of music in the same state.* A case has been mentioned to him by Dr. William Budd, of a patient subject to sudden attacks of temporary suspension of consciousness without convulsion, who, whenever the paroxysm came-on, persisted in the kind of movement in which he was engaged at the moment; and thus on one occasion fell into the water through continuing to walk onwards, and frequently (being a shoemaker by trade) wounded his fingers with the awl in his hand, by a repetition of the movement by which he was endeavouring to pierce the leather.

72. Now in all these forms of *secondarily*-automatic activity, it seems reasonable to infer that the same kind of connection between the excitor and the motor nerves comes to be formed by a process of *gradual development*, as *originally* exists in the Nervous systems of those animals whose movements are *primarily* automatic; this portion of the Nervous mechanism of Man being so constituted, as to *grow-to* the mode in which it is habitually called into play.—Such an idea is supported by all that we know of the formation and persistence of *habits* of Nervo-muscular action. For it is a matter of universal experience, that such habits are far more readily acquired during the periods of Infancy, Childhood, and Youth, than they are after the attainment of

* In playing by memory on a musical instrument, the *muscular* sense (§ 80) often suggests the sequence of movements with more certainty than the *auditory*; and since it is certain that impressions derived from the Muscles may prompt and regulate successional Movements without affecting the Consciousness, there is no such improbability in the above statement as might at first sight appear.

adult age; and that, the earlier they are acquired, the more tenaciously are they retained. Now it is whilst the Organism is growing most rapidly, and the greatest amount of new tissue is consequently being formed, that we should expect such new connections to be most readily established: and, it is then, too, that the nutritive processes most readily take-on that new mode of action (§ 276), which often becomes so completely a "second nature" as to keep-up a certain acquired mode of Nutrition through the whole subsequent life.—It is an additional and most important confirmation of this view, that (as was shown by Dr. Waller) when a Nerve-trunk has been cut-across, the re-establishment of its conductive power which takes-place after a certain interval, is not effected by the re-union of the divided fibres, but by the *development* of a new set of fibres *beyond* the point of section, in the place of the old ones (which undergo a gradual degeneration), the fibres on the *central* side remaining unaltered. And the same may be pretty certainly affirmed of that complete recovery of Nervous power in the hinder part of the body and limbs, which has been shown to take place by M. Brown-Séquard, after they had been entirely paralysed by complete division of the Spinal cord in the back. For if this recovery had been the result of simple *re-union* of the divided surfaces (as in an ordinary cut finger), the restoration of power,—of which no indication can be perceived for some weeks, and which altogether requires several months—would be much more speedy. The length of time required, which corresponds with that needed in Man after severe injuries to the Spine (§ 68), affords clear evidence that the process in these cases is really one of *regeneration;* and this fact, in connection with many others, shows that the Nervous substance is not only more capable of such complete regeneration than any other tissue in the body, but is in a state of more constant and rapid *change.* It will be shown hereafter (§§ 277-282) how intimate is the relation between Mental and Bodily habits; and how the formation and maintenance of both are dependent on this Nutritive reconstruction of the Nervous apparatus.

73. There are many irregular or abnormal Reflex actions, known as *convulsive*, performed through the instrumentality of the Spinal cord, the study of which is peculiarly instructive. These movements are not produced by injuries of the Cerebral hemispheres; but, in the production of them, the *Sensory ganglia* are often associated with the Spinal cord. They may either be (1) simply *reflex*, being the natural result of some extraordinary irritation; or (2) simply *centric*, depending upon an excited condition of the ganglionic centres of the Spinal cord, which occasions muscular movements without any stimulation; or (3) they may depend upon combined action on both principles; the Nerve-centres being in a highly irritable state, which causes very slight irritations (such as would otherwise be inoperative) to excite violent reflex or convulsive movements. The undue excitability of the Spinal cord may have its origin in an abnormal state of the Blood: thus we know that it may be produced by the introduction of certain poisons (as Strychnia) into the circulation; and it is probable that morbid matters generated within the body may have the same effect, and that the convulsive actions which occur in various diseased conditions of the system are generally due in part to that condition. In the case of the convulsions which are not unfrequent during the period of teething, being immediately excited by the irritation which results from the pressure of the tooth as it rises against the unyielding gum, the stimulus would be insufficient to produce the violent result, were it not for a peculiarly excitable state of the Spinal cord, brought about by various causes, amongst which impure air and unwholesome food are the most potent. In like manner, when such an excitable state exists, to which children are peculiarly liable, convulsions may be occasioned by the presence of intestinal worms, of irritating substances, or even simply of undigested matters, in the alimentary canal; and will cease as soon

as they are cleared-out, in the same manner as the convulsions of teething may often be at once checked by the free lancing of the gums. A change to a purer atmosphere is commonly found the most efficacious means of reducing the morbid excitability of the Spinal cord, and thus of diminishing the liability to the recurrence of the Convulsion.

74. The influence of the condition of the Spinal cord itself is manifested in the convulsive diseases known as Tetanus, Epilepsy, and Hysteria.—In *Tetanus* (commonly known as "lock-jaw") there is a peculiarly excitable state of the Spinal cord and Medulla oblongata, not extending to the higher centres. This may be the result of causes altogether internal; the condition exactly resembling that which may be artificially induced by the administration of Strychnia, or by its application to the Cord. Or it may be first occasioned by some local irritation, as that of a lacerated wound; the irritation of the injured nerve being propagated to the Nerve-centres, and establishing the excitable state in them.—In like manner, *Epilepsy*, which consists in a combination of Convulsive actions with temporary suspension of Consciousness, may result from the irritation of local causes, like the convulsions of teething; and may cease, like them, when the sources of irritation are removed. It appears probable from recent researches, that the sudden but temporary suspension of the functions of the Brain, may be due to the spasmodic contraction of the vessels of the Sensorium, induced by the extension of the reflex motor impulse to the "vaso-motor" Nerves (§ 113). Certain forms of Epilepsy, on the other hand, are distinctly traceable to diseased states of the highest Nerve-centres. (See Appendix.)

75. These and other forms of Convulsive disorder, when productive of a fatal result, usually act by suspending the Respiratory movements; the muscles that effect these being fixed by the spasm, which thus prevents the air from passing either in or out; so that suffocation takes place as completely as if the entrance to the air-

passages were closed.—It is remarkable that nearly every one of them may be imitated by *Hysteria;* a state of the Nervous system which is characterized by its peculiar excitability, but in which there is no such fixed tendency to irregular action as would indicate any positive disease; one form of convulsion often taking the place of another, at short intervals, with the most wonderful variety. This state is generally connected with an undue excitability of the *Emotions;* and, from their known influence on the "vaso-motor" Nerves (§§ 113, 565), it seems likely that many of its manifestations are produced through the instrumentality of that system.

76. Proceeding now to the Class of BIRDS, we find a considerable advance in the character of the Brain as compared with that of Reptiles. The Cerebral hemispheres are greatly increased in size; so as to cover-in not merely the Olfactory ganglia, but in great part also the Optic ganglia. The former are of comparatively small size, the organ of Smell in Birds not being much developed: the latter are very large, in conformity with the acuteness of Sight which is their special characteristic. The Cerebellum is of large size, in conformity with the active and varied muscular movements performed by animals of this class; but it consists chiefly of the central lobe, with little appearance of lateral hemispheres. The Spinal cord is still of considerable size in comparison with the Brain; and it is much enlarged at the points whence the legs and wings originate. In the species which have the most energetic flight, such as the Swallow, the enlargement is the greatest where the nerves of the wings come off; but in those which, like the Ostrich, move principally by running on the ground, the posterior enlargement, from which the legs are supplied with nerves, is much the more considerable.

77. It is not a little curious that Birds,—which present so many points of analogy to Insects in structure as well as in mode

of life, as to have been called the "Insects of the Vertebrated series,"—should strongly resemble Insects also in the high development of their *instincts; i.e.*, the marked tendencies they show to particular kinds of movement, at the stimulus and under the guidance of particular sensations, without any experience to direct them. Thus, even the Chick within the egg sets itself free by tapping with its bill (furnished at that time with a sharp horny scale, which soon afterwards falls off) against the shell that encloses it; and, having once penetrated this, carries its chipping in a regular circle round the large end of the shell, which then drops off. In no long time after its escape, it raises itself upon its legs, and soon begins to run about, and to peck at insects, grains, &c., with a very sure aim. Mr. Spalding, who has recently made a series of very interesting observations on this point, found that if he "hooded" Chicks, or put them into a bag immediately on coming forth from the egg, and kept them so for two or three days until they could run about, the first effect of uncovering their eyes was to produce a sort of stunned condition; but recovering from this in a minute or two, they would *immediately* follow the movements of crawling insects, and peck at them with unerring aim. So, he tells us, chickens hatched and kept in the bag for a day or two, when taken out and kept nine or ten feet from a box in which a hen with chicks was concealed, after standing for a minute or two, uniformly set off straight for the box in answer to the call of the hen, though they had never seen her, and had never before heard her voice. This they did, struggling through grass, and over rough ground, when not yet able to stand steadily upon their legs. Even hooded chicks tried to make their way towards the hen, obviously guided by sound alone. So, on the other hand, a turkey only ten days old, which had never in its life seen a hawk, was so alarmed by the note of a hawk secreted in a cupboard, that it fled in the direction opposite to the cupboard with every sign of terror.*

* *Macmillan's Magazine*, Feb., 1873.

78. Now these actions clearly belong to the class which the Physiologist terms *Sensori-motor* or *Consensual ;* being the reflex actions of that higher division of the Nerve-centres, which consists of the *Sensory ganglia* as distinguished from the Cerebrum. The sense of Sight obviously affords the chief direction of the movements of Birds : and that the Sensory ganglia, in these higher Vertebrata, continue to furnish the instrumentality through which Sensations are excited by impressions made on the organs of Sense, and respondent motions are called forth, appears from the effect of experimental removal of the Cerebral hemispheres of Birds, the Sensory ganglia being left intact. For a Bird thus mutilated maintains its equilibrium, and recovers it when it has been disturbed ; if pushed, it walks ; if thrown into the air, it flies. A Pigeon deprived of its Cerebrum has been observed to seek out the light parts of a partially-illuminated room in which it was confined, and to avoid objects that lay in its way ; and at night, when sleeping with closed eyes and its head under its wing, it raised its head and opened its eyes upon the slightest noise.—So, again, the removal or destruction of either pair of these Sensory centres appears to involve the loss of the particular Sense to which it ministers ; and frequently, also, to occasion such a disturbance in the ordinary movements of the animal, as shows the importance of these centres in regulating them. Such experiments have been chiefly made upon the *Optic* ganglia ; the partial loss of which on one side produces temporary blindness in the eye of the opposite side, and partial loss of muscular power on the opposite side of the body ; whilst the removal of a larger portion, or the complete extirpation of it, occasions permanent blindness and immobility of the pupil, and temporary muscular weakness, on the opposite side. This temporary disorder of the Muscular system sometimes manifests itself in a tendency to move on the axis, as if the animal were giddy ; and sometimes in irregular convulsive movements.—Here, then, we have proof of the

necessity of the integrity of this ganglionic centre, for the possession of the sense of Vision; and we have further proof that the ganglion is connected with the Muscular apparatus by motor fibres issuing from it. The reason why the Eye of the *opposite* side is affected, is to be found in the *crossing* of the optic nerves in their course towards the optic ganglia; whilst the influence of the operation on the Muscles of the *opposite* side of the body, results from the like crossing of the motor fibres in their downward course through the Medulla oblongata.—Similar disturbances of movement have been produced by injuries to the organs of Sense themselves, or to the nerves connecting them with the Sensorial centres. Thus the division of one of the "semicircular canals" of the Ear in pigeons and rabbits has been found to occasion constant efforts to move in the plane of that canal. (See Appendix.)

79. Notwithstanding that, in Man, the high development of *Intelligence*, and the exercise of the *Will*, supersede in great degree the operations of *Instinct*, we still find that there are in ourselves certain movements which can be distinguished as neither Volitional nor Excito-motor; being as truly Automatic as the latter, but requiring that the impressions which originate them should be *felt* as Sensations.—As examples of this group, we may advert to the start upon a loud and unexpected sound; the sudden closure of the eyes to a dazzling light, or on the approach of bodies that might injure them, which has been observed to take place even in cases of paralysis, in which the eyelids could not be voluntarily closed; the act of sneezing excited by an irritation within the nostril, and sometimes also by a dazzling light; the semi-convulsive movements and the laughter called forth by tickling; and the vomiting occasioned by the sight or the smell of a loath-some object. So, again, the act of yawning, ordinarily called forth by certain uneasy sensations within ourselves, is also excited by the sight or hearing of the act as performed by another.—Various

phenomena of Disease exhibit the powerful influence of Sensations in producing automatic motions. In *Hydrophobia*, for example the stimuli most effectual in exciting the convulsive movements, are those which act through the nerves of special Sense ; thus the *sight* or the *sound* of water will bring on the paroxysm, and any attempt to *taste* it increases the severity of the convulsions ; and it is further not a little significant, that the suggestion of the *idea* of water will produce the same result (§ 105).—In many *Hysteric* subjects, again, the sight of a paroxysm in another individual is the most certain means of its induction in themselves.—The most remarkable examples, however, of automatic movements depending upon Sensations, are those which we come to perform *habitually*, and as we commonly say *mechanically*, when the attention and the voluntary effort are directed in quite a different channel (§§ 191-194). The man who is walking through the streets in a complete reverie, unravelling some knotty subject, or working-out a mathematical problem, not only performs the movements of progression (which are themselves excito-motor, § 71) with great regularity, but also directs these in a manner which plainly indicates the guidance of Sight. For he will avoid obstacles in the line of his path, and he will follow the course which he has been accustomed to take, although he may have intended to pass along some very different route ; and it is not until his attention is recalled to his situation, that his train of thought suffers the least intermission, so that his Will is brought to bear upon his motions (§ 117).

80. We may recognize the agency of the Sensory ganglia, however, in Man, not merely in their direct and independent operation upon his muscular system, but also in the manner in which they participate in all his Voluntary actions. The existence of a sensation of some kind, in connection with muscular exertion, seems essential to the continuance of the latter. Our ordinary movements are guided by what is termed the *muscular sense ;* that is, by a feeling

of the condition of the muscles, that comes to us through their own afferent nerves. How necessary this is to the exercise of Muscular power, may be best judged-of from cases in which it has been deficient. Thus a woman who had suffered complete loss of sensation in one arm, but who retained its motor power, found that she could not support her infant upon it without constantly *looking* at the child; and that if she were to remove her eyes for a moment, the child would fall, in spite of her knowledge that her infant was resting upon her arm, and of her desire to sustain it. Here, the Muscular sense being entirely deficient, the sense of Vision supplied what was required, so long as it was exercised upon the object; but as soon as this guiding influence was withdrawn, the strongest Will could not sustain the muscular contraction.—Again, in the production of Vocal sounds, the nice adjustment of the muscles of the larynx, which is requisite to produce determinate tones, can only be effected in obedience to a Mental conception of the tone to be uttered; and this conception cannot be formed, unless the sense of Hearing has previously brought similar tones to the mind. Hence it is that persons who are born *deaf* are also *dumb.* They may have no malformation of the organs of Speech; but they are unable to utter distinct vocal sounds or musical tones, because they have not the guiding conception or recalled sensation of the nature of these. By long training, and by efforts directed by the Muscular sense of the larynx itself, some persons thus circumstanced have acquired the power of speech; but the want of sufficiently definite control over the vocal muscles, is always very evident in their use of the organ.—So, again, all the combinations of diverse Muscular actions which take place in the conjoint movements of the eyes, can be shown to be executed by this automatic Mechanism under the guidance of the Visual sense; the mandate to direct the eyes to a given point, being all that is issued by the Will (§ 21).

81. There seems no adequate reason for the belief that the

addition of the Cerebrum in the Vertebrated series *alters* the endowments of the Sensory ganglia on which it is superposed; on the contrary, we everywhere see that the addition of new ganglionic centres, as instruments of new functions, leaves those which were previously existing in the discharge of their original duties. Hence we should be led to regard them as the instruments of Consciousness, even in Man,—each pair of ganglionic centres ministering to that peculiar kind of sensation for which its nerves and the organs they supply are set apart; thus we should consider the Optic ganglia to be the seat of Visual sensations, the Auditory to be the seat of the sense of Hearing, and so on. And we should also consider them as the instruments whereby Sensations, of whatever kind, either originate or direct *instinctive* movements. The mechanism of all such movements, in fact, may be regarded as consisting of that part of the Nervous system which answers to the entire gangliated Cord of Articulated Animals, whose active life may be characterised as almost purely instinctive. And we shall presently see that this automatic Apparatus is as readily distinguishable from the *Cerebrum* (which is the instrument of Intelligence) even in *Man*, as it is in the lower Vertebrata; provided that we study the structure of his Brain under the guidance of Comparative Anatomy.

82. It would be impossible to find a better illustration of the contrast between Instinct and Intelligence as springs of action, than is afforded by the comparison of the habits of Birds in a state of Nature, with those which they acquire when brought into relation with Man. There can be no reasonable doubt that their Architectural constructions, like those of Insects, proceed from an internal impulse, which prompts each individual of a species to build after one particular pattern, to choose a situation suitable to its requirements, and to go in search of materials of a certain kind, though others might be much more easily obtained. But, on the other hand, in the working-out of this design, it is clear that Birds

often profit by experience, and *learn* to use special means when special ends have to be provided for.—The following case, narrated by Mr. Jesse, supplies a very good example of this intelligential modification of the instinctive tendency :—

a. A pair of Jackdaws endeavoured to construct their nest in one of the small windows that lighted the spiral staircase of an old church tower. As is usual, however, in such windows, the sill sloped inwards, with a considerable inclination; and, consequently, there being no level base for the nest, as soon as a few sticks had been laid, and it was beginning to acquire weight, it slid down. This seems to have happened two or three times; nevertheless the birds clung with great pertinacity to the site they had selected, and at last devised a most ingenious method of overcoming the difficulty. Collecting a great number of sticks, they built up a sort of cone upon the staircase, the summit of which rose to the level of the window-sill, and afforded the requisite support to the nest; this cone was not less than six feet high, and so large at its base as quite to obstruct the passage up the staircase; yet, notwithstanding the large amount of material which it contained, it was known to have been constructed within four or five days.—Now as this was a device quite foreign to the natural habit of the bird, and only hit-upon after the repeated failure of its ordinary method of nest-building, the curious adaptation of means to ends which it displayed can scarcely be regarded in any other light, than as proceeding from a *design* in the minds of the individuals who executed it.

The following circumstance, again, which was related to the Writer by a friend who witnessed it, shows how readily some Birds will spontaneously learn to profit by experience in matters which arise out of their relation to Man :—

b. A Wren having built her nest in a rather dangerous situation in the slate-quarries at Penrhyn, was liable to great disturbance from the occasional explosions. She soon learned, however, to take warning by the sound of the bell, which was rung to give notice to the workmen when a blast was about to be made; and would then quit her nest and fly to a little distance, remaining there until the shock of the explosion had passed off. This was noticed by the workmen; and the sagacity of the Wren was made a subject of exhibition to the visitors at the quarries, the bell being frequently rung for the mere purpose of causing her to quit her nest. After a time, however, it was observed that the bird no longer flew away upon the ringing of the bell, but that she remained until she saw whether or no the workmen began to move; if *they* drew off, *she* would go too; but if they remained in their places, she would not stir.—Now this conduct, sagacious as it may appear, is evidently explicable on a very simple hypothesis of the Mental operations of the bird. Observant, from its elevated post, of all that took place in its neighbourhood, the wren in the first instance learnt by experience to *associate* the ringing of the bell with the coming explosion, so as to anticipate the latter on the occurrence of the former. Being frequently disturbed, however, by the demonstration of her sagacity, and driven without occasion from her nest, the bird would perceive that this first association no longer held good; and nothing but a further period of observation was required for the bird to derive a more positive warning from the departure of the workmen, from which she learned by experience that a certain indication of the approaching explosion might be derived. None of those higher processes which enter into our more complex trains of Reasoning, were here required; the mere formation of an *association*, which gives the data for all these, which is the foundation of all knowledge derived from experience, and which appears to be the faculty first called into action in the mind of the Human infant, being quite sufficient to account for it (§ 217).

Another instance in which a Bird, without any direct teaching, learned to perform a particular action altogether foreign to its nature, was related to the Writer by a Swiss friend who had often witnessed the occurrence :—

c. In the town in which he was brought up, was a domesticated Stork, which was accustomed to receive its food every evening about six o'clock, along with the ordinary poultry; and the latter, being usually allowed to roam at large in the streets, were collected together, at the proper time, by a man who went through the town in search of them. The Stork, after having thus learned not to expect its food until the poultry had been all collected, spontaneously accompanied the collector, and assisted him in bringing the fowls together: and after doing this for a considerable time, becoming gradually more and more independent and self-relying, it became quite competent to perform this duty for itself, and was at last intrusted with it, so that it might be seen on any evening, gravely perambulating the town, collecting its flock of poultry, and driving it home, just as a Shepherd's Dog collects the sheep.

So, again, Rooks and other birds which live in the neighbourhood of Man, and are liable to be shot-at, often show in their actions that they distinguish whether a man who approaches them carries a gun, or not; and are said to be able to distinguish a gun from a walking-stick put to the shoulder after the manner of a gun. And it is further noticeable that they distinguish Sunday from other days; flying lower than usual, and sometimes visiting gardens where they would not venture on the days on which they would be liable to be molested. Whether they distinguish the day by some of its outward signs, or are guided by that remarkable power of measuring Time which many animals certainly possess, cannot be stated with certainty.—The following circumstance, of which the Writer is personally cognizant, indicates the acquirement of the same kind of knowledge :—

d. In a Ladies' school formerly kept near Bristol, it was customary for the young people to go into the play-ground for a few minutes every week-day, soon after twelve o'clock, and there to eat their luncheon. The crumbs of bread which they dropped on the ground proved very attractive to the sparrows in the neighbourhood, which would congregate on the walls of the garden a little *before* twelve every day, waiting for the appearance of their young friends, and patiently anticipating the time when the return of the ladies into the school-room would allow them to profit by their leavings. But on Sundays, the habits of the family were altogether different; the visit to the play-ground gave place to attendance on public worship, and the mid-day luncheon to an early dinner; on that day, therefore, the sparrows went without their accustomed meal. But it was obvious that they did not expect it; for it might be observed by any one who happened to remain at home on Sunday morning, that the usual mid-day gathering did not present itself,—the sparrows having evidently learned, not merely to judge very accurately of the approach of noon on the week-days on which they might expect their feed, but also to distinguish the day on which they must dispense with it.

83. The Birds of the *Parrot* tribe are pre-eminent for their educability and apparent intelligence: but this educability chiefly depends upon their great *imitative* power; and their intelligence is really of a very low order, consisting in an exercise of the simple faculty of Association, the manifestations of which are chiefly remarkable as taking the form of vocal utterances. The associations which the Parrot forms between certain vocal sounds and certain visual objects, lead it often to give forth the former under circumstances of singular appropriateness; but it would be quite a mistake to attribute such utterances to any higher intelligence than that of a young child just learning to talk, which repeats the phrases it has learned by imitation, without any distinct idea of their meaning, but sometimes brings them in remarkably *à propos.* A very good illustration of both faculties is afforded by a couple of anecdotes which the Writer remembers to have heard in his youth from his aged friend Mr. Palmer of Bristol:—

In Mr. Palmer's younger days, when Bristol was largely engaged in the African slave-trade, the large grey Parrots (which are far more intelligent than the green) were very numerous in that city, and often created great amusement. There was one which hung outside a shop in the neighbourhood of the quay, and had a remarkable tact in distinguishing sailors; and if a sailor happened to stop before his cage when he was in the middle of singing Handel's 104th psalm (which he performed most correctly), he would break off from "My soul praise the Lord," into "D——n your eyes, you fool, what are you looking at?" The sight of a sailor obviously called forth the phraseology which the bird had been accustomed to associate with that class.—Another Parrot caused no small degree of personal annoyance to Mr. Palmer himself. In his younger days, when an attorney's clerk, he was somewhat given to dandyism, and particularly rejoiced in a very long *queue*. A parrot, which was accustomed to hang outside the window of a house that he passed in his way between his residence and his office, was taught by some waggish boys to salute him with—"There goes the man with the long pig-tail;" and this the parrot learned to sing-out, without any prompting, whenever Mr. Palmer made his appearance; the continual repetition of which remark was so disagreeable to him, that he changed his route, and went through another street, to avoid it.—Now we have no reason to suppose that the bird knew the meaning of what it uttered, or was itself cognizant of the remarkable length of Mr. Palmer's pigtail; it simply learned to distinguish the individual, and to utter the phrase which it had been taught to associate with the sight of him.—On the same simple principles we may explain most, if not all, of what appears most marvellous in the accounts of wonderful Parrots, given to the world from time to time. (See, for example, *Jesse's Gleanings in Natural History*, 5th Edit., p. 220.)

84. It is a fact of no little interest, that Birds which inhabit localities not frequented by Man, know no fear of him, but allow him to approach them closely. Thus, when Mr. Darwin visited the Galapagos islands, he found that mocking-birds and finches, doves, and hawks, would allow him to come near enough to kill them with a stick, and sometimes even to catch them with the hand. The early visitors to the Falkland islands, a century previously, made the same report of the tameness of the birds they found there; and the descendants of these birds still exhibit very little of that apprehension which is shown by the birds of the same species in Tierra del Fuego, where they have been persecuted by man for ages past. The experience of many generations seems to be needed for the acquirement of this fear of man; which, as Mr. Darwin remarked, appears to have the character of an hereditary *instinct*, rather than to proceed from *knowledge* traditionally communicated from one generation to another,—the experience being gradually *embodied*, as it were, in the constitu-

tions, of the Birds, and showing itself, like other congenital tendencies, in the actions they perform without any process of education. Here, then, we have a simple case of that *hereditary transmission of acquired Psychical peculiarities*, which seems to have a large share in the progressive development of the Human Mind (§§ 93, 97).

85. We arrive, lastly, at the Class of MAMMALIA; in which the development of the *Cerebrum* comes to be so predominant, as to

Fig. 10.

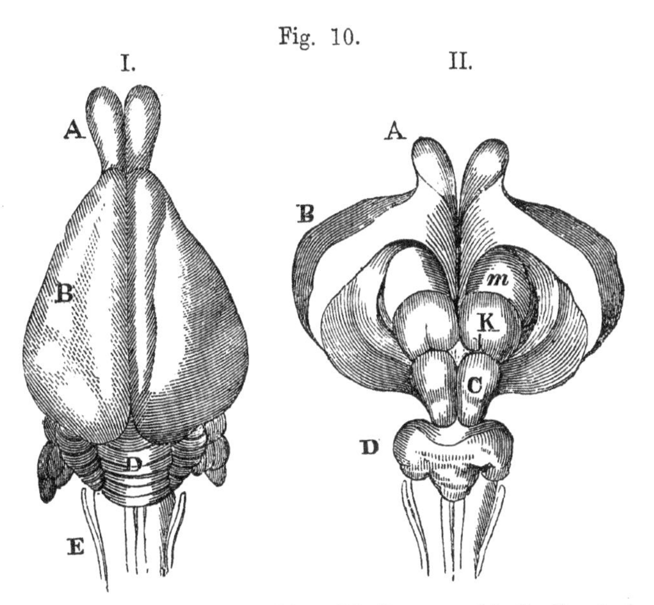

BRAINS OF RODENTS :—I, *Rabbit;* II. *Beaver*, with the Hemispheres drawn apart :—A, Olfactory Ganglia ; B, Cerebral Hemispheres ; C, Optic Ganglia ; D, Cerebellum ; E, Spinal Cord ; K, Thalami Optici ; *m*, Corpora Striata.

mask what has been shown to constitute the fundamental part of the organization of the Brain,—namely the Sensorial tract at its base. Still, among those lower Mammals in which the Brain does not present any great advance upon that of Birds, the Sensorial tract can be at once recognized as something altogether distinct from the Cerebrum, if we simply draw apart the Hemispheres, which, in their natural position, cover it in. Thus, in the Order *Rodentia*, the Cerebrum (B, Fig. 10) is smooth externally, as it is in Birds and Reptiles; it is pointed in front, and is not prolonged sufficiently far forwards to cover the *olfactive* ganglia (A, A); but it is wider behind, and is prolonged so far backwards as completely to cover-in the optic ganglia, and even partly to overlap the Cerebellum (D). But, on drawing its hemispheres apart, we find the *optic* ganglia (C) lying immediately in front of the Cerebellum; whilst in front of these, again, are two pairs of ganglionic masses, known as the *Thalami optici* (K) and the *Corpora striata* (*m*), which may be probably regarded as the terminations of the *sensory* and *motor* columns of the Spinal cord, and as ministering to the sense of Touch, and to the movements immediately related to it. The Thalami have also a connection with the *optic* nerves; and it does not seem improbable that this connection is instrumental in the establishment of that co-ordination between the senses of Sight and of Touch, which is so essential to the formation of trustworthy Perceptions of external objects (§ 167).

86. The large proportion which the Sensory ganglia still bear to the Cerebral hemispheres, and the low development of the latter —as marked by the smoothness of their surface, not less than by their relative size—are in accordance with the predominance of Instinct over Intelligence, which still marks the psychical character of these lower Mammalia, and of which we have a conspicuous example in the *Beaver*. There could scarcely be a better example of the *irrationality* of Instinct, than is afforded by the following account, given by Mr. Broderip, of a Beaver which he kept in his house:—

"The building instinct showed itself immediately it was let out of its cage, and materials were placed in its way; and this before it had been a week in its new quarters. Its strength, even before it was half-grown, was great. It would drag along a large sweeping-brush, or a warming-pan, grasping the handle with its teeth so that the load came over its shoulder, and advancing in an oblique direction till it arrived at the point where it wished to place it. The long and large materials were always taken first, and two of the longest were generally laid crosswise, with one of the ends of each touching the wall, and the other end projecting out into the room. The area formed by the cross-brushes and the wall he would fill up with hand-brushes, rush-baskets, boots, books, sticks, cloths, dried turf, or anything portable. As the work grew high, he supported himself on his tail, which propped him up admirably; and he would often, after laying on one of his building materials, sit up over against it, appearing to consider his work, or, as the country people say, 'judge it.' This pause was sometimes followed by changing the position of the material 'judged,' and sometimes it was left in its place. After he had piled up his materials in one part of the room (for he generally chose the same place), he proceeded to wall up the space between the feet of a chest of drawers which stood at a little distance from it, high enough on its legs to make the bottom a roof for him, using for this purpose dried turf and sticks, which he laid very even, and filling up the interstices with bits of coal, hay, cloth, or anything he could pick up. This last place he seemed to appropriate for his dwelling; the former work seemed to be intended for a dam."

Nothing could be more absurd, from the *reasoning* point of view, than the attempt of the animal to construct a dam where there was no water, or to build up a house where he was already comfortably lodged; but the innate architectural impulse was obviously uncontrolled by any perception of the entire unsuitableness of the work to the conditions under which it was being carried out, under the guidance of a "judgment" which had reference to conditions that did not exist.

87. As we rise through the Mammalian series towards Man, we find not only a marked increase in the *absolute* bulk of the *Cerebral hemispheres*, and a yet greater *relative* excess in their size as compared with the aggregate of that of the Sensory ganglia, but an augmentation of their functional powers beyond all proportion to their size, which is derived from the peculiar manner in which

their ganglionic matter is disposed. In all ordinary ganglia, the Nerve-cells on whose presence their special attributes depend (§ 36), form a sort of *internal nucleus*; but in the Cerebrum they are spread-out on the surface, forming an *external* or *cortical* layer. This layer is covered by a membrane termed the *pia mater*, which is entirely composed of Blood-vessels held together by connective tissue; and thus a copious supply of blood is brought to this important part. But the extent of the cortical layer, and of its contact with the pia mater, is enormously increased by its being thrown into *folds*, so as to produce what is known as the *convoluted surface* of the Hemispheres (Figs. 11-13); for the pia mater everywhere dips-down into the furrows between the convolutions, so as to supply the deepest parts of this plicated ganglionic layer, equally with the most superficial. And thus it comes to pass that the supply of Blood to the Cortical layer is far larger in proportion to the amount of its substance, than it is to any other part of the body. Of the enormous amount distributed to the Brain as a whole (§ 40), by far the greater part goes to the cortical layer of "grey" or "ganglionic" substance; the "white" or "fibrous" structure of the interior, often termed the Medullary substance, which constitutes by far the larger portion of the bulk of the Brain (Figs. 12, 13), receiving comparatively little. It is clear, therefore, that the functional activity of the Cerebrum is immensely augmented by the folding of its Cortical layer; and that its capacity for the production of Nerve-force is marked by the number and depth of its plications, no less than by its absolute size. In the higher orders of Mammalia, the convolutions are well marked; but we do not find them either numerous or complex in their arrangement until we approach Man; and even in the highest Apes they are considerably shallower and less numerous than in the lowest examples of the Human brain. (See Appendix.)

88. The Brain of Man (Fig. 11) differs from that of the animals that most nearly approach him, rather in its large size—as com-

pared alike with the Body generally, and with the Spinal cord,—than in any other character. For in the higher Apes, as in Man, we find that backward development of the Cerebrum into a *posterior lobe,* which makes it cover-in the Cerebellum; whilst in them, as in him, the anterior lobes have a much greater breadth, as well as greater forward extension, than in the lower Mammalia. There is, however, a marked diversity in respect

Fig. 11.

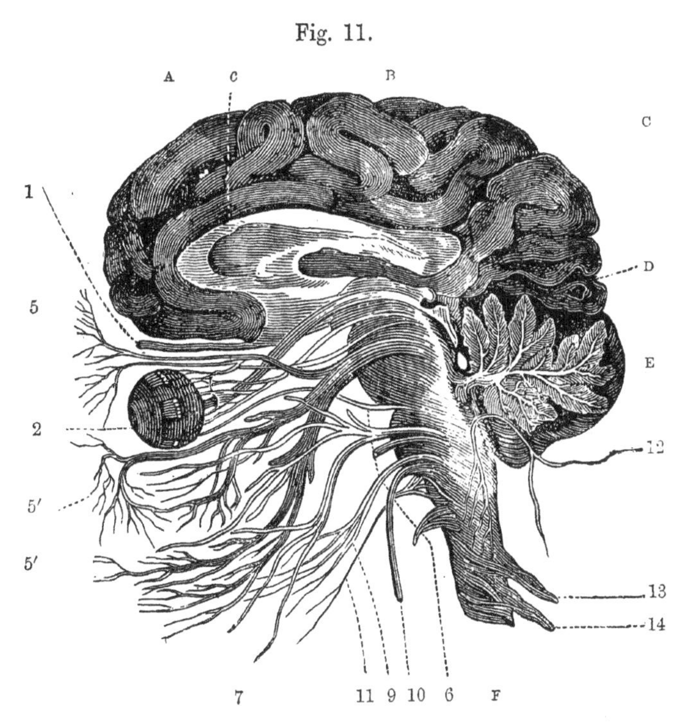

BRAIN AND CEPHALIC NERVES OF MAN, AS SHOWN IN VERTICAL SECTION THROUGH THE MEDIAN PLANE:—A, B, C, anterior, middle, and posterior lobes of the Cerebrum, showing its convoluted surface; D, Optic ganglia; E, Cerebellum; F, Spinal Cord; *c*, corpus callosum;—1, Olfactory bulb; 2, Eye, with Optic nerve; 5, 5′, 5″, Fifth pair of Nerves; 6, Sixth pair; 7, Seventh pair; 9, Glosso-pharyngeal nerve; 10, Par Vagum; 11, Hypoglossal; 12, Spinal Accessory; 13, 14, ordinary double-rooted Spinal Nerves.

of size between the Brains of different Races of men; those of the most civilized stocks, whose powers have been cultivated and improved by Education through a long series of generations, being for the most part considerably larger than those of Savage tribes, or of the least advanced among our own peasantry. So far as can be judged from the few cases which have furnished adequate materials for the determination, the brains of those earliest Races of men, which (like the old "flint-folk") had made but a very slight advance in the arts of life,

Fig. 12.

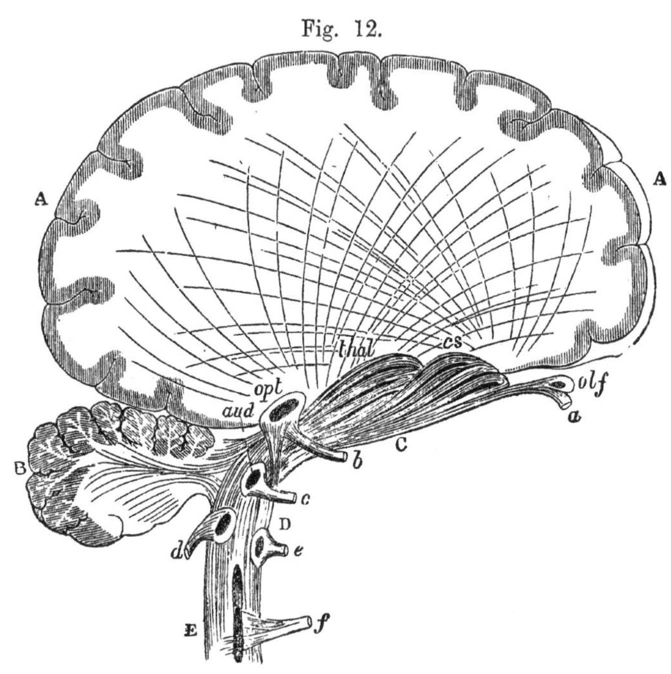

DIAGRAM OF THE MUTUAL RELATIONS OF THE PRINCIPAL ENCEPHALIC CENTRES, AS SHOWN IN *vertical* SECTION :—A, Cerebrum ; B, Cerebellum ; C, Sensori-motor tract, including the Olfactive ganglion *olf*, the Optic *opt*, and the Auditory *aud*, with the Thalami optici *thal*, and the Corpora striata *cs* ; D, Medulla oblongata ; E, Spinal cord ;— *a*, olfactive nerve ; *b*, optic ; *c*, auditory ; *d*, pneumogastric ; *e*, hypoglossal ; *f*, spinal :—radiating fibres of the Medullary substance of the Cerebrum are shown, connecting its cortical layer with the Thalami optici and Corpora striata.

were extremely small. Thus the inference, based on Comparative Anatomy, as to the relation between the development of the

Cerebrum and the predominance of Intelligence over Instinct, seems to hold good when applied to the diversities we encounter in the Human type ; and of this we have a further confirmation in the fact, that where the Cerebrum is so imperfectly developed as to be greatly under the average size, there is a marked deficiency in Intelligence, amounting to absolute Idiocy. The unfortunate

Fig. 13.

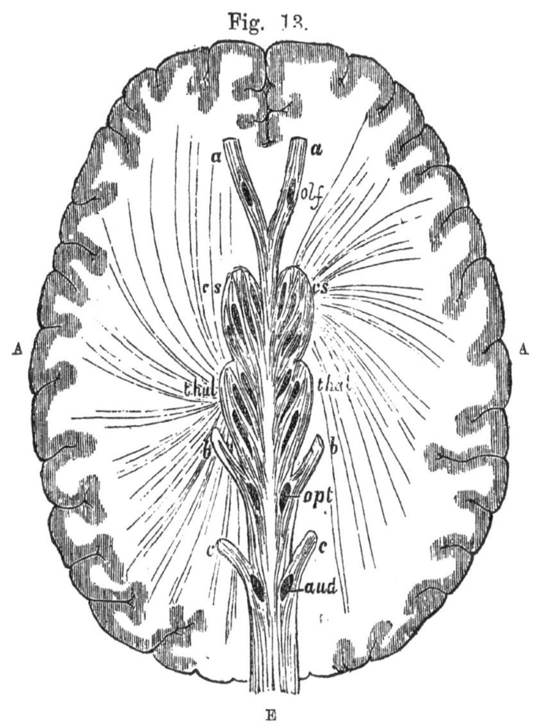

DIAGRAM OF THE MUTUAL RELATIONS OF THE CEREBRUM AND THE SENSORI-MOTOR TRACT, AS SHOWN IN *horizontal* SECTION :—A, A, Cortical layer of Cerebral Hemispheres, connected with Sensori-motor tract by *ascending* fibres (shown on the left side) radiating from *thal*, the Thalami optici, and by *descending* fibres (shown on the right side) converging to *cs*, the Corpora striata ; *olf*, Olfactive ganglia ; *opt*, Optic ganglia ; *aud*, Auditory ganglia ; *a*, *a*, Olfactive nerves ; *b*, *b*, Optic nerves ; *c*, *c*, Auditory nerves ; E, Spinal Cord.

beings thus characterized, are guided almost solely by their Instinctive tendencies, which frequently manifest themselves with a degree of strength that would not have been supposed to exist ;

and Instincts occasionally present themselves, of which the Human being is ordinarily regarded as destitute, and which may be presumed to be *survivals* of those which characterized some lower grade of his development. On the other hand, those who have obtained most influence over the *understandings* of others, have generally been *large-brained* persons, of strong Intellectual and Volitional powers, whose Emotional tendencies have been subordinated to their Reason and Will, and who have devoted their whole energy to the particular objects of their pursuit.—It is very different, however, with those who are chiefly actuated by what is ordinarily termed *genius;* and whose influence is rather upon the *feelings* and *intuitions,* than upon the understandings, of others. Such persons are often very deficient in the power of even comprehending the ordinary affairs of life; and still more commonly, they show a want of judgment in the management of them, being too much under the immediate influence of their Passions and Emotions, which they do not sufficiently endeavour to control by their Intelligent Will. The life of a "genius," whether his bent be towards poetry, music, painting, or pursuits of a more material character, is too often one which cannot be held-up for imitation. In such persons, when the *general* power of the mind is low, the Cerebrum is not usually found of any great size.—The mere comparative size of the Cerebrum, however, affords no accurate measure of the amount of Mental power; for we not unfrequently meet with men possessing large and well-formed heads, whose Psychical capability is not greater than that of others, the dimensions of whose crania have the same general proportion, but are of much less absolute size. Large brains, with deficient activity, are commonly found in persons of what is termed the *phlegmatic* temperament, in whom the general processes of life seem in a torpid and indolent state; whilst small brains and great activity, betoken what are known as the *sanguine* and *nervous* temperaments.

89. It is not only, however, by their size, and by the special development of their Cortical layer, that the Cerebral hemispheres of Man are distinguished from those of the lower Mammalia; for they are further remarkable for the elaborateness of their internal structure, which shows itself especially in the complexity of the arrangement of the nerve-fibres of which the Medullary substance is composed. These may be grouped under three principal divisions. The *first*, which may be distinguished as the *radiating* fibres, connect the different parts of the Cortical layer with the Sensori-motor tract on which the Cerebrum is superposed (Figs. 12, 13); and it is probable that there are two sets of these,—one *ascending* from the *Thalami optici* (which seem to form the terminals of the *sensory* tract of the Axial cord) to the Cortical layer, and conveying to it the result of the Physical changes produced in them by the Sense-impressions which they receive;—the other *descending* from the Cortical layer to the *Corpora striata* (which seem to form the terminals of the *motor* tract of the Axial cord), and conveying to them the Physical results of the changes which take place in itself. These fibres, which bring the instrument of Intelligence and Will into relation with that portion of the Nervous apparatus which furnishes the Mechanism of sensation and of the automatic or instinctive motions, were called by a sagacious old Anatomist, Reil, the *nerves of the internal senses;* and under that name they will be frequently referred-to in this Treatise.—The *second* set of fibres brings the several parts of the Cortical layer of each Hemisphere into mutual communication. The arrangement of these *commissural* fibres is peculiarly complex in Man: one particular group of them is known as the *Fornix*, or great *longitudinal* commissure.—The *third* set of fibres, termed *inter-cerebral*, connects the two Hemispheres together, through the medium of a broad band which is known as the *Corpus callosum*, or great *transverse* commissure (Fig. 11, *c*). This also is much more developed in Man, relatively to the size of his

Cerebrum, than it is in any of the lower Mammalia. It is altogether wanting in Fishes, Reptiles, and Birds; and there is little more than a rudiment of it in Marsupials and Rodents. Cases have occurred in which it has been nearly, or even entirely, deficient in Man; and it is significant that the chief defect in the characters of such individuals has been observed to be a want of forethought, *i. e.*, of power to apply the experience of the past to the anticipation of the future.

90. Thus, then, we see that the Cerebrum is a special organ *superadded* to that automatic Mechanism which constitutes the fundamental and essential part of the Nervous system, even in Man; and which not only supplies the conditions requisite for the maintenance of his Organic functions (§ 32), but ministers to the operations of the Cerebrum itself. For it is through the fibres ascending from the Sensorium to the Cortical layer of the Cerebrum, that the latter derives all that stimulus to its activity, which is furnished by the Ego's consciousness of the changes taking place in the external world;* while it is through the descending fibres that the results of Cerebral change are enabled to produce, through the motor portion of the apparatus, those Muscular movements by which the Mind expresses itself in action. We have now to enquire more closely into the mode in which the Cerebrum is subservient to those higher Mental operations, the capacity for which constitutes the distinguishing characteristic of Man, but to which we may trace very distinct approximations among the lower Mammalia.

91. We have seen that, so far as any Animal is dominated by *instinct*, it is a creature of necessity; performing its instrumental part in the economy of Nature from no design or will of its own, but as an *automaton* executing that limited series of actions for

* Here and elsewhere, the term "external world" is meant to include all that is *external to the Mind itself*,—thus taking-in the changes which occur in the Ego's own Bodily organism.

which its Mechanism fits it: and further, that the highest development of the Instinctive tendencies, with the lowest proportional manifestation of Intelligence, is to be found in *Insects.* On the other hand, that type of Psychical perfection which consists in the highest development of the *reason,* and in the supreme domination of the *will,* to which all the automatic actions—save those which are absolutely essential to the maintenance of the Organic functions—are brought under subjection, is presented in *Man;* who, *in his most elevated phase,* is not only a thinking and reflecting, but a self-determining and self-controlling agent, all whose actions are performed with a definite purpose which is distinctly within his own view, and are adapted to the attainment of that purpose by his own Intelligence. But as, in ascending the Vertebrated series, we observe that the Cerebrum is at first a mere rudimentary organ, and approximates but very gradually to the high development it attains in Man, so do we observe that the Psychical manifestations of its successive types exhibit a greater and yet greater approximation *in kind* to those of which he is capable. And this approximation becomes more obvious, when we compare them, not with those of the Adult, but with those of the Infant and young Child. For whilst the actions of the new-born Infant are entirely automatic, being directly prompted by *present* sensations, it soon becomes obvious that simple *ideas* are being formed as to the objects which excite those sensations, and that the actions begin to be guided by the experience with which these ideas are associated; and this is just what we recognize in studying the actions of such of the lower Vertebrata as we can bring under our observation. In the acts of the higher Mammals, as in those of the Child, we cannot fail to perceive the manifestations of true *reasoning* processes, analogous to those which we ourselves perform; together with the expressions of *emotional* states corresponding more or less closely to our own. Such are especially noticeable in the Dog, the Horse, and the

Elephant; which, having been trained into subservience to Man's requirements, and having come to possess a peculiar sympathetic attachment to him of which other species seem incapable, acquire a peculiar insight into what is passing in *his* mind, which helps to shape their course of action. In so far as that action is based upon the distinct conception of a *purpose*, and is carried-out by the means suggested by their experience as most suitable to its attainment, these animals participate in the rational nature of Man. But there seems no adequate ground for crediting them with that power of *reflecting upon their own Mental states*, which is required for the Intellectual 'processes' of Abstraction and Generalization (§ 227); their most sagacious performances being readily accounted for by the automatic action of *Association* (§ 218).

92. Of all breeds of Dogs, there is none more distinguished for sagacity than the one which has probably been longest associated with Man,—namely the Shepherd's Dog. "The shepherd," says Mr. T. Bell (*British Quadrupeds*, p. 234) "who tends his hundreds or thousands of sheep on the moors and mountain-sides of Scotland and of Wales, or on the extensive and trackless downs of Wiltshire, commits his almost countless charge to the care of his Dogs, with the certainty that their safety and welfare will be surely provided-for by the activity, watchfulness, and courage of these intelligent and faithful guardians. Some of the recorded instances of the almost human sagacity evinced by this valuable race would exceed belief, were they not authenticated by the most credible witnesses. In Scotland, particularly, where the flocks are so liable to be lost in snow-wreaths, these qualities are beyond all price; and are often exhibited in a manner equally affecting and wonderful."—The following is a very remarkable case of this kind, which occurred in the experience of James Hogg, the Ettrick Shepherd, the associate of Walter Scott and Christopher North:—

"He was," quoth the Shepherd, "beyond all comparison, the best dog I ever saw. He was of a surly, unsociable temper, disdaining all flattery, and refused to be caressed; but his attention to his Master's commands and interests will never again be equalled by any of the canine race. When he first came into my possession, he was scarcely a year old, and knew so little of herding, that he had never turned a sheep in his life; but as soon as he discovered that it was his duty to do so, and that it obliged me, I can never forget with what anxiety and eagerness he learned his different evolutions. He would try every way deliberately, till he found out what I wanted him to do; and when once I made him to understand a direction, he never forgot or mistook it again. Well as I knew him, he often astonished me; for when hard pressed in accomplishing the task he was put to, he had expedients of the moment that bespoke a great share of the reasoning faculty."

Mr. Hogg goes on to narrate the following, among other remarkable exploits, in illustration of Sirrah's sagacity. About seven hundred lambs, which were at once under his care at weaning-time, broke up at midnight, and scampered off in three divisions across the hills, in spite of all that the Shepherd and an assistant lad could do to keep them together. "Sirrah," cried the Shepherd in great affliction, "my man, they're a' awa." The night was so dark that he did not see Sirrah; but the faithful animal had heard his master's words—words such as of all others were sure to set him most on the alert; and without any delay, he silently set off in quest of the recreant flock. Meanwhile the Shepherd and his companion did not fail to do all that was in their power to recover their lost charge; they spent the whole night in scouring the hills for miles around, but of neither the lambs nor Sirrah could they obtain the slightest trace. "It was the most extraordinary circumstance," says the Shepherd, "that had ever occurred in the annals of the pastoral life. We had nothing for it (day having dawned), but to return to our master, and inform him that we had lost his whole flock of lambs, and knew not what was become of one of them. On our way home, however, we discovered a body of lambs at the bottom of a deep ravine, called the Flesh Clench, and the indefatigable Sirrah standing in front of them, looking all around for some relief, but still standing true to his charge. The sun was then up; and when we first came in view of them, we concluded that it was one of the divisions of the lambs, which Sirrah had been unable to manage until he came to that commanding situation. But what was our astonishment, when we discovered by degrees that not one lamb of the whole flock was wanting! How he had got all the divisions collected in the dark is beyond my comprehension. The charge was left entirely to himself, from midnight until the rising of the sun; and if all the shepherds in the forest had been there to have assisted him, they could not have effected it with greater propriety. All that I can further say is, that I never felt so grateful to any creature below the sun, as I did to my honest Sirrah that morning."

93. In this and other exercises of Intelligence, we may trace the manifestations of an hereditary transmission of aptitudes for particular kinds of Mental action, which have been originally *acquired* by habit. Dogs of other breeds cannot be taught to herd sheep in the manner which "comes naturally" to the young of the Shepherd's Dog. And it is well known that young Pointers and Retrievers, when first taken into the field, will often "work" as well as if they had been long trained to the requirements of the sportsman. The curious fact was observed by Mr. Knight, that the young of a breed of Springing Spaniels which had been trained for several successive generations to find Woodcocks, seemed to know as well as the old dogs what degree of frost would drive the birds to seek their food in unfrozen springs and rills.—Among the descendants of the Dogs originally introduced into South America by the Spaniards, there are breeds which have learned by their own experience, without any Human training, the best modes of attacking the wild animals they pursue; and since young dogs have been observed to practise these methods the very first time they engage in the chase, with as much address as old dogs, it can scarcely be questioned that the tendency to the performance of them has been embodied in the Organization of the Race, and is thus transmitted hereditarily.—There seems reason to believe that such hereditary transmission is limited to acquired peculiarities which are simply *modifications* of the natural constitution of the Race, and would not extend to such as may be altogether foreign to it.

But the foregoing facts would seem to justify the belief that the like hereditary transmission of acquired aptitudes may take place in Man; and that, in accordance with the far wider range of his faculties, it may become the means of a far higher exaltation of them (§ 97).

94. Whilst, however, we fully recognize the possession, by many of the lower Animals, of an Intelligence comparable (up to a certain point) with that of Man, we find no evidence that any of them have a Volitional power of *directing* their Mental operations, at all similar to his. These operations, indeed, seem to be of very much the same character as those which *we* perform in Reverie or connected Dreams; different "trains of thought" commencing as they are suggested, and proceeding according to the laws of Association until some other disturb them. So long, in fact, as the current of thought and feeling flows on under the sole guidance of Suggestion, and without any interference from the Will, it may be considered as the expression of the *reflex action of the Cerebrum*, called forth, like that of other Nerve-centres, by the stimulus conveyed to it from without; the seat of that activity being its expanded layer of Cortical substance.* This reflex action manifests itself not only in Psychical change, but also in Muscular movements: and these may either proceed from simple Ideas, without any excitement of Feeling, in which case they may be designated *ideo-motor;* whilst, if they are prompted by a Passion or Emotion, they are known as *emotional.* The nature of the response made by the reflex action of the Cerebrum will depend upon the condition of that organ at the time when it receives the impression; and that condition, among the lower Animals, may be regarded as the *resultant*, in each individual, of

* The extension of the doctrine of Reflex action to the Brain was first advocated by Dr. Laycock in a very important Essay read before the British Association in 1844; and published in the "British and Foreign Medical Review" for January, 1845.

the modifications which its inherited Constitution has undergone from the influence of external circumstances.

95. But whilst the Cerebrum of Man, in common with that of the lower Animals, has a reflex activity of its own—which, in the first instance, may be regarded as the direct resultant of his congenital Constitution, modified by early training,—an additional and most important influence subsequently comes into play; namely, the directing and controlling power of the Ego's own *Will*, in virtue of which he can to a great degree *direct* his thoughts and *control* his feelings, and can thus rise superior to circumstances, make the most advantageous use of the intellectual faculties with which he may be endowed, and keep his appetites and passions under subordination to his higher nature. And in proportion as he does this, will he so *shape* his Cerebral mechanism (which, like all other parts of the organism, *grows-to* the manner in which it is habitually exercised), that its automatic responses will be the expressions of the modes of activity in which he has brought it habitually to work,—just as the "trained" Horse automatically does that *of itself*, which it did originally under the will of its master. Thus each Human *Ego*, at any one moment, may be said to be the *general resultant* of his whole Conscious Life; the direction of which has been determined in the first instance by his congenital Constitution, secondly by the education he has received from the Will of others or from the discipline of circumstances, and thirdly by the Volitional power he has himself exercised.

96. It is not only, however, in the possession of this self-determining power, that the Psychical nature of Man is distinguished from that of the animals whose organization most nearly approaches his own; for if his Intellectual and Moral capacity were limited, as narrowly as theirs seems to be, by the Mechanism of his Brain, he could never pass that limit. So far as the lower animals are guided by *Instinct*, the actions of each species are prompted by its own sense of need, and have a direct (though not a self-

designed) adaptation to the supply of them. And these actions we see repeated from generation to generation, with no other variation than may arise from a change of circumstances, which necessitates some modification of the habit. Even where *Intelligence* comes into play, and a *designed* adaptation of means to ends, of actions to circumstances, is made by an individual, the Race does not seem to profit by that experience. And where the influence of Man has been exerted in the domestication of wild animals, it does not appear to produce any permanent improvement in their Psychical characters, but merely developes it in the manner suitable to his own requirements (§ 91); so that when such domesticated Races are left to themselves, they cease in a few generations to show any indication of the training they have received, and relapse into their original wildness. In the Human species, on the other hand, we observe not merely an unlimited *capacity* for Psychical elevation, but an unlimited *desire* to attain it; and this desire serves to stimulate Man not merely to the acquirement of knowledge, and to the application of it in the amelioration of his physical condition, but to the improvement of his Moral nature, by determinately repressing its lower propensities, and by fostering those which he feels to constitute the true nobility of his character.

97. But there is an element in Human nature ranging even beyond this desire and capacity for progress; which, though difficult to define, manifestly interpenetrates and blends-with his whole Psychical character. "The Soul," says Francis Newman, "is that side of our nature which is in relation with the Infinite;" and it is the existence of this relation, in whatever way we may describe it, which seems to constitute Man's most distinctive peculiarity. For it is in the aspiration after a nobler and purer *ideal*, that the highest spring of Human progress may be said to consist; and it is this which is the source of those notions of Truth, Goodness, and Beauty *in the abstract*, which seem peculiar to the higher types of Humanity. Whatever *capacity* for progress may exist among the

lower Races (and this is a question which still remains open to determination by experience), the *desire* for it—as among the lowest part of our own "practical heathen" population—seems altogether dormant. When once thoroughly awakened, however, it "grows by what it feeds on;" and the advance once commenced, little external stimulus is needed, for the desire increases at least as fast as the capacity. In the higher grades of Mental development, there is a continual looking-upwards, not (as in the lower) towards a more elevated Human standard, but at once to something *beyond* and *above* Man and material nature (§ 213). And in proportion as the love of *truth* for its own sake constitutes the incentive of our Intellectual efforts, as the love of *goodness* for its own sake animates our endeavours to bring our own Moral nature into conformity with it, and our love of the highest type of *beauty* withdraws us from all that is low and sensual, are we not only elevating *ourselves* towards our Ideal, but contributing to the elevation of our *race.* For we seem justified by the whole tendency of modern Physiological research, in the belief that alike by the discipline we exert over ourselves, and by the influence we exercise over others, will every effort judiciously directed towards the improvement of our Psychical nature impress itself upon our Physical constitution; and that, by the genetic transmission of such modifications, will the capacity of future generations for yet higher elevation be progressively augmented.

98. It is, in fact, upon the course of our strictly *Mental* operations, that the *Will* exerts its most powerful, and what is commonly regarded as its most direct influence. But it appears to the Writer that this influence is by no means so direct as is commonly supposed; and that observation of our own Psychical phenomena entirely justifies the belief, which Physiological considerations tend to establish,—that the operations of the *Cerebrum* are in themselves as automatic as are those of other Nerve-centres, and that the Volitional control which we exercise over our

thoughts, feelings, and actions, operates through the *selective attention* we determinately bestow upon *certain* of the impressions made upon the Sensorium, out of the *entire aggregate* brought thither by the "nerves of the internal senses" (§ 89). In this point of view, it is the *Sensorium*, not the Cerebrum, with which the Will is in most direct relation; and in order that this doctrine (which lies at the basis of the whole inquiry as to the relation of the Will to motives, and the mode in which it determines our character and actions) may be rightly apprehended, it is necessary here to consider the following Physiological question:—Whether *Cerebral changes are in themselves attended with consciousness*, or whether *we only become conscious of Cerebral changes as states of ideation, emotion, &c., through the instrumentality of the Sensorium*,—that is, of that aggregate of Sense-ganglia, through the instrumentality of which we become conscious of external Sense-impressions, and thus feel *sensations*.

99. The Brain, *as a whole*, has been commonly regarded, alike by Psychologists and by Physiologists, as "the seat of consciousness;"—or, to speak more precisely, as the instrument through which we become conscious of the impressions made by external objects upon our organs of Sense: whilst the Ego has been supposed by Metaphysicians to be *directly* conscious of all Mental operations; or rather, these operations are regarded as "states of consciousness," not in any way requiring material instrumentality. Those Psychologists, however, who recognize the cogency of those considerations which *force* on the Physiologist the conviction that "Brain-change" is a necessary condition of all Mental action, appear generally to take for granted that *all* "Brain-change" must be attended with Consciousness: entirely ignoring the fact that the Brain is an *aggregate* of ganglionic centres having very distinct functions; and that the *Cerebrum*, which in Man is by far the largest of those centres, is *not* the part of the brain which ministers to what may be called the "outer life" of the

Animal, but is the instrument exclusively of its "inner life,"—that is, of those *psychical* operations, of which the sensations received from the outer world constitute the mental *pabulum.* Now this *inner* life seems to have no existence in that vast section of the Animal Kingdom, which is most distinguished by the activity of its *outer* life, viz. the Class of Insects : and taking the Nervous system of that Class as the type of an automatic Apparatus which furnishes all the conditions required for Sensation and Motion, as well as for the working of those fixed or mechanical modes of action which we term *instincts,* we have found that a precisely analogous Automatic apparatus exists through the entire Vertebrated series, that it constitutes almost the whole Nervous system of the Fish, and that it is distinctly recognizable as the fundamental or essential part of that of Man, in whom the vast relative development of the Cerebrum merely indicates a *superaddition* of *new* functions, without affording the least ground to believe that there is any *transfer* to it of the proper attributes of the automatic Apparatus. And it has been shown that this indication is confirmed by the results of the experimental removal of the Cerebrum in Birds (§ 78); which prove that (due allowance being made for the disturbance in the action of other parts of the Brain, necessarily produced by the operation) the Sensori-motor apparatus, which ministers to the outer life, retains its functional activity. Further, it has been positively established, alike by experiments on Animals, and by observation of the phenomena of disease and accident in Man, that the substance of the Cerebrum is itself *insensible;* that is, no injury done to it, or physical impression made upon it, is *felt* by the subject of it (See Appendix). As it is clear, therefore, that the presence of the Cerebrum is not essential to Consciousness, we have next to inquire in what way it seems most likely that the Consciousness is affected by Cerebral changes.

100. When we compare the *anatomical* relation of the Sensorium,

on the one hand to the Cortical layer of the Cerebrum, and on the other to that Retinal expansion of ganglionic matter which is the recipient of Visual impressions, we find the two to be so precisely identical (§ 89), as to suggest that its *physiological* relation to those two organs must be the same. And as we only become *conscious* of the luminous impression by which Nerve-force has been excited in the retina, when the transmission of that nerve-force through the nerve of *external* sense has excited a change in the Sensorium, so it would seem probable that we only become *conscious* of the further change excited in our Cerebrum by the Sensorial stimulus transmitted along its *ascending* fibres, when the reflexion of the Cerebral modification along its *descending* fibres—the nerves of the *internal* senses—has brought it to re-act on the Sensorium. In this point of view, the Sensorium is the one centre of consciousness for Visual impressions on the Eye (and, by analogy, on the other Organs of Sense), and for Ideational or Emotional modifications in the Cerebrum:—that is, in the one case, for *sensations*, when we become conscious of Sense-impressions; and, on the other, for *ideas* and *emotions*, when our consciousness has been affected by Cerebral changes. According to this view, we no more *think* or *feel* with our Cerebrum, than we *see* with our eyes; but the Ego becomes conscious through the same instrumentality of the retinal changes which are translated (as it were) by the Sensorium into visual sensations, and of the Cerebral changes which it translates into Ideas or Emotions. The mystery lies in the *act of translation;* and is no greater in the excitement of *ideational* or *emotional* consciousness by Cerebral change, than in the excitement of *sensational* consciousness by Retinal change.

101. Now although there may seem no *à priori* objection to this view, yet it may be thought to introduce needless complication into what was previously a simple account of the relation of the Brain to Mental phenomena. But this notion of "simplicity"

is really based on ignorance; and when the phenomena of *reproduced* Sensations are carefully considered, they will be found to fit in with it so exactly, as scarcely to admit of being accounted for in any other way. There are many persons who can bring up before the "mind's eye," with extraordinary vividness, the *pictures* of scenes or persons they have been formerly familiar with; while to many who cannot thus recall them volitionally, these pictures present themselves automatically, as in *dreaming* or *delirium.* Thus Dr. Abercrombie relates of Niebuhr, the celebrated Danish traveller, that:—

a. "When old, blind, and so infirm that he was able only to be carried from his bed to his chair, he used to describe to his friends the scenes which he had visited in his early days, with wonderful minuteness and vivacity. When they expressed their astonishment, he told them that as he lay in bed, all visible objects shut out, the pictures of what he had seen in the East continually floated before his mind's eye, so that it was no wonder he could speak of them as if he had seen them yesterday. With like vividness, the deep intense sky of Asia, with its brilliant and twinkling host of stars, which he had so often gazed at by night, or its lofty blue vault by day, was reflected, in the hours of stillness and darkness, on his inmost soul."—*Intellectual Powers,* 5th Edit., p. 130.

The same Author relates the following very remarkable example of the volitional reproduction of a picture formerly impressed on the mental vision, which would be almost too wonderful for belief, if it had not been vouched for by so trustworthy an authority as the late Dr. Duncan, who had himself seen and compared the original picture and the copy reproduced *memoriter*:—

b. "In the church of St. Peter at Cologne, the altar-piece is a large and valuable picture by Rubens, representing the martyrdom of the Apostle. This picture having been carried away by the French in 1805, to the great regret of the inhabitants, a painter of that city undertook to make a copy of it from recollection; and succeeded in doing so in such a manner, that the most delicate tints of the original are preserved with the most minute accuracy. The original painting has now been restored, but the copy is preserved along with it; and even when they are rigidly compared, it is scarcely possible to distinguish the one from the other."—*Op. cit.* p. 131.

102. Now it will not be questioned by any Psychologist, that what were perceived in these two cases were the *ideational representations* or *concepts* of what were formerly presented to the Mind as *objects of sensation;* and it would seem scarcely to admit of question, that *the same Sensorial state* must be excited in the one case as in the other,—that state of the Sensorium which was *originally* excited by impressions conveyed to it by the nerves of the *external* senses, being *reproduced* by impressions brought down to it from the Cerebrum by the nerves of the *internal* senses. In fact, the real complexity lies in supposing that Mental states so closely related as the *perception* of a *present* object, and the *conception* of a *remembered* object, are produced through the instrumentality of two different "seats of consciousness," the *Sensorium* in the one case, and the *Cerebrum* in the other.

103. Still stronger evidence of the same associated action of the Cerebrum and Sensorium, is furnished by the study of the phenomena designated as *Spectral Illusions.* These are clearly *Sensorial* states *not* excited by external objects; and it is also clear that they frequently originate in Cerebral changes, since they represent *creations* of the Mind, and are not mere reproductions of past sensations. The following very interesting experience, which was several years ago communicated to the Writer by the distinguished subject of it (who subsequently published it in fuller detail), affords a striking confirmation, not only of the doctrine here advocated, but also of that further development of it which will be made hereafter under the title of "Unconscious Cerebration" (Chap. XIII.); and it seems to give the clue to the *rationale* o another large class of obscure phenomena, that may now be fairly regarded as results of Physical changes of which we

are unconscious even when our Attention is directed to them (§ 424).

Sir John Herschel stated that he was subject to the involuntary occurrence of Visual impressions, into which Geometrical regularity of form enters as the leading character. These were not of the nature of those ocular Spectra which may be attributed with probability to retinal changes (§ 140); "for what is to determine the incidence of pressure or the arrival of vibrations from without, upon a geometrically devised pattern on the retinal surface, rather than on its general ground?"

"They are evidently not Dreams. The mind is not dormant, but active and conscious of the direction of its thoughts; while these things obtrude themselves on notice, and, by calling attention to them, *direct* the train of thought into a channel it would not have taken of itself."

Even supposing the phenomenon to be the result of a retinal change excited through the Optic nerve, instead of *ab externo*, the question remains—"Where does the pattern itself, or *its prototype in the intellect*, originate? Certainly not in any action *consciously* exerted by the Mind; for both the particular pattern to be formed, and the time of its appearance, are not merely beyond our will or control, but beyond our knowledge. If it be true that the conception of a regular geometrical pattern implies the exercise of thought and intelligence, it would almost seem that in such cases as those above adduced we have evidence of a *thought*, an intelligence, working within our own organization distinct from that of our own personality, in a manner we have absolutely no part in, except as spectators of the exhibition of its results."—*Familiar Lectures on Scientific Subjects*, pp. 406-412.

We have here *not* a reproduction of Sensorial impressions formerly received; but a *construction* of *new* forms, by a process which, if it had been carried on *consciously*, we should have called Imagination. And it is difficult to see how it is to be accounted for in any other way, than by an unconscious action of the Cerebrum; the products of which impress themselves on the Sensorial consciousness, just as, in other cases, they express themselves through the Motor apparatus (§ 425).

104. It may not improbably be in this manner, that a number of those so-called "spiritual" phenomena are produced, in which "subjective" Sensations of various kinds are distinctly felt by persons who are not only wide awake, but are entirely trustworthy on all other matters, though self-deceived as to the reality of the objective sources of their sensations. Having resigned the exercise of their Common Sense *quoad* this particular set of beliefs, and having allowed them to gain a mastery over their ordinary course of thought, there is nothing wonderful in the automatic and unconscious evolution of results corresponding to these beliefs; which results, impressing themselves on the Sensorium, are felt as true sensations. And just as Sir John Herschel *truly saw* as geometrical forms the unconscious constructions of his own Cerebrum, so, it seems probable, may the "spiritualist" *truly see* the strange things he describes as actual occurrences, although they have no foundation whatever in *fact* (§ 147).

105. Another consideration which strongly indicates that the action of Cerebral changes on the Muscular apparatus is exerted through the instrumentality of the Sensorial apparatus, is the identity of the effects often produced by *ideas*, with those produced by sights, sounds, or other Sensations which call forth respondent motions. Thus in a person predisposed to yawn, the verbal suggestion of the notion of yawning is almost as provocative of the act, as the sight or sound of a yawn in another. So, again, a "ticklish" person is affected in the same way by the mental state suggested by the *pointing* of a finger, as by the actual contact. And so in a hydrophobic patient, the same paroxysm is excited by the *idea* of water suggested by words or pictures, as by the actual sight or sound of it. So far, then, from being a source of additional complexity, the doctrine of the *singleness* of the Sensorial nerve-centre, through the instrumentality of which we become conscious alike of Sense-impressions and of Cerebral changes, and from which the Motor

impulses to respondent action immediately proceed, will be found (the Writer believes) to lead to a real simplification in the interpretation of a large class of phenomena occupying the border ground between Physical and Psychical action.

106. That the different portions of the *Cerebrum* should have different parts to perform in that wonderful series of operations by which the Brain as a whole becomes the instrument of the Mind, can scarcely be regarded as in itself improbable. But no determination of this kind can have the least scientific value, that is not based on the facts of Comparative Anatomy and Embryonic Development. In ascending the Vertebrate series, we find that this organ not only increases in relative size, and becomes more complex in general structure, but undergoes progressive additions which can be defined with considerable precision. For the Cerebrum of Oviparous Vertebrata is not a miniature representative of the entire Cerebrum of Man, but corresponds only with its "anterior lobe;" and is entirely deficient in that great transverse commissure, the *corpus callosum* (§ 89), the first appearance of which, in the Placental Mammals, constitutes "the greatest and most sudden modification exhibited by the brain in the whole Vertebrated series" (Huxley). It is among the smooth-brained *Rodentia* that we meet with the first distinct indication of a "middle lobe," marked off from the anterior by the "fissure of Sylvius;" this lobe attains a considerably greater development in the *Carnivora;* but even in the *Lemurs* it still forms the hindermost portion of the Cerebrum. The "posterior lobe" makes its first appearance in *Monkeys;* and is distinctly present in the anthropoid *Apes.* The evolution of the Human Cerebrum follows the same course. For in the *first* phase of its development which presents itself during the second and third months, there is no indication of any but the anterior lobes; in the *second,* which lasts from the latter part of the third month to the beginning of the fifth, the middle lobes make their appearance; and it is not until the latter

part of the fifth month that the *third* period commences, characterized by the development of the posterior lobes, which sprout, as it were, from the back of the middle lobes, and remain for some time distinctly marked off from them by a furrow. The exact mutual confirmation afforded by these two sources of knowledge seems fatal to the ordinary Phrenological doctrine, which locates in the posterior part of the Cerebrum those Instincts and Propensities which Man shares with the lower Animals; while it would lead us to regard the *posterior* lobes as the instruments of those higher forms of Ideational activity by which Man is especially distinguished, and the *anterior* and *middle* lobes as the instruments of those simpler Ideational states which are the *most general* forms of Mental activity, being most directly excited by Sensorial suggestions. And it seems probable that evidence to this effect may be derived from a careful comparison of the Cerebral convolutions in different animals; the researches of Leuret, Gratiolet, and others having made it clear that notwithstanding the apparent indefiniteness of their distribution in Man, a distinct plan shows itself in their arrangement in each Family (this being simpler in the smaller members of it, and more complex in the larger), and that certain identities are traceable between the fundamental convolutions in representatives of different Families. The lower Quadrumana, for example, present a sort of sketch of the plan on which the convolutions are arranged in the higher Apes; and whilst the study of the latter gives the key to the complex arrangement of the convolutions in the Human Cerebrum, that of the former enables the Simioid plan to be correlated with that of inferior types. (See Appendix.)—One remarkable localization of function to which recent Pathological enquiry has been thought to point, will be considered hereafter (§ 355).

SECTION 4. *General Summary:—Functional Relations of the Ganglionic Centres of the Cerebro-Spinal System of Man:—Sympathetic System.*

107. It was well remarked by Cuvier, that the different tribes of Animals may be said to be so many "experiments ready prepared for us by Nature; who adds to, or takes from, the aggregate of their organs, just as we might wish to do in our laboratories, showing us at the same time in their actions the results of such addition or subtraction." And to no part of the organization of Animals is this view more applicable, than it is to the Nervous apparatus; for the different Ganglionic centres which are combined in the *Cerebro-spinal system* of Man and the higher Vertebrates, have such an intimate structural relation to each other, and so much more frequently act consentaneously than separately, that, notwithstanding the abundant evidence of the diversity of their respective endowments, there is considerable difficulty in the determination of their special functions; since the destruction or removal of any one portion of the Nervous system, not only puts a stop to the phenomena to which that portion is directly subservient, but so deranges the general train of nervous activity, that it often becomes impossible to ascertain, by any such method, what is its real share in the entire performance.—Under the guidance of Comparative Anatomy, however, we are enabled to recognize the following Ganglionic centres as essentially distinct in function, however intimately connected in structure:—

I.—The *Spinal Cord*, consisting of a tract of ganglionic matter enclosed within strands of longitudinal fibres, and giving-off successive pairs of nerves which are connected at their roots with both of these components. This obviously corresponds with the gangliated Ventral cord of the *Articulata;* chiefly differing from it in the *continuity* of the ganglionic tract which occupies its interior. And each segmental division of it, which serves as the centre of Reflex action for its own pair of nerves, may be considered, like each ganglion of the ventral cord of the Articulata, as a repetition of the single "pedal" ganglion of those Mollusca which have but one instrument of locomotion.

II.—The *Medulla Oblongata*, or prolongation of the Spinal Cord within the skull: which consists of a set of strands that essentially correspond with the cords passing round the œsophagus in Invertebrated animals, and connecting the cephalic ganglia with the first ganglion of the trunk; although, as the whole Cerebro-spinal axis of the Vertebrata lies *above* the alimentary canal (the trunk being supposed to be in a horizontal position), there is no divergence of these strands to give it passage. Interposed among them, however, are certain collections of ganglionic matter, which serve as the centres for the reflex movements of Respiration and Deglutition, corresponding with the separate respiratory and stomato-gastric ganglia found in many Invertebrated animals.—This incorporation of so many distinct centres into one system, would seem destined in part to afford to all of them the protection of the Vertebral column; and in part to secure that consentaneousness of action, and that ready means of mutual influence, which are peculiarly requisite in beings in whom the activity of the Nervous system is so predominant. Thus the close connection which is established in the higher Vertebrated animals, between the Respiratory and the general Motor apparatus, is obviously subservient to the use which the former makes of the latter in the performance of its functions; whilst, on the other hand, the control which their Cephalic centres possess over the actions of the Respiratory ganglia, enables the Will to regulate the inspiratory and expiratory movements in the manner required for the acts of Vocalization.

III.—The *Sensory Ganglia*, comprehending that assemblage of ganglionic masses lying along the base of the skull in Man, and partly included in the Medulla Oblongata, in which the nerves of the special Senses,—Taste, Hearing, Sight, and Smell,—have their central terminations. With these may probably be associated the two pairs of ganglionic bodies known as the Corpora Striata and Thalami Optici; into which may be traced the greater proportion of the fibres that constitute the various strands of the Medulla Oblongata, and which seem to stand in the same kind of relation to the nerves of Touch or common sensation, that the Olfactive, Optic, Auditory, and Gustative ganglia bear to *their* several nerve-trunks.—These Ganglia, the aggregate of which constitutes the *Sensorium*, are the centres of Reflex movements prompted by the impressions brought to them by the several nerves of sense.

The foregoing together constitute the Automatic Apparatus which ministers to our purely *animal* or *outer life*, namely, the functions of Sensation and Locomotion; and which also sustains the movements that are necessary for the maintenance of our Organic functions. To this apparatus is superadded :—

IV.—The *Cerebrum*,—the instrument of our *Psychical* or *inner life;* of which organ, although it is so enormously developed in Man as apparently to supersede the Sensorial centres, scarcely a trace exists in the lowest Vertebrates; and the relative proportion borne by which to the Sensorial centres, in regard alike to size and to complexity of structure, corresponds closely with the degree of predominance which the *Intelligence* possesses over the Animal Instincts. Much of its action, however, may still be purely *automatic* in its nature; for so long as the current of Thought and Feeling flows-on in accordance with the direct promptings of suggestion, and without any interference from Volition, it may be considered as a manifestation of the reflex activity of the Cerebrum, which takes the form of a *mental instinct.* This reflex activity manifests itself not only in the Psychical operations themselves, but also in Muscular movements: and these, when they proceed from simple Ideas, without any excitement of feeling, may be designated as *ideo-motor;*

whilst, if they spring from a Passion or Emotion, they are termed *emotional.* The *mental* Instincts, however, are by no means as invariable in the different individuals of the same species, as are the *animal* Instincts of that inferior part of the Nervous apparatus which is more closely connected with the maintenance of the Organic life: the particular changes which any given suggestion will excite in each individual, being partly determined by *original constitution,* and partly by *acquired habits;* and the hereditary constitution being itself determined to a large extent by the acquired habits of the ancestral Race.—There seems a strong probability that there is *not* (as was formerly supposed) a direct continuity between even all or any of the Nerve-fibres distributed to the body, and those of the Medullary substance of the Cerebrum. For whilst the nerves of special sense have their own ganglionic centres, it cannot be shown that the nerves of common Sensation have any higher destination than the *thalami optici.* So, the Motor fibres which pass-forth from the Brain, though commonly designated as *cerebral,* cannot be certainly said to have a higher origin than the *corpora striata.* And there is strong reason to believe that the Cerebrum has no communication with the external world, otherwise than by its connection with the Sensori-motor apparatus; and that even the movements which are usually designated *voluntary* (or more correctly *volitional*), are only so as regards their original *source,* —the stimulus which *immediately* calls the Muscles into contraction being still supplied from the automatic centres.

v.—Wherever a Cerebrum is superimposed upon the Sensory ganglia, we find another ganglionic mass, the *Cerebellum,* superimposed upon the Medulla oblongata. The development of this organ bears a general, but by no means a constant, relation to that of the Cerebrum; for in the lowest Fishes it is a thin lamina of nervous matter on the median line, only partially covering-in the fissure between the two lateral halves of the upper part of the

Spinal cord (§ 64); whilst in the higher Mammalia, as in Man, it is a mass of considerable size, having two lateral lobes or hemispheres, in addition to its central portion (Fig. 11). The direct communication which the Cerebellum has with both columns of the Spinal cord, and the comparatively slight connection which it possesses with the higher portions of the Brain, justify the supposition that it is rather concerned in the regulation and co-ordination of the Muscular movements, than in any proper Psychical operations; and though its precise function is still unknown, that general conclusion seems in harmony with our best knowledge on the subject. (See Appendix.)

108. Now although every segment of the Spinal cord and every pair of the Sensory ganglia, may be considered, in common with the Cerebrum, as an independent centre of Nervous power, yet this independence is only manifested when these organs are separated from each other; either structurally—by actual division, or functionally—by partial suspension of activity. In their state of perfect integrity and complete functional activity, they are for the most part (at least in Man) in such subordination to the Cerebrum, that they only minister to *its* actions; except in so far as they are subservient to the maintenance of the Organic functions, as in the automatic acts of breathing and swallowing. The impressions which call forth these and similar movements, ordinarily excite them by the direct reflex action of the lower centres, without passing-on to the Cerebrum; so that we only perceive them when we specially direct our attention to them, or when they exist in unusual potency. Thus we are ordinarily unconscious of that internal need for air, by which our movements of Respiration are prompted; and it is only when we have refrained from breathing for a few seconds, that we experience a sensation of uneasiness which impels us to make forcible efforts for its relief. Notwithstanding, however, that the Cerebrum is unconcerned in the

ordinary performance of those automatic movements, yet it can exert a certain degree of control over many of them, so as even to suspend them for a time; but in no instance can it carry this suspension to such an extent, as seriously to disarrange the Organic functions. Thus, when we have voluntarily refrained from breathing for a few seconds, the inspiratory impulse so rapidly increases in strength with the continuance of the suspension, that it at last overcomes the most powerful effort we can make for the repression of the movements to which it prompts. That the Will should have a certain degree of control over such movements, is necessary in order that they may be rendered subservient to various actions which are necessary for the due exercise of Man's Psychical powers; but that they should not be left dependent upon its exercise, and should even be executed in opposition to it, when the wants of the system imperatively demand their performance, constitutes an essential provision for the security of Life against the chance of inattention or momentary caprice, as well as for maintaining it during the unconsciousness of sleep.

109. In that action and reaction, however, between the Mind and all that is outside it, in which the Conscious Life of every Human *Ego* consists, the whole Cerebro-spinal system participates. For in virtue of the peculiar arrangement of the Nervous apparatus, every excitor *impression* travels in the *upward* direction, if it meet with no interruption, until it reaches the Cerebrum, without exciting any reflex movements in its course. When it arrives at the Sensorium, it makes an impression on the consciousness of the individual, and thus gives rise to a *sensation;* and the change there induced, being propagated onwards to the Cerebrum, becomes the occasion of further changes in its cortical substance, the downward reflexion of whose results to the Sensorium gives rise to the formation of an *idea.* If with this idea any pleasurable or painful *feeling* should be associated, it assumes the character of an *emotion;* and either as a simple or as an emotional

idea, it becomes the stimulus to further Cerebral changes, which, when we become conscious of them, we call *Intellectual operations.* These may express themselves either *directly* in respondent Movements; or *indirectly*, by supplying motives to the Will; which may exert itself either in producing or in checking Muscular movement, or in controlling or directing the current of Thought and Feeling.

110. But if this ordinary *upward* course be anywhere interrupted, the impression will then exert its power in a *transverse* direction, and a reflex action will be the result; the nature of this being dependent upon the part of the Cerebro-spinal axis at which the ascent had been checked. Thus if the interruption be produced by division or injury of the Spinal cord, so that its lower part is cut-off from communication with the Cephalic centres, this portion then acts as an independent centre; and impressions made upon it, through the afferent nerves proceeding to it from the lower extremities, excite violent reflex movements, which, being thus produced without sensation, are designated as *excito-motor.*—So, again, if the impression should be conveyed to the Sensorium, but should be prevented by the removal of the Cerebrum, or by its state of functional inactivity, or by the direction of its activity into some other channel, from calling-forth Ideas through the instrumentality of that organ, it may re-act upon the Motor apparatus by the reflex power of the Sensory ganglia themselves. Such actions, being dependent upon the prompting of Sensations, are *sensori-motor* or *consensual.*—But further, even the Cerebrum responds automatically to impressions fitted to excite it to reflex action, when from any cause the Will is in abeyance, so that its power cannot be exerted either over the muscular system or over the direction of the thoughts and feelings. Thus in the states of Reverie, Dreaming, Somnambulism, &c., whether spontaneous or artificially induced (Chaps. XIV.—XVI.), *ideas* which take full possession of the mind, and from which it cannot free itself, may

excite respondent *ideo-motor* actions; as happens also when the force of the Idea is morbidly exaggerated, and the Will is not suspended, but merely weakened, as in many forms of Insanity (Chap. XIX).

111. The general views here put-forth in regard to the independent and connected actions of the several primary divisions of the *Cerebro-spinal* apparatus, may perhaps be rendered more intelligible by the following Table; which is intended to represent (1) the

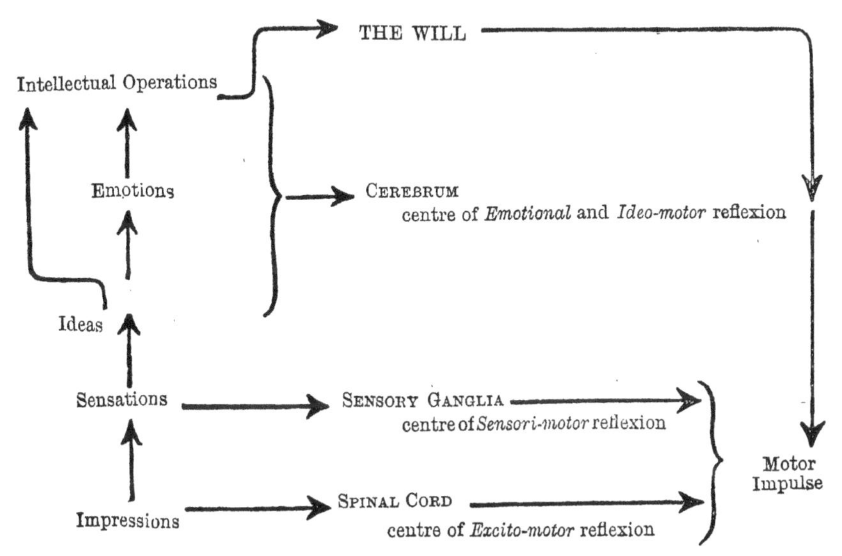

ordinary course of operation, when the whole is in a state of complete functional activity, and (2) the character of the Reflex actions to which each part is subservient, when it is the highest centre that the impression can reach.

112. The Cerebro-spinal system is intimately blended with another set of Ganglionic centres and Nerve-trunks, scattered through different parts of the body, but mutually connected with each other, which is termed the *Sympathetic* system.* The principal ganglionic

* This System has been sometimes termed the *ganglionic* system, on account of the connection of its nerve-trunks with isolated and scattered centres, in striking contrast with the continuity and apparent fusion of the ganglionic

centres of this system are the two great *Semilunar Ganglia*, which lie in the abdominal cavity near the spine, and from which there issues a radiating series of trunks and branches, constituting the great *Solar plexus*, to be distributed to the muscular walls of the Intestinal canal along its whole length, and also to the various Glandular organs in connection with it. There are two other principal though smaller systems of ganglia and nerves; one in connection with the Heart and great bloodvessels, constituting the *Cardiac plexus;* and the other in connection with the organs of reproduction and other viscera contained in the cavity of the pelvis, constituting the *Hypogastric plexus.* These plexuses communicate by connecting cords with a chain of *Prevertebral* ganglia, which lies along the front of the Spine on each side; and this, in its turn, communicates with the *Spinal* ganglia, which are ganglionic enlargements on the posterior roots of the Spinal nerves (§ 62). There are also several small ganglionic centres in the Head; which have the same kind of connection with its ordinary sensory and motor nerves, that the prevertebral ganglia have with those of the trunk. In virtue of these communications, the trunks and branches of the *Sympathetic* system contain, in addition to their own fibres (which are distinguished from the Cerebro-spinal by not possessing the double border that is given by the "white substance of Schwann," § 36), a great number of fibres derived from the Cerebro-spinal system. On the other hand, the Cerebro-spinal nerves contain Sympathetic fibres, sometimes in considerable quantity.

113. The Sympathetic system is sometimes designated the

centres of the Cerebro-spinal system. And an analogy has even been drawn between the chain of *prevertebral* ganglia of the Sympathetic, and the Ventral cord of Articulated animals. But this analogy entirely fails when we look at the distribution of the two sets of nerves, and the functions to which they respectively minister; since it is perfectly clear from such comparison, that it is the Spinal cord of Vertebrata which really represents the Ventral cord of Articulata, as a series of locomotive or Pedal ganglia.

nervous system of Organic Life, to distinguish it from the Cerebro-spinal system, which is the *nervous system of Animal Life.* Its *motor* action is exerted upon the Muscular walls of the Alimentary canal, of the Heart and Arteries, of the Gland-ducts, Uterus, and other organs; as has been experimentally proved by irritating these trunks immediately after the death of an animal. But as the very same contractions may be excited by irritating the roots of those Spinal nerves from which the several Sympathetic plexuses receive fibres, there is reason to believe that the motor endowments of the Sympathetic system are chiefly dependent on its connection with the Cerebro-spinal. And this seems to be especially the case with that very important division of the Sympathetic which is distributed on the walls of the Arteries, now known as the *Vaso-motor* system. For the real centre of this system, which has for its function to regulate the supply of blood to different parts, by its action on the calibre of the Arteries, appears to lie in the *Medulla oblongata*; from which also proceed certain nerve-fibres (included in the trunks of the *Pneumogastric* or *Par vagum*), which have a special influence on the movements of the Heart, and which probably regulate their rate in accordance with that of Respiratory action.—No motor power can be exerted through the Sympathetic system by any act of *Will;* but the muscular actions of many of the parts just enumerated are greatly affected by *Emotional* states; and this is particularly the case in regard to those of the Heart and Arteries. Thus we continually see the action of the heart quickened by Emotional excitement; whilst a violent "shock" to the feelings may seriously reduce it (as in fainting), or may even completely paralyse it. Of the action of the Emotions, through the same channel, on the Blood-vessels, we have a familiar example in the phenomenon of blushing; and this is only one of (probably) a vast number of changes thus induced, some of which have a very important influence on our Mental operations (§§ 356, 472). The blood-vessels that supply some of the Glands most directly affected by Mental states—such as the Lachrymal, the Salivary, and the Mammary,—seem to receive their supply of vaso-motor nerves direct from the Cerebro-spinal System ; and it is by the influence of those nerves in determining the calibre of their arteries, that the quantity of the Secretion is regulated ; as in the sudden flow of Tears, of Saliva, or of Milk. Those portions of the Glandular apparatus, on the other hand, the amount of whose secretions is affected, not so much by mental conditions, as by states of other parts of the Visceral apparatus, are supplied by the Sympathetic exclusively, or nearly so.

114. Whatever *Sensory* endowments are possessed by the parts supplied by the Sympathetic system, must be referred to the same connection with the Cerebro-spinal system. In the ordinary condition of the body, there is no evidence of the possession of any such endowments ; for the organs exclusively supplied by the Sympathetic system perform their functions without our consciousness, and no sign of pain is given when the Sympathetic nerve-trunks are irritated. But in diseased conditions of those organs, violent pains are often felt in them ; and experiment shows that whilst slight irritations of the healthy organs call forth no indications of suffering, such indications are manifested when the impression is made stronger. It is clear, therefore, that the effect of such impressions, when unusually strong, must be transmitted to the Sensorium ; and the reason why they do not ordinarily proceed thither, is probably because the *excitor* impulse is usually expended in calling forth reflex *movement* through the Sympathetic ganglia themselves. There is a remarkable tendency to *radiation* in such impulses, in virtue of the extraordinary intercommunication between different parts of the Sympathetic system ; and it is in this manner that those "morbid sympathies" between remote organs are established, which have a very important share in the phenomena of disease.

115. There is considerable evidence, moreover, that the

Sympathetic system has a modifying influence on the *Nutrition* of the body, and on the *quality* (as well as on the quantity) of the *Secretions*. And it seems probable that this is exerted through the proper fibres of the Sympathetic, rather than through those of the Cerebro-spinal system. This influence has been especially studied in the case of the Fifth Pair (Fig. 11), which is the nerve of common Sensation for the head and face, and which contains a great number of Sympathetic fibres that have their centre in a large ganglion on its sensory root. For it has been found that if its *trunk* be divided after having passed through the ganglion, a disorganizing inflammation of the Eye always follows; whilst if the *roots* be divided, so as to cut off all the Cerebro-spinal fibres from *their* centre, whilst the fibres proceeding from the Sympathetic ganglion can still go on to the eye, the derangement of its nutrition is either wanting altogether, or is greatly diminished in intensity.—There can be no doubt whatever, that though the *Will* can exert no modifying influence on the Nutritive operations, yet that these are very much affected by *mental states;* and especially by the persistence of that which may be termed *expectant attention*, whose remarkable action will form the subject of special consideration hereafter (Chap. XIX.).

NOTE.

Since the foregoing Chapter was in type, Dr. Ferrier has obtained a very remarkable series of Experimental results, by the application of Faradic Electricity to the Cortical substance of the Cerebrum, and to other Ganglionic centres of the Brain, in different animals. As these results—so far as they have been yet made public—do not appear to the Writer in any way inconsistent with the views set forth in the preceding pages, but, on the contrary, serve to confirm and extend them, he has thought it preferable to leave in its original shape the expression of the opinions at which he had arrived long previously, and had recorded in nearly the same words; deferring to the Appendix an account of Dr. Ferrier's experiments, with the inferences which they seem in the Writer's judgment to warrant.

CHAPTER III.

OF ATTENTION.

116. It has been the Writer's object in the preceding Chapter, not only to explain the general structure and working of the Nervous mechanism, but also to indoctrinate the Reader with that idea of its *reflex* activity, which we derive from experiment on the lower Animals, and from observation of the phenomena of disease or injury in Man. For the information we obtain from the study of the *lower* centres, in regard to that form of reaction which manifests itself directly in Muscular motion, furnishes the key to the study of that reflex activity of the *higher* centres, which expresses itself in states of Consciousness, — namely, in the production of Sensations, the formation of Ideas, and the excitement of Emotions; these states of activity being either the *excitors* of other Cerebral changes of the like kind, or discharging themselves (so to speak) by operating downwards on the Muscular apparatus.—It seems desirable, at our very entrance upon the enquiry into the action of these *higher* centres (which is, in fact, the Physiology of the *Mind*), to take special note of the *active* as distinguished from the *passive* state of recipiency for impressions which are brought to the Sensorium, whether by the nerves of the *external* or by those of the *internal* senses (§ 89); in virtue of which we *fix our Attention* either on something that is going on *outside* us, or on something that is going on *within* us, instead of being affected by each impression exactly in proportion to its strength. For it is in the power which the Will possesses over the *direction* of this active recipiency, or *Attention*, that the capacity of the Ego, alike for the systematic acquirement of Knowledge, for the

control of the Passions and Emotions, and for the regulation of the Conduct, mainly consists. In studying the working of the "Mechanism of Thought and Feeling," therefore, we must first trace the effect of Attention on each principal form of Mental activity.

117. The augmented recipiency of the *Sensorium* for some particular kind of impression, involves—apparently as its direct consequence—a proportionate reduction, or even an entire suspension, of its recipiency for impressions of other kinds. The Philosopher who is walking in a crowded thoroughfare, may have his attention so completely engrossed by an internal "train of thought," that he takes no heed whatever of what is going on around him, so long as this does not interfere with his onward progress; his vision having been *passively* exercised merely in directing his Muscular movements, and none of its impressions having gone up further than the Sensorium, the *activity* of which has been limited for the time to its Cerebral side.

The Writer was informed by Mr. John S. Mill, when his "System of Logic" was first published, that he had thought-out the greater part of it during his daily walks between Kensington and the India House; and himself more than once met Mr. Mill in Cheapside, at its fullest afternoon tide, threading his way among the foot-passengers with which its narrow pavement was crowded, with the air of a man so deeply absorbed in his own contemplations that he would not recognize a friend, and yet not jostling his fellows or coming into collision with lamp-posts.

On the other hand, the Countryman who comes up to London for the first time, may have his attention so attracted by the novelties he sees at every step, as to be led with difficulty to discuss a matter of business with the friend with whom he is walking. But suppose the Philosopher's course to be checked by some unusual obstruction, — such as a procession, or a street-accident,—the activity of his Sensorium is diverted from its Cerebral (or Intellectual) to its Sensational side; in other words, his attention is given to what is passing *outside* himself, rather than to what is passing *within* himself; his train of thought is completely interrupted; and he cannot recover it, until his attention is no longer occupied by the difficulty of making his way onward, which has temporarily diverted it. On the other hand, the nature of the communication which the Countryman receives from his friend, may be of a kind so powerfully to interest him, whether pleasureably or painfully,—as, for example, his inheritance of a fortune, or the success of a commercial speculation; or, on the other hand, a serious loss of property, or the adverse decision of a law-suit,—that from the moment he receives the news, he takes no note of the novelties which previously attracted him so strongly; but gives his whole attention to the particulars which his friend has to communicate.

118. Now this state of *active* as compared with *passive* recipiency, —of *Attention* as compared with mere *Insouciance*—may be either *volitional* or *automatic;* that is, it may be either *intentionally* induced by an act of the Will, or it may be produced *unintentionally* by the powerful *attraction* which the *object* (whether external or internal) has for the Ego. Hence, when we *fix* our Attention on a particular object by a determinate act of our own, the *strength of the effort* required to do so is greater, in proportion to the attractiveness of *some other* object. Thus, the Student who is earnestly endeavouring to comprehend a passage in "Prometheus," or to solve a Mathematical problem, may have his attention grievously distracted by the sound of a neighbouring piano, which *will* make him think of the fair one who is playing it, or of the beloved object with whom he last waltzed to the same measure. Here the Will may do its very utmost to keep the attention fixed, and may yet be overmastered by an involuntary attraction too potent for it; just as if a powerful electro-magnet were to snatch from our hands a piece of iron which we do our very utmost to retain within our grasp. Or, again, when "the thoughts begin to

wander" through fatigue of Brain, a powerful effort of the Will may be needed to keep them fixed on the completion of a task which the Ego has determined to execute, until the strongest Volition can no longer resist the imperious demand of the Physical mechanism for repose. Yet even then, the attractiveness of some new object (the coming-in, for example, of an anxiously-desired book, or the unexpected arrival of a friend charged with important news) shall produce not only a complete awakening of the attention, but an irresistible diversion of it into a new channel.

119. The power of the *Will* over the state of *attention* is therefore not unlimited; and its degree varies greatly in different individuals. In the young Child, as among the lower Animals, the Attention seems purely *automatic*, being solely determined by the *attractiveness of the object;* and the diversion of it from one object to another simply depends upon the relative force of the two attractions. It is this automatic fixation of the attention on the Sense-impressions received from the external world, that enables the Infant to effect that marvellous combination of visual and tactile perceptions, which guides the whole subsequent interpretation of its phenomena (§ 167). When an attractive object is presented to it, which it grasps in its little hands, carries to its lips, and holds at different distances, earnestly gazing at it all the while, it is learning a most valuable lesson; and the judicious Mother or Nurse will not interrupt this process, but will allow the infant to go on with its examination of the object as long as it is so disposed.—During the earlier stage of Childhood, it is mainly the attractiveness which the *changes* going on in the world around have for the observing faculties, which leads to the employment of them in connection with Ideational activity; the child wanting to know the *meaning* of what it sees, breaking open its toys to find out what makes them move, and asking the "why" of everything that excites its curiosity. In this stage, it is of great importance that the child should be led to limit his enquiries to some one

object, until he has made himself acquainted with all that he can learn of its characters; and here a judiciously-devised system of "object lessons" answers the double purpose of communicating information and of cultivating the habit of fixity of the attention, which, at first purely *automatic*, gradually comes to be under the control of the Ego.

120. So soon, however, as the work of systematic instruction commences, other influences come into play. It is the aim of the Teacher to fix the attention of the Pupil upon objects which may have in themselves little or no attraction for it; and in this stage, the direct operation of *motives* becomes very apparent. The "unconscious influence" which the Parent or Nurse has acquired by Habit (§ 290), the desire of approbation or reward, or the fear of punishment, first call forth the *effort* which is required to keep the Attention steadily fixed, even for a short time, upon some unattractive object, and to resist the solicitations of a new toy or a game of play. And in this early stage, all experience shows the advantage of *moderating* this effort, by giving to the object to which the Attention is to be directed, such attractiveness as it may be capable of, and by not requiring the attention to be too long sustained. Thus a picture-alphabet, with jingling rhymes, will often do what a simpler and severer method of "teaching the child its letters" fails to accomplish; and the "multiplication table" is much sooner acquired by being put into rhyme and sung in the march of an Infant-school, than when presented in the repulsive nakedness of 2 × 2 = 4. Those "strong-minded" Teachers who object to these modes of "making things pleasant," as an unworthy and undesirable "weakness," are ignorant that in this stage of the child-mind, the Will—that is, the power of *self*-control—*is* weak; and that the primary object of Education is to encourage and strengthen, not to repress, that power. Great mistakes are often made by Parents and Teachers, who, being ignorant of this fundamental fact of child-nature, treat as *wilfulness* what is

in reality just the contrary of Will-fullness; being the direct result of the *want* of Volitional control over the automatic activity of the Brain. To punish a child for the want of obedience which it *has not the power* to render, is to inflict an injury which may almost be said to be irreparable. For nothing tends so much to prevent the healthful development of the Moral Sense, as the infliction of punishment which the child *feels to be unjust;* and nothing retards the acquirement of the power of directing the Intellectual processes, so much as the Emotional disturbance which the feeling of injustice provokes. Hence the determination often expressed to "break the will" of an obstinate child by punishment, is almost certain to strengthen these reactionary influences. Many a child is put into "durance vile" for not learning "the little busy bee," who simply *cannot* give its small mind to the task, whilst disturbed by stern commands and threats of yet severer punishment for a disobedience it cannot help; when a *suggestion* kindly and skilfully adapted to its automatic nature, by directing the turbid current of thought and feeling into a smoother channel, and *guiding* the activity which it does not attempt to *oppose*, shall bring about the desired result, to the surprise alike of the baffled teacher, the passionate pupil, and the perplexed bystanders.

121. The habit of Attention, at first purely automatic, gradually becomes, by judicious training, in great degree amenable to the Will of the Teacher; who encourages it by the suggestion of appropriate motives, whilst taking care not to overstrain the child's mind by too long dwelling upon one object. Even at a very early period, there will be found marked differences among individuals, as to their power of sustained attention: some being distracted by every passing occurrence; whilst others have not much difficulty in keeping their minds fixed upon an object, for a sufficient length of time to enable them to learn all that the exercise of their senses can teach them; while with others, again, the difficulty lies in the *transference* of their attention from one

object to another, so that, when the Teacher thinks that the Pupil's mind is being exercised on B, he is still "ruminating" upon A. And thus many children require special modifications of this disciplinary process; the "bird-witted" being encouraged to *fix* their attention, whilst those in whom the opposite tendency predominates should be exercised in *mobilizing* it. These opposite tendencies are noticeable in after life, and give a marked direction to the Intellectual character. Many a "dull" boy is supposed to be stupid, when he is simply introspective; his attention being given rather to the ideas which are passing through his mind, than to what is going on around him. On the other hand, many a "quick-witted" boy gets a reputation for cleverness which he does not deserve; his mind being keenly alive to all that is passing outside, so that he rapidly takes-in new impressions, but loses their traces *as* quickly, one set of impressions superseding another before any have had time to fix themselves.

122. As the power of determinately fixing the Attention gains strength, only requiring adequate motives for its exercise, the influence of a *system* of discipline by which each individual feels himself borne along as if by a Fate, still more that of an Instructor possessing a strong Will guided by sound judgment (especially when united with qualities that attract the affection, as well as command the respect, of the pupil, § 290, III.), greatly aid him in learning to use that power. As Archbishop Manning has truly said (*Contemporary Review*, Feb. 1871), "During the earlier period of our lives, the potentiality of our intellectual and moral nature is elicited and educed by the Will of others. . . . . . Our 'plagiosus Orbilius' did for our brain in boyhood, what our developed Will, when we could wield the ferule, did for it in after life." With the general progress of Mental development, the direction of the Attention to *ideas* rather than to sense-impressions, which was at first difficult, becomes more and more easy; its *continuous* fixation upon one subject becomes so completely habitual, that it is often

less easy to break the continuity than to sustain it; and the time at last arrives, when the *direction* of that Attention is given by the individual's *own* Will, instead of by the will of another.

123. It will serve to help us in the study of the manner in which volitional Attention operates in the higher spheres of Thought and Emotion, if we first study its action in the reception of Sense-impressions.— When we *wish* to make ourselves thoroughly acquainted with a Landscape or a Picture, we *intentionally* direct the axes of our eyes to each part of it successively, and study that part in its details until we have formed a composite conception of the whole. Whilst we do this, the determinate fixation of the Attention upon any one part weakens the impression made by all the rest; so that of what lies within the Visual range at any one moment, nothing is distinctly seen, except the limited spot at which we are fixedly looking. Again, the practised Microscopist, whilst applying one of his eyes to his instrument, and determinately giving his whole Attention to the visual picture he receives through it, can keep his other eye open, without being in the least disturbed by the picture of the objects on the table which must be formed upon its retina, but which he *does not see*, unless their brightness should *make* him perceive them.— So in the act of *listening*, we are not only distinctly conscious of sounds so faint that they would not excite our notice but for the volitional direction of the Attention; but we can single out these from the midst of others by a determined and sustained effort, which may even make us quite unconscious of the rest, so long as that effort is kept up. Thus, a person with a practised "musical ear" (as it is commonly but erroneously termed, it being not the *ear*, but the *brain*, which exerts this power), whilst listening to a piece of music played by a large orchestra, can single out any one part in the harmony, and follow it through all its mazes; or can distinguish the sound of the weakest instrument in the whole band, and follow its strain through the whole performance.

And an experienced Conductor will not only distinguish when some instrumentalist is playing out of tune, but will at once single out the offender from the midst of a numerous band.

124. The contrast between the *volitional* and the *automatic* states of Attention is particularly well shown in the effects of *painful* impressions on the Nervous system. It is well known that such impressions as would ordinarily produce severe pain, may for a time be *completely unfelt*, through the exclusive direction of the Attention elsewhere; and this direction may either depend (*a*) upon the *determination of the Ego*, or (*b*) upon the *attractiveness of the object*, or (*c, d, e,*) on the combination of both.

a. Thus, before the introduction of Chloroform, patients sometimes went through severe operations without giving any sign of pain, and afterwards declared that *they felt none*; having concentrated their thoughts, by a powerful effort of abstraction, on some subject which held them engaged throughout.

b. On the other hand, many a Martyr has suffered at the stake with a calm serenity that he declared himself to have no difficulty in maintaining; his entranced attention being so *engrossed* by the beatific visions which presented themselves to his enraptured gaze, that the burning of his body *gave him no pain* whatever.

c. Some of Robert Hall's most eloquent discourses were poured forth whilst he was suffering under a bodily disorder which caused him to roll in agony on the floor when he descended from the pulpit; yet he was entirely unconscious of the irritation of his nerves by the calculus which shot forth its jagged points through the whole substance of his kidney, so long as his soul continued to be "possessed" by the great subjects on which a powerful effort of his Will originally fixed it.

d. The Writer has himself frequently begun a lecture, whilst suffering neuralgic pain so severe as to make him apprehend that he would find it impossible to proceed; yet no sooner has he, by a determined effort, fairly launched himself into the stream of thought, than he has found himself continuously borne along without the least distraction, until the end has come, and the attention has been released; when the pain has recurred with a force that has over-mastered all resistance, making him wonder how he could have ever ceased to feel it.

e. A similar experience in the case of Sir Walter Scott is thus recorded by his biographer:—"John Ballantyne (whom Scott, while suffering under a prolonged and painful illness, employed as his amanuensis) told me that though Scott often turned himself on his pillow with a groan of torment, he usually continued the sentence in the same breath. But when dialogue of peculiar animation was in progress, spirit seemed to triumph altogether over matter,—he arose from his couch, and walked up and down the room, raising and lowering his voice, and as it were acting the parts. It was in this fashion that Scott produced the far greater portion of the Bride of Lammermoor,—the whole of the Legend of Montrose,—and almost the whole of Ivanhoe."—(*Lockhart's Life of Scott*, chap. xliv.) See also § 352 *a*, for a curious sequel to the foregoing.

125. These facts throw considerable light upon a question which will hereafter come to be considered, whether Cerebral changes by which Intellectual results are evolved may not go on *without our consciousness* (§ 417). For there are Metaphysicians who fully admit the *automatic* nature of the operations referred to, but at the same time assert that, as they are truly *Mental*, we cannot be really *unconscious* of them, but merely do not *remember* them, in consequence either of the occupation of our attention by another train of thought, or of the severance of the connection between our *sleeping* and our *waking* consciousness. But this assertion does not constitute proof. In the case of the Physical impressions that produce the sense of pain, we have ample evidence that they *must* have been *made;* and the only question is as to their having been *felt.* "Did Robert Hall, for example, really *feel* the pain which he declared that he did *not* feel?" If it be replied that he did, but that he did not remember it, it may be further inquired, "What is the evidence of his having felt it?" His consciousness and memory said that *he did not;* and what higher evidence is attainable? No doubt, if his attention had been for a

moment withdrawn from the subject of his discourse, the pointed calculus in his kidney would have made its presence most distressingly perceptible; but there is no more evidence that pain was *consciously felt, though not remembered,* whilst he was preaching, than that he felt it when a large dose of opium procured for him the refreshment of sound sleep. When Damiens, worn out by his protracted sufferings, slept on the rack (§ 471), enjoying a remission of suffering until awoke by some new and more exquisite torture, did *he feel* his pain? It would be a mere gratuitous assumption to say that he *must* have felt it, because the organic condition was present that would make him feel it if he were awake; since the presence of this organic condition goes for nothing, unless there be a receptive condition on the part of the Sensorium. And there seems just as much evidence that this receptivity may be entirely suspended *quoad* any one set of impressions (whether internal or external) by the *complete engrossment of the attention* upon another, as that it may be suspended altogether in Sleep or Coma (see § 488).

126. Now, just as the Organic impressions which make themselves felt in *pain,* when the sensorium is receptive of them, may exist *without consciousness* if the sensorium be otherwise engaged, so (it appears to the Writer) may it be affirmed—and on precisely the same evidence—that the Organic changes which are concerned in the automatic production of Thought, and of which we become conscious as *ideas* when the Sensorium takes cognizance of them, may go on *without consciousness* if the sensorium be otherwise engaged. The affirmation that such automatic changes *cannot* take place without the consciousness of them, is, as Sir William Hamilton has pointed out (§ 418), a mere *petitio principii;* and may be regarded as a "survival" of those older notions of the essential independence of Mind and Body, which a truly philosophical Psychology can no longer accept as consistent with the fundamental facts of our composite nature.

127. It is to the habitual direction of the Attention to some particular kind of Sense-impressions, that we are to attribute the *increase in discriminative power*, which is specially remarkable in the case of such as suffer under deprivation of other Senses. This s most frequently seen in the case of the Touch, which may be brought by practice to such wonderful acuteness, that some blind persons can read from raised print not much larger than that of an ordinary folio Bible, by merely passing the point of the finger along the lines; whilst by attending to minute differences which ordinarily pass entirely unnoticed, they can not only distinguish persons among whom they are living, but can also recognize such as have not been near them for months or even years previously, by the mere contact of their hands. (Thus Laura Bridgeman unhesitatingly recognized the Writer's brother, after the lapse of a year from his previous interview with her, by the "feel" of his hand.) It is well known that an extraordinary acuteness of Touch is possessed by the weavers of the finest of those textile fabrics for which India is celebrated; and as this manufacture, like others, is handed down in the same families, it does not seem improbable that a special *aptitude* for it, originally acquired by the experience of the individual, may be transmitted hereditarily with progressive improvement.—The like improvement is also occasionally noticed in regard to the Smell, which may acquire an acuteness rivalling that of the lower animals; and this not only in the blind, but among races of Men whose existence depends upon such discriminative power. Thus we are told by Humboldt that the Peruvian Indians in the darkest night can not merely perceive through their scent the approach of a stranger whilst yet far distant, but can say whether he is an Indian, European, or Negro. And it is said that the Arabs of the Sahara can recognise the smell of a fire thirty or forty miles distant.—In the same manner, the sense of Taste may be trained to the recognition of differences which would ordinarily pass unnoticed; of which we have an example in the Wine-taster

who can tell the vineyard by which any particular choice wine was yielded, and the year of the vintage which produced it; a not less striking case being furnished by the Tea-taster, the delicacy of whose sense is said to be seldom preserved for more than a few years.—The familiar case of the Seaman who makes out the distinct "loom of the land," where a landsman can discern nothing but an indefinite haze above the horizon-line, illustrates the improvement in the Visual sense of individuals, which arises from the habitual direction of the Attention to a particular class of impressions. But the possession of this faculty, also, seems occasionally to be an attribute of Race; the power of descrying objects at vast distances being (it is asserted) hereditarily possessed by the Mongols of Northern Asia and the Hottentots of Southern Africa, both of which races habitually dwell on vast plains, that seem to stretch without limit in every direction. As no dweller among them seems able to acquire the same visual power by any amount of individual experience, and as even half-breeds do not possess the aptitude in a degree by any means equal to that which characterizes the men of pure race, it seems probable that, as in the cases already referred to, the power acquired by habitual Attention in the first instance has become fixedly hereditary, improving with habitual use in successive generations.

128. Whilst, however, we give full credit to the cumulative effect of Hereditary transmission, in cases in which the same habit is kept up by force of circumstances through successive generations, there is adequate evidence that an extraordinary increase in the discriminative power of individuals may be brought about by the *concentration* of the Attention upon the Sensorial impression received through the organ of sense, rather than upon an improvement produced by practice in the organ itself. For the same exaltation often shows itself without any practice at all, in that curious form of Somnambulism (natural or induced), in which the Attention is entirely engrossed by the particular thought or feeling which

may be before the consciousness at the moment (§§ 494, 498). And it is not a little remarkable that this exaltation should show itself especially in the *muscular sense*, to the indications of which we ordinarily give very little heed. Thus, the Writer has repeatedly seen Hypnotized patients write with the most perfect regularity, when an opaque screen* was interposed between their eyes and the paper; the lines being equidistant and parallel, and the words at a regular distance from each other. He has seen, too, an algebraical problem thus worked out, with a neatness which could not have been exceeded in the waking state.—But the most curious proof of the exaltation of this Muscular sense, which conveys to the mind of the Somnambule that exact appreciation of distance and relative position for which we ordinarily trust to our Vision (§ 192), is derived from the manner in which the writer will sometimes carry back his pen or pencil to dot an *i* or cross a *t*, or to make a correction in a letter or word. Mr. Braid had one patient (in whom the sense of Smell also was remarkably exalted, § 498), who could thus go back and correct with accuracy the writing on a whole page of note-paper; but if the paper was moved from the position it had previously occupied on the table, all the corrections were on the *wrong* points of the paper as regarded the *then* place of the writing, though on the *right* points as regarded its *previous* place. Sometimes, however, he took a fresh departure (to use a nautical phrase) from the upper left-hand corner of the paper; and all his corrections were then made in their *right* position, notwithstanding the displacement of the paper. "This," says Mr. Braid, "I once saw him do, even to the double-dotting a vowel in a German word at the bottom of the

* This is a far more satisfactory test than bandaging the eyes. It is impossible to see through a slate, a music book, or a piece of pasteboard; but those who have carefully experimented on the asserted *clairvoyance* of Mesmerized "subjects," know well that the best-arranged bandages may be shifted, by the working of the muscles of the face, sufficiently to permit the use of the eyes in certain directions.

page, a feat which greatly astonished his German master, who was present at the time."

129. The effects of Attention in either augmenting or diminishing the intensity of Sensations, are manifested, not only in regard to those which are excited by external Impressions, but also in respect to those which originate within the body. Every one is aware how difficult it is to remain perfectly at rest, especially when there is a particular motive for doing so, and when the attention is strongly directed to the object. This is experienced whilst a Photographic likeness is being taken, even when the position is chosen by the individual, and a support is adapted to assist him in retaining it; and it is still more strongly felt by the performers in "Tableaux Vivans," who cannot keep up the effort for more than three or four minutes.—On the other hand, when the Attention is strongly directed to an entirely different object (as, for example, in listening to an eloquent sermon or an interesting lecture), the body may remain perfectly motionless for a much longer period, the Sense-impressions which would otherwise have induced the individual to change his position, not being felt; while no sooner is the discourse ended, than a simultaneous movement of the whole audience takes place, every one then becoming conscious of some discomfort, which he seeks to relieve. This is especially the case in regard to the Respiratory sense: for it may generally be observed that the usual reflex movements do not suffice for the perfect aëration of the blood, and that a more prolonged inspiration, prompted by an uneasy feeling, takes place at intervals; but under such circumstances as those just alluded to, this feeling is not experienced until the Attention ceases to be engaged by a more powerful stimulus, and then it manifests itself by the deep inspirations which accompany, in almost every individual, the general movement of the body.—Sensations may even be called into existence, as Sir H. Holland has pointed out, by the deter-

minate direction of the Attention to particular parts of the body :

a. "The Attention by an effort of Will concentrated on the sensorium, creates certain vague feelings of tension and uneasiness, caused possibly by some change in the circulation of the part; though it may be an effect, however difficult to conceive, on the nervous system itself. Persistence in this effort, which is seldom indeed possible beyond a short time without confusion, produces results of much more complex nature, and scarcely to be defined by any common terms of language." "Stimulated attention will frequently give a local sense of arterial pulsation where not previously felt, and create or augment those singing and rushing noises in the ears, which probably depend on the circulation through the capillary vessels." "A similar concentration of consciousness on the region of the stomach, creates in this part a sense of weight, oppression, or other less definite uneasiness; and, when the stomach is full, appears greatly to disturb the due digestion of the food." The state and action of the bladder and the bowels are much influenced by the same cause. A peculiar sense of weight and restlessness approaching to cramp, is felt in a limb to which the attention is particularly directed. So, again, if the attention be steadily directed to almost any part of the surface of the body, some feeling of itching, creeping, or tickling will soon be experienced.—(*Chapters on Mental Physiology*, pp. 18—24.)

Evidence will hereafter be adduced, that this direction of the Attention changes the local action of the part; so that, if habitually maintained, it may produce important modifications in its Nutrition. In this way it often happens that a *real* malady supervenes upon the *fancied* ailments of those, in whom the want of healthful occupation for the mind leaves it free to dwell upon its own Sensations; whilst, on the other hand, the strong *expectation* of benefit from a particular mode of treatment, will often *cure* diseases that involve serious organic change (Chap. XIX.). Hence it seems probable that in the cases just cited, as in others to be presently noticed (§ 140), the Sensations really originate in an impression on the nerves of the part to which they are referred.

130. The difference between *volitional* and *automatic* Attention, again, is well shown by the difference between an observant and an unobservant person; still more by the phenomena of that state which is strangely misnamed "absence of mind." One man is designated as observant, whose Will prevents his attention from being so far enchained by the attractiveness of any one object, whether a Sense-perception or an internal train of Thought, as to interfere with the reception of other impressions; whilst another is spoken of as unobservant, who, by allowing his attention to remain engrossed by one object, whether a Sense-perception or a train of Thought, is kept from bestowing a legitimate share of it upon the other impressions which he receives through either his external or his internal senses. The state of *Abstraction* is only an intensified condition of this last form of exclusive *un*-volitional attention (§ 445).

131. The effect of Attention in the limitation and intensification of our *external* sense-impressions, is exerted also upon those Cerebral operations of which we become conscious as states of Thought and Emotion, and which may be conveniently distinguished as *internal* sense-impressions. For these, like the preceding, may excite no more than a *passive* cognizance of them; whilst, on the other hand, our attention may be *actively* directed to them. And the result of this direction is similar: for the Mental state, of whatever nature it may be, upon which the attention is fixed, becomes intensified to such a degree, as to exclude for the time the cognizance of other impressions; whilst it acquires a special power of suggesting other Mental states.—This direction of the attention to states of Cerebral activity, may, like its direction to impressions received through the organs of Sense, be either *automatic* or *volitional*. When it is automatic, the Mind is engrossed for the time by some Idea or Emotion, in virtue of the intensity with which it has been called

up, or of the peculiar hold which it has upon our nature; and it may remain thus fixed, until this Mental state shall have given-rise to some other, or shall have expended its force in bodily action, or until the attention has been determinately detached from it by an exertion of the Will. But volitional Attention consists in the fixation of the Mental gaze, by a purposive *effort,* upon some single state, or on some class of Ideas or Feelings, which the Ego desires to make the special object of his contemplation; and it is by means of this *selecting* power, and of the tendency of the Mental state thus intensified to call-forth other states with which it has pre-formed links of association, that the Will acquires that *directing* power over the current of Thought and Feeling, which characterizes the fully developed Man (§§ 25-28).—Thus it is in the degree of Attention which we bestow upon certain classes of Ideas presented to us by suggestion, that our power of determinately *using* our Minds in any particular mode consists; and hence we see the fundamental importance of early learning to *fix* our attention, and to resist all influences which would tend to distract it. And this is essential, not merely to the advantageous employment of our Intellectual powers in the acquirement of Knowledge, but also to the due regulation of our Emotional nature: for it is by fixing the Attention upon those states of *feeling* which we desire to intensify, and, conversely, by withdrawing it from those we desire to repress (which is most easily effected by choosing *some other* object that exercises a healthful attraction for us), that we can encourage the growth of what we recognize as worthy, and can keep in check what we know to be wrong or undesirable.—This part of the subject will be more fully treated hereafter (Chaps. VI—IX).

CHAPTER IV.

OF SENSATION.

132. Sensation is that primary change in the condition of the conscious *Ego*,* which results from some change in the *Non-ego* or External World,—this last term including the Bodily organism itself; for it is through the instrumentality of a certain part of the Nervous apparatus, that the change in the Non-ego is enabled to affect the Ego. A physical impression made upon an afferent nerve, is propagated by it to its Ganglionic centre forming part of the *Sensorium;* and according to the particular centre which is thus affected, will be the nature of the Sensation produced. Thus impressions on the Olfactive, Optic, or Auditory nerves excite sensations of Smell, Sight, or Hearing, in virtue of their transmission to the Olfactive, Optic, or Auditory ganglia respectively. This is proved by the fact that *similar Impressions* will produce *entirely diverse Sensations*, according as they are made on one or another of the nerves of Sense. Electric stimulation does this most effectively, producing in each Sensory nerve the change which is necessary to call forth the particular affection of the Consciousness to which it ministers; so that, by proper management, the Ego may be made conscious at the same time of flashes of light, of distinct sounds, of a phosphoric odour, of a peculiar taste, and of a feeling of pricking, all excited by the same stimulus, the effects of which are modified by the respective peculiarities of the instru-

* Some Physiologists, it is true, have spoken of *Sensation without Consciousness;* but it seems very desirable, for the sake of clearness and accuracy, to limit the application of the word to the *mental* change; especially since the term "impression" serves to designate that change in the state of the Nervous system, which is its immediate antecedent.

ments through which it operates. So pressure, which produces through the nerves of common Sensation the feeling of resistance, is well known to occasion, when exerted on the Eye, the sensation of light and colours; and when made with some violence on the Ear, to produce a ringing sound. It is not so easy to excite sensations of Taste and Smell by mechanical irritation; and yet, as was shown by Dr. Baly, a sharp light tap on the papillæ of the tongue excites a Taste which is sometimes acid, sometimes saline. The sense of nausea may be easily produced, as is familiarly known, by mechanical irritation of the fauces.—But although there are some stimuli which can produce sensory impressions on all the nerves of Sense, those to which any one is specially fitted to respond, produce little or no effect upon the rest. Thus the Ear cannot distinguish the slightest difference between a luminous and a dark object. A tuning-fork, which, when laid upon the Ear whilst vibrating, produces a distinct musical tone, excites no other sensation when placed upon the Eye, than a slight jarring feeling, which is a modification of common not of visual sensation. The most delicate Touch cannot distinguish a substance which is sweet to the Taste, from one which is bitter; nor can the Taste (if the communication between the mouth and the nose be cut-off) perceive anything peculiar in the most strongly odorous bodies.—It may hence be inferred that no nerve of *special* Sense can take-on the function of another, any more than it can minister to common Sensation (§ 38).

133. The first stage in the Mental operation excited by a Sense-impression, is the *localization* of the Sensation; and this is clearly an *automatic* action, in regard to which it is impossible to say with certainty whether it is *primary* or *secondary*, a *congenital Intuition*, or an *acquired Instinct.* The latter view is perhaps the more probable; for though the young Infant cries when it feels pain, it does not show by any sign that it refers that pain to any particular seat; and we ourselves often wake out of sleep with a feeling of

discomfort or distress, which we are not at first sufficiently wide-awake to refer to a local origin. Yet the fully-developed Consciousness unhesitatingly refers Sense-impressions to the *origins* of the nerves that convey them to the Sensorium; those of any special Sense to the particular organ of that sense, and those of common Sensation to the part in which the afferent nerve-trunks have their roots. There is, as Professor Huxley has phrased it, "an *extradition* of that consciousness which has its seat in the Brain, to a definite point of the body; which takes place without our volition, and may give rise to ideas which are contrary to fact." Thus after amputations, the patients are for some time affected with sensations (probably excited by irritation at the cut ends of the nerves), which they refer to the fingers or toes of the lost limbs; and flashes of light are often experienced when the Eye has been completely extirpated, as also when its structure has been destroyed by disease. The effects of the Taliacotian operation afford a curiously-illustrative example of this principle; for until the flap of skin from which the new nose is formed, obtains vascular and nervous connections in its changed situation, the sensation produced by touching it is referred to the forehead.

134. There are cases, however, in which Sensations are referred to parts quite remote from those on which the impressions are made that give rise to them. Thus, disease of the hip-joint is often first indicated by pain in the knee; various disorders of the liver occasion pain under the left scapula; attention is often drawn to disease of the heart by shooting pains along the arms; the sense of nausea is more commonly excited by conditions of the stomach, than by the direct contact of the nauseating substance with the tongue or fauces; the sudden introduction of ice into the stomach will cause intense pain in the supra-orbital region; and the same pain is frequently occasioned by the presence of acid in the stomach, and may be very quickly relieved by its neutralization with an alkali. It will be seen that in most of these cases, it is

impossible to refer the sensations to any direct nervous connection with the parts on which the impressions are made; and they can scarcely be otherwise accounted-for, than by supposing that these impressions produce Sensorial changes, which are referred to other parts in virtue of some *central* track of communication with them, analogous to that through which reflex movements are excited. There are circumstances, indeed, which seem to render it not improbable, that just as the impression brought by the afferent nerves to the central organs, calls forth a reflex Motion by exciting the nerve-force of a motor nerve, it may produce a reflex Sensation by a like excitation of a sensory nerve. Certain it is, that, after a long continuance of some of these reflex sensations, the organs to which they are referred often themselves become diseased, although previously quite healthy; this perversion of their normal action being not improbably due to that habitual direction of the Attention to the part, which is prompted by the habitual Sensation (§ 129).

135. It has already been pointed out (§ 41) that, for the production of Sensations, each part of the Nervous apparatus must be in a state of activity, which can only be maintained by the constant Circulation of blood;—this being specially needed at what may be considered the *origins* of the Sensory nerves in the general surface of the Skin and in the organs of special Sense, and at their *terminations* in the Ganglionic centres. An enfeeblement of the circulation where impressions are first *received*, diminishes their strength, as we see in the numbness produced by an obstruction to the flow of blood through the main artery of a limb; and a like enfeeblement of the circulation in the Ganglionic centre through the instrumentality of which we are rendered *conscious* of the physical impression, produces a corresponding torpor of Sensibility. The local action of Cold, in like manner, produces numbness, not only by retarding the blood-circulation, but also (it would appear) by directly lowering the conducting power of the Nerve

itself; for if cold be applied to an afferent nerve-trunk *in its course*, complete insensibility is produced in all the parts from which it receives fibres. So, local *anœsthesia* or want of sensibility may be produced by the action of Ether or Chloroform on the nerves of the part.—On the other hand, in that first stage of local Inflammation in which the capillary circulation is unduly active, and the heat of the part is augmented, there is an unusual susceptibility, or "tenderness," which renders ordinary impressions productive of pain.

136. The like diminution or exaltation of Sensibility may arise from states of the Sensorium. Thus in *Sleep* there is a want of receptivity for *ordinary* Sense-impressions; though *extraordinary* impressions will make themselves felt, recalling the sleeper to consciousness. In the profound *Coma* of apoplexy or of narcotism, on the other hand, complete suspension of Sensorial activity is produced, in the one case by continued pressure within the skull (which probably acts by disturbing the circulation), in the other by the direct action of the poison on the nerve-substance; whilst the torpor which is produced by continued exposure to severe Cold, is attributable to the congestion of the veins of the brain, which results from the contraction called-forth by the cold in the vessels of the general surface of the body. On the other hand, there are states of general *exaltation* of Sensibility, which obviously depend upon affections of the Sensorial centres. Thus the first stage of Inflammation of the Brain is characterized by an extreme susceptibility of this kind; the most ordinary impressions of light, sound, &c., giving rise to sensations of unbearable violence. The presence of certain poisons, as lead, in the blood, sometimes induces the condition termed *hyperœsthesia*, or excessive sensibility; though it more commonly induces local *anœsthesia*, or want of sensibility.—In all these cases it is perfectly clear to the Physiologist, that the degree in which Sensations are *felt*, entirely depends upon the condition of the Mechanism by

the instrumentality of which Physical Impressions are received and are translated into states of Consciousness.

137. It is no less certain, however, that the intensity of Sensations is greatly affected by the degree in which the recipient Mind is directed towards them; and this may operate in regard either to Sensory impressions generally, or to those of some particular class. Of the former we have a characteristic example in what is known as the *hysterical* condition; in which the patient's Attention is so fixed upon her own bodily state, that the most trivial impressions are magnified into severe pains; while there is often such an extraordinary acuteness to sounds, that she overhears a conversation carried-on in an undertone in an adjoining room, or (as in a case known to the Writer) in a room on the second floor beneath. There is here, doubtless, a peculiar Physical susceptibility to Nervous impressions, which is to a certain degree remediable by medical treatment; but much depends upon the diversion of the patient's Attention from her own fancied ailments; and we here see the importance of the *self*-determining power of the Will, which, if duly exercised, can substitute a *healthful* direction of the Mental activity, for the *morbid* imaginings to which the patient has previously yielded herself.*—The transition is easy from the cases in which there is an exaggeration of *real* Sensations, to those in which there is an actual *production* of sensations not originating in any external impressions, by an *expectation* generated in the Mind itself (§ 147).

138. The vividness of Sensations usually depends rather on the degree of *change* which they produce in the system, than on

* This condition is by no means peculiar to Females; although, from the greater impressibility of the Nervous system, and the lower development of Volitional power, by which the Sex is ordinarily characterized, it is more common among them than in males. It is often fostered, from a very early date, by the habit in which injudicious Parents and Nurses indulge, of *fixing* the Child's attention on any little hurt or ache, instead of *withdrawing* it by the *counter*-attraction of some object of interest. (See § 269.)

the *absolute amount* of the impressing force; and this is the case with regard alike to special and to ordinary sensations. Thus, our sensations of Heat and Cold are entirely governed by the previous condition of the parts affected; as is shown by the well-known experiment of putting one hand into hot water, the other into cold, and then transferring both into tepid water, which will seem *cool* to one hand, and *warm* to the other. Every one knows, too, how much more we are affected by a warm day at the commencement of summer, than by an equally hot day later in the season. The same is the case in regard to Light and Sound, Smell and Taste. A person going out of a totally dark room into one moderately bright, is for the time painfully impressed by the light, but soon becomes habituated to it; whilst another, who enters it from a room brilliantly illuminated, will consider it dark and gloomy. Those who are constantly exposed to very loud noises, become almost unconscious of them, and are even undisturbed by them in illness; and the medical student well knows that even the effluvia of the dissecting-room are not perceived, when the Sensorium has been habituated to impressions they produce: although an intermission of sufficient length would, in either instance, occasion a renewal of the first unpleasant feelings, when the individual is again subjected to the impression.—Thus there seems reason to believe that Sensorial changes of frequent occurrence, produce a modification in the nutrition of the Sensorium itself, which *grows-to* them, as it were, just as the Nervous system generally may be considered as growing-to "nervine stimulants" habitually taken-in (§ 155); for not only would the production of such a modification be quite in accordance with the general phenomena of Nutrition,* but we can scarcely other-

* We have a remarkable exemplification of this, in the *tolerance* which may be gradually established in the system for various Poisons, especially for such as particularly affect the Nervous substance, such as Opium or Alcohol. It seems impossible to explain this tolerance on any other hypothesis, than that of the

wise explain the progressive formation of that connection between Sensorial changes and Motor actions, which gives rise to the "secondarily automatic" movements (§ 191). Hence it seems reasonable to attribute that diminution in the force of Sensations which is the ordinary consequence of their *habitual* recurrence, to the want of such a *change* in the condition of the Sensorium, as is needful to produce an impression on the Consciousness; the effects which they at first induced being no longer experienced in the same degree, when the structure of that part has accommodated itself to them.—But the same principle does not apply to those impressions to which the *attention* is habitually directed; for these lose none of their power of exciting Sensations by frequent repetition; on the contrary, they are so much *more* readily recognized, that they affect the Consciousness under circumstances in which the Ego is insensible to much stronger impressions of other kinds (§ 480).

139. *Subjective Sensations.*—The designation "subjective" is commonly given to all those Sensations which arise out of either bodily or mental states, whose existence is not consequent upon any "objective" or external change. But, strictly speaking, it should be limited to those which arise from the workings of the Ego's own Mind; since those which are produced by Physical impressions made on the nerves *within* his Body, just as truly belong to the *Non-ego*, as do those made by impressions operating from *without*. Thus, for example, when incipient Inflammation of a part produces the sense of Heat in it, exactly resembling that which would be excited by the proximity of a heated body, it can scarcely be doubted that the Physical impression on the afferent nerves *of the part*, exciting Sensorial change, is the same in both cases: such a Sensation, therefore, is no more truly "subjective" in the one case than in the other. But when a "biologized"

alteration of the Nutrition of the tissue by repeated doses, so that no further change can be produced by the quantity originally taken.

subject is made to believe that a body he holds in his hands is unbearably hot, and throws it down accordingly (§ 458), the *sensorial* change is produced by the Mental suggestion; the Sensation, which is only *referred* to the locality by the mental preconception, being the creation of the Ego himself.

140. Of those so-called "subjective" Sensations which have their origin in *local* changes that produce impressions on the nerves of the parts to which they are referred, we have examples in the flashes of light which are symptomatic of disease of the Retina or of the Optic nerve; and in the ringing in the ears, which, while sometimes due to a disordered condition affecting the nervous apparatus within the ears themselves, appears more frequently to arise from an affection of the Auditory nerve in its course by the pulsations of a neighbouring artery. And it is probable that the persistence of a bad taste or of an unpleasant odour, having no source *outside* the body, is often to be attributed to analogous *local* changes *within* it. — On the other hand, there is probably no kind of Sensation that may not be produced by physical conditions of the Sensorium, which have *not* been induced by impressions transmitted thither by the afferent nerves, but which arise from morbid changes, either in its own substance, or in the blood which circulates through it. For subjective sensations are among the commonest indications of incipient Brain-disease; and they are especially noticeable as results of the presence of *poisons* in the blood, whose action is specially exerted on the Cephalic nerve-centres.

141. We have now to consider, however, that class of *truly* "subjective" Sensations, of which the origin is to be looked-for, neither in local impressions on the nerves of the external senses, nor in abnormal affections of the Sensorium; but in impressions transmitted to the Sensorium by the "nerves of the internal senses," which convey to it the results of changes taking place in that cortical layer of the Cerebrum which we have seen reason

to regard as the instrument of the higher Psychical operations (§ 100).

142. Every one is familiar with the fact that Sensations formerly experienced are *reproduced* in Dreaming, with a vividness and reality quite equal to that with which his consciousness was originally impressed by the actual objects. And this not unfrequently happens also in the waking state; in which we are able distinctly to trace-out the causation of this reproduction, in the suggestive action of pre-formed Ideational associations (§§ 101, 103). Of these associative actions, it cannot be reasonably doubted that the *Cerebrum* is the instrument; and the mechanism by which they occasion the reproduction of Sensations, has been already explained to be (according to the Writer's view) the transmission to the Sensorium, along the nerves of the *internal* senses, of an impression equivalent to that which it originally received through the nerves of the *external* senses (§§ 99—105).—But if Cerebral (ideational) states can *reproduce* Sensations, they can also *produce* them; and as this fact is of fundamental importance in our interpretation of a large class of phenomena to which attention has been drawn of late years under the designations "Odylism," "Spiritualism," &c., it will be desirable to adduce the proofs of it in some detail. For nothing is more common than to hear the advocates of these doctrines appealing to "the evidence of their senses" as conclusive in regard to the actual occurrence of the phenomena which they believe themselves to have witnessed; in utter ignorance of the fact that *nothing is more fallacious* than that evidence, when the Mind is previously "possessed" by an idea of what the Sense-impressions are to be. Of this we have an apposite illustration in the well-known exclamation of Dr. Pearson, "Bless me, how heavy it is," when he first poised upon his finger the globule of Potassium produced by the battery of Sir H. Davy; his preconception of the coincidence between metallic lustre and high specific gravity, causing him to feel *that* as ponderous,

which the unerring test of the balance determined to be lighter than water. The excitement of the peculiar sensation of tickling in a "ticklish" person, by a threatening movement that suggests the idea, and of that of creeping or itching by the mention of bed-infesting insects to those who are peculiarly liable to their attacks, are familiar instances of the same fact.

143. In the two following cases related by Professor Bennett, the effect of the Idea was not limited to the production of the Sensations, but extended itself to the consequences which would have followed those sensations, if their supposed cause had been real:—

a. "A clergyman told me, that some time ago suspicions were entertained in his parish, of a woman who was supposed to have poisoned her newly-born infant. The coffin was exhumed, and the Procurator-fiscal, who attended with the medical men to examine the body, declared that he already perceived the odour of decomposition, which made him feel faint, and in consequence he withdrew. But, on opening the coffin, it was found to be empty; and it was afterwards ascertained that no child had been born, and consequently no murder committed."

b. The second case is yet more remarkable. "A butcher was brought into the shop of Mr. Macfarlan, the druggist, from the market-place opposite, labouring under a terrible accident. The man, on trying to hook-up a heavy piece of meat above his head, slipped, and the sharp hook penetrated his arm, so that he himself was suspended. On being examined, he was pale, almost pulseless, and expressed himself as suffering acute agony. The arm could not be moved without causing excessive pain; and in cutting-off the sleeve, he frequently cried out; yet when the arm was exposed, it was found to be quite uninjured, the hook having only traversed the sleeve of his coat." —(*The Mesmeric Mania* of 1851.)

No evidence *could* be stronger than that afforded by the almost pulseless condition of the subject of the second of these cases, as to the *reality* of the severe pain which he experienced; and yet this pain entirely arose from his Mental conviction that the hook had penetrated the flesh of his arm.

144. Nearly thirty years ago, the scientific world was startled by the announcement made by Baron von Reichenbach (who had previously attained considerable reputation as a chemist) of the discovery of "a new Imponderable,"—a peculiar Force existing in nature and embracing the Universe, distinct from all known forces —to which he gave the name *Odyle.* This force could only be recognised by the effects it produced on certain "sensitive subjects;" who could *see,* it was averred, flames streaming from the poles of magnets, could *smell* odours issuing from them, and could *feel* sensations of warmth or coolness when magnets were drawn over any part of the surface of the body; some of them being also similarly affected by crystals; and one, in particular, by almost *any substance whatever*, so that she saw (in the dark) flames issuing from nails or hooks in a wall, or streaming from the finger-ends of human beings. Experienced Physicians, however, at once recognized in Baron Reichenbach's descriptions, the influence of the *ideas* with which these "sensitives" had become "possessed;" the phenomena being only, under another form, the manifestations of a tendency with which they were previously familiar in Hysterical and Hypochondriacal patients. Hence there was to them nothing in the least surprising in the fact, that such persons, *placed in a perfectly dark room for two hours*, could be brought to see a multitude of luminous phenomena, could hear varied sounds, could smell odours, and could touch intangible things, quite independently of any "Odyle" whatever, by the mere *suggestion* of what they were to experience. And although Reichenbach himself considered that he had taken adequate precautions to exclude the conveyance of any suggestion of which his "sensitives" should be conscious, yet those who were familiar with the extraordinary receptivity for Sense-impressions which is a special characteristic of Hysterical subjects (§ 137), could readily discern the modes in

which such suggestions would reach the "sensitives," without any intention on the part of the operator.

145. The very fact that *no manifestation of this supposed Force could be obtained, except through the conscious Human being,** was quite sufficient to convince every philosophic investigator, that he had to do, not with a new Physical Force, but with a peculiar phase of Psychical action, by no means unfamiliar to such as had previously studied the influence of the Mind upon the Body. From this point of view, Reichenbach's researches were accepted as an important contribution to Mental Physiology; and this estimate of their character was entirely confirmed by the inquiries of Mr. Braid, of Manchester, who was early led to the adoption of it by the experience he had already gained in a parallel line of investigation, the results of which had thrown great light on the phenomena of Mesmerism (Chaps. XIV, XV). For he found that whatever Sensations were producible by the agency of magnets, crystals, &c., the very same sensations occurred when the "subjects" *believed* that such agency was being employed, although nothing whatever was really being done; and further, that the character of the Sensations experienced by the "subjects" depended very much on the Ideas they had been led to form of them, either by their own mental action, or by the suggestion of others. The following are a few examples of the results obtained by Mr. Braid, of many of which the Writer was himself a witness:—

"A lady, upwards of fifty-six years of age, in perfect health, and wide awake, having been taken into a dark closet, and desired to look at the poles of the powerful horse-shoe magnet of nine elements, and describe what she saw, declared, after looking a considerable

* Thus although certain of the "sensitives" felt their hands powerfully attracted towards a magnet, yet, as Reichenbach himself confessed, when the magnet was poised in a delicate balance, and the hand was placed above or beneath it, *the magnet was never drawn towards the hand.* The "attraction," therefore, although *real* to the "subject" of it, was generated by the Idea in the "sensitive's" own Mind (§448).

time, that she saw nothing. However, after I told her to look attentively, and she would see fire come out of it, she speedily saw sparks, and presently it seemed to her to burst forth, as she had witnessed an artificial representation of the volcano of Mount Vesuvius at some public gardens. Without her knowledge, I closed down the lid of the trunk which contained the magnet, *but still the same appearances were described as visible.* By putting leading questions, and asking her to describe what she saw from *another* part of the closet (where there was nothing but bare walls), she went on describing various shades of most brilliant coruscations and flame, according to the leading questions I had put for the purpose of changing the fundamental ideas. On repeating the experiments, similar results were repeatedly realised by this patient. On taking this lady into the said closet after the magnet had been removed to another part of the house, she still perceived the same visible appearances of light and flame when there was nothing but the bare walls to produce them; and, two weeks after the magnet was removed, when she went into the closet by herself, the mere association of ideas was sufficient to cause her to realize a visible representation of the same light and flames. Indeed such had been the case with her on entering the closet, ever since the few first times she saw the light and flames. In like manner, when she was made to touch the poles of the magnet when wide awake, no manifestations of attraction took place between her hand and the magnet; but the moment the idea was suggested that she would be held fast by its powerful attraction, so that she would be utterly unable to separate her hands from it, such result was realized; and, on separating it by the suggestion of a new idea, and causing her to touch the *other* pole in like manner, predicating that *it* would *exert no attractive power* for the fingers or hand, such negative effects were at once manifested.—I know this lady was incapable of trying to deceive myself, or others present; but she was self-deceived and spell-bound by the predominance of a pre-conceived idea, and was not less surprised at the varying powers of the instrument than others who witnessed the results."—(*The Power of the Mind over the Body*, 1846, p. 20.)

146. Other "subjects" taken by Mr. Braid into his dark closet, and unable to see anything in the first instance, when told to look

steadily at a certain point (though there was no magnet there), and assured that they would see flame and light of various colours issuing from it, very soon declared that they saw them; and in some of them, "individuals of a highly concentrative and imaginative turn of mind," the same sensations could be called up in open daylight. The following was an experiment made *with* and *without* the magnet, upon the sensations of the general surface; the "subject" being a young gentleman twenty-one years of age:—

a. "I first operated on his right hand, by drawing a powerful horse-shoe magnet over the hand, without contact, whilst the armature was attached. He immediately observed a sensation of cold follow the course of the magnet. I reversed the passes, and he felt it *less cold*, but he felt no attraction between his hand and the magnet. I then removed the cross-bar, and tried the effect with both poles alternately, but still there was no change in the effect, and decidedly no proof of attraction between his hand and the magnet.—In the afternoon of the same day I desired him to look aside and hold his hat between his eyes and his hand, and observe the effects when I operated on him, whilst he could not see my proceedings. He very soon described a recurrence of the same sort of sensations as those he felt in the morning, but they speedily became more intense, and extended up the arm, producing rigidity of the member. In the course of two minutes this feeling attacked the other arm, and to some extent the whole body; and he was, moreover, seized with a fit of involuntary laughter, like that of hysteria, which continued for several minutes—in fact, until I put an end to the experiment. His first remark was, 'Now this experiment clearly proves that there must be some intimate connection between mineral Magnetism and Mesmerism; for I was most strangely affected, and could not possibly resist laughing during the extraordinary sensations with which my whole body was seized, as you drew the magnet over my hand and arm.' I replied that I drew a very different conclusion from the experiments, as *I had never used the magnet at all*, nor held it, nor anything else, near to him; and that the whole proved the truth of my position as to the extraordinary power of the Mind over the Body."—(*Op. cit.*, p. 15.)

Phenomena of the same kind were found to be producible without the use of a Magnet at all :—

b. "Another interesting case of a married lady, I experimented with in presence of her husband, was as follows. I requested her to place her hand on the table, with the palm upwards, so situated as to enable her to observe the process I was about to resort to. I had previously remarked, that by my drawing something slowly over the hand, without contact, whilst the patient concentrated her attention on the process, she would experience some peculiar sensations in consequence. I took a pair of her scissors, and drew the bowl of the handle slowly from the wrist downwards. I had only done so a few times, when she felt a creeping, chilly sensation, which was immediately followed by a spasmodic twitching of the muscles, so as to toss the hand from the table, as the members of a prepared frog are agitated when galvanized. I next desired her to place her *other* hand on the table, in like manner, but in such a position, that by turning her head in the opposite direction she might not see what was being done, and to watch her sensations in that hand, and tell us the result. In about the same length of time, similar phenomena were manifested as with the other hand, although in this instance *I had done nothing whatever*, and was not near her hand. I now desired her to watch what happened to her hand, when I predicted that she would feel it become *cold;* and the result was as predicted; and *vice versâ*, predicting that she would feel it become intensely *hot*, such was realized. When I desired her to think of the tip of her nose, the predicted result either of heat or cold was speedily realized in that part.

"Another lady, twenty-eight years of age, being operated on in the same manner, whilst looking at my proceedings, in the course of half a minute, described the sensation as that of the blood rushing into the fingers; and when the motion of my pencil-case was from below upwards, the sensation was that of the current of blood being reversed, but less rapid in its motion. On resuming the downward direction, the original feeling recurred, still more powerfully than at first.—This lady being requested now to look aside, whilst I operated, *realized similar sensations*, and that whilst *I was doing nothing*.

"The husband of this lady, twenty-eight and a half years of age, came into the room, shortly after the above experiment was finished.

She was very desirous of my trying the effect upon him, as he was in perfect health. I requested him to extend his right arm laterally, and let it rest on a chair with the palm upwards, to turn his head in the opposite direction, so that he might not see what I was doing, and to concentrate his attention on the feelings which might arise during my process. In about half-a-minute he felt an *aura* like a breath of air passing along the hand; in a little after, a slight pricking, and presently a feeling passed along the arm, as far as the elbow, which he described as similar to that of being slightly electrified:—*all this, while I had been doing nothing*, beyond watching what might be realized. I then desired him to tell me what he felt NOW,—speaking in such a tone of voice, as was calculated to lead him to believe I was operating in some different manner. The result was that the former sensations ceased; but when I requested him once more to tell me what he felt *now*, the former sensations recurred. I then whispered to his wife, but in a tone sufficiently loud to be overheard by him, observe now, and you will find his fingers begin to draw, and his hand will become clenched,—see how the little finger begins to move, and such was the case; see the next one also going in like manner, and such effects followed; and finally, the entire hand closed firmly, with a very unpleasant drawing motion of the whole flexor-muscles of the fore-arm. I did nothing whatever to this patient until the fingers were nearly closed, when I touched the palm of his hand with the point of my finger, which caused it to close more rapidly and firmly. After it had remained so for a short time, I blew upon the hand, which dissipated the previously existing mental impression, and instantly the hand became relaxed. The high respectability and intelligence of this gentleman rendered his testimony very valuable; and especially so, since he was not only wide awake, but had never been either mesmerised, hypnotised, or so tested before.—(*Op. cit.*, pp. 15—17.)

147. The results thus obtained by experiment, being at the same time consistent with ordinary Medical experience, and accordant with Physiological probability, have an adequate claim to acceptance as *Scientific facts;* and it is obvious that, if the principle be once admitted that *real Sensations are producible by Mental states,*

this principle furnishes the key to the explanation of a large number of those "spiritualistic" experiences, in which objects are affirmed to be actually seen and felt, that only exist in the Imagination of the "subjects" of them. It has been no less happily than philosophically said by the Laureate, that "Dreams are true while they last;" but we become conscious in our waking state of the "objective" *unreality* of what was for the time *real* to us, by its discordance with that general resultant of our waking experiences which we call "Common Sense." (Chap. XI.) Occasionally we are puzzled to answer the question, "Did this really happen, or did I dream it?"—our perplexity arising from the fact, that the "trace" of what passed in our dream equals in vividness that which would have been left by the actual occurrence, and that there is nothing inconsistent with our experience, in the idea that it *might* have happened.—Now when a number of persons who are "possessed" with the current ideas in regard to Spiritualistic manifestations, sit for some time in a dark room in a state of "expectant attention," it is conformable to all scientific probability that they should *see* luminous manifestations, should *smell* flowers, should *feel* the contact of spirit-hands or the crawling of live lobsters, or should *hear* musical sounds or the voices of departed friends,—just as they are prompted to do by their own course of thought, or by the suggestions of others; the correction of these dreamy imaginings, by bringing common sense and scientific knowledge to bear upon them, being just what the votaries of the doctrine referred-to scornfully repudiate.

148. Very nearly connected with the foregoing, are the phenomena of *Spectral Illusions;* which, like the creations of dreaming or delirium, are the products of the excitement of *Sensorial* activity by *Cerebral* change, operating through the "nerves of the *internal* senses;" the essential difference being that as the Sensorium is not closed to external impressions, these mental images mingle with the sensations called forth by objective realities. A simple but

very illustrative case of this kind, which occurred in the experience of Sir John Herschel, has already been cited (§103); and it is probable that we are to regard in the same light that Spectrum of the Sun, which Sir Isaac Newton was able to recall by going into the dark and directing his mind intensely, "as when a man looks earnestly to see a thing which is difficult to be seen," and which, after a frequent repetition of this process, came (he says) to return "as often as I began to meditate on the phenomena, even though I lay in bed at midnight with my curtains drawn." For although phenomena of this class are often regarded as *ocular* spectra produced by *retinal* change, their reproduction by *mental* states seems to place them in the same category as the visual sensations which are distinctly reproduced by Memory, that is, by *cerebral* change (§102). In fact, there is such a gradational transition from the one state to another, that it seems clear that they have a common origin. Thus Dr. Abercrombie (*Inquiries concerning the Intellectual Powers,* 5th Edit., p. 382) mentions the case of a gentleman who was all his life haunted by Spectral figures, and could call up any at will, by directing his attention steadily to some conception of his own mind, which might either consist of a figure or a scene that he had seen, or might be a composition of his own imagination: but although possessing the faculty of producing the illusion, he had no power of banishing it; so that when he had called-up any particular figure or scene, he could not say how long it might continue to haunt him. In this case the Sensorial state produced by Cerebral action so closely resembled the impression produced by the actual object, that, on meeting a friend in the street, the subject of it could not satisfy himself whether he saw the real individual or the spectral figure, save by touching his body, or by hearing the sound of his footsteps,—the correction being here supplied by other Sense-impressions. In certain instances the unreality of these phantasms, however vivid, is recognized by the Intellect, from the consideration of the circumstances under which they occur: as in the well-known case of Nicolai, who, when suffering from intermittent fever, saw coloured pictures of landscapes, trees, and rocks, resembling framed paintings, but of half the natural size; so long as he kept his eyes closed, they underwent constant changes, some figures disappearing while new ones showed themselves; but as soon as he opened his eyes, the whole vanished. The following is another case of this kind, in which the same Sensorial condition as in dreaming or delirium was accompanied by an Intellectual recognition of its objective unreality:—

"We knew a gentleman of strong mind, and a most accomplished Scholar, who was for many years subject to such phantasms, some sufficiently grotesque; and he would occasionally laugh heartily at their antics. Sometimes it appeared as if they interrupted a conversation in which he was engaged; and then, if with his family or intimate friends, he would turn to empty space, and say, 'I don't care a farthing for ye; ye amuse me greatly sometimes, but you are a bore just now.' His spectra, when so addressed, would to his eye resume their antics, at which he would laugh, turn to his friend, and continue his conversation. In other respects he was perfectly healthy, his mind was of more than ordinary strength, and he would speak of his phantoms, and reason upon their appearance, being perfectly conscious that the whole was illusive."

149. It is a curious confirmation of the view here advocated as to the distinctness between the parts of the Brain which are the instruments of *sensorial* and of *ideational* states respectively, and of the immediate relation of the former to the *motor* apparatus, that the presence of a Spectral illusion will often operate in directing movement, even though there be an intellectual consciousness that there is no objective cause for it, and that the movement is consequently inappropriate. A lady nearly connected with the Writer, having been frightened in childhood by a black cat, which sprang-up from beneath her pillow just as

she was laying her head upon it, was accustomed for many years afterwards, whenever she was at all indisposed, to see a black cat on the ground before her; and although intellectually aware of the spectral character of the appearance, yet she could never avoid lifting her foot as if to step over the cat, when it seemed to be lying in her path.*

150. When, as the cases last cited, the Spectral image distinctly represents an external object, it must be regarded as not merely a Sensorial but as a Perceptional illusion. A far more frequent cause of false *perceptions*, however, lies in the *misinterpretation* of real Sense-impressions, under the influence of pre-existing Ideational states,—a subject which will be more fitly considered hereafter (§ 186).

151. *Relation of Sensations to other Mental States.*—It is through the medium of Sensation, that we acquire a knowledge of the Universe external to us, by the Psychical operations which its changes excite in ourselves. The Psychologist of the present day views Matter entirely through the light of his own consciousness:—his idea of matter in the abstract being that it is a "something" which has a permanent power of exciting Sensations; his idea of any property of matter being the mental Representation of some kind of sense-impression he has received from it; and his idea of any particular kind of matter being the Representation of the whole aggregate of the sense-perceptions which its presence has called up in his mind.

"Thus when I press my hand against this table, I recognise its unyieldingness through the conjoint medium of my sense of Touch, my Muscular sense, and my Mental sense of Effort, to which it will be convenient to give the general designation of the Tactile sense; and I attribute to that table a *hardness* which resists the effort I

* A very interesting collection of cases of Spectral Illusions will be found in Dr. Abercrombie's Treatise, and in Sir B. Brodie's "Psychological Inquiries;" see also Sir John Herschel's own experiences in his "Familiar Lectures on Scientific Subjects," pp. 403-5.

make to press my hand into its substance, whilst I also recognise the fact that the force I have employed is not sufficient to move its mass. But I press my hand against a lump of dough; and finding that its substance yields under my pressure, I call it *soft*. Or again, I press my hand against this desk; and I find that although I do not thereby change its *form*, I change its *place;* and so I get the Tactile idea of *Motion*. Again, by the impressions received through the same Sensorial apparatus, when I lift this book in my hand, I am led to attach to it the notion of *weight* or ponderosity; and by lifting different solids of about the same size, I am enabled, by the different degrees of exertion I find myself obliged to make in order to sustain them, to distinguish some of them as *light*, and others as *heavy*. Through the medium of another set of Sense-impressions (which some regard as belonging to a different category), we distinguish between bodies that *feel* 'hot' and those that *feel* 'cold'; and in this manner we arrive at the notion of differences of Temperature. And it is through the medium of our Tactile sense, without any aid from Vision, that we first gain the idea of *solid form*, or the three dimensions of Space.

"Again, by the extension of our Tactile experiences, we acquire the notion of *liquids*, as forms of matter yielding readily to pressure, but possessing a sensible weight which may equal that of solids: and of *air*, whose resisting power is much slighter, and whose weight is so small that it can only be made sensible by artificial means. Thus, then, we arrive at the notions of *resistance* and of *weight* as properties common to all forms of Matter; and now that we have got rid of that idea of Light and Heat, Electricity and Magnetism, as 'imponderable fluids,' which used to vex our souls in our scientific childhood, and of which the popular term 'electric fluid' is a 'survival,' we accept these properties as affording the practical distinction between the 'material' and the 'immaterial.'

"Turning, now, to that other great portal of Sensation, the Sight, through which we receive most of the messages sent to us from the Universe around, we recognise the same truth. Thus it is agreed alike by Physicists and Physiologists, that *colour* does not exist *as such* in the object itself; which has merely the power of reflecting or transmitting a certain number of millions of undulations in a second; and these only produce that affection of our consciousness which we call Colour, when they fall upon the retina of the living percipient. And if there be that defect either in the retina or in the apparatus behind it, which we call 'colour-blindness' or 'Daltonism,' some particular hues cannot be distinguished, or there may even be no power of distinguishing any colour whatever. If we were all like Dalton, we should see no difference, except in form, between ripe cherries hanging on a tree, and the green leaves around them: if we were all affected with the severest form of colour-blindness, the fair face of Nature would be seen by us as in the chiaroscuro of an engraving of one of Turner's landscapes, not as in the glowing hues of the wondrous picture itself."—(*Address to the Meeting of the British Association*, 1872.)

152. If it were possible for a Human being to come into the world with a Brain perfectly prepared to be the instrument of Psychical operations, but with *all* the inlets to Sense-impressions closed, we have every reason to believe that the Mind would remain dormant, like a seed buried deep in the earth. The attentive study of cases in which there is congenital deficiency of one or more Sensations, makes it evident that the Mind is utterly incapable of forming any definite Ideas in regard to those properties of objects, of which those particular sensations are adapted to take cognizance. Thus the man who is born blind, can form no conception of colour; nor the congenitally-deaf, of musical tones. And in those lamentable cases in which the sense of Touch is the only one through which Ideas can be called-forth, the Mental operations necessarily remain of the simplest and most limited character,—unless the utmost attention be given by a judicious instructor, to the development of the Intellectual faculties, and the cultivation of the Moral feelings, through that restricted class of ideas which there *is* a possibility of exciting.*

153. The activity of the Mind, then, is just as much the result of its consciousness of external impressions, by which its faculties

* Of the extent to which this *may* be accomplished, the well-known case of Laura Bridgeman affords a most remarkable exemplification.

are called into play, as the life of the Body is dependent upon the appropriation of nutrient materials, and the constant influence of external forces. But there is this difference between the two cases,—that whilst the Body continually requires *new* materials and a continued action of external agencies, the Mind, when it has been once called into activity, and has become stored with Ideas, may remain active, and may develope new relations and combinations amongst these, after the complete closure of the Sensorial inlets by which new ideas can be excited *ab externo.* Such, in fact, is what is continually going-on in the state of Dreaming; but examples yet more remarkable are furnished in the vivid conceptions which may be formed of a landscape or a picture, from oral description, by those who have once enjoyed sight; or in the composition of music, even such as involves new combinations of sounds, by those who have become deaf, as in the well-known case of Beethoven. The mind thus feeds, as it were, upon the store of Ideas which it has laid-up during the activity of its Sensory organs: and not only are those impressions which it *consciously* retains, worked-up into a never-ending variety of combinations and successions of ideas, thus continuing to afford new sources of mental activity, even to the very end of life; but impressions of which the Mind, though once conscious of them, seems even to itself to have entirely lost the traces, may recur spontaneously, and influence its trains of thought, at periods long subsequent to their reception. (Chap. X.)

154. With particular Sensations are connected *feelings of Pain or Pleasure,* which cannot (for the most part at least) be explained upon any other principle than that of the necessary association of these feelings, by an original law of our nature, with the sensations in question. As a general rule, it may be stated that the *violent* excitement of *any* sensation is disagreeable, even when the same sensation in a moderate degree may be a source of extreme pleasure. This is the case alike with those impressions which are

communicated through the organs of Sight, Hearing, Smell, and Taste, as with those that are received through the nerves of Common sensation; and the association of painful feelings with such violent excitement, serves to stimulate the individual to remove himself from what would be injurious in its effects. Thus, the pain resulting from violent pressure on the cutaneous surface, or from the proximity of a heated body, gives warning of the danger of injury, and excites mental operations destined to remove the part from the influence of the injurious cause; and this is shown by the fact, that loss of sensibility is frequently the indirect occasion of severe lesions,—the individual not receiving the customary intimation that an injurious process is taking place.* Thus, violent inflammation of the membrane lining the air-passages has resulted from the effects of ammoniacal vapours incautiously introduced into them during a fainting-fit,—the patient not receiving that notice of the irritation, which, in the active condition of his Nervous

* The following case, recorded in the "Journal of a Naturalist," affords a remarkable instance of this general fact. The correctness of the statement having been called in question, it was fully confirmed by Mr. Richard Smith, the late senior Surgeon of the Bristol Infirmary, under whose care the sufferer had been:—"A travelling man, one winter's evening, laid himself down upon the platform of a lime-kiln, placing his feet, probably numbed with cold, upon the heap of stones newly put on to burn through the night. Sleep overcame him in this situation; the fire gradually rising and increasing, until it ignited the stones upon which his feet were placed. Lulled by the warmth, the man slept on: the fire increased until it burned one foot (which probably was extended over a vent-hole) and part of the leg above the ankle entirely off, consuming that part so effectually that a cinder-like fragment was alone remaining,—and still the wretch slept on! and in this state was found by the kiln-man in the morning. Insensible to any pain, and ignorant of his misfortune, he attempted to rise and pursue his journey; but missing his shoe, requested to have it found; and when he was raised, putting his burnt limb to the ground to support his body, the extremity of his leg-bone, the tibia, crumbled into fragments, having been calcined into lime. Still he expressed no sense of pain, and probably experienced none; from the gradual operation of the fire, and his own torpidity during the hours his foot was consuming. This poor drover survived his misfortunes in the hospital about a fortnight; but the fire having extended to other parts of his body, recovery was hopeless."

system, would have prevented him from inhaling the noxious agent.

155. The feelings of Pain or Pleasure which unaccustomed sensations excite, are often exchanged for each other when the system is *habituated* to them; this is especially the case in regard to impressions communicated through the organs of Smell and Taste. There are many articles in common use among mankind,—such as Tobacco, Alcoholic liquors, &c., the use of which cannot be said to produce a natural enjoyment, since they are at first unpleasant to most persons; and yet they first become tolerable, then agreeable; and at last the want of them is felt as a painful privation, and the stimulus must be applied in an increasing degree in order to produce the usual effect. These all belong to the class of "nervine stimulants;" and it can scarcely be questioned that the result of their continual employment is to produce a modification of the *nutrition* of the Nervous system, which engenders a Physical want when they are withheld, comparable to that of Hunger or Thirst (§ 138).

156. On the same level with the simple feelings of Pleasure and Pain which are associated with particular Sensations, but distinct from these in the manner in which they affect us, are those *general* feelings of personal *well-being*, or of its reverse *malaise*, which, whilst so intimately connected with states of the Bodily system as to be producible by them alone, are also the rudimentary forms of those higher Psychical states which we term *emotions*. These feelings, in their lowest stage of development, are purely "subjective;" the individual being simply conscious of them, and not referring them to any external source. There are many persons who are so keenly susceptible of both, that they pass their whole lives in an alternation between *cheerfulness* and *depression*: the former state being favoured by freedom from anxiety, by the healthful activity of all the organic functions, by a bright sun and a dry bracing atmosphere; whilst the latter is immediately induced by

mental disquietude, by a slight disorder of digestion or excretion, or by a dull oppressive day. And a concurrence of favourable conditions may even exalt this *Cœnæsthesis* (or self-feeling) into *exhilaration* or absolute *joy;* whilst the combined influence of those of the opposite kind may produce *gloom,* which may be exaggerated almost to *despair.* We shall hereafter see cogent reasons for regarding these conditions as purely Physical (§§ 535, 552). The condition of "the spirits" (as these Mental affections are commonly designated) most to be desired, however, is that of *tranquil comfort;* which is far more favourable than the alternation of extremes, to healthful activity and to sustained energy, alike of Body and of Mind. And this may be cherished by cultivating the habit of Volitional self-control (§ 271), whereby any tendency to undue exhilaration is moderated, and excessive depression is resisted by a determinate effort not to yield to it.

157. Similar states of Consciousness may be excited by causes purely Psychical; and although we are then accustomed to designate them as Emotions, yet their nature and their seat are probably the same in the one case as the other. The simple *feeling* which we experience from a piece of "good" or of "bad news," is so nearly allied to the Pleasure or Pain we experience in connection with Sensations, that we may fairly regard the instrumentality by which we become conscious of it as *sensorial* rather than Cerebral; the state of the Sensorium being affected, in the one case, by impressions conveyed to it by the "nerves of the *internal* senses," just as it is in the other by those brought to it by the nerves of the *external* senses. It often happens, moreover, that the impression thus made upon the "emotional sensibility" is more persistent than the ideational state which gave rise to it; for after some disagreeable occurrence, or the receipt of ill-tidings, we feel an abiding consciousness of discomfort or distress, although we determinately keep from our Mental view the recollection of the unpleasant idea, in order that we may not be disturbed by dwelling

too painfully on it. It may often be observed, moreover, that when the passions have been stirred in states of Somnambulism, Hypnotism, &c., a disturbed *cœnæsthesis* is carried-on into the ordinary state, although the "subject" is altogether unconscious of the nature or causes of the emotional excitement (§ 491).—There are few other forms of emotional sensibility which are so completely subjective as the foregoing; most of them having reference to some object which is felt to be external to Self, and therefore belonging to the next category (§ 189). But we seem justified in referring to this group, as being nearly allied to the foregoing, though scarcely capable of being grouped together with them, *the sense of enjoyment in activity*, and its converse *the sense of tedium in inactivity* (commonly known as *ennui*); both of which are purely subjective states, and are obviously manifested by the lower Animals,—chiefly, however, in connection with their *bodily* functions,—whilst in Man it is the want of *mental* occupation that is the chief source of Ennui.*

* The Writer would here express his obligations to his friend Dr. Noble, of Manchester, for many valuable suggestions in regard to the diversified forms of "Emotional Sensibility," and its relations to Sensational, Perceptional and Ideational states respectively —See his "Elements of Psychological Medicine" 2nd Ed., 1855), and his subsequent treatise on "The Human Mind" (1858).

CHAPTER V.

OF PERCEPTION AND INSTINCT.

158. NEITHER the operations of the Intellectual Powers, nor the higher Emotional states, are *immediately* called-forth by Sensations; for in that stage of consciousness we merely recognize the fact that certain changes have occurred in our own "subjective" state, and do not refer these changes to any external or "objective" source. Of such a limitation, we occasionally meet with examples among the phenomena of Sleep, and in some of the conditions resulting from the use of Anæsthetic agents: for if we fall asleep whilst suffering from bodily pain, we may entirely lose all perception of the cause of that pain, and yet remain conscious of a perturbed state of feeling, which may affect the course of our dream; and when a surgical operation is performed in a state of incomplete anæsthesia, it is obvious that pain is felt without any distinct consciousness of its source, and the patient may subsequently describe his state as an uneasy dream. Such, it is probable, is the condition of the infant at the commencement of its Psychical life. "If," as has been well remarked by Dr. J. D. Morell (*Philosophy of Religion*, p. 7), "we could by any means transport ourselves into the mind of an Infant before the Perceptive consciousness is awakened, we should find it in a state of absolute isolation from everything else in the world around it. Whatever objects may be presented to the Eye, the Ear, or the Touch, they are treated simply as *subjective feelings*, without the Mind's possessing any consciousness of them *as objects* at all. To it, the inward world is *everything*, the outward world is *nothing*."—However difficult it may be, under the influence of our

life-long experience, to dissociate any Sensation of which we are cognizant, from the notion of its external cause—since, the moment the feeling is experienced, and the Mind is directed to it, the *object* from which it arises is immediately suggested,—yet nothing is more certain than that all of which we are *primarily* conscious in any case whatever, is a certain internal or subjective state, a change in our previous Consciousness; and that the *mental recognition* of the object to which that change is due, is dependent upon a higher process, to which the name of *Perception* is now generally accorded. We may recognize the manifestation of this process in the Child, as it advances beyond the first few months of its helplessness. "A sight or a sound," remarks Dr. Morell (*Op. cit.*), "which at first produced simply an involuntary start, now awakens a smile or a look of recognition. The mind is evidently struggling *out of itself*; it begins to throw itself into the objects around, and to live in the world of outward realities." A similar transition, more rapidly effected, may be distinguished in ourselves, during the passage from Sleep, or from the insensibility of a swoon, to the state of wakeful activity; when we are at first conscious only of our own sensations, and gradually come to the knowledge of our condition as it relates to the world around, and of the position and circumstances, new and strange as they may be, in which we find ourselves.

159. Now the *apprehension*, or the formation of an elementary Notion, of the *outness* or *externality* of the cause of a Sensational change, is an operation which the Mind seems necessarily to perform, when it has attained a certain stage of development; instinctively or intuitively making a definite distinction between the *self* and the *not-self*, the *subject* and the *object*. We do not infer the existence of objective realities by any act of the Reason; in fact, the strict application of logical processes tends rather to shake than to confirm the belief in the External World; but our Minds being at first subjectively impressed by the qualities of matter,

we gradually learn to interpret and combine the impressions they make upon our consciousness, so as to derive from them a more or less definite *notion* of the object (§ 151). Some of these Notions are so simple, and so constantly excited by certain Sensations, that we can scarcely do otherwise than attribute their formation to original and fundamental properties of the Mind, called into activity by the sensations in question; thus, the notion of *hardness* seems to connect itself from the first with the sense of absolute resistance, the notion of *direction* with the consciousness of diversity of parts in the visual picture. Such perceptions are said to be *instinctive* or *original*. In other cases, however, the notions are connected with the sensations by *habit* alone; the connection being the result of the association which gradually establishes itself between them, so that a certain sense-impression invariably calls up a certain notion of an object answering to it. And thus it may happen that a *wrong* interpretation may be put upon the sensational state, merely through some change in the conditions under which it has been habitually received; as in the following very simple experiment:—If the middle finger of either hand be crossed behind the fore-finger, so that its extremity is on the thumb-side of the latter, and the ends of the two fingers thus disposed be rolled over a marble, pea, or other round body, a sensation will be produced, which, if uncorrected by reason, would cause the mind to believe in the existence of two distinct bodies; this being due to the impression being made at the same time upon two spots, which, in the ordinary position of the fingers, are at a considerable distance from each other.

160. There can be no doubt that, during the period of Infancy, a very rapid and energetic process of self-education is going on; the *whole mind*, so far as it is yet developed, being concentrated upon its Perceptive activity: and when once a complete interpretation has thus been attained of any particular group of Sensations, it so immediately occurs to the consciousness whenever

those sensations may be renewed, as to have all the directness of an *original* perception. Thus it is very difficult, at later periods of life, to discriminate the Perceptions which are really *instinctive*, from those which have been *acquired* during Infancy.

It would be wrong to draw inferences on this point from the actions of the lower Animals; for in those cases in which the young are dependent from the first on the exertion of their own powers, it is obvious that they have a larger range of Instinctive perceptions than is possessed by those which derive their early sustenance from their parents. Many of them, for example, manifest a guiding appreciation of direction and distance, which Man can only gain by long experience. Thus, a Fly-catcher just come out of its shell, has been seen to peck-at and capture an Insect, with an aim as perfect as if it had been all its life engaged in learning the art.*—Still more remarkable is the perception that guides the actions of a little Fish, the *Chœtodon rostratus*, which shoots-out drops of fluid from its prolonged snout, so as to strike Insects that happen to be near the surface of the water, thus causing them to fall-in, and to be brought within its reach. Now by the refraction of light, the real place of the Insect in the air will not be that at which it appears to the Fish in the water, but will be a little *below* its apparent place; and to this point the aim must be directed. The difference between the real and the apparent place, moreover, will not be constant; for the more perpendicularly the rays enter the water, the less will be the variation; and, on the other hand, the more oblique the direction, the greater will be the difference.

161. It has been recently maintained that the want, both of the apprehension of *distance*, and of the power of *directing* the Muscular movements so as at once to lay hold of an object, which every one who carefully observes the actions of the Human Infant must recognize, is a mere matter of *development;* the human infant coming into the world in a less advanced condition

* See the experiments of Mr. Spalding on this subject, detailed in his paper in "Macmillan's Magazine" for February, 1873, to which reference has already been made (§ 77).

than the young of many other animals, which are able to run about and seek their own sustenance from the first. But the Writer has strong personal reasons for asserting that such is not the fact. Having been introduced into the Medical profession by an eminent Surgeon of Bristol (the late Mr. J. B. Estlin), who had a large Ophthalmic practice in the West of England and South Wales, he had the opportunity of seeing many cases of congenital Cataract cured by operation; the condition of these children being exactly parallel in respect of Vision, to that of Mr. Spalding's hooded chicks. Generally speaking, the operation was performed within the first twelve months; but he distinctly remembers two cases, in one of which the subject was a remarkably sturdy little fellow of three years old, whilst the other was a lad of nine. In the latter, however, there had been more visual power before the operation, than in the former; and he therefore presents the well remembered case of Jemmy Morgan as the basis of his assertion, that the acquirement of the power of visually guiding the muscular movements is *experiential* in the case of the Human Infant:—

Jemmy had most assuredly come to that stage of his development, which would justify the expectation that if he *had* his Sight he would *at once* use it for his guidance, supposing the power of doing so to be congenital, for his father being a farmer a few miles out of Bristol, he was accustomed to go about by himself in the farm-yard, where he made friends with every one of its inhabitants, and picked up from the labourers a very improper accomplishment,—that of swearing most horribly. He was so strong, that it was necessary for the performance of the operation that his body should be bound down upon a table, and that each of his limbs and his head should be held by a separate assistant. The Writer remembers that he had charge of his head, which he found it impossible altogether to prevent him from rolling from side to side; whilst his roars and curses seem even now ringing in the Writer's ears. The operation, performed with consummate dexterity,—the handle of the cataract-needle being left by Mr. Estlin to "play" between his fingers, as Jemmy's head *would* move in spite of the strongest efforts to restrain it,—was entirely successful. In a few days both pupils were almost clear; and it was obvious from his actions that he had distinct Visual Perceptions. But though he clearly recognised the *direction* of a candle or other bright object, he was as unable as an Infant to apprehend its *distance;* so that when told to lay hold of a watch, he *groped* at it, just like a young child lying in its cradle. It was *very gradually* that he came to use his Sight for the guidance of his movements: and when going about the house at which he was staying in Bristol, with which he had familiarized himself before the operation, he generally *shut his eyes*, as if puzzled rather than aided by them. When he came up to Mr. Estlin's house, however, he would show that he was acquiring a considerable amount of visual power; and it was his favourite amusement there, to blow about with his breath a piece of white paper on the surface of a dark mahogany table, round and round which he would run, as he wafted the paper from one side to another, shouting with glee at his novel exploit. Nevertheless, when he returned *home* to his father's house and farm-yard, his parents (very intelligent people) remarked that he was for some time obviously puzzled by his Sight, *shutting his eyes* as he went about in his old way; though whenever he went to a *new* place, he was obviously aided by it. But it was several months before he came to trust to it for his guidance, as other children of his age would do.—Jemmy's case was very carefully observed, both by Mr. Estlin and the Writer, with full knowledge of the interest attaching to such observations; and every fact the Writer has here stated remains as distinctly impressed on his mind at the distance of more than forty years, as if it had only happened yesterday,—the image of Jemmy, in his red frock, and with his still redder legs, being more vivid than any other reminiscence of his early professional life.

162. This formation of *acquired* Perceptions, and their gradual assumption of the *immediate* character of those which belong to our original constitution (§§ 167, 168), bear a striking analogy to the process by which habitual Movements come to be linked-on to the Sensations that prompt them, so as at last to be automatically performed, although originally directed by the Will. And it can scarcely be regarded as improbable, that, in the one case as in the

other, the Nervous Mechanism *grows-to* particular modes of activity (§ 138); so that successions of change are uniformly excited by particular stimuli, which were not provided-for in its original construction. Such a view harmonizes well with the fact, that such associations, alike between Sensations and respondent *movements*, and between Sensations and respondent *ideas*, are formed much more readily during the period of Childhood and Adolescence, than they are after the full measure of development has been attained; and that they are much more durable in the former case than in the latter. Throughout the whole Constitution of Man, as well Physical as Mental, we witness a marked capacity of adaptation to a great variety of circumstances; and by the self-education directed by those circumstances, he gradually *acquires* those modes of action, which in other Animals are originally and uniformly prompted by their Instinctive tendencies. It will be shown hereafter (§ 275) to be accordant with the general laws of Nutrition, that such habitual modes of action should express themselves in the formation of the Nervous mechanism, so as to *develope* in it arrangements corresponding to those which it elsewhere originally possessed; and that such arrangements, when once formed, should be kept-up through life, provided that they are not allowed to pass into disuse. Hence, although placed at a disadvantage in comparison with other Animals during the earlier periods of his life, Man is enabled ultimately to attain to a far wider range of Perceptive appreciation than that to which *they* are limited; there being, in fact, no class of Sense-impressions, from which, by habitual Attention to them (§ 127), he may not draw information of a far more precise and varied nature, than they seemed at first to be capable of affording.

163. We have seen that, for the production of a Sensation, a *conscious* state of Mind is all that is required; whilst, on the other hand, for the exercise of the Perceptive power, a certain degree of *Attention* is requisite; or, in other words, the Mind must be *directed*

towards the sensation. And thus it happens that, when the Mind is either inactive, or is completely engrossed by some other subject of thought, the Sensation may neither be perceived nor remembered, notwithstanding that we have evidence, derived from the respondent movements of the body, that it has been felt. Thus a person in a state of imperfect Sleep may start at a loud sound, or may turn-away from a light shining on his face; being conscious of the Sensation, and acting automatically upon it, but forming no kind of appreciation of the externality of its source. And, in like manner, a person in a state of profound Abstraction (§ 443) may perform many automatic movements, which cannot (so far as we know) be excited otherwise than through the medium of Sensation; and yet the exciting sensations are neither *perceived* by him at the time, nor are they afterwards remembered; so that when he is aroused from his reverie, he may be astonished to find himself in circumstances altogether different from those under which he passed into it. Sometimes, however, the Sense-impression may excite a sort of imperfect Perception, which is subsequently remembered and completed. For example, the Student who does not hear the repeated strokes of the clock, when his mind is entirely given-up to his object of pursuit, may have a sort of vague consciousness of them if his attention be less completely engrossed by his studies; and although the sounds may not suggest at the moment any distinct idea of the passage of time, yet, when he subsequently gives his attention to the Sensorial impression, he may remember to have heard the clock strike, and may even be able to retrace the number of strokes.* When the Attention

* It is curious that, in so retracing a number, we are often assisted by mentally reproducing the succession of strokes; *imagining* their recurrence, until we feel that we have *counted-up* to the impression that was left upon our Sensorium.—In the same way, if asked how many stairs there are in a staircase which we are in the habit of using, we may not be able to name the number; yet, when actually ascending or descending, we are conscious that we have arrived at the top or the bottom, by the completion of that series of Sensorial changes which has become habitual to us.

is directed, however, to the sonorous impressions (as when we are *listening* for the striking of the clock), or when it is not so closely fixed on any other object as to prevent it from being attracted by the Sense-impressions, the sounds are not only recognized as proceeding from an external source, which is a simple act of Perception, but the sensations we perceive are discriminated from all others of like nature; and it is by this kind of mental intensification of the perceptive change to which they give rise, that the sensations themselves are impressed with so much additional force on our consciousness, as to seem extraordinarily increased in acuteness. Although we are accustomed to see this chiefly in cases where some particular kind of perceptive acuteness has been acquired by *habit* (§ 127), yet we may learn from certain phenomena of Somnambulism (both spontaneous and artificial) that nothing more is needed, than that concentration of the *whole mind* upon the Sensorial indications, which is the natural state of the Infant (§ 498).

164. Taking as the basis of the knowledge possessed by Man of any object external to him (and therefore of the External World generally), *first*, a subjective Sensation called forth by the presence of that object; *secondly*, the recognition of the *externality* of the cause of that sensation; and *thirdly*, the formation of a *notion* respecting the quality of the object which called it forth,—we have next to inquire into the mode in which such elementary Notions or Cognitions (which are afterwards to be combined into the composite Idea of the object) are generated. How far *any* of them are original or intuitive, is a question which has been much discussed by Psychologists: some maintaining that *all* such Notions are generalizations based on experience; whilst others regard them as the products of Intuition, that is, as mental instincts, of which no other account can be given, than that such cognitions are formed—like the sensations that excite them—from "a law of our nature." Psychologists of both schools agree in considering the

formation of these elementary Notions, like the performance of movements prompted or guided by them, as an *automatic* action; the difference in their views consisting in this,—that Intuitionalists regard this action as *primarily* automatic, while Psychologists of the Experience-school regard it as *secondarily* automatic. As it appears to the Writer that Physiology can throw considerable light on this question, he will here examine in some detail the mode in which those *visual* perceptions are formed, whereon we mainly depend for our ordinary guidance.

165. One of the most elementary of our Visual cognitions is the *Sense of Direction,* whereby we recognize the relations of the points from which luminous rays issue, and thus see objects *erect,* though their pictures on the retina are *inverted.* Some Psychologists have gone so far as to assert that Infants really see all objects inverted, and only acquire the true notion of their position in reference to themselves, by the corrective experience gained by touching and handling them. But this is a pure assumption, founded on an entirely erroneous notion of the nature of Sensation. For it supposes that we look at the picture formed on the retina, by the "mind's eye" placed behind it, just as we look at the picture formed by a camera with the bodily eye; whereas the fact is unquestionable, that Sensation is a state of consciousness excited by the transmission to the Sensorium (through the optic nerve) of the impression produced by the picture on the retina; and as we know nothing whatever of the mode in which the Physical change is *translated,* so to speak, into the Mental, there is no reason why it should be *less natural* for the retinal impression to suggest to the mind the notion of the *real* position of an object, than to call up a representation corresponding to its *inverted* picture. As a matter of fact, it is found that persons who have for the first time acquired Sight by operation, at an age when they can describe their sensations, are able to recognise the *direction* of any luminous object, though quite incapable of appreciating its

distance. And it appears from the experiments of Dr. Serre,* that the luminous spectra produced when pressure is made upon the eye in a dark room, are seen in a direction which has in each case a constant and definite relation to the part of the retina that is affected by the pressure, either directly, or secondarily through its transmission to the opposite side of the globe. By an extensive series of observations on the relation of the positions of these primary and secondary *phosphènes,* both to each other and to the seat of compression, Dr. Serre has been able to deduce the important conclusion, that the lines joining these spectra and the spots of the retina by the affection of which they are produced, all pass through a common "centre of direction," which is situated nearly in the middle of the crystalline lens. And from these facts it seems a legitimate conclusion, that our sense of the relative directions of external objects, which affect different points of our retina by their luminous rays, is primarily derived from a kind of *extradition* of the Visual sensation, corresponding to that which takes place in the case of our Tactile (§ 132). The Writer quite agrees, however, with Professor Bain, that in our ordinary use o Vision we are greatly guided as to the recognition of *relative* direction, by the Muscular Sense called forth in the movements we give to our eyeballs, when we transfer our gaze from one point of a visual picture to another.

166. The recognition of the *singleness* of the object which forms a simultaneous visual picture on both eyes, has been very generally regarded by Physiologists as necessarily arising, by "a law of our nature," from a certain structural relation between what have been termed "corresponding points" on the two retinæ; "double vision," or the recognition of two distinct images, taking place whenever, through a want of harmony in the action of the muscles, the axes of the two eyes do not converge in the object looked-at. But this view of the case is inconsistent with the fact, that if such

* "Essai sur les Phosphènes," Paris, 1853.

abnormal conditions should become permanent (as in squinting) the vision after a time becomes "single" again, notwithstanding that the pictures are formed on parts of the retinæ which do not correspond. Further, if the Muscular irregularity be rectified by surgical means, so that the axes of the two eyes can be again brought into convergence in the object looked at, *double vision recurs for a time*, although the images are now formed upon the original "corresponding points." It is also a fact well known to Ophthalmic Surgeons, that if an opaque spot has been formed in the centre of the cornea, or an artificial pupil has been made at the margin of the iris, so that the most distinct vision is gained when the axis of one eye is directed, not to the object, but to some other point, such direction will become habitual; yet although, when the two eyes work together, there is a decided squint, there is no "double vision." Since it is clear from these facts, that the recognition of the singleness of the object of sensation is the result of *experience*, in the cases in which it supersedes a temporary double vision, it may be fairly so regarded in the case of the Infant; more particularly since observation shows that the convergence of its eyes upon the object looked-at, is, in the first instance, by no means so immediate or exact as it subsequently becomes. And, further, it is obvious that if (as seems not improbable) there *is* some structural arrangement which conduces to singleness of vision when the images are thrown on the originally corresponding points of the two retinæ, such mechanism must have developed itself *de novo*, whenever single vision is the result of the habitual conjoint use of two eyes whose axes do *not* converge so as to meet in the object.

167. A like process of *experiential acquirement* of Perceptional cognitions having the immediateness and trustworthiness of the Sensations on which they are based, is demonstrable in regard to those notions of *form* which we derive from the Visual sense alone, when it has been educated by *co-ordination* with the Tactile. It

may now be affirmed with certainty, that Sight originally informs us only of what can be represented in a Picture—that is, light and shade, and colour; and it may be affirmed, with equal certainty, that the notions of form which we obtain through the sense of Touch (when exercised in combination with muscular movements, of which the "muscular sense" renders us cognizant) are originally unrelated to those derived from Sight; so that when a blind adult first acquires vision, objects with which he (or she) possesses the greatest tactile familiarity, are not recognized by its means, until the two sets of Sense-impressions have been co-ordinated by repeated experience. The best evidence of this kind is derived from observations made upon persons born blind, to whom sight has been communicated by an operation, at a period of life which enabled them to give an accurate description of their sensations:—

a. The case long ago recorded by Cheselden still remains one of the most interesting of these. The youth (about twelve years of age), for some time after tolerably distinct vision had been obtained, saw everything *flat* as in a picture, simply receiving the consciousness of the impression made upon his retina; and it was some time before he acquired the power of judging, by his sight, of the real forms and distances of the objects around him. An amusing anecdote recorded of him, shows the complete want which there is in Man of any *original* or *intuitive* connection between the ideas formed through visual and through tactile sensations. He was well acquainted with a Dog and a Cat by *feeling*, but could not remember their respective characters when he *saw* them; and one day, when thus puzzled, he took up the cat in his arms, and felt her attentively, so as to associate the two sets of cognitions, and then, setting her down, said, "So puss, I shall know you another time."

The same indication, moreover, is obviously afforded by the case of Jemmy Morgan already cited (§ 161).—In a recently-recorded case in which sight was imparted by an operation to a young woman who had been blind from birth, it was interesting to contrast the rapidity and accuracy of her Tactile perception, which was highly educated, with the slow, laborious process by means of which she arrived at a conception of the shape and nature of an object through the medium of her newly-acquired and imperfectly-educated Vision :—

b. "I found," says the operator, "that she was never able to ascertain what an object really was by Sight alone, although she could correctly describe its shape and colour; but that after she had once instructed one sense, through the medium of the other, and compared the impressions conveyed by touch and sight, she was ever after able to recognise the object without touching it. In this respect her memory was very perfect: I never knew her fail in a single instance, though I put this power frequently to the test of experiment. It was curious to place before her some very familiar object that she had never compared in this way, such as a pair of scissors. She would describe their shape, colour, glistening metallic character, but would fail in ascertaining what they really were, until she put a finger on them, when in an instant she would name them, and laugh at her own stupidity, as she called it, in not having made them out before." —(See Critchett, in *Medico-Chirurgical Transactions*, vol. xxxviii.)

168. Still more remarkable is the acquirement of those Perceptions of *solid form* or *Relief*, which we derive, as Sir C. Wheatstone's admirable investigations have shown, from the Mental combination of the dissimilar perspectives that are projected by solid objects upon our two retinæ. When we bring to our right and left eyes respectively, by means of the Stereoscope, pictures corresponding to those which would be formed on their two retinæ by the actual object if placed before them at a moderate distance, the resulting perception of the solidity of the image seems as necessary and immediate as if it were the product of an original Intuition; and this perception is strong enough to assert itself, in spite of our intellectual knowledge that we are looking at two plane surfaces. Now, although it may be inferred from the actions of many of the lower Animals, that the perception of the relative

distances of near objects or parts of an object (which constitutes the basis of the conception of solidity) is in their case intuitive, it may be affirmed, as a conclusion beyond reasonable doubt, that this also is *acquired* by the Human infant during the earliest months of its life, by a co-ordination of its muscular and visual sensations; which enables the automatic mechanism to adopt the dissimilarity of position between corresponding points in the two pictures, as the measure of their relative distances. The self-education of this Perceptive faculty which goes on during the first few months of infantile life, is the basis of our subsequent Visual knowledge of the External World, as it seems to be for the most part also of the primary belief in its objective reality (§ 159).

169. In this Visual recognition of the solid form of an object by the mental combination of its two dissimilar perspectives, we have an exercise of *judgment*, the decision of which may be as implicitly trusted (at least under ordinary circumstances) as if it were authoritatively delivered by a congenital faculty, but which really rests on a basis of Experience. It is scarcely conceivable that the Infant consciously asks itself the question, "What do I see?" But there can be little doubt that, in the earlier stages of its experience, it is incapable (like the newly-seeing adult) of distinguishing between a picture and the solid object which it represents; and that the essential condition of a judgment—the possibility of the opposite, or of something else—therefore exists for it. But with every consentaneous exercise of the visual, tactile, and muscular sensations, during the Infant's gaze at an object grasped in its hands and carried to different distances by the motion of its arms, there is a new co-ordination, which helps to supply the deficiency in the sum of all that preceded; and this process is repeated until the complement of the whole serves as the basis of the cognition, which we thenceforth rightly characterize as "self-evident."

170. It is not a little remarkable that even that Visual percep-

tion of Solidity, which is based on the Binocular combination of dissimilar perspectives, may, under certain circumstances, be antagonized by a higher experience, so as to be for a time, or even permanently, excluded. The very ingenious Pseudoscope contrived by Sir C. Wheatstone, effects a lateral reversal of the perspective projections of actual objects on the two retinæ, corresponding to that which would be made by "crossing" the pictures in the Stereoscope; and thus, in viewing through it any solid object, we ought at once, if the visual perception were a necessary product (as Sir David Brewster maintained) of the geometrical relations of the two images, to see all its projections and depressions reversed,—the exterior of a basin, for example, being changed into a concave interior; and the projecting rim on which it rests, into a deep furrow. But this "conversion of relief" is generally resisted, for a time at least, by the preconception of the actual form which is based on habitual experience; and it only takes place immediately, in cases in which the converted form is at least as familiar to the mind as the actual form. Thus, when we look with the Pseudoscope at the *interior* of a mask, or at a hollow *mould* of a plaster bust, the mental representation of the image in relief is at once called-up. But when we look pseudoscopically at the *face* of a plaster bust, or at the *outside* of a mask, it is only after a lengthened gaze that such "conversion of relief" occurs; the mind being so much more familiar with the actual form, that the mental image of the interior of a mould or mask is not called-up, until the visual perception has overcome, as if by continued pressure, the resistance of the preconception; and for this a considerable *time* is often required. In the case of the *living human face*, however, it seems that no protraction of the Pseudoscopic gaze is sufficient o bring about a "conversion of relief"; the perceptive consciousness (probably here under the complete domination of the Intellectual) refusing to entertain the notion of an *actual visage* having the form of the interior of a mask.

171. The notion of *Solidity* or *projection in three dimensions*, as distinct from a representation of an object on a plane surface, may, under certain circumstances, be derived from a *single flat picture*, no less strongly than from the combination of the two dissimilar perspectives of the object. It has long been known that if we gaze steadily at a picture, whose perspective projection, lights and shadows, and general arrangement of details, are such as accurately correspond with the reality which it represents, the impression it produces will be much more vivid when we look with *one* eye only, than when we use both; and that the effect will be further heightened, when we carefully shut out the surroundings of the picture, by looking through a tube of appropriate size and shape. This fact has been commonly accounted-for in a very erroneous manner. "We see more exquisitely," says Lord Bacon, "with one eye than with both, because the vital spirits thus unite themselves the more and become the stronger;" and other writers, though in different language, agree with Bacon in attributing the result to the *concentration* of the visual power, when only one eye is used. But the fact is, that when we look with *both* eyes at a picture within a moderate distance, we are *forced* to recognize it as a flat surface; whilst, when we look with only *one*, our Minds are at liberty to be acted-on by the suggestions furnished by the perspective, chiaroscuro, &c.; so that, after we have gazed for a little time, the picture may begin to start into relief, and may even come to possess the solidity of a model. The completeness of this illusion will essentially depend upon the exactness with which the picture represents the real "projection" of its object upon a flat surface. It is very rarely that pictures painted by human hands "come out" after this fashion in a degree at all comparable to sun-pictures; for the obvious reason that the Photograph represents not merely the exact perspective of the scene or object, but the actual chiaroscuro as it was at any one moment, with a fidelity which the Artist, who requires *time* for his work, cannot possibly

equal, since the shadows on the object are so constantly changing as he proceeds, that he can scarcely by any possibility avoid a departure from strict truth in his combinations.—The nearest approach to sun-pictures in respect to the truthfulness of the chiaroscuro, is presented by pictures painted by artificial light, the uniformity of which can be maintained while the "sitting" lasts.

a. The Writer possesses three Photographs, two of *bassi-relievi*, and one of an *alto-relievo*, by Lucca della Robia; which, when looked at with one eye in the manner now described, give rise to a feeling of *solidity* so vivid, that it is almost impossible not to credit it.

b. As the shadows are strong in all these Photographs, the illusion is promoted by causing the light by which they are viewed, to fall on them in the direction corresponding to that in which it fell on the originals, when (so to speak) they sat to the photographic camera for their portraits; but this is by no means a necessary condition, the effect being produced with nearly the same vividness in diffused day-light.—If, indeed, a strong light be seen to proceed from the *opposite* side, so that the direction of the shadows in the Photograph is reversed with reference to it, a picture may be turned (as it were) inside-out, so as no longer to present the relievo, but its hollow mould, *provided that the Mind will readily accept the conversion.* This effect the Writer can produce most effectively with a beautiful Photograph of a large American Trilobite imbedded in its rocky matrix; for according to the direction in which the light is allowed to fall upon it, the surface of the back of the Trilobite appears to project, or is turned into a concave reverse, representing the "cast" of that surface in the surrounding rock,—the effect in either case being aided by a Mental predetermination as to which view shall be seen.

c. In the beautiful Medallion-engravings (produced by mechanical means) that were in fashion some years ago, the like illusion could be produced; the same picture being caused to represent either a cameo or an intaglio, by such a disposition as made its lights and shadows correspond with those which would have been thrown from the source of illumination, had the rays fallen on an actual cameo or an actual intaglio.

d. It is remarkable that the effect of this mode of viewing Photographic pictures is not limited to bringing out the solid forms of objects; for other features are thus seen in a manner more true to the reality, and therefore more suggestive of it. This may be noticed especially with regard to the representation of *still water*, which is generally one of the most unsatisfactory parts of a Photograph; for although, when looked at with *both* eyes, its surface appears opaque, like white wax, a wonderful depth and transparence are often given to it by viewing it with only *one.*—And the same holds good also in regard to the characters of *surfaces* from which light is reflected,—as bronze or ivory; the material of the object from which the Photograph was taken being recognized much more certainly when the picture is looked at with *one* eye, than when *both* are used, unless in stereoscopic combinations.

172. The superiority of Monocular to Binocular vision depends in these cases upon the freedom with which the Mind is left to interpret the picture after its own fashion, when no longer forced to view it as a flat surface; and the interpretation is here so obviously based on experience, which gives to every incident of the picture a *suggestive* power of its own, as to destroy the force of any argument that might be erected on the immediateness and uniformity of the perception of Relief derived from the binocular combination of two dissimilar perspectives, as to the *original* or *intuitive* character of that perception. For it thus becomes clear that this combination is only *one* out of *several* modes of suggestion by which that notion is formed; whilst the phenomena of the Pseudoscope show that the notion is by no means *necessarily* called up by the visual impressions which *ordinarily* produce it.

173. But further, it is not a little curious that an actual *conversion* of relief may be produced by a Mental preconception, when we look at certain solid forms with *one* eye only. For just as the want of power to appreciate distance monocularly, enables us to invest a pictorial representation with the attribute of solid form, so is a solid form represented to the mind as a flat picture; and to this picture we may *mentally* give a solid form *the very*

opposite of that which it really possesses. Of this fact, which is of no small importance in elucidating the nature of *false perceptions* (§ 186), the following are illustrations :—

a. It has long been known that when a *seal* is looked-at through a Microscope, it will appear sometimes projecting as a cameo, sometimes excavated as an intaglio; this "conversion of relief" taking place alike with the engraven stone or its waxen impression. That it is due, not (as some have supposed) to an optical change effected by the Microscope, but simply to the limitation of the visual impression to a single eye, which deprives the judgment of the positive guidance whereon it ordinarily relies, is clearly proved by the fact that no such conversion can be produced under a properly constructed Binocular Microscope,—a seal, like every other object, being represented in its true projection; while it is readily effected in regard to larger objects of a suitable nature, without the intervention of any optical instrument. Thus, as Sir D. Brewster pointed out in his "Natural Magic," if we take the intaglio mould of a bas-relief, and look steadily on it for a time with one eye, excluding surrounding objects as much as possible from our attention, we may distinctly see the bas-relief as if projecting. "After a little practice," he says, "I have succeeded in raising a complete hollow mask of the human face, the size of life, into a projecting head."

b. If instead of a plaster mould, we take a common pasteboard mask (such as is sold in every toy-shop), and paint the inside, which is usually left in the rough, so that the colours of its different parts may imitate, as closely as possible, those of the corresponding parts of its exterior, and the inside or hollow surface of the mask be then held at arm's length from the eye, with the light so arranged that no shadow falls anywhere upon it,—not only will the image of the projecting face very readily present itself, but it will be difficult for an observer who has once caught this, to see the mask as it really is, even by a determinate effort. The illusion is the more complete, if his view be limited to the mask itself, and he be brought to the proper point of sight without being aware of what he is to see; so that, of a large number of persons on whom the Writer has tried this experiment, almost all have at once pronounced that they were looking at the projecting surface of the mask, and have only been convinced to the contrary by the conjoint use of both eyes.

The facility with which these conversions, and others of like nature, occur to the "mind's eye," may be readily shown to depend upon the degree of readiness with which, in virtue of our previous habits and experiences, the visual picture *suggests* the real form, or its converse.

c. In the case of a seal, the hollow mould and its projecting cast are objects almost equally familiar; hence the representation of either may offer itself, and the one may be substituted for the other by a slight effort of the volitional power of Conception. The conversion of the hollow mask into the projecting face is, to most persons, still more easy, because they are more accustomed to the life-like features of the plaster-model, than they are to the concave mould which has no similitude in nature; whilst, on the other hand, the Writer has not found it possible to convert the face of a bust into the likeness of a hollow mask by the simple monocular gaze, however long continued, even with the aid of the strongest effort so to conceive it.—When a seal is looked at in a Microscope, or larger objects of the same kind are seen through an inverting Telescope, the "conversion of relief" is aided by the fact that the optical inversion of the images has caused the relation of the shadows to the known source of the light to be also reversed, so that they fall as they would do if the cameo were really replaced by the intaglio, or the intaglio by the cameo.

174. Another singularly interesting demonstration of the inability of monocular vision to afford any *certainly-true* idea of solid form, was given by Sir C. Wheatstone in the first of his two remarkable memoirs on Binocular Vision :—

If we hold up at arm's length a small skeleton-cube made of wire or ebony-beading, and look at it with one eye whilst placing it in a variety of positions by turning it between the fingers, so long as the Mind perceives the cube, its various perspective projections are interpreted by it as so many different representations of one object, all of them suggesting the same primitive form. But as certain of these perspective projections might be given by an object of very different shape, it will probably happen that in some position of the cube one of these dissimilar figures will suggest itself to the mind; and, if this new conception be fixed by a steady gaze for a short time, it will take such possession of the Mind, that some effort is required to bring back the original conception, so long, that is, as the position of the cube remains unchanged. But if, whilst the Mind is thus possessed with the false idea, the cube be again made to turn between the fingers, the series of successive projections then presented not being reconcilable with the converse form, either the Mind reverts to the original conception of the cube as the only one with which they are consistent, or (if this should not be adopted) the skeleton-figure seems to be continually undergoing a change of shape, *as if its sides were hinged together* and fell into new inclinations with every new position given to the object.—(*Philosophical Transactions*, 1838.)

175. Thus our Perception of solid form, when only *one* eye is used, is clearly a matter of *judgment*, determined by the tendency of the Mind to interpret the visual picture according to its previous familiarity with the forms which that picture may represent: its choice between two or more of these being quite involuntary, when one is decidedly more familiar to the mind than another; but being to a certain extent under volitional control, when they present themselves with equal or nearly equal readiness, through the power possessed by the Will of *fixing the attention* upon either one, to the exclusion of the others.—In ordinary Binocular Vision of a moderately near object, on the other hand, there is no wavering; we feel that there *can* be no mistake. There is but one solid form that *can* furnish the two dissimilar perspective projections. Hence that form presents itself to our Minds, independently of any previous acquaintance with it, as the *necessary resultant* of the combination of those pictures; and this is the case even with pairs of pictures which differ in a degree that is itself quite inconsistent with our experience, provided that the *resultant* suggested by their combination is *conformable to our experience.* Thus, when we look at an actual Landscape, the perspective views we receive through the right and left eyes respectively, of every part of it save the fore-ground, are so nearly the same, as to convey no sugges-

tion of their relative distances; that suggestion being here conveyed by other differences, as of size, distinctness, and the like (§ 180). But the two Photographs of such a landscape taken for the Stereoscope, represent it as seen from two points of view sufficiently remote from each other, to produce that degree of dissimilarity in the pictures to which we are accustomed in looking at a *near* object; and thus the idea of the relative distances of the different parts of the landscape, is suggested with all the force derived from that difference. So, the Photographs of the Moon which are taken at the extremes of her "libration in longitude," are sufficiently dissimilar to one another to "pair" in the Stereoscope, and thus to bring out not only the solid form of her globe, but even the projection of some of the principal craters, with unmistakeable effect. And further, a most striking effect is sometimes produced by the Stereoscopic combination of a pair of pictures, of which neither by itself suggests any idea of the scene it represents. But in all these cases, the result of the combination is *consistent* with our previous experience, or, at any rate, is *not inconsistent* with it. As we intellectually know that the Moon is really globular, though her face—as ordinarily seen by us—*looks* flat, we are quite prepared for the acceptance of the suggestion made by the Stereoscopic view of it. And when a dark patch upon the apparent face of an ice-cliff is carried back stereoscopically to a remote distance beyond, we interpret it as representing a far-off village seen through an arch in the cliff,—a view readily conceivable by our minds, though we may have never actually seen it.

176. However different in kind, then, may be the visual data on which our ordinary Monocular and Binocular perceptions are based, the mental operation by which we build upon them is essentially the same in both cases. For the binocular view of an object, like the monocular, does nothing else than *suggest* to the Mind the conception of a certain solid form; and that the adoption of this conception depends much more upon the antecedent condition of

the *mind*, than it does upon the purely *optical* relations of the two retinal pictures, is rendered quite certain (in the Writer's opinion) by the application of the pseudoscopic test (§170). For according to the theory of *intuitive cognition*, based on the *optical* differences of the two pictures, *every thing* at which we look with the Pseudoscope ought to be at once "turned inside-out." But a large proportion of the objects on which we try its converting powers, are proof against them; those only being readily *metamorphosed*, whose new forms can be readily *conceived*. And the percipient Mind will not admit too strange a novelty; it obstinately clings to so much of the reality, as is recognized by its previous Tactile experience to be the necessary interpretation of the Visual impression ordinarily received from the object; and it can only accept such modifications as are capable of being fitted-on to the results of that experience.

177. Our visual recognition of solid form is aided by suggestions of another kind, which are furnished by that measurement of the relative *distances* of the different points of an object, which we make by bringing the axes of the eyes into convergence upon those points successively; the degree of such convergence being indicated to us by the "muscular sense" that originates in the state of the Muscles we put in action to produce it. Here, again, it is obvious from what has been already stated, that our interpretation of that sensation is *acquired*, not intuitive; and it will be presently shown that, under certain circumstances, an increase or diminution of the angle of convergence rather suggests change of *size*, than change of *distance* (§ 182). How much of our right estimation of the relative distances of objects not far removed from the eye, depends upon the conjoint use of *both* eyes, is made evident by the fact that if we close *one* eye, we find ourselves unable to execute with certainty any actions which require the guidance of that estimation,—such as threading a needle, or passing a crooked stick through a suspended ring. And it has sometimes happened that

persons who have lost the sight of one eye, have been first made aware of its want by their inability to execute such actions.

178. It will now be apparent how, when one eye is closed, we lose that power of *certainly* distinguishing a flat picture from a projecting relievo, or either from a concave mould, which we derive from the conjoint use of both organs. We can make no mistake in our *binocular* estimation of such objects, provided their dimensions and distances be such as to make their two retinal projections appreciably different, and to require a sensible difference in the convergence of the optic axes as they are successively directed to different points. We are *forced* to see that a picture is nothing but a plane-surface, that the outside of a mask represents the actual features of a human face, and that the hollow mould is the concave reflex of the cast which has been turned out of it, so long as these objects are within a few feet distance, and are seen by both eyes at once. And thus it becomes evident that the remarkable converting power of *monocular* vision, by which a single picture may be raised into stereoscopic relief, and cameos and intaglios be mistaken for each other (§§ 171-173), is—however interesting as a Psychological phenomenon—really a mark of imperfection in the visual sense when thus exercised.

179. That such is the true view of the case, appears further from this; that we are liable to be thus deceived in regard to the very same objects, even when we look at them with *both* eyes, provided that they are removed to a sufficient distance to render the difference of their retinal projections inappreciable, and to prevent the relative distances of their parts from being measured through the sense of convergence.

The large Architectural pictures formerly exhibited in the Diorama often gave such an impression of projection, that almost everyone who saw them would have been ready to affirm that a particular column or statue *must* have been painted on a different surface from the rest, like a detached part of a scene in a theatre;—until, on slightly moving the head from side to side, the absence of any alteration in its apparent position made it evident that it must be on the same plane with the adjacent parts. The perplexing vividness of this deception was due, as is now well known, to the early possession by MM. Daguerre and Niepce of one form of the Photographic art; which enabled them to impart to their architectural pictures a truthfulness previously unattainable, and therefore gave to these pictures an extraordinary power of *suggesting* the solid forms of the objects they represented.—Many of the apartments in the Louvre are decorated with cornices which so vividly represent projecting forms, as to be generally mistaken for them by such as see them for the first time; and visitors to the Bourse of Paris will recollect the large allegorical paintings in its interior, which are so executed, and so disposed, as very strongly to suggest to those who only view them from a distance the perception of high relief.

180. Our estimate of the distance of *remote* objects is clearly a matter of *judgment* based on experience; being chiefly founded upon their apparent size, if their actual size be known or guessed; or, if we have no knowledge of this, and our view does not range over the intervening space, upon that modification of their distinctness of colour and outline, which is known to Artists as "aërial perspective." Hence this estimate is liable to be greatly affected by varying states of the atmosphere: the same mountain-peak, for example, being apparently brought much nearer than it is in reality, when an extraordinary clearness of the air enables all its features to be distinctly seen; and carried to a much greater apparent distance, when a slight haziness of the air softens them all down. This alteration has a very curious effect upon our appreciation of the *sizes* of distant objects (§ 181).

181. Of the relative *sizes* of objects, our estimate is partly based on the sizes of the pictures formed of them on our retina, or, in other words, on the "visual angles" they subtend; and partly on our appreciation of their distances,—the *apparent* size of an object seen under a given visual angle being estimated as larger or smaller than the reality, according as we suppose it to be more or

less distant than it really is. Thus the apparent *height* of mountains is so greatly affected by our estimate of their *distance*, that, according to the varying atmospheric conditions which modify the latter (§ 180), the same mountain may appear much higher or much lower than it really is; its height being *under*-estimated when the peak is made to seem very near, and *over*-estimated when its apparent distance is exaggerated,—just as, when we are walking across a common in a fog, a child dimly seen through it seems to have the stature of a man, and a man that of a giant. In the case of a near object, however, we are not liable to any such error; since, if we truly appreciate its *distance* in the mode already described (§ 177), the appreciation of its *size* can be derived with certainty from the dimensions of its visual picture.

182. The appreciation of size, like that of solid form, is a matter of *immediate judgment*: but there is strong evidence that in this, as in the preceding case, the power of forming that judgment has been *acquired by experience*. Much light has been thrown upon this as upon other phenomena of Binocular Vision, by the ingenious experimental researches of Sir Charles Wheatstone. A simple modification of his mirror-stereoscope enables the observer to vary the *distances* of the pair of pictures from his eyes, without altering the angle of their convergence; and, conversely, to alter the *angle of convergence of the optic axes*, without altering the distance of the pictures. Now in the first case, the *perceived dimensions* of the pictures change—diminishing as their distance increases, and *vice versâ*,—in accordance with the change in the *actual* dimensions of the retinal pictures; the effect being very much like that of the phantasmagoria. But in the second case, a most remarkable change takes place in the *perceived dimensions* of the pictures, although the *actual* dimensions of their retinal pictures remain unaltered. For when the optic axes are made to converge upon them more and more, as they would do if they were fixed upon a single picture gradually brought very near the eyes, *the apparent*

size of the pictures undergoes a most remarkable reduction; whilst, if the arms of the stereoscope are so turned, that the optic axes, instead of being moderately convergent, are brought into parallelism, or even into slight divergence, *the apparent dimensions of the pictures undergo a not less remarkable increase.* (Phil. Trans., 1852).

183. It does not seem possible to account for this fact in any other way, than by supposing that the percipient Mechanism has been developed in conformity with the experience gained during the early part of Infantile life; in which objects held in the hand are brought nearer to, or removed further from, the eyes, the axes of which are steadily directed to them at varying angles of convergence. The *identity* of the object being recognized throughout, the two sets of changes are brought into *mutally corrective action,* but when either of them takes place without the other, the Mechanism evolves a wrong result. If the *angle of convergence* remain unaltered, changes in the size of the retinal images produce corresponding changes in the *apparent* size of the picture; whilst if the *size of the retinal image* remain unaltered, changes in the angle of convergence, acting on the Mechanism in the same manner that a change of distance would do, cause (as it were) an *organic expectation* that the size of the retinal image will vary accordingly; and, as it does *not* change, it is instinctively interpreted exactly like the image of a mountain or other distant object, which is made to seem *larger* by an *increase,* and *smaller* by a *diminution* of its apparent distance.—It is a curious illustration of the same principle, that if we take up such a position at a Railway-station as to see a train approaching "end on," it seems to *swell-out* as it approaches our stand-point; the retinal image being rapidly enlarged, without any such corresponding indication of diminished distance, as would serve to account (so to speak) to our percipient Mechanism for that enlargement.

184. Every *acquired* visual Perception, then, may be regarded as the *resultant of our whole previous experience* relating to the

object of it; such resultant, however, not being worked out by a process of conscious reasoning, but being the reflex action of the nervous Mechanism of the Ego, which has *formed itself* in accordance with that experience, so as to acquire powers of reaction of a far higher kind than it originally possessed. The "self-evidence" of the truthfulness of the Perception is of the same kind, therefore, as that of the Sensation which has called it forth; the Mental affection being in each case the *immediate* and *invariable* response of the organization to the impression made upon it. But whilst that response, in the case of the deliverances of our *sensational* consciousness, is given by our *original* constitution, it is given in the case of our *perceptional* consciousness by our *acquired* constitution; in which are embodied those results of primary experience, which are common to every normally-constituted Human being.

185. The power of immediate and acute Perception is one eminently capable of being increased by habitual Attention. We are here concerned not so much with that exaltation of the *discriminating consciousness* of Sense-impressions, which has been already noticed (§ 127); as with the augmentation of the power of *taking cognizance* of the objects that excite Sensations, which depends upon a rapid exercise of that higher faculty by which those sensations are interpreted. It would be easy to adduce many examples of the improvement of this faculty by practice; so that individuals who have cultivated it in particular modes, derive from ordinary Sense-impressions an amount of information which they could scarcely have been supposed capable of conveying. The following, however, will suffice:—

a. It has long been known that individuals among the Deaf-and-Dumb have acquired the power of "lip-reading"; that is, of so interpreting the visible movements of the mouth and lips of a speaker, as to apprehend the words he utters, no less accurately than if they were heard. And it has been latterly proposed to make this a matter of systematic instruction; so that every deaf-mute should be enabled to understand what is said, without the aid of the "sign-language" or the "finger-alphabet."—It appears, however, that it is not every one who is capable of acquiring this power; and it is still questionable whether it can be even *generally* attained by any amount of practice. But that it should have been even *exceptionally* acquired, shows the extraordinary improvability of the Perceptive faculty.

b. The celebrated conjuror, Robert Houdin, relates in his Autobiography the mode in which he prepared himself and his son for the performance of the trick which he termed "second sight;" the success of it mainly depending upon the rapidity with which the information given by Sense-impressions could be apprehended and interpreted, and the accuracy with which (for a short time at least) they could be remembered;—In the first instance, Houdin put down a single domino, and required his son to name the total number of points *without counting them*, which each could readily do. *Two* dominoes were then tried; and, after a little practice, the total number of points on both was correctly named by each *at the first glance*. The next day the lesson was resumed, and they succeeded in naming the points on *four* dominoes at a single glance; on the following day those of *six;* and, at length, they found themselves able to give, without counting, the sum of the points on *twelve* dominoes.—This result having been attained, they applied themselves to a far more difficult task, over which they spent a month. The father and son passed rapidly before a toy-shop, or any other displaying a variety of wares; and each cast an attentive glance upon it. A few steps further on, each drew paper and pencil from his pocket, and tried which could enumerate the greater number of the objects momentarily seen in passing. The son surpassed the father in quickness of apprehension, being often able to write down forty objects, whilst his father could scarcely reach thirty; yet, on their returning to verify his statement, he was rarely found to have made a mistake.

The following remarkable proof of the efficacy of this training may be best given in Houdin's own words:—

c. "One evening, at a house in the Chaussée d'Antin, and at the end of a performance which had been as successful as it was loudly applauded, I remembered that while passing through the next room to the one we were now in, I had begged my son to cast a glance at the library, and remember the titles of some of the books, as well as the order they were arranged in. No one had noticed this rapid examination.

"'To end the second-sight experiment, Sir,' I said to the master of the house, 'I will prove to you that my son can read through a wall. Will you lend me a book?'

"I was naturally conducted to the library in question, which I pretended now to see for the first time; and I laid my finger on a book.

"'Emile,' I said to my son, 'what is the name of this work?'

"'It is Buffon,' he replied, quickly.

"'And the one by its side?' an incredulous spectator hastened to ask.

"'On the right or the left?' my son asked.

"'On the right,' the speaker said, having a good reason for choosing this book, for the lettering was very small.

"'The Travels of Anacharsis the Younger,' the boy replied. 'But,' he added, 'had you asked the name of the book on the left, Sir, I should have said Lamartine's Poetry; a little to the right of this row, I see Crébillon's works; below, two volumes of Fleury's Memoirs;' and my son thus named a dozen books before he stopped.

"The spectators had not said a word during this description, as they felt so amazed; but when the experiment had ended, they all complimented us by loud plaudits." — (*Autobiography of Robert Houdin*, p. 206.)

186. *False Perceptions.*—It has been shown (§ 148) that the action of *ideational* states upon the Sensorium can modify or even produce *sensations*. But the action of pre-existing states of Mind is still more frequently shown in modifying the *interpretation* which we put upon our sense-impressions. For since almost every such interpretation is an act of *judgment* based on experience, that judgment will vary according to our Mental condition at the time it is delivered; and will be greatly affected by any domi-

nant idea or feeling, so as even to occasion a complete misinterpretation of the objective source of the sense-impression, as often occurs in what is termed "absence of mind" (§ 445). The following case, mentioned by Dr. Tuke as occurring within his own knowledge, affords a good example of this fallacy:—

a. "A lady was walking one day from Penryn to Falmouth, and her mind being at that time, or recently, occupied by the subject of drinking-fountains, thought she saw in the road a newly erected fountain, and even distinguished an inscription upon it, namely—

"If any man thirst, let him come unto me and drink."

Some time afterwards, she mentioned the fact with pleasure to the daughters of a gentleman who was supposed to have erected it. They expressed their surprise at her statement, and assured her that she must be quite mistaken. Perplexed with the contradiction between the testimony of her senses and of those who would have been aware of the fact had it been true, and feeling that she could not have been deceived ("for seeing is believing"), she repaired to the spot, and found to her astonishment that no drinking fountain was in existence—only a few scattered stones, which had formed the foundation upon which the suggestion of an expectant imagination had built the super-structure. The subject having previously occupied her attention, these sufficed to form, not only a definite erection, but one inscribed by an appropriate motto corresponding to the leading idea."—(*Influence of the Mind upon the Body*, p. 44.)

So it is mentioned by Sir Walter Scott, in his "Demonology and Witchcraft," that having been engaged in reading with much interest, soon after the death of Lord Byron, an account of his habits and opinions, he was the subject of the following illusion:—

b. "Passing from his sitting-room into the entrance-hall, fitted up with the skins of wild beasts, armour, &c., he saw right before him, and in a standing posture, the exact representation of his departed friend, whose recollection had been so strongly brought to his imagination. He stopped, for a single moment, so as to notice the wonderful accuracy with which fancy had impressed upon the bodily eye the peculiarities of dress and posture of the illustrious poet. Sensible, however, of the delusion, he felt no sentiment save that of wonder at the extraordinary accuracy of the resemblance; and stepped onwards towards the figure, which resolved itself, as he approached, into the various materials of which it was composed. These were merely a screen occupied by great-coats, shawls, plaids, and such other articles as are usually found in a country entrance-hall. Sir Walter returned to the spot from which he had seen this product of what may be called imagination proper, and tried with all his might to recall it by the force of his Will, *but in vain*—a good illustration of the slight influence of volition over sensation, compared with that of a vivid Mental image or idea acting upon the Sensorial centres, and distorting or moulding into other forms the impressions received from objects of sense."—(*Op. cit.*, p. 45.)

187. Moreover, if not only a single individual, but several persons, should be "possessed" by one and the same idea or feeling, the same misinterpretation may be made by all of them; and in such a case the concurrence of their testimony does not add the least strength to it.—Of this we have a good example in the following occurrence cited by Dr. Tuke as showing the influence of a "dominant idea" in falsifying the perceptions of a number of persons at once:—

d. "During the conflagration at the Crystal Palace in the Winter of 1866-67, when the animals were destroyed by the fire, it was supposed that the Chimpanzee had succeeded in escaping from his cage. Attracted to the roof, with this expectation in full force, men saw the unhappy animal holding on to it, and writhing in agony to get astride one of the iron ribs. It need not be said that its struggles were watched by those below with breathless suspense, and, as the newspapers informed us, 'with sickening dread.' But there was no animal whatever there; and all this feeling was thrown away upon a tattered piece of blind, so torn as to resemble, to the eye of fancy, the body, arms, and legs of an ape!"—(*Op. cit.*, p. 44.)

Another example of a like influence affecting several individuals simultaneously in a similar manner, is mentioned by Dr. Hibbert in his well-known Treatise on Apparitions :—

b. A whole ship's company was thrown into the utmost consternation, by the apparition of a cook who had died a few days before. He was distinctly seen walking a-head of the ship, with a peculiar gait by which he was distinguished when alive, through having one of his legs shorter than the other. On steering the ship towards the object, it was found to be a piece of floating wreck.

Many similar cases might be referred-to, in which the Imagination has worked-up into "apparitions" some common-place objects, which it has invested with attributes derived from the previous Mental state of the observer; and the belief in such an apparition as a reality, which usually exists in such cases, unless antagonized by an effort of the reason, constitutes a *delusion.*—The delusions of Insanity usually have their origin in a perverted state of *feeling;* which begins by imparting a *false* colouring to *real* occurrences; and then, if not checked or diverted, goes on to suggest Ideas having no foundation in fact, which are accepted as realities on account of the incapacity of the disordered mind to bring them to the test of Common Sense (§ 562). And there are many persons quite sane upon ordinary matters, and even (it may be) distinguished by some special form of ability, who are yet affected with what the writer once heard Mr. Carlyle term a "diluted Insanity"; allowing their minds to become so completely "possessed" by "dominant ideas," that their testimony as to what they declare themselves to have witnessed—even when several individuals concur in giving exactly the same account of it—must be regarded as utterly untrustworthy. Of this we have examples at the present time, alike in the asserted appearances of the Virgin, and in the marvels of "Spiritualism;" while the same lesson is taught by the records of the prevalent delusions of past ages, and pre-eminently by those of Witchcraft.

188. *Instinctive Feelings.*—The attainment of that grade of

Mental development which enables us to apprehend the objective reality of external things, seems to make us capable of experiencing certain *feelings* in regard to them, which are nearly akin to those that are immediately associated with Sensations (§ 154), but constitute the germs (so to speak) of higher forms of consciousness. Thus the *æsthetic* sense of the beautiful, of the sublime, of the harmonious, &c., seems in its most elementary form to connect itself immediately with the Perceptions which arise out of the contact of our Minds with external Nature. "All those," says Dr. J. D. Morell, "who have shown a remarkable appreciation of form and beauty, date their first impressions from a period lying far behind the existence of definite ideas or verbal instruction. The germs of all their Æsthetic impressions manifested themselves, first of all, as a spontaneous Feeling or Instinct, which, from the earliest dawn of reason, was awakened by the presentation of the phenomena which correspond objectively with it in the Universe." These primitive feelings exist in very different intensity in different individuals; and it is where they have most strongly manifested themselves at a very early period of life (the sense of Harmony, for example, in the infant Mozart, § 206), that we can see how fundamental a part of our nature they constitute, although they may be but faintly shadowed-forth in a large part of Mankind. They are peculiarly susceptible of development, however, by appropriate culture; under the influence of which they not merely grow-up in the individual, but manifest themselves with increased vigour and more extended range in successive generations (§§ 201—203).

189. The same may be said of those simple forms of *Emotional* sensibility (§ 157), which, being no longer purely subjective, require as a condition of their existence that they shall relate to an external object. This is pre-eminently the case with all those which are termed *emotions of sympathy:* thus, the Perception of the pain or distress of another tends to call forth (except in individuals of a

peculiarly unsympathetic temperament) a corresponding affection in the percipient Self; and the opposite state of cheerfulness or mirth has a like tendency to affect those who are brought into contact with it, provided that there be nothing positively antagonistic in their own condition. But further, the Perception of enjoyment in others calls-forth a respondent *gladness* in ourselves; whilst the perception of suffering tends to excite in ourselves that feeling of sorrow which we term *pity;* and either of these feelings may be experienced, even when we do not ourselves share in the state of elevation or depression which excited them.—More closely connected with the foregoing than is commonly conceived, is that sense of the *humorous*, which attaches itself to certain manifestations of character presented to us in the actions of others; that *sympathy with human nature* in which the former have their source, being the foundation of the latter also; and thus it happened that those writers who have the strongest power of exciting our sense of Humour, are usually distinguished also by their mastery of the Pathetic. To the sense of the humorous, that of the *ludicrous* is obviously related. Both these, however, when excited by operations of the Intellect, instead of by external objects, belong to a different category (§ 404). The same may be said of the sense of *wonder;* which in its simplest form may be connected with our Sense-perceptions, but which is more commonly experienced in regard to the Ideas which they excite.—Another group of Instinctive feelings belonging to the same category, is that which may receive the general designation of *attractions* and *repulsions.* These are the elementary states of those Emotions which involve a distinct *idea* of the object which attracts or repels, and which then assume the forms of *desires* and *aversions* (§ 261); but it is in this form that they seem to act in the lower Animals and in young Children, whose minds are not yet fully developed into the stage of Ideational consciousness. The various terms *like* and *dislike*, *partiality* and *distaste*, *love* and *hatred*, which we use to signify the modes in which we ourselves feel affected by external objects, indicate the existence of this elementary form of Emotional sensibility in connection with the Perceptive consciousness.—There are other Emotional states, some of them rising to the intensity of *Passions*, which seem to belong to this category; but the examples already cited are sufficient to illustrate the principle, that the *elementary forms of Emotion* belong to the *Perceptional* stage of consciousness.

190. So, too, there seems to lie in this Perceptional stage of Mental activity, the germ which, in a higher phase of development, is evolved into the *Moral* sense. Experience shows, as Dr. J. D. Morell justly remarks, "that an Instinctive *apprehension* of 'right' and 'wrong,' as attached to certain actions, precedes in the child any distinct *comprehension* of the language by which we convey Moral truths. Moreover, the power and the purity of Moral feeling not unfrequently exist even to the highest degree, amongst those who never made the question of Morals in any way the object of direct thought, and may perchance be unconscious of the treasure they possess in their bosoms." Of these elementary Moral feelings, those of the lower Animals which associate most closely with Man are obviously capable. The *sense of duty* towards a being of a higher nature, which shows itself in the *actions* of the young Child towards its Parent or Nurse, long before any Ideational comprehension of it can have been attained, is exactly paralleled by that of the Dog or the Horse towards its Master. "Man," as Burns truly said, "is the God of the Dog." It is the substitution of the *superior* for the *inferior* directing principle, the distinct Intellectual comprehension of it, and the Volitional direction of the Attention to it, which constitutes the essential difference between the most conscientious effort of the enlightened Christian, and the honest and self-sacrificing response to his sense of Duty, which is seen in the Horse that falls down dead from exhaustion after putting forth his utmost power at the behest of his rider, or

in the Dog who uses his utmost skill and intelligence in seeking and collecting his master's flock (§ 92).—The elementary form of the Religious sense appears to connect itself, not merely with that simple apprehension of a Power external to ourselves which comes to us from the recognition of its manifestations, but with those feelings of Awe and Solemnity which are directly excited by objects of sublimity, grandeur, and mystery. Its higher development, however, requires an Ideational exercise of the Mind; and with this are connected Emotional states of a more elevated character (§§ 213—215).

191. *Instinctive Movements.*—It has been already shown that the Instinctive actions of the lower Animals may be regarded as constituting the *direct and immediate response* of their Nervous Mechanism to the impressions made upon it; and that there is reason to believe that, in some instances at least, this mechanism has shaped itself in accordance with the manner in which it has been habitually called into activity (§§ 84, 93). Now there are, perhaps, no movements in Man of a higher character than those immediately related to the maintenance of his Organic Functions (§ 32), which *originally* have this character; but there is a very large class in which the immediate response *comes* to be made, in consequence of the habitual "training" of the Automatic mechanism to a certain sequence of movements, under the direction of a certain sequence of Sense-perceptions.

192. One of the most universal of these *secondarily automatic* actions is that of *walking erect;* for which the whole Human organization is so obviously adapted, that it seems probable that every Child would acquire the habit *proprio motu,* without either teaching or example. But this acquirement depends upon the establishment of a very complicated set of relations between Sense perceptions and respondent Muscular contractions; in virtue of which the latter come to be instinctively prompted by the former. Thus the effort needed for the mere *support* of the body is

ordinarily kept up by the "muscular sense;" of which, indeed, we only become cognizant when our attention is directed to it; but the necessity for which is evidenced by the fact, that if the *sensory* nerve of a limb be paralysed, the contraction of its muscles cannot be sustained by the strongest exertion of the Will, unless the Sight be used to replace the lost Feeling (§ 80). The existence of this partial paralysis may sometimes be recognized by the persistent *looking-downwards* of those who suffer from it; for if, whilst walking, they were to withdraw their eyes from their feet, their legs would at once give way under them. In the ordinary *balancing* of the body, our movements are still prompted in great degree by the Muscular sense; and this is alone sufficient to the blind, as it is to the seeing man when walking in the dark. It frequently happens, indeed, that Vision, instead of aiding and guiding, brings to us sensations of an antagonistic character; and our movements then become uncertain, from the loss of that power of control over them, which the harmony of the two sensations usually affords. Thus a person unaccustomed to look down heights, *feels* insecure at the top of a tower or a precipice, although he *knows* that his body is properly supported; for the void which he sees below him contradicts (as it were) the Muscular sense by which he is made conscious of its due equilibrium. So, again, although any one can walk along a narrow plank which forms part of the floor of a room, or which is elevated but little above it, without the least difficulty, and even without any consciousness of effort, yet if that plank be laid across a chasm, the bottom of which is so far removed from the eye that the Visual sense gives no assistance, even those who have braced their nerves against all Emotional distraction, feel that an effort is requisite to maintain the equilibrium during their passage over it; that effort being aided by the withdrawal of the eyes from the depth *below*, and the fixation of them on a point *beyond*, which at the same time helps to give steadiness to the movements, and distracts the mind from the sense of its danger. On the other hand, the sufficiency of the Muscular sense, when the Mind has no consciousness of the danger, and when the Visual sense neither affords aid nor contributes to distract the attention, is remarkably illustrated by the phenomena of Somnambulism; forthe sleep-walker traverses, without the least hesitation, the narrow parapet of a house, crosses narrow and insecure planks, clambers roofs, &c.—But how soon a new co-ordination of this kind can be acquired, is shown (as Mr. H. Mayo pointed-out) by what happens to a landsman on first going to sea. "It is long before the passenger acquires his 'sea-legs.' At first, as the ship moves, he can hardly keep his feet; the shifting lines of the vessel and surface of the water unsettle his Visual stability; the different inclinations of the planks he stands-on, his Muscular sense. In a short time, he learns to disregard the shifting images and changing motions, or acquires facility in adapting himself (like one on horseback) to the different alterations in the line of direction in his frame." And when a person who has thus learned by habit to maintain his equilibrium on a shifting surface, first treads upon firm ground, he feels himself almost as much at fault as he did when he first went to sea: and it is only after being some time on shore, that he is able to resume his original manner of walking. Indeed, most of those who spend the greater part of their time at sea, acquire a peculiar gait, which becomes so habitual to them, that they are never able to throw it off.

193. Not less universal, in the ordinarily constituted Human being, is some definite form of *Vocalization;* requiring a very exact and complicated co-ordination between the Respiratory movements, and those of the Larynx, the Tongue, and the Lips, which is ordinarily directed by the sense of Hearing. This co-ordination is acquired, in the first instance, under the guidance of the Sounds actually heard; but, when subsequently called into action volitionally, it depends on the presence of a mental conception

(or internal sensation) of the tone to be uttered,—save in those cases in which a special training has brought "deaf-mutes" to regulate the action of their organs of Speech by the guiding sensations originating in the muscles themselves (§ 80). It is very rarely that a person who has once enjoyed the sense of Hearing, afterwards becomes so *completely* deaf, as to lose all auditory control over his vocal organs. An example of this kind, however, was communicated to the public by a well-known Author, as having occurred in himself; and the record of his experiences contains many points of much interest:—

The deafness was the result of an accident occurring in childhood, which left him for some time in a state of extreme debility; and when he made the attempt to speak, it was with considerable pain in the vocal organs. This pain probably resulted from the unaccustomed Muscular effort which it was necessary to make, when the usual guidance was wanting; being analogous to the uneasiness we experience when we attempt to move our eyes with the lids closed. His voice at that time is described as being very similar to that of a person born deaf-and-dumb, but who has been taught to speak. With the uneasiness in the use of the vocal organs, was associated an extreme mental indisposition to their employment; and thus, for some years, the voice was very little exercised. Circumstances afterwards forced it, however, into constant employment; and great improvement subsequently took place in the power of vocalization, evidently by attention to the indications of the Muscular sense. It is a curious circumstance fully confirming this view, that the words which had been in use previously to the supervention of the deafness, were still pronounced (such of them, at least, as were kept in employment) as they had been in childhood; the muscular movements concerned in their articulation being still guided by the original Auditory conception, in spite of the knowledge derived from the information of others that such pronunciation was erroneous. On the other hand, all the words subsequently learned were pronounced according to their spelling; the acquired associations between the Muscular sensations and the written signs being in this case the obvious guide. —(See Dr. Kitto's *Lost Senses*, vol. i., chaps. 2, 3.)

194. The extraordinary *adaptiveness* of the Organism of Man, is shown in his power of acquiring a vast number of more *special* actions, which have no direct relation to his bodily wants, but minister to requirements of his own creation. These often become, by a process of prolonged "training," not less *automatic* than the act of walking; as is shown by the fact that, when once set going, they will continue in regular sequence, not only without any Volitional exertion, but whilst the Attention is wholly directed elsewhere. Thus a Musical performer will play a piece which has become familiar by repetition, whilst carrying on an animated conversation, or whilst continuously engrossed by some train of deeply interesting thought; the accustomed sequence of movements being directly prompted by the *sight* of the notes, or by the remembered succession of the *sounds* (if the piece is played from memory), aided in both cases by the guiding sensations derived from the Muscles themselves. But further, a higher degree of the same "training" (acting on an Organism specially fitted to profit by it) enables an accomplished Pianist to play a difficult piece of music at sight; the movements of the hands and fingers following so immediately upon the sight of the notes, that it seems impossible to believe that any but the very shortest and most direct track can be the channel of the Nervous communication through which they are called forth. The following curious example of the same class of *acquired aptitudes*, which differ from Instincts only in being prompted to action by the Will, is furnished by Robert Houdin:—

With a view of cultivating the rapidity of visual and tactile Perception, and the precision of respondent Movements, which are necessary for success in every kind of "prestidigitation," Houdin early practised the art of juggling with balls in the air; and having, after a month's practice, become thorough master of the art of keeping up *four* balls at once, he placed a book before him, and, while the balls were in the air, accustomed himself to read without hesitation. "This," he says, "will probably seem to my readers very extraordinary; but I shall surprise them still more when I say that I have just amused myself with repeating this curious experiment. Though thirty years have elapsed since the time I was writing, and though I have scarcely once touched my balls during that period, I can still manage to read with ease while keeping *three* balls up."—(*Autobiography*, p. 26.)

This last fact appears to the Writer to be one of peculiar significance, for it seems to justify the conclusion, that even a most complex series of actions which essentially depends on guiding perceptions, may be performed by the automatic mechanism, without any other Volitional action than that which "starts" it, when once this mechanism has been developed by the *habitual exercise* originally imposed on the Nerve-centres by the Will. And further, it shows that this mechanism, having been originally so shaped *at an early period of life*, is kept up by Nutritive action, even though not called into use (§ 276); just as the "traces" of our early mental acquirements are persistently retained in our organism, long after we have lost the conscious Memory of them (§ 339).

195. To the same category as Instinctive movements, may be referred those movements of *expression*, which are automatically prompted by states of *feeling* connected with the Perceptional consciousness. These Movements are often more powerfully significant than any verbal language can be; for they convey the immediate experiences of the percipient mind, which have not been (and are often incapable of being) evolved into ideas, and thence translated into words (§ 198); and they are immediately or instinctively apprehended by other minds. It may be noticed that long before Children have attained to any comprehension of verbal language, they intuitively interpret the expressions of emotion, and are sympathetically affected by them; as seems the case, too, with regard to such of the lower Animals as habitually associate with Man, and thus acquire that sympathy with his emotional nature,

which enables them to recognize its manifestations. And they often reveal the state of Mind of the individual even *more truly* than his spoken words; being less under the control of his Will, which may use his Speech rather to conceal than to make known his thoughts.*

* The subject of the Movements of Expression being too large to be here discussed in detail, the reader who seeks further information upon it may be referred to the recent Treatise of Mr. Darwin, by whom it is handled with hi usual ability.

CHAPTER VI.

OF IDEATION AND IDEO-MOTOR ACTION.

SECTION I.—*Of Ideation Generally.*

196. IN ascending the scale of Psychical activity, we find the operations of the *Intelligent Mind* becoming more and more independent of the Sensorial changes which first excited them. It has been shown that in the first or *sensational* stage, the Consciousness is engrossed with *self,* not being as yet awake to the existence of any external cause for the *subjective* change it experiences; whilst in the second or *perceptive* stage, in which that objective cause is apprehended as something *not-self,* the Mind is entirely given-up to the contemplation of it, and recognizes its properties as the sources of the various affections it experiences. Some of these affections relate to *knowledge,* whilst others partake more of the nature of *feeling;* but in all of them the percipient mind is brought face to face, as it were, with the object perceived; and the knowledge which comes to us from this direct relation, whether through our *original* or our *acquired* intuitions, has a certainty to which no other kind of knowledge can lay claim. But it is not until the Mind attains a still higher kind of activity, that it forms that distinct *mental representation,* or *idea,** of the object, which stands altogether apart from our

* The Writer does not think it expedient to enter into the inquiry which has been the subject of so many abstruse and laboured Metaphysical discussions, as to whether our fundamental Ideas originate altogether *without,* or altogether *within,* the Mind; or partly without, and partly within. It will be sufficient for him to express his own conviction, that the latter is the view at which any Psychological inquirer *must* arrive, who looks at the subject from the Physiological side. An Idea can no more correctly be designated a "transformed sensation," than a Sensation could be designated "a transformed impression," or Muscular Con-

immediate experience, and assumes the character of an independent Intellectual reality. Thus *Ideation* forms, as it were, the climax of that reaction between the external world and the intelligent *Ego*, of which *sensation* and *perception* constituted the lower stages; and looking at the *Cerebrum* (as we seem justified in doing) as the instrument of that activity, we see how its operations, prompted in the first instance by changes in the Sensorium, may come to be entirely independent of them, by that singular power of recording ideational changes, which constitutes the Physiological basis of Memory (§ 344). And in all the higher intellectual operations, it is by its own ideational activity, rather than by sensorial promptings, that the further action of the Cerebrum is sustained.—In forming these "mental representations," the Mind is determined by the nature and intensity of the various affections of its consciousness which have been excited by the object; and as these depend in part upon the original constitution of the Cerebrum, and in part upon the mode in which its activity has been habitually exercised, it follows that the ideas of the same object or occurrence which are formed by different individuals, may be widely discrepant. This, indeed, continually proves to be the case; and we cannot have a better example of the fact, than is afforded by the variety of modes in which the same face or landscape shall be depicted by different Artists, each expressing in his peculiar "manner" that representation of the object which his Mind has formed. As Carlyle has well said, "The eye sees what it brings the power to see."

traction could be called a "transformed stimulation." The one is antecedent; the other is consequent. Just as an electrical or chemical stimulus, applied to a Muscle, calls it into contraction, so does the sensational stimulus, acting on the Cerebrum, excite the changes which give rise to the Ideational form of consciousness. On the other hand, to affirm that ideas are either "innate," or are in any way generated by the mind itself without original excitement by sensations *ab extra*, is a position so entirely inconsistent with experience, as not to bear any careful scrutiny.—For a concise view of the various doctrines which have been propounded on this subject, see Dr. J. D. Morell's "Elements of Psychology," pp. 269 *et seq.*

How much more the Artist's pencil is guided by his *mental* than by his *sensorial* view of certain objects, has been recently pointed out by Mr. Hamerton, who states it as a fact that every Landscape-painter *represents* mountains much higher than he *sees* them; as is shown by the comparison of his drawings either with photographs, or with tracings taken by a perspective apparatus (*Thoughts about Art*, p. 62).—Another departure from *visual* truth, for the purpose of producing *ideal* truth, is made by every Artist in his pictorial representation of the perpendicular lines of a building as vertical and parallel; notwithstanding that, as projected upon his retina, they converge towards a vanishing point in the sky.*

197. The influence either of preconceived *notions*, or of the *feelings* by which the Mind is habitually pervaded, may be continually recognized by the observant, as modifying the ideas which every one forms of what is presented to his observation: and it is by an exaggeration of such influences, in such as allow themselves to become "possessed" by "dominant ideas" without bringing them to the test of Common Sense, that those *mis*-representations come to be accepted as realities, which have the same source as the delusions of Insanity; differing from them only in their degree of fixity and intensity, and in the kind of influence which they exert over the conduct (§§ 187, 561).—This want of conformity between the *ideal* and the *actual* is peculiarly apt to arise in the minds of those who live too much in the former and too

* The Writer's statement on this point has been called in question, on the ground that a perspective projection *on a vertical plane* shows perpendicular lines as vertical and parallel. But when we are looking at a lofty building, like the west front of York Minster, we do not direct our eyes horizontally, but look towards a point some way up, so that *the retinal plane becomes oblique;* and what our visual picture really is under such circumstances, is proved by the unerring test of Photography. For, in taking a picture of a lofty building, the Photographer tilts his camera upwards, so that the plane of the picture becomes oblique; and in every Photograph thus taken, the perpendiculars of the building most unmistakably converge. Now when a pair of such Photographs, taken stereoscopically, are so viewed in the Stereoscope that their planes are brought into parallelism to that on which they were taken (which is easily done by *sloping* the pictures, so that their upper edges are brought nearer to the eye), the perpendicularity of the verticals is restored.

little in the latter; for in proportion as the Mind dwells too exclusively upon its own conceptions, and refrains from bringing these into contact with the realities of every-day life, do aberrations which would speedily be checked by experience, progressively gain a preponderating influence, until at last they may acquire the character of settled delusions, and may altogether upset the balance of the Intellect.

198. The whole tendency of the Ideational activity of the Mind, being thus to separate the "representation" formed by itself, from the restraints of outward experience, so as to make it a distinct and intelligible object of contemplation, it is requisite for the perfection of this *objectifying* process, that we should possess some mode of *signifying* our ideas, so that they may at the same time be made clear and distinct to ourselves, and be rendered intelligible to other minds. This may be accomplished by means of *signs* visible to the eye, or transmissible through the touch; or by means of *spoken language*, in which certain combinations of sounds are made to symbolize ideas.

The deaf-and-dumb are trained to communicate with each other, not merely by the "finger-language," by which words are *alphabetically* spelled, but also by the "sign-language," by which ideas are conveyed through the much more direct medium of *single signs.* These signs, though partly conventional, are made to conform as nearly as possible to the *natural* expressions of ideas; and are usually acquired very quickly by the deaf-and-dumb, whose want of other modes of utterance forces into activity a mode of expressing their ideas and emotions, which is unnecessary to those who have the command of language, and is consequently but little exercised by them. Young Children, however, who associate much with the deaf-and-dumb, very readily acquire this sign-language, and will often prefer the continued use of it to the acquirement of *spoken* language. —The inquiries of Mr. E. B. Tylor ("Researches into the Early History of Mankind," chaps. ii., iii.) have shown that the sign-language is very generally used among the least civilized Races; and that it presents such a remarkable uniformity among different

Families of Mankind, that it must be regarded as the most *natural* and *direct* mode in which ideas can be expressed.

The *range* of such signs, however, is necessarily very limited; and every Family of Mankind has substituted for them a set of *arbitrary* sounds, which are not only much more perfect in themselves as instruments for the expression of ideas, but are capable of being made to convey (by means of that wonderful apparatus of *articulation* with which Man is provided) an unlimited variety of meanings. In proportion as, by inflexion and combination, a verbal Language is capable of readily and precisely embodying the results of the Intellectual processes, in that proportion can these results be *objectified* by each individual, and be thus made the basis of further operations; and in the same proportion can they be clearly presented to the minds of others, and be employed by them for the same purpose. Thus whilst the structure of the Language of any people is to a certain extent a measure of its mental development, it comes to exert a most important influence over the further progress and direction of that development; different languages being in their very nature adapted for the expression, both of different *classes*, and of different *relations*, of Ideas, and having very different capacities for further development.

Some have maintained that Words which are used to designate external objects are the signs of those objects, and that such words form a class distinct from that of the words which stand as signs of abstract ideas. It is true that to the Child first learning the use of language, as among the lower Races of Mankind, every such noun is originally a *proper name*, standing as the symbol of the *individual* object with which it has become associated. But the Child is very early led by the familiar experiences of its nursery, to apply such words as chair, table, bed, to *classes* of objects, and thus to appreciate their significance as symbols of generalized or abstract ideas. And when that process has been accomplished in a few instances, the child's intellect soon extends it to others (its chief activity in this stage of its development being directed to the expansion and multiplication of its ideas); and thus—except in the case of *proper names* which are only applicable to individuals—*all* words come really to express *generalized representations* of the objects to which they refer. If, for example, we attempt to define the most familiar object, such as a house, a table, or a basket, by any verbal description, we find it extremely difficult to frame a definition that shall include *all* houses, *all* tables, *all* baskets; notwithstanding that our *idea* of a house, of a table, or of a basket, is sufficiently precise to enable us to say at once with regard to any particular object, whether it *does*, or *does not*, fall under one of these categories.

Hence Words do not appeal directly to the Intuitions of other minds, but must be comprehended by translation through their Ideational consciousness; signifying to each one the ideas he is prepared by his previous habits of thought to attach to them.

Thus every branch of Knowledge has its own Language, the terms of which, even when identical with words in ordinary use, can only convey their full and peculiar signification to those who have already gained an extensive acquaintance with the department of thought to which they relate. So, in rendering from one Language into another, great difficulty is continually experienced in the choice of words which shall convey in the translation the precise ideas signified in the original; the difficulty being greater in proportion to the diversity between the habits of thought of the two nations respectively. We can scarcely have a more "pregnant instance" of the obstruction thus created to the transmission of ideas through language, by the peculiarity of Scientific Terminology, in combination with diversity of National habitudes of thought, than is presented in the attempt to bring the abstract refinements of German Metaphysics within the comprehension of a "common-sense" English mind.

It is from their purely ideational significance, that, as expressions of *feeling*, words are often less potent than tones or gestures, which directly appeal to the emotional sensibility of the Percipient. And it is a striking testimony to the correctness of the view to be hereafter advocated in regard to the composite nature of the Emotions (§ 260), that they are most strongly excited by *language* that appeals to their *ideational* component, uttered in a *tone* and *manner* that calls forth the associated *feeling*.

199. There are certain Ideas which spring up in the Mind during the course of its own operations, whenever it *attends* to these; presenting themselves so universally, being so little subject to modification by peculiarities of individual character (whether original or acquired), and being so unhesitatingly recognized as "necessary" Truths, either when they spontaneously occur to ourselves, or are presented to our acceptance by others, that they take rank as *Primary Beliefs*, or *Fundamental Axioms*. Such are :—

I. The belief in our own *present existence*, or the faith which we repose in the evidence of Consciousness; this idea being necessarily associated with every form and condition of Mental Activity.

II. The belief in our *past existence*, and in our *personal identity* so far as our Memory extends (§ 364); with this, again, is connected the general Idea of *Time*.

III. The belief in the *external and independent existence* of the causes of our Sensations, leading to the recognition of the External World as distinct from the Ego; out of this arises the general idea of *Space*.

IV. The belief in the existence of an *efficient Cause* for the changes that we witness around us, which springs from the recognition of our own conscious agency in the production of such changes; whence is derived our idea of *Power*.

V. The belief in the *Uniformity of the Order of Nature*, or in the *invariable sequence* of similar effects to similar causes, which also springs from the perception of external changes, and is the foundation of all applications of our own experience, or of that of others, to the Conduct of our lives, or to the extension of our Knowledge.

VI. The belief in *our own free will*, involving the general idea of Volitional agency; which is in like manner a direct result of our recognition of a self-determining power (§ 5) within ourselves.

200. Again, those Axioms or first truths upon which the whole fabric of Geometry rests (such as "Things which are equal to the same thing, are equal to one another"), are statements of *universal fact, necessarily true under all circumstances;* which we unhesitatingly accept as such, because *any statement inconsistent with them would be inconceivable.* And so every step of a Mathematical or a Logical demonstration, which is based on such fundamental axioms, derives its validity from the fact, that either the *contrary* or *anything else* than the fact asserted is "unthinkable." Where each step is thus necessarily true to our Minds, the final Q. E. D. carries with it the same authority. So, too, the deliverances of our "Common Sense" (§ 378) derive their trustworthiness from what we consider the "self-evidence" of the propositions affirmed. Hence it is evident that "the only foundation of much of our belief, and the only source of much of our knowledge, is to be found in the Constitution of our own Minds."

201. The origin of these Primary Beliefs is one of the great Philosophical problems of our day, which has been discussed by Logicians and Metaphysicians of the very highest ability as leaders of opposing Schools, with the one result of showing how much can be said on each side.—By the *Intuitionalists* it is asserted that the tendency to form them is an intellectual instinct inborn in Man, an original part of his Mental organization; so that they grow up spontaneously in his mind as its faculties are gradually unfolded and developed, requiring no other experience for their genesis, than that which suffices to call these faculties into exercise. But by the advocates of the doctrine which regards *Experience* as the basis of all our knowledge, it is maintained that the primary beliefs of each individual are nothing else than generalizations which he forms of such experiences as he has either himself acquired, or has consciously learned from others; and they deny that there is any original or intuitive tendency to the formation of such beliefs, beyond that

which consists in the power of retaining and generalizing experiences.—A careful study, however, of the manner in which those Beliefs grow-up in our minds, seems to supply a means of reconcilement between these opposing doctrines. Even the generalization of actual experiences requires a certain preparedness of Intellect; and we can readily trace the growth of this in the Child, whose mind, like that of the untutored savage, dwells minutely on the *particular*, long before any idea of the *general* occurs to it; whilst a far higher development is required for it to pass from the general to the *universal*, to extend its conceptions from the experiential sphere of the *actual* to the imaginary range of the *possible*. And this development can only take place in a Mind which is continually acquiring new experiences; these being as necessary a *pabulum* to the mental organism, as *food* is to the bodily. But as the growth of the Body and the increase of its capabilities are dependent, not on the accumulation, but on the assimilation, of the food it has ingested, even so it is not in the accumulation of experiences, but in the increase of its capacity to deal with them, that the growth of the Mind essentially consists; of which capacity one most essential feature is the power of *direct apprehension* of truth. And in view of the many considerations hereafter to be adduced, no Physiologist can deem it improbable that the Intuitions which we recognise in our own Mental constitution have been thus acquired by a process of gradual development in the Race, corresponding to that which we trace by observation in the Individual.—That the great Master of the Experiential school, Mr. J. S. Mill, was latterly tending towards the acceptance of this view, will hereafter appear (p. 486).

The doctrine that the Intellectual and Moral Intuitions of any one Generation are the embodiments in its Mental constitution of the experiences of the Race, was first explicitly put forth by Mr. Herbert Spencer, in whose Philosophical Treatises it will be found most ably developed. But it had been distinctly foreshadowed as regards the

Instincts of animals (which are only lower forms of Man's intellectual Intuitions) by Sir John Sebright, Mr. T. A. Knight, and M. Roulin; of whose observations a summary has been given by the Writer in the "Contemporary Review," January, 1873. Sir John Sebright went so far as to express it as his decided conviction "that by far the greater part of the propensities which are generally supposed to be instinctive, are not implanted in animals by Nature, but are the results of long experience, acquired and accumulated through many generations, so as, in the course of time, to assume the characters of instinct." And in the Fourth and Fifth Editions of his "Human Physiology," published respectively in 1852 and 1855, the Writer had distinctly expressed his belief that the Cerebrum of Man *grows-to the modes of thought in which it is habitually exercised; and that such modifications in its structure are transmissible hereditarily.* (See § 838 of the Fourth Edition, and §§ 629, 630, of the Fifth Edition.) He here refers to this fact, merely to show that the *general doctrine* above enunciated (which he believes to have been held also by other Physiologists who had made Psychology their study, such as Sir H. Holland, Sir B. Brodie, and Dr. Laycock), is much older than Mr. Herbert Spencer.

202. We have an illustration of this progress in the fact of continual occurrence, that Conceptions which prove inadmissible to the minds of one generation, in consequence either of their want of Intellectual power to apprehend them, or of their pre-occupation by older habits of thought, subsequently find a universal acceptance, and even come to be approved as "self-evident." Thus the First Law of Motion, divined by the genius of Newton, though opposed by many Philosophers of his time as contrary to all experience, is now accepted by common consent, not merely as a legitimate inference from experiment, but as the expression of a necessary and universal truth. And the same axiomatic value is extended to the still more general doctrine, that Energy of *any* kind, whether manifested in the "molar" motion of masses, or consisting in the "molecular" motion of atoms, *must* continue under some form or other without abatement or decay; that which all admit in

regard to the indestructibility of Matter, being accepted as no less true of Force, namely, that as *ex nihilo nil fit*, so *nil fit ad nihilum.**

203. But, it may be urged, the very conception of these and similar great truths is in itself a typical example of Intuition. The men who divined and enunciated them stand out above their fellows, as possessed of a Genius which could not only combine but create, of an Insight which could clearly discern what Reason could but dimly shadow forth. Granting this freely, it may yet be shown that the Intuitions of individual Genius are but specially-exalted forms of endowments which are the general property of the Race at the time, and which have come to be so in virtue of its whole previous culture.—This appears readily capable of proof in the case of two forms of Mental activity, the tendency to which occasionally manifests itself so remarkably in individuals as a *congenital aptitude*, that it must be considered as embodied in their Constitution; and which are yet so completely the products of *culture*, that we are able to trace pretty clearly the history of their development. These are the Ideas which relate to *Number*, and those which relate to *Music*.

204. There can be no reasonable question that the definite Ideas which we now form of *numbers* and of the *relations of numbers* are the products of Intellectual operations based on experience. There are Savages at the present time, who cannot count beyond five; and even among races that have attained to a considerable proficiency in the arts of life, the range of numerical power seems extremely low. In Eastern nations generally, it would appear that *definite conceptions* of Number are more limited than their language implies; for in their descriptions of what they have themselves witnessed, they are in the habit of using what to *our*

* This is the form in which the doctrine now known as that of the "Conservation of Energy" was enunciated by Dr. Mayer, in the very remarkable Essay published by him in 1845, entitled "Die organische Bewegung in ihrem Zusammenhange mit dem Stoffwechsel."

"matter-of-fact" apprehension are ludicrous exaggerations in regard to numbers, although these descriptions would probably not convey any erroneous ideas to those for whom they were intended.* Although the ancient Greeks developed the science of Arithmetic up to a certain point, they were incapacitated from carrying it further by the clumsiness of their mode of expressing large numbers; which made it necessary for their higher computations to use symbols borrowed from Geometry—the science of Space; as when they spoke of the *square* or the *cube* of a number. It was the introduction into Europe, from India, of what we are accustomed to call the "Arabic notation," that gave an entirely new development to Arithmetical science; the essential features of this notation being the combination of the *local* value of each of the figures representing any number, with the *decimal* multiplication in the value given to them by their position. The science of Arithmetic, as at present existing, may be regarded as the accumulated *product* of the intellectual ability of successive generations; each generation building up some addition to the *knowledge* which it has received from its predecessor. But it can scarcely be questioned by any observant person, that an *aptitude* for the apprehension of numerical ideas has come to be embodied in the congenital Constitution of races which have long cultivated this branch of knowledge; so that it is far easier to teach Arithmetic to the child of an educated stock, than it would be to a young Yanco of the Amazons, who, according to La Condamine, can count no higher than *three*, his name for which is *Poettarrarorincoaroac.*

205. The most satisfactory evidence of the existence of a *numerical intuition*, or congenital aptitude for recognising the relations of Numbers, is furnished by the not unfrequent display of

* A very interesting example of this tendency was presented by the "Journal of Two Parsee Shipbuilders," who visited this country about forty years ago, and published their experiences for the benefit of their countrymen.

this faculty among Children; for, as the Writer is informed by a friend who has a large field of observation among Primary Schools in which "mental arithmetic" is cultivated, it often happens that individuals who have received very little instruction surpass their fellows in the quickness and accuracy of their replies to numerical questions proposed to them, though they cannot be brought to explain the processes by which they have worked-out their results. More remarkable instances, however, are presented by the occasional display of very extraordinary Arithmetical ability on the part of individuals, who, having received very little instruction, have not only anticipated, but have gone far beyond, any power derivable from instruction, in almost immediately arriving at the answers to questions, which, according to ordinary Arithmetical methods, would involve long computations of a very elaborate character. The case of Zerah Colburn, the son of an American peasant, is especially remarkable among these, not only for the immediateness and correctness with which he gave the answers to questions resolvable by simple but prolonged computation,—such as the product of two numbers, each consisting of two, three, or four figures; the exact number of minutes and seconds in a given period of time; the raising of numbers up to high powers; or the extraction of the square and cube roots;—but, still more, for his power of at once answering questions to which no rules known to Mathematicians would apply. It was when the lad was under six years of age, and before he had received any instruction either in writing or in arithmetic, that he surprised his father by repeating the products of several numbers; and then, on various arithmetical questions being proposed to him, by solving them all with facility and correctness. Having been brought over to London in 1812, at the age of eight years, his powers were tested by several eminent Mathematicians; among them Francis Baily, from whose account of him the following examples are selected:—

He raised any number consisting of *one* figure progressively to the *tenth* power; giving the results (by actual multiplication, and not by memory) *faster than they could be set down in figures* by the person appointed to record them. He raised the number 8 progressively to the *sixteenth* power; and in naming the last result, which consisted of fifteen figures, he was right in every one. Some numbers consisting of *two* figures he raised as high as the *eighth* power; though he found a difficulty in proceeding when the products became very large.

On being asked the *square root* of 106929, he answered 327, *before the original number could be written down.* He was then required to find the *cube root* of 268,336,125; and with equal facility and promptness he replied 645.

He was asked how many *minutes* there are in 48 years; and before the question could be written down, he replied 25,228,800, and immediately afterwards he gave the correct number of *seconds.*

On being requested to give the *factors* which would produce the number 247483, he immediately named 941 and 263, which are the *only two* numbers from the multiplication of which it would result. —On 171395 being proposed, he named 5×34279, 7×24485, 59×2905, 83×2065, 35×4897, 295×581, and 413×415.—He was then asked to give the factors of 36083, but he immediately replied that it had none, which is really the case, this being a prime number.—Other numbers being proposed to him indiscriminately, he always succeeded in giving the correct factors, except in the case of prime numbers, which he generally discovered almost as soon as proposed. The number 4,294,967,297, which is $2^{32} + 1$, having been given to him, he discovered (as Euler had previously done) that it is not the prime number which Fermat had supposed it to be, but that it is the product of the factors $6,700,417 \times 641$. The solution of this problem was only given after the lapse of some weeks; but the method he took to obtain it clearly showed that he had not derived his information from any extraneous source.

When he was asked to multiply together numbers both consisting of more than three figures, he seemed to decompose one or both of them into its factors, and to work with these separately. Thus, on being asked to give the square of 4395, he multiplied 293 by itself, and then twice multiplied the product by 15. And on being asked to tell the square of 999,999, he obtained the correct result,

999,998, 000,001, by twice multiplying the square of 37037 by 27. He then of his own accord multiplied that product by 49; and said that the result (viz. 48,999,902,000,049) was equal to the square of 6,999,993. He afterwards multiplied this product by 49; and observed that the result (viz. 2,400,995,198,002,401) was equal to the square of 48,999,951. He was again asked to multiply this product by 25; and in naming the result (viz. 60,024,879,950,060,025) he said that it was equal to the square of 244,999,755.

On being interrogated as to the method by which he obtained these results, the boy constantly declared that he did not know *how* the answers came into his mind. In the act of multiplying two numbers together, and in the raising of powers, it was evident (alike from the facts just stated, and from the motion of his lips) that *some* operation was going forward in his mind; yet that operation could not (from the readiness with which the answers were furnished) have been at all allied to the usual modes of procedure, of which, indeed, he was entirely ignorant, not being able to perform on paper a simple sum in multiplication or division. But in the extraction of roots and in the discovery of factors of large numbers, it did not appear that any operation *could* take place: since he gave answers *immediately*, or in a *very few seconds*, which, according to the ordinary methods, would have required very difficult and laborious calculations; and *prime* numbers cannot be recognized as such by any known rule.

It is remarked by Mr. Baily that the same faculty, improved by cultivation, appears to have been possessed by the illustrious Euler; who had not only a most extraordinary *memory* for numbers—to the extent, it is said, of being able to recall the first six powers of any number under 100,—but also a kind of *divining power*, by which "he perceived, almost at a glance, the factors of which his formulæ were composed; the particular system of factors belonging to the question under consideration; the various artifices by which that system might be simplified and reduced; and the relation of the several factors to the conditions of the hypothesis."—This power of *divining* truths in advance of existing knowledge, is the special attribute of those Mathematicians who have done most for the development of their science. A notable instance of it is furnished by the celebrated formula devised by Newton for the solution of equations; for although its correctness was proved experientially by the results of its application in every conceivable variety of case, its *rationale* seems to have been unknown to Newton himself, and remained a puzzle to succeeding Mathematicians, until discovered by the persevering labours of Professor Sylvester, who is himself specially distinguished for the possession of this highest form of Mathematical genius.—That such a power as Zerah Colburn's should exist in a child who had never been taught even the rudiments of Arithmetic, seems to point (as Mr. Baily remarks) to the existence of *properties of numbers* as yet undiscovered, somewhat analogous to those on which the system of Logarithms is based. And if, as he grew older, he had become able to make known to others the methods by which his results were obtained, a real advance in knowledge might have been looked for. But it seems to have been the case with him, as with George Bidder and other "calculating boys," that with the *general* culture of the mind, this *special* power faded away.

206. The development of the Science and Art of *Music* has been even more recent. Whatever may have been the advances made in early times towards the "scale" of notes which all civilized Races now accept as the basis alike of Melody and of Harmony, it is pretty certain that the ancients cultivated Melody (or the *succession* of notes) exclusively, and that Harmony (or the combination of *simultaneous* tones) is of quite modern origin, the first indications of such combination not showing themselves until the Middle Ages. It was not, indeed, until the 16th century, that the system of *counterpoint*, or the arrangement of separate "parts" in harmony, was developed; and although this rapidly attained a high degree of perfection, as regards both Vocal and Instrumental music, the art of *orchestration*—that is, the use, either in combination or in contrast, of Instruments of different capacities and

qualities of tone, so as enormously to increase the range and variety of musical effects—is the product of the 18th century. Now whilst, as in the case of Number, the Musical science of any given period is the general expression of the accumulated *knowledge*, based on *experience*, of those who had devoted themselves to its culture in previous generations, there have arisen, from time to time, individuals in whom there has obviously been not merely a congenital aptitude for the *acquirement* of the Musical knowledge previously attained, but a power of *anticipating*, without any experience of their own, the results at which their predecessors had arrived, and then of *creating* forms of Musical thought entirely new, which have served as standards or models for those who have come after them.—No more remarkable instance of this kind could be adduced, than that which is presented by the short but brilliant career of Mozart :* and this will also furnish illustrations of the *spontaneous* working of Genius of the highest order, trained and disciplined by the most thorough Culture (§§ 232, 400).

The father of Mozart was not only an excellent performer on the violin (for which instrument he produced a Method that was long esteemed the best of its kind), but was well skilled in the Theory of Music, and wrote in various styles with no inconsiderable success. Of his seven children, only two survived the period of infancy; Anna Maria (born Aug. 29, 1751), and Wolfgang (born Jan. 27, 1756), who was four years and a half younger than his sister. That the girl inherited considerable musical ability, appears from the fact that at seven years old she was her father's pupil on the clavier (the early form of pianoforte), at which her progress was great and uniform; that when on the musical tour which she made with her brother, her performance was considered only less wonderful than his; and that she finally gained the highest reputation that any female performer on a keyed instrument had at that time acquired. She seems, however, to have been altogether destitute of the *inventive* faculty by which her brother was pre-eminently distinguished.

* The materials of the following sketch are chiefly derived from the admirable "Life of Mozart," by Edward Holmes, London, 1845.

At the time that his sister was commencing clavier practice, Wolfgang, then three years old, "was a constant attendant on her lessons; and already showed, by his fondness for striking thirds, and pleasing his ear by the discovery of other harmonious intervals, a lively interest in Music. At four he could always retain in memory the brilliant solos in the Concertos which he heard; and now his father began, half in sport, to give him lessons. The musical faculty seems to have been intuitive in him; for in learning to play, he learned to compose at the same time: his own nature discovering to him some important secrets in melody, rhythm, symmetry, and the art of setting a bass. To learn a minuet, he required half an hour, for a longer piece an hour; and having once mastered them, he played them with perfect neatness and in exact time. His progress was so great, that at four years of age, or earlier, he composed little pieces, which his father wrote down for him." From four to six years old, he was continually exercising himself in this manner, and acquired great experience in design and modulation; so that there could be no longer a doubt of the extraordinary precocity of his Musical genius.

"His desire of knowledge was great on all subjects; but in Music he astonished his teacher, not so much by an avidity for information, as by the impossibility of telling him anything which he did not know before. At the age of six years, Mozart knew the effect of sounds as represented by notes, and had overcome the difficulty of composing unaided by an instrument. Having commenced composition without recourse to the clavier, his powers in mental music constantly increased, and he soon imagined effects of which the original type existed only in his brain."

An incident which occurred at Wassenburg when the boy, not yet eight years old, first tried an organ with pedals, is thus narrated in one of his father's letters:—"To amuse ourselves, I explained the pedals to Wolfgang. He began immediately *stante pede* to try them, pushed the stool back, and preluded standing and treading the bass, and really as if he had practised many months. Every one was astonished; this is a new gift of God, which many only attain after much labour." This is the more remarkable, as not merely the execution, but the style suitable to it, must have been new to the juvenile musician.

When young Mozart was nearly fourteen, his father took him to

Italy for about sixteen months; and this tour seems to have had a considerable influence on his musical development. "In a country which was pre-eminently the seat of excellence in the fine arts, and where to excite admiration was proportionably difficult, his progress was a perpetual ovation. Under these circumstances, his genius was in a state of peculiar exaltation; *for sympathy*, it is to be observed, *was the atmosphere of his artistic existence;* and he could neither play nor compose to his own satisfaction, without the consciousness of being enjoyed and appreciated. But the stamp of his great individuality as a dramatic musician was not as yet visible." In his Church music he seems to have followed the dictates of his artistic feelings and musical science; but in writing for the Theatre he at first aimed chiefly at gaining success by consulting the taste of his audience.—"In reviewing the numerous instrumental Compositions of Mozart's youth, we are struck with the effort he made to master his ideas. The Quartett and Symphony productions of this period show many beautiful thoughts not yet turned to due account, but which he resumed and more fully developed in subsequent compositions. Thus his memory in after-life became a perfect storehouse of melodies and subjects which had long been floating n his imagination, and which his exquisite tact and judgment enabled him instantly to apply. We find this particularly in his Operas and Symphonies."

It was in 1780, when Mozart was in his twenty-fifth year, that he produced the Opera of Idomeneo; the first of that series of great Dramatic works, which have retained a permanent place to the present time in the estimation of all true Musicians. Up to this period, in attempts at dramatic composition, he had followed existing models; but in "Idomeneo" he asserted his independence of them, and developed modes and powers of musical expression, which took the most cultivated musicians by surprise, and have ever since been accepted as true and appropriate. "Youthful fire and invention," says Mr. Holmes, "were never so happily tempered by consummate experience." The performers, who had brought tone and facility of execution upon their instruments to great perfection, but had never been *animated* by what they played, "were awakened by the magic touch of genius to a new life in their art; they found themselves discoursing in an unheard and rapturous language; and the effect upon them was one of intoxication and enchantment."—Though it is usual to assign to Haydn that development of the powers and capacities of the different instruments of the Orchestra, which unquestionably constitutes the distinguishing feature of the Music of this epoch, yet competent critics maintain that the basis of this development was clearly laid in "Idomeneo," which was produced several years before the great Symphonies of Haydn. The position to which Mozart was at once raised by its production, as the greatest of dramatic composers, was made still more glorious by the immortal works that followed it, "Le Nozze di Figaro" and "Don Giovanni:" of the former of which it has been said that "while all the comic operas coeval with it are lost, not a note of that composition has faded, so that when reproduced it still finds as many enthusiastic admirers as a Comedy of Shakspere;" while the latter still "stands alone in dramatic eminence, combining the labour of the greatest melodist, symphonist, and master of dramatic expression ever united in the same individual."

But even these grand works, each of which occupied only a few weeks in its composition, constitute only a small part of the productions poured forth from the pen of Mozart, which seemed to be an inexhaustible fountain of music of the most varied character. There are scarcely any of the "unconsidered trifles" which he briefly gave forth, sometimes as the mere overflowings of his inventive faculty, that do not bear the stamp of his genius; while every one of those which he purposely elaborated with all the resources of his art, such as his Quartetts and Quintetts, his Symphonies, and above all the "Requiem," would of itself, if it stood alone, have marked an era in Musical history. "These works," it has been said, "show the variety of powers that Mozart brought to composition: the great organist and contrapuntist—the profound master of harmony and rhythm, are there—but taste and imagination ever preside. The quality of these productions can, in fact, only be estimated by the attempts which musicians have been making ever since to attain some credit in the same path." Like other works in advance of their time, however, they were not at first appreciated. The Six Quartetts dedicated to Haydn, for example,—in which Mozart, making use of the constructive skill which he had learned from the works of the same kind previously produced by his great contemporary, advanced beyond him in the invention of new harmonic combinations,—were repudiated by many musicians as full of unauthorised innovations; the Italians, in particular, imputing to mistakes of the engraver what *they* regarded as grievous blemishes, though *now* accepted as the greatest beauties of these fascinating compositions.

The most remarkable evidence of the *fertility* and *versatility* of Mozart's creative power, is furnished by the closing part of his history. It was soon after he commenced the "Zauberflöte," that, in an interval of depression which marked the commencement of the malady that terminated his life before the attainment of his thirty-sixth year, he composed the "Ave Verum;" a short strain of calm but elevated devotion, which has nowhere its equal for its combination of expressive beauty, religious feeling, and scientific skill. He then resumed the "Zauberflöte," and had nearly finished it, when he undertook the "Requiem;" having, as he told his wife, a desire to produce a work in which he could develope the elevated and the pathetic in Church music to the highest degree. The "Zauberflöte" was put aside, and the "Requiem" was begun; but he had not proceeded far, when his further progress was interrupted by a commission to compose the opera "La Clemenza di Tito" for the Coronation of the Emperor Leopold at Prague; and though this was completed within the wonderfully short space of *eighteen days*, he astonished his friends at whose house he was staying, by also producing the beautiful Quintett in the first act of the "Zauberflöte," the subject of which had come into his mind while he was playing a game of billiards, and had been at once noted down in a memorandum-book of "musical ideas" which he carried with him. On his return to Vienna, he completed and produced the "Zauberflöte;" and then, while stricken down by mortal disease, resumed the "Requiem," which he did not live entirely to complete, but in which, according to the judgment of all cultivated Musicians, there is a more wonderful combination of sublimity with pathetic beauty, than is to be found in any other Ecclesiastical composition, whether of earlier or later date.

207. In each of the foregoing cases, then, we have a typical example of the possession of an extraordinary congenital aptitude for certain forms of Mental activity; which showed itself at so early a period as to exclude the notion that it could have been

acquired by the experience of the individual; and which, in the case of Mozart, led its possessor far beyond the accumulated experience of his predecessors. To such congenital gifts we give the name of *Intuitions*; and it can scarcely be questioned that, like the *Instincts* of the lower Animals, they are the expressions of constitutional tendencies embodied in the organism of the individuals who manifest them. But whilst extraordinary in *degree*, they were not so in *kind*; for Zerah Colburn's faculty for numbers only placed him on the level of those who had previously attained the same results; and the creations of Mozart's genius, even when they passed the previous boundaries of musical thought, were soon appreciated by those who had already reached them. And it can scarcely be conceived that a Zerah Colburn could suddenly arise in a race of savages who cannot count above five; or that an infant Mozart could be born amongst a tribe, whose only musical instrument is a tom-tom, whose only song is a monotonous chant.

208. Again, by tracing the *gradual genesis* of some of those Intellectual Ideas which we now accept as "self-evident,"—such, for example, as that of the "Uniformity of Nature"—we are able to recognize them as the expressions of certain tendencies, which have progressively augmented in force in successive generations, and now manifest themselves as Mental Instincts that penetrate and direct our ordinary course of thought (§ 199). Such instincts constitute a precious heritage, which has been transmitted to us with ever-increasing value through the long succession of preceding generations; and which it is for us to transmit to those who shall come after us, with all that further increase which our higher culture and wider range of knowledge can impart.

209. In a similar light we are probably to rank those elementary notions of Truth, Beauty, and Right, which present themselves to our consciousness in connection with certain Ideational conditions respectively adapted to excite them; the first being associated especially with the operations of the Reason, the second with those

of the Imagination as directed by the Æsthetic sense, and the third with the determination of the Will in the regulation of conduct under the guidance of the Moral sense.—*Truth* may be defined to be an apprehension of the relations of things as they actually exist; and the conception of Truth, which is originally based upon Sensational Ideas, comes to be also applied to those which are purely Intellectual.—The notion of *Beauty*, the germ of which, as we have seen (§ 188), exists in the Perceptive consciousness, is one that is very difficult to define; but it seems to consist, when fully developed, in the conformity of an external object to a certain ideal standard, by which conformity a pleasurable feeling is produced. That ideal standard is a work of the Imagination, and is generated (by a kind of automatic process, § 412) by the elimination of all those elements which we recognize as inferior, and by the intensification and completion of all those which we regard as excellent. Hence according to the Æsthetic judgment which every individual pronounces as to these particulars, will be his idéal of Beauty; and although this judgment is subject to so wide a variation, that the uselessness of disputing about matters of Taste has become proverbial, yet a gradual approximation to agreement shows itself among those who are distinguished by the possession of a high measure of the Æsthetic sense, and who have cultivated it by the intelligent study of what, by common consent, are regarded as the noblest works of art. In fact, it is from the careful scrutiny of the products of the highest Genius (§ 409), that the *rules of art*, alike in Poetry and Music, in Painting and Sculpture, are derived. The notion of Beauty extends itself also to the pure conceptions of the Intellect; and thus we may experience the sense of Beauty in the recognition of a Truth. We experience the sense of Beauty, too, in witnessing the conformity of conduct to a high standard of Moral excellence; which excites in our minds a pleasure of the same order as that which we derive from the contemplation of a noble work of Art.—The notion of

Right, which is purely Ideational, connects itself with Voluntary action. We have no feeling of approval or disapproval with respect to actions that are necessarily connected with our Physical well-being; but in regard to most of those which are left to our choice, it is impossible to feel indifferent; and the sphere of operation of this principle becomes widened, in proportion as the mind dwells upon the notion of Moral obligation which arises out of it. Then, too, the idea of Right is brought to attach itself to *thoughts* as well as to *actions;* and this, not merely because the right regulation of the thoughts is perceived to be essential to the right regulation of the conduct, but also because *whatever* we can govern by the Will may present itself to the Mind in a Moral aspect.

210. It has been usually considered by Moralists and Theologians, that *Conscience*, or the *Moral sense*, is an "autocratic" faculty, which unmistakeably dictates *what is right* in each individual case, and which should consequently be unhesitatingly obeyed as the supreme and unerring guide. Now this view of the case is attended with practical difficulties, which make it surprising that it can ever have been entertained. For it must be obvious to every one who carefully considers the matter, that whilst *a notion of right as distinguished from wrong*, attaching itself to certain actions, is as much a part of the Moral nature of every individual, as the *feeling of pleasure* or *pain* attaching itself to certain states of consciousness is of his Sensational nature, yet the determination of *what* is right, and *what* is wrong, is a matter in great degree dependent upon race, education, habits of thought, conventional associations, &c.; so that the Moral standard of no two men shall be precisely alike, while the moral standards of men brought up under entirely different circumstances shall be of the most opposite nature. (Without having recourse, for an illustration of this position, to the strange estimates of right and wrong which prevail amongst Savage nations, it may be sufficient to

refer to the different views which used to be *conscientiously* entertained on the question of Slavery, by high-minded, estimable, and Christian men and women in different parts of the American Union.) Moreover, in what have been designated as "cases of conscience," the most enlightened Moralist may have a difficulty in deciding what is the right course of action, simply because the Moral sense finds so much to approve on both sides, that it cannot assign a preponderance to either. Thus, individuals in whose characters the love of *truth* and of *justice* and the *benevolent* affections are the prominent features, and who would shrink with horror from any violation of these principles of action for any selfish purpose whatever, are sorely perplexed when they are brought into collision with each other; a strong motive to tell a falsehood (for example) being presented by the desire to protect a defenceless fellow-creature from unmerited oppression or death.

211. It is evident, then, that the determination of *what* is *right* and *what* is *wrong* in any individual case, must be a matter of *judgment*; the rule of Moral action being based on a comparison of the *relative nobility of the motives* which impel us to either course, and being decided by the preference which is accorded to one motive or combination of motives above another. As Mr. Martineau has well said,* "Every Moral judgment is *relative;* and involves a comparison of (at least) two terms. When we praise what *has been* done, it is with the co-existent conception of *something else that might have been* done; and when we resolve on a course as *right*, it is to the exclusion of some other that is *wrong*." If it be asked, how are the relative values of these motives to be decided, the answer must be sought in the Moral consciousness of Mankind in general; which is found to be more and more accordant in this respect, the more faithfully it is interpreted, the more habitually it is acted-on, and the more the whole Intelligence is expanded and enlightened.—It is this tendency towards universal agree-

* "Prospective Review" for November, 1845, pp. 587-9.

ment, which shows that there is really as good a foundation for Moral Science in the Psychical nature of Man, as there is for that of Music in the pleasure which he derives from certain combinations and successions of Sounds.

212. On the other hand, as we cannot attach any Moral character to the actions of Animals that are performed under the direction of a blind undiscerning Instinct, leaving them no choice between one course and another, neither can we attach it to those which are executed by Human beings, even when possessed of their full Intelligence, who are dominated by impulses which they have it not in their power to restrain (§ 264); nor, again, to those performed by individuals whose Moral sense has either been never awakened, or has been so completely misdirected by early education, that their standard of right and wrong is altogether opposite to that which the enlightened Conscience of Mankind agrees in adopting (§ 8). But, although there are doubtless many cases in which Criminal actions are committed under the impulse of passions (such as anger, lust, &c.) which the individual has not *at the moment* the power to control, so that he must be absolved from Moral responsibility *quoad* the immediate impulses to those particular actions, yet these impulses too frequently derive all their force from the habit of yielding to their promptings in lesser matters, which gradually gives them a "dominance," such as the Will (weakened by want of exercise in the habit of self-restraint) is unable to resist (§ 287). Hence the Criminal *action* is to be regarded as but the expression of a long previous course of Criminal *thought*, for which, in so far as he could have otherwise directed it, the individual may legitimately be held responsible,—just as he is for actions committed in the state of Intoxication, in which he has temporarily lost, by his own voluntary act, the power of self-control.

213. Closely connected with many of the foregoing Tendencies to Thought, and arising in most minds from some or other of them

by the very nature of our Psychical constitution, are those Ideas which relate to the Being and Attributes of the Deity. There is, in fact, no part of man's Psychical nature, which does not speak to him, when it is rightly questioned, of something beyond and above himself. The very perception of *finite* existence, whether in Time or Space, leads to the idea of the Infinite. The perception of *dependent* existence leads to the idea of the Self-existent. The perception of *change* in the Universe around leads to the idea of an unseen Power as its cause. The perception of the *order* and *constancy* underlying all those diversities which the surface of Nature presents, leads to the idea of the Unity of that power. The recognition of Intelligent Will as the source of the power we ourselves exert, leads to the idea of a like Will as operating in the Universe. And our own capacity for Reasoning, which we know not to have been obtained by our individual exertions, is a direct testimony to the Intelligence of the Being who implanted it.—So are we led from the very existence of our Moral feelings, to the conception of the existence of attributes, the same in kind, however exalted in degree, in the Divine Being. The sense of Truth implies its actual existence in a being who is Himself its source and centre; and the longing for a yet higher measure of it, which is experienced in the greatest force by those who have already attained the truest and widest view, is the testimony of our own souls to the Truth of the Divine Nature. The perception of Right, in like manner, leads us to the Absolute lawgiver who implanted it in our constitution; and, as has been well remarked, "all the appeals of innocence against unrighteous force are appeals to eternal justice, and all the visions of moral purity are glimpses of the infinite excellence." The aspirations of the more exalted Moral natures after a yet higher state of holiness and purity (§ 97), can only be satisfied by the contemplation of such perfection as no merely Human being has ever attained; and it is only in the contemplation of the Divine Ideal that they meet their appropriate object. And the sentiment of Beauty, especially as it rises from the *material* to the *spiritual*, passes beyond the noblest creations of Art and the most perfect realization of it in the outward life, and soars into the region of the Unseen, where alone the Imagination can freely expand itself in the contemplation of such beauty as no objective representation can embody.—And it is by combining, so far as our capacity will admit, the ideas which we thus derive from reflection upon the facts of our own consciousness, with those which we draw from the contemplation of the Universe around us, that we form the justest conception of the Divine Being of which our finite minds are capable. We are led to conceive of Him as the absolute, unchangeable, self-existent,—infinite in duration,—illimitable in space,—the highest ideal of Truth, Right, and Beauty,—the all-Powerful source of that agency which we recognize in the phenomena of Nature,—the all-Wise designer of that wondrous plan, whose original perfection is the real source of the uniformity and harmony which we recognise in its operation,—the all-Benevolent contriver of the happiness of His sentient creatures,—the all-Just disposer of events in the Moral world, for the evolution of the ultimate ends for which Man was called into existence. In proportion to the elevation of our own spiritual nature, and the harmonious development of its several tendencies, will be the elevation and harmoniousness of our conception of the Divine; and in proportion, more particularly, as we succeed in raising ourselves towards that ideal of perfection which has been graciously presented to us in the "well-beloved Son of God," are the relations of the Divine Nature to our own *felt* to be more intimate. And it is from the consciousness of our relation to God, as His creatures, as His children, and as independent but responsible fellow-workers with Him in accomplishing His great purposes, that all those ideas and sentiments arise, which are designated as Religious, and which constitute that most exalted portion of our nature.

214. The pervading consciousness of that relation expresses itself in the notion of *Duty*; which attaches itself to every action as to which the Ego may believe that the Divine Will has been expressed. But the dictates of this sense will vary with the ideas entertained respecting the Divine character and requirements; and actions may be sincerely regarded as an acceptable sacrifice by one class of religionists, which are loathed as barbarous and detestable by another. Moreover, the difficulty which attends the determination of what is Morally right (§ 210), often occurs in regard to Religious duty; each of two or more possible modes of action being recommended by its conformity to the Divine law on certain points, whilst it seems opposed to it on others.

Thus if a man who might be urged to conceal a Political refugee in immediate danger of capture, were to refuse to do so merely on the fear of unpleasant consequences to himself, he would be justly branded with the character of a cold-hearted coward; but if his refusal should proceed from the conviction that the Divine Law requires the preference of rigid Truthfulness over every other motive, and that, by concealing the suppliant, he should be forced into a violation of that law, he cannot be blamed even by those who believe that the Law of Compassion "written upon our hearts" is at least equally imperative.—Similar difficulties beset the upholders of the non-resistance creed, which teaches that *love* is the all-powerful principle in the Moral world, and that it should entirely supersede all those lower impulses of our nature which lead us to oppose force to force, and to resist an unjust and unprovoked assault. Here, again, we might readily understand and sympathise with those, who consider that the fear of personal suffering does not warrant our doing a severe injury to another in warding-off a threatened attack; but when the question comes to be, not of *self*-defence, but of protection to *others* who are helpless dependents upon our succour, and who are bound to us by the closest ties of natural affection, we feel that the comparative nobility of the latter motive warrants actions which our individual peril might scarcely justify.

215. But as in Morals, so in Religion, does it become increasingly

obvious, that the more elevated are the ideas of Mankind in regard to the character and will of the Deity, the more do they approach to a general accordance in regard to what constitutes Religious duty; and the complete coincidence which is thus found to exist between the dictates of the Christian law and the highest principles of pure Morality, should prevent one set of motives from ever coming into antagonism with the other.—The *Conscience* of the religious man, indeed, may be said to be the *resultant* of the combination of his Moral sense with the idea of Duty which arises out of his sense of relation to the Deity. With the former are closely associated all those emotions and dispositions, which render him considerate of the welfare of his fellow-men, as of his own; and with the notion of duty to God are closely united the desire of His favour, the fear of His displeasure, the aspiration after His perfection, all which act like other motives in deciding the Will. Their relative force on any occasion, as compared with that of the lower propensities and sensual desires, greatly depends on the degree in which they are *habitually* brought to influence the mind; and it is in its power of fixing the attention on those higher considerations which ought to be paramount to all others, and of withdrawing it from the lower, that the Will has the chief influence in the direction of the conduct according to the dictates of virtue (Chap. IX.).

SECTION 2.—*Succession of Ideas:—Laws of Thought.*

216. The conscious Mind, when not engrossed in Sensational or Perceptive acts, is incessantly occupied in *thinking*, with or without the accompaniment of *feeling*; its whole inner life being *a succession of ideas and emotions*, only suspended by Sleep and Death, or interrupted by the concentration of its attention on Sense-impressions. Now whatever difference of opinion there may be in regard to the degree in which the ordinary Laws of Causation are applicable to Mental phenomena (in other words, as to how far

each state of consciousness may be considered as *determined* by its antecedents), all are agreed that there *are* certain "Laws of Thought," expressive of the *uniformities of succession* which are observable in Mankind in general; whilst there are others which are characteristic of Races and Individuals; arising either from peculiarities in *original* constitution, or from the special direction which its congenital activities have acquired, or from both combined. It is not so much, however, the presence or absence of particular attributes, as their *proportional development*, that differentiate Minds from one another; and it is the habitual predominance (whether original or acquired) of particular *sequences of thought and feeling*, determined by that proportion, which constitutes the Character of each race or individual. Thus we find the *Intellectual* character to consist in the predominance of certain Faculties, which, as we shall presently see, are only designations of particular *modes* of intellectual activity; and a knowledge of these enables us to predicate, to a certain extent, the nature of the result at which any individual Mind will arrive, by its exercise upon a given subject previously thought-out by others. So, again, the *Moral* character will depend upon the relative predominance which may exist in the individual nature, of those Emotional tendencies, which not merely furnish a large share of the governing motives of the conduct, but which also contribute in a very important measure to the habitual direction of the thoughts: and in proportion to the completeness of our knowledge of the Moral character of any individual, will be our power of predicting the manner in which he will act under any particular contingency.

217. But these *uniformities of succession* are predicable only of the *automatic* activity of the Mind: and our own consciousness tells us that there is something in our Psychical nature, which is beyond and above this automatic exercise of our powers; and that the direction of our thoughts is placed, within certain limits, under

the control of the Will (§ 25). These limits, like those of the automatic activities, are partly universal, and partly peculiar to the individual. It is a universal fact that the Will *cannot originate* anything; but that it *has* a power of *selecting* any one out of several objects that present themselves either simultaneously or successively before the mental vision, and of so *limiting* and *intensifying* the impression which that particular object makes upon the consciousness, that all others shall be (for the time) non-existent to it. On the other hand, the *degree* in which this Volitional power is possessed by different individuals, is subject to wide variation. In some it is weak from the beginning, and no training seems effectual in developing it to a degree of full efficiency. In others it shows itself very early in a "masterful" disposition, which aims to bring *others* under subjection to itself; and here the aim of the Educator should be to cultivate *self*-mastery, by showing how much nobler is "he that ruleth his spirit, than he that taketh a city." It often happens, however, that strong *passions* are mistaken for strong *Will;* and that an entirely wrong method of discipline is adopted with a view to "*break* the Child's will," when what is really needed is to *direct* its Mental action aright (§ 120). Not unfrequently a strong Volitional power originally exists, but lies dormant for want of being called into exercise (§ 8); and here it is that judicious training can work its greatest wonders.

218. *Laws of Association.*—The most powerful agency in the automatic determination of the succession of our Mental states, is undoubtedly that tendency which exists in all minds that have attained the ideational stage of development, to the *association of ideas;* that is, to the formation of such a connection between two or more Ideas, that the presence of one tends to bring the other also before the consciousness,—or, in other words, that the one *suggests* the other. Certain Laws of Association, expressive of the conditions under which this connection is formed, and the

mode in which it acts, have been formularized by Psychologists;* of these the most important will be now specified.

219. Two or more states of consciousness, habitually existing together or in immediate succession, tend to cohere, so that the future occurrence of any one of them restores or revives the other; this is designated the *law of contiguity.* It is thus (to take a simple illustration) that the impressions made upon our Sensational consciousness by natural objects, which are usually received through two or more senses at once, become compacted into those composite notions, which, however simple they may appear, really result from the intimate combination of many distinct states of ideation. Thus our notion of the *form* of an object is made-up of separate notions derived from the visual and muscular senses respectively; our notion of the character of its *surface,* from the combination of impressions received through the visual and tactile senses; and with both of these our notion of *colour,* as in the case of an orange, may be so blended, that we do not readily conceive of its characteristic form and surface, without also having before our minds the hue with which these have been always associated in our experience. So, again, the external aspect of a body suggests to our minds its internal arrangement and qualities, such as we have before found them invariably to be; thus, to use the preceding illustration, the shape and colour of the orange bring before our consciousness its fragrant odour and agreeable taste, as well as the internal structure of the fruit. And our idea of "an orange" must be considered as the aggregate of all the preceding notions.—Not only the different ideas excited by one object, but those called-up by objects entirely dissimilar, may thus come to be associated, provided that the mind has been accustomed

* In the writings of Prof. Bain will be found the fullest and ablest exposition yet given of the Laws of Association, with copious illustrations of their operation drawn from a great variety of Mental phenomena, by which the Writer has profited in the outline here given.

to the presentation of them in frequent contiguity one with the other. Such conjunctions may be natural, that is, they may arise out of the "order of nature;" or they may be artificial, being due to human arrangements; all that is requisite is, that they should have sufficient permanence and constancy to habituate our minds to the association.—Of this Law of Contiguity, moreover, we have a most important example in the association which the mind early learns to form between successive *events*, so that when the first has been followed by the second a sufficient number of times to form the association, the occurrence of the first suggests the *idea* of the second; if that idea be verified by its occurrence, a definite *expectation* is formed; and if that expectation be unfailingly realized, the idea acquires the strength of a *belief*. And thus it is that we come to acquire that part of the notion of "cause and effect," which rests upon the "invariability of sequence;" and to form our fundamental conception of the "uniformity of nature" (§ 199, v). It is by the same kind of operation, again, that we come to employ Words as the symbols of ideas, for the convenience of intercommunication and reference (§ 198); a certain number of repetitions of the sound, concurrently with the sight of the object, or the suggestion of the notion of that object, being sufficient to establish the required relation in our minds.—Of the large share which this kind of action has in the operations of Memory and Recollection, evidence will be hereafter given (Chap. X).

220. But a not less important *tendency of thought*, and one whose operation is even more concerned in all the higher exercises of our Reason, is that which may be designated the *law of similarity;* and which expresses the general fact that any present state of consciousness tends to revive previous states which are *similar* to it. It is thus that we instinctively invest a new object with the attributes we have come to recognise in one we have previously examined, if the new object bears such a resemblance to it, that the sight of the second suggests those ideas which

our minds have connected with the first. Thus, we will suppose a man to have once seen and eaten an orange; when he sees an orange a second time, although it may be somewhat larger or smaller, somewhat rougher or smoother, somewhat lighter or darker in hue, he recognizes it as "an orange," and mentally assigns to it the fragrance and sweetish acidity of the one which he had previously eaten. But if, instead of being yellow, the fruit were green, he would doubt its being an orange; and if assured that it still was, but had not come to maturity, he would no longer expect to find it sweet; the notion of acidity being suggested to his mind by his previous experience of other green and unripe fruit.—It is in virtue of this kind of action, that we extend those elementary notions which are primarily excited by sensation, to new objects. Thus, the idea of roundness (like other notions of form) is originally based on the combination of the tactile and visual sensations, and must be first acquired by a process of considerable complexity; but when once derived from the examination of a single object, it is readily extended to other objects of the same character.—So, again, it is by the operation of this mental tendency, that we recognize similarity where it exists in the midst of difference, and separate the points of agreement from those of discordance; and this, again, not merely as regards objects which are before our consciousness at the same time or in close succession, but also with regard to all past states of consciousness. It is thus that we *identify* and *compare*, that we lay the foundations of classification, and that we recover all past impressions which have anything in common with our present state of consciousness.—The intensity of this tendency, and the habitual direction which it takes, vary extremely in different individuals. Some have so great an incapacity for recognizing *similarity*, that they can only perceive it when it is in marked prominence, their minds taking much stronger note of *differences*; whilst others have a strong bias for the detection of resemblances and analogies, and discover them where ordinary minds cannot recognize them. Some, again, address themselves to the discovery of similarity among objects of sense, whilst others study only those ideas which are the objects of our internal consciousness; and it is in the detection of what is essentially similar among the latter, that all the higher operations of the Intellect essentially consist. Even here we find that some are contented with superficial analogies, whilst others are not satisfied until they have penetrated by analysis to the depths of the subject, and are able to compare its "fundamental idea" with others of like kind.—It may be remarked that this mode of action of the mind is in some degree opposed to the preceding; for whilst *contiguity* leads to the arranging of ideas as they happen to present themselves in actual juxtaposition, and thus to induce a routine which is often most unmeaning (§ 285), *similarity* breaks through juxtaposition, and brings together from all quarters objects which have an Ideational likeness.

221. It is this habit of mind, which is of essential value in all the sciences of *classification* and *induction.* Thus, in the formation of generic definitions to include the characters which a number of objects have in common, their subordinate or specific differences being for a time left out of view, we are entirely guided by the recognition of similarity between the objects we are arranging; and the same is the case in the formation of all the higher groups of Families, Orders, and Classes, the points of similarity becoming fewer and fewer as we proceed to the more comprehensive groups, whilst those of difference increase in corresponding proportion. The sagacity of the Naturalist is shown in the selection of the *best* points of resemblance, as the foundation of his Classification; the value of characters being determined, on the one hand by their constancy, and on the other by their degree of coincidence with important features of general organization or of physiological history.—In the determination of Physical laws by the process of Induction, the process is somewhat of the same kind; but the

similarities with which we have here to do, are not, as in the preceding case, objective resemblances, but exist only among our subjective ideas of the *nature* and *causes* of the phenomena brought under our consideration. Thus, there is no obvious relation between the fall of a stone to the Earth, and the motion of the Moon in an elliptical orbit around it; but the penetrating mind of Newton detected a relation of *common causation* between these two phenomena, which enabled him to express them both under one Law. It was by a like Intellectual perception of similarity, that Franklin was led to determine the identity of lightning with the spark from an electrical machine. And it would be easy to show that it has been in their extraordinary development of this power of recognizing *causative* similarity, leading to a kind of intuitive perception of its existence, where no adequate ground could be assigned by the Reason for such a relationship, that those men have been eminent, who have done the most to advance Science by the process of inductive generalization.

222. The same kind of Mental activity is also exercised in the contrary direction: namely, in that application of general laws to particular instances, which constitutes *deductive reasoning;* and in that extension of generic definitions to new objects, which takes place upon every discovery of a new species. We may trace it, again, even in the extension of the meaning of words so as to become applicable to new orders of ideas, in consequence of the resemblances felt to exist between the latter and the ideas of which the words were previously the symbols;—as in the application of the word "head," which primarily designated the most elevated part of the human body, in such phrases as the "head of a house," the "head of a state," the "head of an army," the "head of a mob," in each of which the idea of superiority and command is involved;—or in the phrases the "heads of a discourse," or the "heads of an argument," in which we still trace the idea of authority or direction;—or in the phrases the "head of a

table," the "head of a river," in which the idea of superiority or origin comes to be locally applied;—or in the "head of a bed," or "head of a coffin," in which we have the more distinct local association with the position of the head of Man. Of the foregoing applications (the presence of which in Languages of entirely different families indicates their origin in wide spread identities of *modes of thought*), those first cited belong to the nature of a *metaphor*, which has been defined to be "a simile comprised in a word;" and the judicious use of metaphors, which frequently adds force as well as ornamental variety to the diction, is most seen amongst those who possess a great power of bringing together the *like* in the midst of the *unlike*.

223. Every effort, in fact, to trace-out unity, consistency, and harmony, in the midst of the wonderful and (at first sight) perplexing variety of objects and phenomena amidst which we are placed, is a manifestation of this tendency of the Human mind: and, when conducted in accordance with the highest teachings of the intellect, or guided by that insight which in some minds supersedes and anticipates all reasoning, it enables us to rise towards the comprehension of that great Idea of the Universe, which we believe to exist in the Divine Mind in a majestic simplicity of which we can here but faintly conceive, and of which all the phenomena of Nature are but the manifestations to our consciousness.—With this purely Intellectual operation, there is frequently associated a peculiar feeling of *pleasure*, which constitutes a true Emotional state. There are few who devote themselves to the pursuit of Science, who do not experience this pleasure, either from the detection of new relations of similarity by their own perception of them, or in the recognition of them as developed by others. It is, however, much more intense in some minds than in others; and according to its intensity, will it act as a *motive* in the prosecution of the search for Truth amidst discouragements and difficulties. But *all* discoveries of identifica-

tion, where use and wont are suddenly broken through, and a common feature is discerned among objects previously looked-on as entirely different, produce a flash of agreeable *surprise*, and the kind of sparkling cheerfulness that arises from the sudden lightening of the burden. And it is in this, that our enjoyment of *wit* seems essentially to consist (§ 402).

224. Although the single relations established between ideas, either through Contiguity or through Similarity, may suffice for their mutual connection, yet that connection becomes much stronger when two or more such relations exist consentaneously. Thus, if there be present to our minds two states of consciousness, each of them associated, either by contiguity or similarity, with some third state that is past and "out of mind" at the time, the compound action is more effective than either action would be separately; that is, although the suggestions might be separately too weak to revive the past state of consciousness, they reproduce it by acting together. Of this, which has been termed the *law of compound association*, we have examples continually occurring to us in the phenomena of Memory; but it is especially brought into operation in the volitional act of Recollection (§ 372).

225. Another mode in which the Associative tendency operates, is in the formation of aggregate conceptions of things that have never been brought before our consciousness by sense-impressions. This faculty, which has been termed *constructive association*, is the foundation of Imagination (§ 396); and it is exercised in every other mental operation in which we pass from the *known* to the *unknown*. When we attempt to form a conception which shall differ from one that we have already experienced, as a matter of objective reality, by the introduction of only a single new element,—as when we imagine a brick building replaced by one of stone, in every respect similar as to size and form,—we substitute in our minds the idea of stone for that of brick, and associate it by the principle of contiguity with those other ideas,

of which that of the whole building is an aggregate. So, again, if we conceive a known building transferred from its actual site to some other already known to us, we dissociate the existing combinations, and keep-together the ideas which were previously separated, until their contiguity has so intimately united them, that the picture of the supposed combination may present itself to the mind exactly as if it had been a real scene which we had long and familiarly known. By a further extension of the same power, we may conceive the elements to be varied, as well as the mode of their combination; and thus we may bring before our consciousness a *representation* in which no particular has ever been present to our minds under any similar relations, and which is, therefore, entirely new to us *as a whole*, notwithstanding that, when we decompose it into its ultimate elements, we shall find that each of these has been previously before our consciousness. Such a representation, by being continually dwelt-on, may come to have all the force and vividness of one derived from an actual sensory impression; and we can scarcely conceive but that the actual state of the Sensorium itself must be the same in both cases, though this state is induced in the one case by an act of mind, and in the other by objective impressions (§ 100).—A very common *modus operandi* of this "constructive association," is the realization of a landscape, a figure, or a countenance, from a pictorial representation of it. Every picture *must* be essentially defective in some of the attributes of the original, as, for example, in the representation of the *projection* of objects; and all, therefore, that the picture can do, is to *suggest* to the mind an idea, which it completes for itself by this constructive process, so as to form an aggregate which may or may not bear a resemblance to the original, according to the fidelity of the picture, and the mode in which it acts upon the mind of the individual. Thus, a mere sketch shall convey to one person a much more accurate notion of the object represented, than a more finished picture shall give to another; because from practice in this kind of mental reconstruction, the former recognizes the true meaning of the sketch, and fills it up in his "mind's eye;" whilst the latter can see little but what is actually before his bodily vision, and interprets as a literal presentation that which was intended merely as a suggestion. And it is now generally admitted, that in all the higher forms of representative Art, the aim should be, *not* to call into exercise the faculty of mere objective *realization*, but to address that higher power of *idealization*, which invests the conception suggested by the representation with attributes more exalted than those actually possessed by the original, yet not inconsistent with them. It depends, however, as much on the mind of the individual addressed, as on that of the Artist himself, whether such conceptions shall be formed; since by those who do not possess this power, the highest work of Art is only appreciated, in so far as it enables them to realize the object which it may represent.

226. Having thus pointed out what may be considered the most elementary forms of Mental action, we shall briefly pass in review those more complex operations which may be regarded as in great part compounded of them. The capacity for performing these is known as the *Intellect* or the *reasoning power;* and the capacities for those various forms of intellectual activity, which it is convenient to distinguish for the sake of making ourselves more fully acquainted with them, are termed "intellectual faculties." It appears to the Writer, however, to be a fundamental error to suppose, that the entire Intellect can be split-up into a certain number of faculties; for each faculty that is distinguished by the Psychologist, expresses nothing else than a *mode of activity* in which the whole power of the Mind may be engaged at once,—just as the whole power of the locomotive steam-engine may be employed in carrying its load forwards or backwards, according to the *direction* given to its action. It is the direction of the attention to *external* objects, for example, that constitutes the "faculty" of *obser-*

vation; which is simply that form of activity, in which the Mind is occupied by the Sense-impressions it is receiving, either from a number of sources at once, or from a more limited area, the impressions in the latter case being proportionally intensified (§ 123). On the other hand, it is the direction of our attention to what is passing *within* us,—not merely intensifying the Mental state, but separating and bringing it forward as an object of contemplation, —which is designated as *reflection,* but is more appropriately termed *introspection.* In each of these *the whole Mind* may be so completely engaged, that the two activities cannot go on simultaneously (§ 117). So, again, in that reproduction of past states of consciousness which we term *memory,* and, still more in that volitional recall of them which constitutes *recollection,* we have the *whole mind* at work in certain definite sequences expressed by the "laws of association."

227. Upon the various Ideational states, which are either directly excited by Sense-impressions, or are reproduced by Memory, and are sequentially connected in "trains of thought" by suggestions arising out of pre-formed associations, all acts of *Reasoning* are founded. These consist, for the most part, in the aggregation and collocation of ideas, the decomposition of complex ideas into more simple ones, and the combination of simple ideas into general expressions; in which processes are exercised the faculty of *comparison,* by which the relations and connections of ideas are perceived,—that of *abstraction,* by which we mentally isolate from the rest any particular quality of the object of our thought, —and that of *generalization,* by which we recognize the common properties we have abstracted, as composing a distinct notion, that of some *genus* in which the objects are comprehended. These operations, when carefully analyzed, seem capable of reduction to this one expression,—namely, the fixation of our Attention either on some particular *classes of ideas,* from among those which suggestion brings before our consciousness, or on some particular

relations of those ideas; and this fixation may depend, as already shown, either on the peculiar attractiveness which these ideas or relations have for us (the constitution of individual minds varying greatly in this respect), or on the determination of our own Will. All these faculties are exercised in the act of *judgment;* which is a summary expression of the entire process—how simple or how complex soever—by which we arrive at a decision either as to the absolute or probable truth or falsehood of any proposition, or as to the moral or prudential bearing of any course of action. —There is strong reason to believe that these processes may be performed *automatically* to a very considerable extent, without any other than a *permissive* act of Will. It is clearly by such automatic action, that the before-mentioned "fundamental axioms" or "secondary intuitions" (§ 199) are evolved; and there is not one of the operations above described which may not be performed quite involuntarily, especially by an individual who is naturally disposed to it. Thus, to some persons, the tendency to *compare* any new object of consciousness with objects that have been previously before the mind, is so strong as to be almost irresistible; and this, or any other original tendency, is strengthened by the habit of acting in conformity with it. So, again, the tendency to *abstract* is equally strong in the minds of others, who instinctively seek to separate what is fundamental and essential in the properties of objects, from what is superficial and accidental; and their attention being most attracted by the former, they readily recognize the same characters elsewhere, and are thus as prone to combine and generalize, as others are to analyse and distinguish.

228. It is only, in fact, when we *intentionally* divert the current of thought from the direction in which it was previously running, —when we *determine* to put our minds in operation in some particular manner,—and make a *choice of means* adapted to our end (as in the act of recollection (§ 370), by *purposely* fixing our attention upon one class of objects to the exclusion of others,—that we can be

said to use the Will in our intellectual processes; and this exercise of it is shown, by the analysis of our own consciousness, to be less constant than is commonly supposed. Thus we may imagine a man sitting-down at a fixed hour every day, to write a treatise upon a subject which he has previously thought-out: after that first effort of Will by which his determination was made, the daily continuance of his task becomes so habitual to him, that no fresh exertion of it is required to bring him to his desk; and, unless he feel unfit for his work, or some other object of interest tempt him away from it, so that he is called-upon to decide between contending motives, his will cannot be fairly said to be brought into exercise. It may need, perhaps, some Volitional fixation of his attention upon the topics upon which he had been engaged when he last dropped the thread, to enable him to recover it, so as to commence his new labours in continuity with the preceding; but when once his mind is fairly engrossed with the subject, this developes itself before his consciousness according to his previous habits of mental action; ideas follow one another in rapid and continuous succession, clothe themselves in words, and prompt the movements by which those words are expressed in writing; and this automatic action may continue uninterruptedly for hours (§ 236 *a*), without any tendency of the mind to wander from its subject, the Will being only called into play when the feeling of fatigue or the distraction of other objects renders it difficult to keep the attention fixed upon that which has previously held it by its own attractive power (§ 315).—The converse of this condition is experienced, when some powerful interest tends to draw-off the attention elsewhere, and the thoughts are found to wander continually from the subject in hand; or when, from the undue protraction of mental exertion, the physical condition is such, that the thoughts no longer develope themselves consecutively, nor shape themselves into appropriate forms of expression. In either of these cases, the Intellectual powers can only be kept in action upon the pre-determined subject, by a strong *effort* of the will: of this effort we are conscious at the time, and feel that we need to put forth even a greater power than that which would be required to generate a large amount of physical force through the muscular system; and we subsequently experience the results of it, in the feeling of excessive *fatigue* which always follows any such exertion.

229. The more carefully the actions of early Childhood are observed, the more obvious does it become that they are solely prompted by ideas and feelings which *automatically* succeed one another, in uncontrolled accordance with the laws of suggestion. This principle has already been referred to (§ 120); but the following illustrations of it, which show that a Child very early comes to adapt the expression of its wants, or the communication of its ideas, to the receptivity of the person addressed,—and this not by intention, but in accordance with an *acquired intuition* based on its everyday experience,—may be here appropriately introduced:—

a. Dr. Kitto, whose experience of *entire loss* of the sense of Hearing has formed the subject of a very interesting Autobiography, tells us that his children, in their successive infancies, would begin to imitate the finger-language *whenever they saw him*, even whilst they were yet in arms, and could have had no true cognizance of his peculiar condition.—(*The Lost Senses*, vol. i., p. 97.)

b. The following case, originally recorded by Dr. C. B. Radcliffe, has been found by the subsequent enquiries of the Writer to be one of very common occurrence.—A child of English parents residing in Germany, being under the care of a German nurse, had acquired the power of speaking on ordinary matters either in German or English, without confusing the words or idioms; but yet seemed invariably compelled to reply in the language used by the person he was addressing. Thus, in conveying a message to his German nurse, he delivered it in German, though he had received it the moment previously in English; but on returning to the English family in the parlour, if asked what the maid had said, he answered in English as often as the question was proposed in English; and even though pressed to give the words he had heard in the nursery, he still continued to do the same, without seeming to be aware of the difference. But if the question was put to him in German, the answer was in German; there being the same inability to reply in English, as there had previously been to give a German answer to the English question.—(*Philosophy of Vital Motion*, p. 137).

c. In another instance known to the Writer, the child of a French father resident in England, and of an English-speaking mother, who was growing-up to speak to his father in French and to his mother in English, was taken by his father to spend the summer in Switzerland, where he never heard anything but French spoken, and for several weeks himself spoke French exclusively. One day, as the father and child were walking together, they met some English friends, who addressed the boy in English, but could get no reply from him, though he answered them at once in French when they spoke to him in that language. The father feared that the boy had already lost his *mother* tongue; but on returning home the lad at once found it again, telling his mother in English of all that had happened to him abroad.

These two cases, though in some respects dissimilar, are obviously referable to the same principle; for the result was determined in each by the automatic action of the Mind, in accordance with the laws of association. In the former case, the language of each answer was suggested by that in which the question was put; whilst in the latter, it was determined in the first instance by the *last acquired* habit, and in the second by the recurrence of the circumstances under which the *original* habit had been formed.

230. Even in the adult, the predominance of the *automatic* activity of the Mind over that which is regulated by the *will*, is often seen as a result of a want of balance between the two; arising either from the excessive *force* of the former, or from the unusual *weakness* of the latter. We have an example of it in the loose rambling talk of persons who have never schooled themselves to the maintenance of a coherent train of thought, but are perpetually "flying off in a tangent,"—sometimes at a mere sensorial suggestion (conveyed

by the sound or the visual conception of a word), sometimes at the prompting of an ideational association of a most irrelevant kind. A most truthful portraiture of a low type of this order of mind is presented in the "Mrs. Nickleby" of Dickens: while, in real life, we have had a most striking exemplification of its most exalted form in Coleridge, whose talk was just as disjointed as Mrs. Nickleby's, though relating to the highest instead of the most trivial subjects. His career, indeed, affords so remarkable a "study" to the Psychologist who takes as his guiding idea the relation between *automatic activity* and *volitional direction*, that the principal features of it will be here brought under review.

231. There was probably no man of his time, or perhaps of any time, who surpassed Coleridge in the combination of the reasoning powers of the Philosopher with the imagination of the Poet and the inspiration of the Seer; and there was perhaps not one of the last generation, who has left so strong an impress of himself in the subsequent course of thought of reflective minds engaged in the highest subjects of human contemplation. And yet there was probably never a man endowed with such remarkable gifts, who accomplished so little that was worthy of them,—the great defect of his character being the want of Will to turn his gifts to account; so that, with numerous gigantic projects constantly floating in his mind, he never brought himself even seriously to attempt to execute any one of them. It used to be said of him, that whenever either natural obligation or voluntary undertaking made it his *duty* to do anything, the fact seemed a sufficient reason for his *not* doing it. Thus, at the very outset of his career, when he had found a bookseller (Mr. Cottle) generous enough to promise him thirty guineas for poems which he recited to him, and might have received the whole sum immediately on delivering the manuscript, he went on, week after week, begging and borrowing for his daily needs in the most humiliating manner, until he had drawn from his patron the whole of the promised purchase-money, without supplying

him with a line of that poetry which he had only to *write down* to free himself from obligation.—The habit of recourse to nervine stimulants (alcohol and opium) which he early formed, and from which he never seemed able to free himself, doubtless still further weakened his power of Volitional self-control; so that it became necessary for his welfare, that he should yield himself to the control of others. The character of his Intellect was eminently *speculative.* He tells us, in his "Biographia Literaria," that even before reaching his fifteenth year, he had bewildered himself in metaphysics and theological controversy; that nothing else pleased him; and that, in especial, *history and particular facts had no interest for him.* This complete isolation of his mind from all the *realities* of life, except the friendships to which he was held by personal sympathy, marked his character throughout; what he would himself have called its *subjective* side having so great a predominance, that he seldom seemed to care to bring his ideas to the test of conformity with *objective* facts. All accounts of Coleridge's habits of thought, as manifested in his conversation (which was a sort of *thinking aloud*), agree in showing that his train of mental operations, once started, went on *of itself,*—sometimes for a long distance in the original direction, sometimes with a divergence into some other track, according to the consecutive suggestions of his own mind, or to new suggestions introduced into it from without.

a. The Writer once heard a very characteristic instance of this, from a gentleman who had obtained an introduction to him when he was domiciled with the Gillmans at Highgate. After presenting his credentials, his visitor expressed a hope that he was better, having heard that he had been ill. "Yes," said Coleridge, "I am better, but I should be better still if I did not *dream* so much. These subjective states are very curious." And then he discoursed for two hours continuously on "subject" and "object;" or, as Carlyle graphically tells us, on what "he sang and snuffled into 'om—m—mject,' and 'sum—m—mject,' with a kind of solemn shake or quaver as he rolled along."

b. How little he thought of his listeners, when he was once fairly launched, is proved by the following account of his habits, narrated to the Writer by a friend who was a school-girl at Highgate at the time of Coleridge's residence there. Being accustomed to walk every day in the "Grove," at an hour when the girls were at play there, he would sometimes draw one of the children to him, and begin by caressing and coaxing her to talk to him; but very soon the conversation would pass into the accustomed monologue, altogether beyond the comprehension of the poor child, who was like the "wedding-guest" under the spell of the "ancient mariner," vainly endeavouring to free herself that she might resume her sport. Thus "old Coley," as the school-girls irreverently nicknamed him, became the terror of the children of the neighbourhood, who learned sedulously to keep out of his way.

c. Charles Lamb's story of his having cut off the button by which Coleridge was holding him one morning, when he was going in to London by the Enfield stage; of his leaving Coleridge in full talk, with the button in one hand, and sawing the air with the other; and of his finding him discoursing in exactly the same attitude when he came back to Enfield in the afternoon,—is, of course, a ludicrous exaggeration; but it conveys, like other "myths," a true idea of the degree in which Coleridge was habitually "possessed" by the train of thought that happened to be passing through his mind at the time.

In fact, Coleridge's whole life might almost be regarded as a sort of waking dream. The composition of the poetical fragment "Kubla Khan" *in his sleep*, as told in his "Biographia Literaria," is a typical example of automatic mental action.

d. He fell asleep whilst reading the passage in "Purchas's Pilgrimage" in which the "stately pleasure-house" is mentioned; and, on awaking, he felt as if he had composed from two to three hundred lines, which he had nothing to do but to write down, "the images rising up as things, with a parallel production of the correspondent expressions, without any sensation or consciousness of effort." The whole of this singular fragment, as it stands, consisting of fifty-four lines, was written as fast as his pen could trace the words; but having been interrupted by a person on business, who stayed with him above an hour, he found, to his surprise and mortification, that, "though he still retained some vague and dim recollection of the general purport of the vision, yet, with the exception of some eight or ten scattered lines and images, all the rest had passed away, like the images on the surface of a stream into which a stone had been cast; but, alas! without the after-restoration of the latter."

In the wonderfully graphic description of Coleridge's appearance and style of discourse, given by Carlyle ("Life of John Sterling," Chap. VIII.), it is necessary to bear in mind the essential difference, one might almost say the contrariety, between the characters of the "subject" and his pourtrayer: the "history and particular facts" which had "no interest" for the one, being the favourite mental food for the other; while the purely speculative problems in which Coleridge delighted (parodied by his friend Charles Lamb, in the question "How many angels can dance on the point of one needle?"), would have been regarded by Carlyle as altogether futile.

e. "Coleridge's whole figure and air," says Carlyle, "good and amiable otherwise, might be called flabby and irresolute; expressive of weakness under possibility of strength. He hung loosely on his limbs, with knees bent, and stooping attitude. In walking he rather shuffled than decisively stept; and a lady once remarked he never could fix which side of the garden-walk would suit him best, but continually shifted in corkscrew fashion, and kept trying both.

"Nothing could be more copious than his talk; and, furthermore, it was always virtually or literally of the nature of a monologue; suffering no interruption, however reverent; hastily putting aside all foreign additions, annotations, or most ingenuous desires for elucidation, as well-meant superfluities which would never do. Besides, it was talk not flowing any whither like a river, but spreading every whither in inextricable currents and regurgitations like a lake or sea; terribly deficient in definite goal or aim—nay, often in logical intelligibility; *what* you were to believe or do, on any earthly or heavenly thing, obstinately refusing to appear from it. So that, most times, you felt logically lost, swamped, near to drowning, in this tide of ingenious vocables, spreading out boundless as if to submerge the world.

"He began anywhere. You put some question to him, made some suggestive observation; instead of answering this, or decidedly setting out towards answering it, he would accumulate formidable apparatus, logical-swim bladders, transcendental life-preservers, and other precautionary and vehiculatory gear, for setting out; perhaps did at last get under weigh, but was swiftly solicited, turned aside by the glance of some radiant new game on this hand or that, into new courses, and ever into new, and before long into all the universe, where it was uncertain what game you would catch, or whether any. His talk, alas! was distinguished, like himself, by irresolution; it disliked to be troubled with conditions, abstinences, definite fulfilments; loved to wander at its own sweet will, and make its auditor and his claims and humble wishes a mere passive bucket for itself.

"Glorious islets, too—balmy, sunny islets—islets of the blest and the intelligible! I have seen rise out of the haze, but they were few, and soon swallowed in the general element again.

"Eloquent, artistically expressive words you always had; piercing radiances of a most subtle insight came at intervals; tones of noble pious sympathy, recognisable as pious, though strangely coloured, were never wanting long; but, in general, you could not call this aimless, cloud-capt, cloud-based, lawlessly meandering human discourse of reason by the name of 'excellent talk,' but only of 'surprising;' and were reminded bitterly of Hazlitt's account of it: 'Excellent talker, very—if you let him start from no premises, and come to no conclusion.'"

It was by the brilliance and subtlety of those occasional flashes of thought which Carlyle designates as "islets of the blest and the intelligible," and by the profound suggestiveness of those fragmentary writings which constitute all he ever executed of his colossal project of a system of Mental Philosophy in its widest meaning, that Coleridge exerted that influence over the thinkers of the succeeding generation, which no one acquainted with its Intellectual history can question.

232. So, again, the artistic life of Mozart, from his infancy to his

death, presents a typical example of the spontaneous or automatic production of Musical conceptions; which, under the skilful training he received from his father, developed themselves into creations of the very highest order, whose number, considering the early age at which he died (less than thirty-six years) is nothing less than marvellous. In fact, whether we estimate Mozart by the spontaneity, the productiveness, or the variety of his inventive power, as attested by the multitude of those "things of beauty" he called into existence, every one of which will be "a joy for ever,"—or by that wonderful divining faculty which enabled him, as a *boy*, to anticipate almost everything that was then known in Music, and as a *man*, to advance, in every style he took up, far beyond his greatest predecessors in each department,—or by the permanent impress he has left upon his Art, not merely in furnishing the most perfect models for the study of those who especially cultivate it, but in elevating that general appreciation of the highest order of beauty, which only the works of a consummate Artist can call forth,—Mozart certainly stands alone among Musicians, and deserves to rank as a typical example of *genius*. Mozart, like Coleridge was a man whose Will was weak in proportion to the automatic activity of his mind; and it is probable that if he had not been under the guidance, in the first instance, of a judicious father, and afterwards of an excellent wife, to both of whom he had the good sense to submit himself, his career would have been comparatively inglorious. For his lively sensibility made him the sport of every kind of impulse, so that he could neither keep firm to a resolution, nor resist a temptation: and hence he would never of his own accord have subjected himself to the discipline which his father imposed upon him, and without which he could not have been anything else than a "musical prodigy;" nor would he have had the motive which his conjugal affection supplied, for the steady application that was required for the elaboration of his greatest works. Hence his life becomes a most interesting study to the

Psychologist, no less than to the Musician. Of the general features of his career, a sketch has already been given (§ 206); we shall now endeavour to trace-out the manner in which he worked; and of this we fortunately have a pretty full account from himself, in a letter to a friend:—

"You say you should like to know my way of composing, and what method I follow in writing works of some extent. I can really say no more on the subject than the following, for I myself know no more about it, and cannot account for it. When I am, as it were, completely myself, entirely alone, and of good cheer, say, travelling in a carriage, or walking after a good meal, or during the night when I cannot sleep; it is on such occasions that my ideas flow best and most abundantly. *Whence* and *how* they come, I know not, nor can I force them. Those ideas that please me I retain in my memory, and am accustomed (as I have been told) to hum them to myself. If I continue in this way, it soon occurs to me how I may turn this or that *morceau* to account, so as to to make a good dish of it, that is to say, agreeably to the rules of counterpoint, to the peculiarities of the various instruments, &c.

"All this fires my soul, and, provided I am not disturbed, my subject enlarges itself, becomes methodised and defined, and the whole, though it be long, *stands almost complete and finished in my mind*, so that I can survey it like a fine picture, or a beautiful statue, at a glance. Nor do I hear in my imagination the parts *successively*, but I hear them, as it were, all at once (*gleich alles zusammen*). What a delight this is, I cannot tell! All this inventing, this pondering, takes place in a pleasing lively dream. Still the actual hearing of the *tout ensemble* is after all the best. What has been thus produced I do not easily forget, and this is perhaps the best gift I have my Divine Maker to thank for.

"When I proceed to write down my ideas, I take out of the bag of my memory, if I may use that phrase, what has previously been collected into it in the way I have mentioned. For this reason, the committing to paper is done easily enough; for everything is, as I said before, already finished; and it rarely differs on paper from what it was in my imagination. At this occupation I can therefore suffer myself to be disturbed; for whatever may be going on around me, I write and even talk, but only of fowls and geese, or of Gretel or Barbel or some such matters.* But why my productions take from my hand that particular form and style that makes them *Mozartish*, and different from the works of other composers, is probably owing to the same cause which renders my nose so, or so large, so aquiline, or in short, makes it Mozart's, and different from that of other people. For I really do not study or aim at any originality; I should in fact not be able to describe in what mine consists, though I think it quite natural that persons who have really an individual appearance of their own, are also differently organised from others, both externally and internally. At least I know that I have not constituted myself, either one way or the other."—(Holmes's *Life of Mozart*, p. 318.)

An interesting pendant to this remarkable self-analysis is supplied by Mozart's answer to the question asked him by a lad of twelve

* The story of the production of the Overture to "Don Giovanni" affords so admirable an illustration of the above description, that, though often told, it ought not to be omitted here:—On the very evening before the first performance of this Opera, not a note of the Overture had been written; and Mozart was giving himself up to social enjoyment at the house of Dussek, for whose wife (a finished singer) he had been that day composing a highly scientific scena, peculiarly suited to her style. About midnight he retired to his apartment, desiring his wife to make him some punch, and to stay with him to keep him awake while he wrote. She accordingly began to tell him fairy tales and odd stories, which made him laugh till the tears came. The punch, however, occasioned such a drowsiness, that he could only go on while his wife was talking; as soon as she ceased, he dropped asleep. The efforts which he made to keep himself awake, the continual alternation of sleep and watching, so fatigued him, that his wife persuaded him to take some rest, promising to awake him in an hour's time; but he slept so profoundly, that she suffered him to repose for two hours. At five in the morning she awoke him; and by seven o'clock, the hour at which he had appointed the music-copiers to come to him, the Overture was finished. The commencement of the evening performance was delayed, because the copiers had not completed their work; and the parts were brought into the orchestra with the notes still covered with the sand which had been used to dry up the ink.—Of course the Overture had to be performed without any rehearsal; and Mr. Holmes thinks it not unlikely that this *tour de force* was intended by Mozart as a compliment to the Prague Orchestra. It is clear that the Overture must have previously *evolved itself in all its completeness* in his creative imagination; since the mere writing it down must have engrossed the whole of the time within which it was committed to paper. As a mere feat of Memory, its production was therefore most marvellous; to say nothing of the transcendant merits of the work itself, which none but a Mozart could have produced.

years old, who already played the pianoforte very skilfully. "Herr Kapellmeister, I should very much like to compose something. How am I to begin?" "Pho-pho," said Mozart, "you must wait." "You," said the boy, "composed much earlier." "But," replied Mozart, "I asked nothing about it. If one has the spirit of a composer, *one writes because one cannot help it.*"—What can be a better description of the exuberant automatic activity of his Musical faculty? When he was "in the vein" for composition, it was difficult to tear him from his desk; and when he was in the mood to improvise upon the pianoforte, either alone or in the society of a friend, sitting down to the instrument in the evening, he commonly pursued the train of his musical thoughts till long after midnight.

That, notwithstanding the exuberance of his own creative power, Mozart constantly disciplined it by the most sedulous study, and that he could, without being chargeable with imitation, *assimilate* (so to speak) into his own Musical constitution all that he found suitable in the works of others as *pabulum* for his genius, is one of its most remarkable features. "It is a very great error," he wrote to a friend, "to suppose that my art has become so exceedingly easy to me. I assure you there is scarcely any one who has worked at the study of composition as I have. You could hardly mention any famous composer, whose writings I have not diligently and repeatedly studied throughout." And, in this self-education, as Mr. Holmes remarks, "whatever of striking, new, or beautiful he met with in the works of others, left its impression on him; and he often reproduced these effects, not servilely, but mingling his own nature and feeling with them, in a manner not less surprising than delightful." Thus no musician more thoroughly appreciated, than did Mozart, the surpassing greatness, in his own particular walk, of Handel. "Handel," he said, "understands effects better than any of us; when he chooses, he strikes like a thunderbolt." Mozart's "additional accompaniments" to the "Messiah" show how thoroughly impregnated he had become with the *feeling* which pervades that immortal work; so as to be able to fill up,—with the rich colouring of an instrumentation, and with the telling effects of harmonies, that Handel could not have devised,—the grand outlines traced by the Master's hand, in such a manner that none but a pedant could take exception to difference of style. Let any competent listener compare the effect produced on his own Musical sense by that wonderful song, "The people that walked in darkness," as accompanied *with* and *without* the wind-instrument parts added by Mozart; and he must acknowledge how admirably they carry out the "groping" sentiment of the air itself, which in Handel's score is merely repeated in octaves by the stringed instruments. It would seem as if Mozart had made this air (as it were) *a part of himself*, and that these accompaniments then *evolved themselves* in what was *to him* its most natural form of complete expression.

Only second to Mozart's creative genius, was his *executive skill;* which enabled him to render his own Musical ideas, and to express his own feelings, on the pianoforte, in a way which, by the judgment of his ablest contemporaries, it would be impossible even now to surpass. The testimony of Haydn, in particular, who was a frequent guest at the parties at which Mozart was wont to introduce his new compositions to his friends, and who constantly showed the most cordial appreciation of the genius of one whom less disinterested men depreciated as a dangerous rival, is singularly emphatic. "Mozart's playing," he said, "I can never forget." Doubtless, this executive skill was partly due to the early and excellent training he received from his father; and was partly the result of the animating influence of the genius which thus found expression. But looking to the fact of no unfrequent occurrence in every department of invention, that genius has often to *struggle* for its expression, and looking also to the peculiar *mobility* of Mozart's physical constitution, which showed itself in a variety of ways, it can scarcely be doubted that his was a case in which there was that complete harmony between his bodily and his mental organisation, which enables each to minister in the highest degree to the requirements of the other.—If the self-discipline which Mozart so admirably exercised in the culture of his Musical gifts, had been carried into his Moral nature, so as to restrain the impulses of his ardent temperament within due bounds, and to prevent him from consuming the energy of his frail body in the pursuit of exhausting pleasures, the world might have profited by a still higher development of his genius, and a still larger bequest of treasures of pure and elevated enjoyment.

233. It may be well to contrast with these examples, two others drawn from the careers of men by whom inferior endowments were turned to their best account, under the direction of a steady Will. Coleridge's brother-in-law, Southey, was as honourably distinguished by his strong sense of duty, and strict fidelity to his engagements and resolutions, as Coleridge himself was lamentably notorious for the reverse. Although few of Southey's poems may retain a lasting celebrity, yet his prose writings will always be models of excellence in composition; and he had his powers under such complete command, that he never failed (save from physical incapacity) to perform those promises which are too often made by men of genius "only to be broken," and never shrank from what he felt to be a task of disagreeable drudgery, when once he had undertaken it. And it is specially worthy of note, that even his poetic faculty seemed to be so far under his command, that he could fix how many lines he should write per day, so as to complete a poem of a given length within a specified time. Though poetry of the *highest* eminence cannot be thus "made to order," few in Southey's day could have produced what would have equalled his, either in opulence of imagination, or in splendour and appropriateness of diction. "His mind," it has been said, "although a teeming, was not an inventive or creative one. It returned manifold the seed deposited in it, but communicated to it comparatively little of any new nature or quality." What his poetry wanted, was the true vitality of the *mens divinior;* and this it was not in Southey's nature to impart.

234. The contrast between Haydn and Mozart, as regards both their artistic genius and their personal character, is scarcely less striking. Haydn's musical ability, like Mozart's, manifested itself in childhood; but it received a far less complete training. Being

early thrown, however, in a great degree upon his own resources, he displayed that steadiness of purpose which mainly contributed to his subsequent distinction, in acquiring from books, and from such chance instruction as he could obtain, the theoretical knowledge which he felt that he needed for succeeding as a Composer. By giving a few lessons in music, and occasionally performing in the orchestra, he managed to supply himself with what, frugally husbanded, served to provide him with absolute necessaries, and to enable him to maintain a decent appearance; and having thus gradually acquired a reputation as an able Musician, at the age of twenty-nine (before attaining which Mozart had produced some of his master-pieces), he was appointed Maestro di Capella to Prince Esterhazy, in whose service he remained to the end of his long life of seventy-seven years.

a. Comfortably settled in the palace of Eisenstadt, in Hungary, enjoying in moderation his favourite diversions of hunting and fishing, and relieved from care for the future, Haydn there composed the long series of works in various styles which he produced before his visit to London at nearly sixty years of age; which visit was the immediate occasion of his bringing out his "Twelve Grand Symphonies," and indirectly (by the impression which his hearing of Handel's music made upon him) prompted the composition of the "Creation," which he produced in his sixty-fifth year. During the whole period of his residence with Prince Esterhazy, he may be said to have been educating himself, under peculiar advantages, for those great works of his advanced life, on which his reputation now chiefly rests. He had a full and choice band living under the same roof with him, at his command every hour in the day; he had only to order, and they were ready to try the effect of any piece, or even of any passage, which, quietly seated in his study, he might commit to paper. Thus at leisure he heard, corrected, and refined whatever he conceived; and never sent forth his compositions, until they were in a state to fearlessly challenge criticism.

There can be no question of Haydn's inferiority to Mozart in *creative* power; but the steadiness of his application to his art, and the advantage he possessed in being constantly able to test his productions by actual trial, enabled him ultimately to attain a place among the first of modern Musicians, which Mozart had reached at a bound. He did not possess enough of the emotional temperament to succeed in dramatic composition; and his Operas have been long forgotten. But his *forte* lay in the *development* of musical ideas, and in the *construction* of elaborate Orchestral combinations; so that he is commonly regarded as "the father of modern orchestral music." As already pointed out (§ 206), however, Mozart is fully entitled to share in this distinction; his marvellous *intuition* having directly led him to anticipate many of those effects, which Haydn was engaged in elaborating by successive steps. It is worthy of note, as showing the different temperaments of these two illustrious contemporaries, that while Mozart's musical ideas were almost always in free flow (§ 400)—their character changing with the mood in which he happened to be—those of Haydn seemed only to come when he *set himself* to compose, which he usually did at a fixed hour every day, in this respect strongly resembling Southey.

b. It is related of him that, when he sat down to compose, he always dressed himself with the utmost care, had his hair nicely powdered, and put on his best suit. Frederick II. had given him a diamond ring; and Haydn declared that, if he happened to begin without it, he could not summon a single idea. He could write only on the finest paper; and was as particular in forming his notes, as if he had been engraving them on copper-plate. After all these minute preparations, he began by choosing the theme of his subject, and fixing into what keys he wished to modulate it; and he varied, as it were, the action of his subject, by imagining to himself the incidents of some little adventure or romance.—Haydn had strong religious feeling; and when, in composing, he found his imagination at fault, or was stopped by some difficulty which appeared insurmountable, he rose from the pianoforte and began to run over his rosary, and was accustomed to say that he never found this method fail.

This last fact is a "pregnant instance" of the principle of action which we shall hereafter have to consider (Chap. XIII.);—that, namely, of the working of a mechanism *beneath the consciousness*, which, when once set going, runs on of itself; and which is more likely to evolve the desiderated result, when the *conscious* activity of the mind is exerted in *a direction altogether different.*

Section 3.—*Ideo-Motor Action.*

235. Although it has been usual to designate by the term *voluntary* all those muscular movements which take-place as the result of mental operations, save when they are the expression of *emotional* states, yet a careful analysis of the sources from which many of even our ordinary actions proceed, will show that the Will has no direct participation in producing them; and that they are, psychologically speaking, the direct manifestations of Ideational states excited to a certain measure of intensity, or, in physiological language, *reflex actions of the Cerebrum.* This mode of operation has been already shown (§§ 94—111) not only to be fully conformable to the general plan of the activity of the Nervous system, but even to complete or fill-up a part of it which would otherwise be left void; and we shall find that it takes account of a great number of phenomena which had not previously been included under any general category, and which, when thus combined and generalized, form a most interesting and remarkable group, well deserving of attentive study.—It is, of course, when the Intellect is in a state of exalted (though it may be aberrant) activity, but when the directing power of the Will is suspended or weakened, that we should expect to see the most remarkable manifestations of this "reflex" power of the Cerebrum; and such is the condition of the Somnambulist who *acts* his dreams (§ 492), and of the Biologized subject who *acts* his reverie (§ 451). In

each case, the mind is "possessed" by a succession of Ideas, which may either be spontaneously evolved by its own operations, or may be directly suggested through the senses, or may be the products of associative action called forth by the promptings which it receives from without. In whatever mode the Ideas have been generated, it is the essential characteristic of these states, that the mind is *entirely given-up* to whatever may happen to be before it at the time; which consequently exerts an uncontrolled directing power over the actions, there being no antagonistic agency to keep it in check.

236. We may range under the same category all those actions performed by us in *our ordinary course of life*, which are rather the automatic expressions of the ideas that may be dominant in our minds at the time, than prompted by distinct volitional efforts (§ 228). Of this kind, the act of expressing the thoughts in Language, whether by speech or by writing, may be considered as a good example: for the attention may be so completely given-up to the choice of words and to the composition of the sentences, that the movements by which the words and sentences already conceived are uttered by the voice or traced on paper, no more partake of the truly Volitional character, than do those of our limbs when we walk through the streets in a state of abstraction (§ 16). And it is a curious evidence of the influence of Ideas, rather than of the agency of the Will, in producing them, that, as our conceptions are a little in advance of our speech or writing, it occasionally happens that we mis-pronounce or mis-spell a word, by introducing into it a letter or syllable of some other whose turn is shortly to come; or, it may be, the whole of the anticipated word is substituted for the one which ought to have been expressed. Now it is obvious that there could be neither any consciously-formed intention of breaking the regular sequence, nor any volitional effort to do so; and the result is evidently the automatic expression of the Idea represented by the anticipated word, which interferes with the working

out of that which we have previously given it in charge to our automaton to execute.—An interesting example of this familiar phenomenon (which, like many other mental phenomena, has not attracted the notice it merits, simply because it *is* so familiar) is given us by the amanuensis to whom Sir Walter Scott dictated his "Life of Napoleon Buonaparte" :—

a. "His thoughts flowed easily and felicitously, without any difficulty to lay hold of them or to find appropriate language ; which was evident by the absence of all solicitude (*miseria cogitandi*) from his countenance. He sat in his chair, from which he rose now and then, took a volume from the book-case, consulted it, and restored it to the shelf,—all without intermission in the current of ideas, which continued to be delivered with no less readiness than if his mind had been wholly occupied with the words he was uttering. It soon became apparent to me, however, that he was carrying on two distinct trains of thought, one of which was already arranged, and in the act of being spoken, while at the same time he was in advance, considering what was afterwards to be said. This I discovered by his sometimes introducing a word which was wholly out of place—*entertained* instead of *denied*, for example,—but which I presently found to belong to the next sentence, perhaps, four or five lines further on, which he had been preparing at the very moment that he gave me the words of the one that preceded it."—(*Life of Sir Walter Scott*, Chap. lxxiii.)

237. It is the *dominant Idea*, then, which really *determines* these movements, the Will simply *permitting* them ; and the more completely the volitional power is directed to other objects, the more completely automatic are the actions of this class. They may, indeed, come to be performed even without the consciousness, or at least without the *remembered* consciousness, of the agent ; as we see in the case of those who have the habit of "thinking aloud," and who are subsequently quite surprised on learning what they have uttered. The one-sided conversation of some persons, who are far more attentive to their own trains of thought, than they are to what may be expressed by others, and who are

allowed to proceed with little or no interruption, is often a sort of "thinking aloud."—This was pre-eminently the habit of Coleridge, whose whole life was little else than a waking dream, and whose usual talk has been shown to have been the outpouring of his "dominant ideas." (See § 231.)—The following case, recently communicated to the Writer, shows how strongly the *mode of expression* of our ideas is influenced by habit; and how, after the chain would seem to have been completely broken, it may come to renew itself when the circumstances recur under which it had been formed :—

b. A Military Officer, who had seen much hard service at a time when a command was scarcely ever given without the accompaniment of an oath, and who had thus acquired the habit of continual swearing, determined, on retiring into private life, to do his best to forego this practice; and by keeping a constant check upon himself, with the assistance of the friendly monitions of others, he entirely succeeded. After the lapse of many years, however, he found himself called upon to perform some Military duty; and, in the discharge of it, he used much of the bad language to which he had formerly accustomed himself. A friend who happened to notice this, having afterwards expressed his regret that he should have relapsed into his old habit of swearing, the Officer assured him (and he was a man whose word could be implicitly relied on) that *he was not at the time in the least degree conscious of uttering an oath, and that he had not the slightest recollection of having done so.*

238. Much attention has recently been given to a set of Involuntary movements, which, however diverse the circumstances under which they occur, all have their source in one and the same mental condition,—that of *expectant attention* :—the whole Mind being "possessed" with the idea that a certain action will take place, and being eagerly directed (generally with more or less of emotional excitement) towards the indications of its occurrence. This is a very curious subject of inquiry, and one on which adequate scrutiny has scarcely yet been bestowed; the phenomena

which are referable to the principle of action here enunciated, having been very commonly explained by the agency of some other hypothetical Force.—Thus, if a button or ring be suspended from the end of the finger or thumb, in such a position that, when slightly oscillating, it shall strike against a glass tumbler, it has been affirmed by many who have made the experiment, that the button continues to swing with great regularity, striking the glass at tolerably-regular intervals, until it has sounded the hour of the day, after which it ceases for a time to swing far enough to make another stroke. This certainly does come to pass, in many instances, without any *intention* on the part of the performer; who may be really doing all in his power to keep his hand perfectly stationary. Now it is impossible, by any voluntary effort, to keep the hand absolutely still, for any length of time, in the position required; an involuntary tremulousness is always observable in the suspended body; and if the *attention* be fixed upon the part, with the *expectation* that the vibrations will take a determinate direction, they are very likely to do so.*—Their persistence in this direction, however, *only takes place so long as they are guided by the visual sensations;* a fact which at once points to the real spring of their performance. When the performer is impressed with the conviction that the hour *will* be thus indicated, the result is very likely to happen; and when it has once occurred, his confidence is sufficiently established to make its recurrence a matter of tolerable certainty. On the other hand, the experiment seldom succeeds with sceptical subjects; the "expectant idea" not having in them the requisite potency. That it is through the Mind that these movements are regulated, however involuntarily, appears evident from these two considerations:—first, that if the performer be

* This was long since pointed out by M. Chevreul, who investigated the subject in a truly philosophic spirit. See his letter to M. Ampère, in the "Revue des Deux-Mondes," Mai 1833; and his more recent treatise "De la Baguette Divinatoire, du Pendule dit Explorateur, et des Tables Tournantes," Paris, 1854.

entirely ignorant of the hour, the strokes on the glass do not indicate its number, except by a casual coincidence ;—and second, that the division of the entire period of the earth's rotation into twenty-four hours, and the very nomenclature of these hours, being entirely arbitrary and conventional, no other *modus operandi* can be imagined. For example, the button which strikes *eleven* at night in London, should strike *twenty-three* in Italy, where (as in the astronomer's Observatory) the cycle of hours is continued through the whole twenty-four; and if an Act of Parliament were to introduce the Italian horary arrangement into this country, all the swinging buttons in her Majesty's dominions would have to add twelve to their number of post-meridiem strokes; all which would doubtless come to pass, if the experimenters' *expectation* of the result were sufficiently strong. These phenomena, in which no hypothetical "odylic" or other "occult" agency can be reasonably supposed to operate, are here alluded-to only for the sake of illustrating those next to be described, which have been imagined to prove the existence of a new Force in Nature.

239. If "a fragment of anything, of any shape," be suspended from the end of the fore-finger or thumb, and the Attention be intently fixed upon it, regular oscillations will be frequently seen to take place in it; and if changes of various kinds be made in the conditions of the experiment, corresponding changes in the direction of the movements will very commonly follow.

a. The public mind was directed to these facts, about the year 1850, by Dr. Herbert Mayo; who, having brought himself to accept Baron Reichenbach's "Odyle" as a "new force in Nature," accepted these oscillations as a manifestation of it, and gave to this suspended body the designation of "odometer." After varying his experiments in a great variety of modes, Dr. Mayo came to the conclusion that the direction and extent of the oscillations were capable of being altered, either by a change in the nature of the substances placed beneath the odometer, or by the contact of the hand of a person of the opposite sex, or even of the experimenter's other hand, with that from which the odometer was suspended, or by various other changes of the like nature. And he gradually reduced his results to a series of definite Laws, to which he seems to have imagined them to be as amenable, as are the motions of the heavenly bodies to the law of Gravitation.—(*The Truths contained in Popular Superstitions*, 3rd edition, 1851, Letter XII.)

b. Other observers, however, who were induced by Dr. Mayo's earlier experiments to take-up the subject, and who worked it out with like perseverance and good faith, framed a very different code; so that it at once became apparent to those who knew the influence which "expectant attention" exerts in determining involuntary muscular movements, that this was only another case of the same kind; and that the cause of the change of direction in each case lay in the Idea that some such change would result from a certain variation in the conditions of the experiment. Hence the general conclusions which each experimenter works out for himself, so far from being entitled to rank as "laws of Odylic force," are merely expressions of what has been passing (though perhaps almost unconsciously to himself) *in his own mind.*—The truth of this *rationale* was proved by the results of a few very simple variations in the conditions of the experiment. When it was tried upon *new* subjects, who were entirely devoid of any expectant idea of their own, and who received no intimation, by word, sign, or look, of what was anticipated by others, the results were found to have no uniformity whatever. And even those who had previously been most successful in this line of performance, *found all their success vanish, from the moment that they withdrew their eyes from the oscillating body*, its movements thenceforth presenting no regularity whatever.—Thus it became obvious that the definite direction which the oscillations previously possessed, was due, not to any Magnetic, Electric, or Odylic force of which the operator was the medium, but to the influence directly exercised by his Ideas over his muscles, under the guidance of his visual sense.

240. Now this will occur, notwithstanding the strong Volitional determination of the experimenter to maintain a complete immobility in the suspending finger. And it is very easily proved that, as

in the preceding case, the movements are guided by his Visual sensations, and that the impulse to them is entirely derived from his *expectation* of a given result. For if he be ignorant of the change which is made in the conditions of the experiment, and should *expect or guess something different from that which really exists*, the movement will be in accordance with his Idea, not with the reality :—

a. Rather more than twenty years ago, when no inconsiderable portion of the British public was amusing itself with swinging buttons and rings from its finger-ends, the attention of Scientific men was invited by Mr. Rutter of Brighton to the fact, that a very definite series of movements of a like kind was exhibited by a ball suspended from a metallic frame *which was itself considered a fixture*, when the finger was kept for a short time in contact with it; and that these movements varied in direction and intensity, according as the operator touched other individuals with his disengaged hand, laid hold with it of bodies of various kinds, or altered his condition in various other modes. These experiments appeared to many persons of great general intelligence, to indicate some new and mysterious agency not hitherto recognised in our philosophy; for even among those who might be disposed to attribute the oscillations of a button suspended from the *finger*, to the involuntary movements of the hand itself, some were slow to believe that the simple *contact* of the finger with a frame of *solid metal* could produce the like vibrations through such a medium. Yet there were certain troublesome sceptics, who persisted in asserting that this was but pro another case of "expectant attention;" and such it was soon proved to be.

b. The mode in which the *dénouement* took place, however, was not a little curious. Among Mr. Rutter's disciples was a Homœopathic Physician at Brighton, Dr. H. Madden; who conceived the notable idea of testing the value of the indications of the Magnetometer (as it was called), by questioning it as to the characters of his remedies, in regard to which he was of course himself "possessed" with certain foregone conclusions. Globules in hand, therefore, he consulted its oscillations, and found that they corresponded exactly with his idea of what they *ought* to be; a medicine of one class producing *longi-*

tudinal movements, which were at once exchanged for *transverse* when a medicine of opposite virtues was substituted for it. In this way Dr. Madden was systematically going through the whole Homœopathic Pharmacopœia; when circumstances led him to investigate the subject *de novo*, with a precaution which had never occurred to him as requisite in the first instance, but of which the importance is obvious to every one who holds the real clue to the mystery;—namely, that he *should not know* what were the substances on which he was experimenting, the globules being placed in his hand by another party, who should give him no indication whatever of their nature. From the moment that he began to work upon this plan, the whole aspect of affairs was altered. *The results ceased altogether to present any constancy.* Oscillations at one time transverse, at other times longitudinal, were produced by the very same globules; whilst remedies of the most opposite kinds frequently gave no sign of difference. And thus, in a very short time, Dr. Madden was led to the conviction, which he avowed with a candour that was very creditable to him, that the whole system which he had built-up had no better foundation than *his own anticipation of what the results should be.*—(*Lancet*, Nov. 15, 1851.)

241. This case—which seems so easily disposed of by the phrase "all humbug," or "all imagination,"—is, in truth, neither the one nor the other; but a singularly complete and satisfactory example of the general principle, that, in certain individuals, and in a certain state of mental concentration, the *expectation* of a result is sufficient to determine,—without any voluntary effort, and even in opposition to the Will (for this may be honestly exerted in the attempt to keep the hand perfectly unmoved),—the Muscular movements by which it is produced. It is obvious, too, that the unconscious rhythmical motion of the hand constituted the *vera causa* of the vibrations of the magnetometer: a fact which will not surprise any one who knows how difficult it is to prevent the tremors of a Telescope or a Microscope, by the most careful construction of its supporting frame-work; or who bears in mind that the form of the great speculum of Lord Rosse's telescope, weighing

five tons, having a thickness of eight inches, and composed of the hardest known combination of metals, is perceptibly altered (as is demonstrated by the immediate impairment of the distinctness of its reflection) by a moderate pressure of the hand against its back. Moreover, as Dr. Madden justly remarked, the arrangement of Mr. Rutter's apparatus was such as to admit of the greatest sensible effect being produced by the smallest amount of imparted motion; and every modification of it which increased its immobility, decreased in the same proportion its apparent sensibility to the so-called "magnetic currents." It was further ascertained that no definite vibrations took place, unless the pendulum was *watched;* showing that, as in the preceding cases, the guidance of the visual sense was required to determine their direction. It is a curious example, however, of the hold which the belief in the "occult" has upon the Imagination, that, notwithstanding the complete proof thus given of the dependence of these vibrations upon the unconscious movements of the operator himself, the vague hypothesis of "human electricity" long continued to be entertained by Mr. Rutter and his disciples; just as the Spiritualists of the present day will not accept Faraday's demonstration (§ 245) that tables are really "turned" and "tilted" by the pressure of the hands placed on them,—refusing to submit the question to the test of Physical experiment, because (as they say) it *cannot* negative *their own conviction* that they are exerting no pressure whatever.

242. It is clearly on the very same Physiological principle, that we are to explain the mysterious phenomena of the "Divining rod;" which have been accepted as true, or rejected as altogether fabulous, according to the previous habits of thought of those who have given their attention to the subject.—That the end of a hazel-fork, whose limbs are grasped firmly in the hands of a person whose good faith can scarcely be doubted, frequently points upwards or downwards without any *intentional* direction on his own

part, and often thus moves when there is *metal* or *water* beneath the surface of the ground at or near the spot, is a fact which is vouched-for by such testimony that we have scarcely a right to reject it; and when we come to examine into the conditions of the occurrence, we shall find that they are such as justify us in attributing it to a state of *expectant attention*, which (as we have seen) is fully competent to induce muscular movement. For, in the first place, as not above one individual in forty, even in the localities where the virtues of the divining-rod are still held as an article of faith, is found to succeed in the performance of this experiment, it is obvious that the agency which produces the deflection—whatever be its nature—must operate by affecting the *holder* of the rod, and not by attracting or repelling the *rod* itself. And when experiments are carefully made with the view of determining the nature of this agency, they are found to indicate most clearly that the state of expectant attention, induced by the *anticipation* of certain results, is fully competent to produce them. For the mere act of holding the rod for some time in the required position, and of attending to its indications, is sufficient to produce a tendency to spasmodic contraction in the grasping muscles, notwithstanding a strong effort of the Will to the contrary; and when, by such contractions, the limbs of the fork are made to approximate-towards or to separate-from each other, the point of the fork will be caused to move either upwards or downwards, according to the position in which it is held. If, when the muscles have this tendency to contract, occasioned by their continued restraint in one position, the mind be possessed with the expectation that a certain movement will ensue, that movement will actually take-place, even though a strong effort may be made by the Will to prevent any change in the condition of themuscles. And a sufficient ground for such expectation exists, on the part of those who are "possessed" by the idea of the peculiar powers of the divining-rod, in the belief, or even in the surmise, that *water* or *metal* may lie beneath particular points of the surface over which they pass.

a. Thus Dr. H. Mayo, notwithstanding his belief in the existence of an "Od-force" governing the movements of the divining-rod, admitted that he found in the course of his experiments, that when his "diviner" knew which way he expected the fork to move, it invariably answered his expectations; but when he had the man blindfolded, the results were uncertain and contradictory. Hence he became certain that several of those in whose hands the divining-rod moves, set it in motion, and direct its motion (however unintentionally and unconsciously) by the pressure of their fingers, and by carrying their hands near-to or apart-from each other.—(*Op. cit.*, Letter I.)

b. The following statement of the results obtained by a very intelligent friend of the Writer, who took up the inquiry some years ago, with a strong prepossession (derived from the assurances of men of high scientific note) in favour of the reality of the supposed influence, but yet with a desire to investigate the whole matter carefully and philosophically for himself, will serve as a complete illustration of the doctrine enunciated above:—Having duly provided himself with a hazel-fork, he set out upon a survey of the neighbourhood in which he happened to be staying on a visit; this district was one known to be traversed by Mineral Veins, with the direction of some of which he was acquainted. With his "divining-rod" in his hand, and with his attention closely fixed upon his instrument of research, he walked forth upon his experimental tour; and it was not long before, to his great satisfaction, he observed the point of the fork to be in motion, at the very spot where he knew that he was crossing a metallic lode. For many less cautious investigators, this would have been enough; but it served only to satisfy this gentleman that he was a favourable subject for the trial, and to stimulate him to further inquiry. Proceeding in his walk, and still holding his fork *secundum artem*, he frequently noticed its point in motion, and made a record of the localities in which this occurred. He repeated these trials on several consecutive days, until he had pretty thoroughly examined the neighbourhood, going over some parts of it several times. When he came to compare and analyse the results, he found that there was by no means a satisfactory accordance amongst them; for there were many spots over which the rod had moved on one occasion, at which it had been obstinately stationary on others, and *vice versâ;* so that the constancy of a physical agency seemed altogether wanting. Further, he found that whilst some of the spots over which the rod had moved, were those *known* to be traversed by Mineral Veins, there were many others in which its indications had been no less positive, but in which those familiar with the Mining Geology of the neighbourhood were well assured that no veins existed. On the other hand, the rod had remained motionless at many points where it *ought* to have moved, if its direction had been affected by any kind of terrestial emanation.—These facts led the experimenter to a strong suspicion that the cause existed in *himself* alone; and by carrying out his experiments still further, he ascertained that he could not hold the fork in his hand for many minutes consecutively, concentrating his attention fixedly upon it, without an alteration in the direction of its point, in consequence of an involuntary though almost imperceptible movement of his hands; so that in the greater number of instances in which the rod exhibited motion, the phenomenon was clearly attributable to this cause; and it was a matter of pure accident whether the movement took place over a Mineral Vein, or over a blank spot. But further, he ascertained on a comparison of his results, that the movement took-place more frequently where he knew or suspected the existence of mineral veins, than in other situations; and thus he came, without any knowledge of the theory of *expectant attention*, to the practical conclusion that the motions of the Rod were produced by his own Muscles, and that their actions were in great degree regulated automatically by the Ideas which possessed his mind.

The same instrument appears to have been used, even from a very early period, by those who were supposed to possess "a spirit of divination," for the purpose of giving replies to questions by its movements, precisely after the fashion of the "talking tables" of our own day; the hands of the operators (where they really believed in their power, and were not impostors) being automatically impelled to execute the appropriate movements of the rod, either by their consciously-formed idea of what the answer should be,* or by Cerebral changes which excite reflex movements that give expression to them, without themselves rising into the "sphere of consciousness" (§§ 424, 425).

243. No difficulty can be felt by any one who has been led by the preceding considerations to recognize the principle of *Ideo-motor action*, in applying this principle to the phenomena of "Table-turning" and "Table-talking;" which, when rightly analysed, prove to be among the very best examples of the "reflex action of the Cerebrum," that are exhibited by individuals whose state of mind can scarcely be considered as abnormal. The *facts*, when stripped of the investment of the marvellous with which they have too commonly been clothed, are simply as follows:—A number of individuals seat themselves round a table, on which they place their hands, with the *idea* impressed on their minds that the table will move in a rotatory direction; the direction of the movement, to the right or to the left, being generally arranged at the commencement of the experiment. The party sits, often for a considerable time, in a state of expectation, with the whole attention fixed upon the table, and looking eagerly for the first sign of the anticipated motion. Generally one or two slight changes in its place herald the approaching revolution; these tend still more to excite the eager attention of the performers, and then the actual "turning" begins. If the parties retain their seats, the revolution only continues as far as the length of their arms will allow; but not unfrequently they all rise, feeling themselves obliged (as they assert) to *follow* the table; and from a walk, their pace may be accelerated to a run, until the table actually spins-round so fast that they can no longer keep-up with it. All this is done, not merely without the least consciousness on the part of the performers that they are exercising any force of their

* See Chevreul, Op. cit., première partie.

own, but for the most part under the full conviction that they are not.

244. Now the *rationale* of these and other phenomena of a like kind, is simply as follows. The continued concentration of Attention upon a certain idea gives it a *dominant* power, not only over the mind, but over the body; and the muscles become the involuntary instruments whereby it is carried into operation. In this case, too, as in that of the divining-rod, the movement is favoured by the state of muscular tension, which ensues when the hands have been kept for some time in a fixed position. And just as in the case of the victims of the Dancing Mania (§ 259), it is by the continued influence of the "dominant idea" that the performers are impelled to *follow* (as they believe) the revolution of the table, which they really *maintain* by their continued propulsion. However conscientiously they may believe that the "attraction of the table" carries them along with it, instead of the table being propelled by an impulse which originates in themselves, yet no one feels the least difficulty in withdrawing his hand, if he really *wills* to do so. But it is the characteristic of the state of mind from which these Ideo-motor actions proceed, that the Volitional power is for the time in abeyance; the whole mental power being absorbed (as it were) in the high state of tension to which the Ideational consciousness has been wrought-up.

245. The demonstration that the table is really moved by the hands placed upon it, notwithstanding the positive conviction of the performers to the contrary, was first afforded by the very ingenious "indicator," devised by Professor Faraday, which is constructed as follows:—

A couple of boards of the size of a quarto sheet of paper, a couple of small rulers or cedar-pencils, a couple of india-rubber bands, a couple of pins, and a strip of light wood or cardboard eight or ten inches long, constituted its materials. The rulers being laid on one of the boards, each at a little distance from one of its sides and parallel to it, the other board was laid upon the rulers, so that it would roll on them from side to side; and its movements were restrained, without being prevented, by stretching the india-rubber bands over both boards, so as to pass above and beneath the rulers. One of the pins was fixed upright into the lower board close to the middle of its farther edge, the corresponding part of the upper being cut away at that part, so that the pin should not bear against it; the second pin was fixed into the upper board, about an inch back from the first; and the strip of wood or cardboard was so fixed on these pins, as to constitute a lever of which the pin on the lower board was the fulcrum, while motion was imparted to the short arm of it by the pin on the upper board. Any lateral motion given to the upper board by the hands laid upon it, would thus cause the index-point of the long arm of the lever to move through a long arc in the opposite direction; the amount of that motion being dependent on the ratio between the long and the short arms of the lever.

The first point tested by Faraday, in the spirit of the true Philosopher, was whether the interposition of his "indicators" between the hands of the operators and the table in any way interfered with the movements of the latter; and he found, by tying the boards together, and taking off the index, that no such interference was observable, the table then going round as before. When, however, the upper board was free to move, and each performer fixed his (or her) eyes upon the index, so as to be made cognizant by its movement of the slightest lateral pressure of the hands, any communication of motion to the table was usually kept in check; but if the table did go round under this condition, its motion was always preceded by a very decided movement of the index in the opposite direction. And the same indication was given when the index was hidden from the operator, but was watched by another person; any movement shown by the table under that condition being always preceded by a considerable motion of the index in the opposite direction. And thus it may be considered as demonstrated that as the table never went round unless the "indicator" showed that lateral muscular pressure had been exerted in the direction of its movement, and as it always did go round when the "indicator" showed that such lateral pressure was adequately exerted, *its motion was solely due to the unconscious muscular action of the performers.**

246. A sufficient explanation of these wonders, then, being found in the known principles of Mental Physiology, it is against all the rules of Philosophy to assume that any other force is concerned in their production. Yet experience has shown that when the Common Sense of the public once allows itself to be led away by the *love of the marvellous*, there is nothing too monstrous for its credulity. And the greatest difficulty in this case was to convince the performers that the movement of the table was really due to the impulse which it received from their hands: their conviction being generally most positive, that *as they were not conscious* of any effort, the table *must* have been propelled by some other agency, and that their hands were drawn along by its attraction. So resolutely was this believed, that when the table was intentionally prevented from moving by the determined pressure of the hands of one of the parties, so that those of another —automatically moving in the expected direction—slid over its surface, instead of carrying the table with them, the fact, instead of being received as evidence that the hands *would* have moved the table had it been free to turn, was set down to a "repulsive" influence exerted by the table on the hands! It might have been thought that Common Sense would teach, that, if half-a-dozen persons lay their hands on a table, any movements which it executes are to be fairly attributed to muscular force communicated by them, until proof shall have been

* See his memorable letter on Table Turning, in the "Athenæum," of July 2, 1853.—It would be well that experimenters on "Psychic Force" should profit by the admirable models set before them in this Letter, and in the Treatise "De la Baguette Divinatoire" of M. Chevreul, by two of the greatest Masters of Experimental Science.

given to the contrary; and that the *absence of conscious effort* on the part of the performers is no valid proof to the contrary, since it is within the experience of every one that *muscular movements are continually being executed without such effort*,—as in the case of a man who continues to walk, to read aloud, or to play on a musical instrument, whilst his whole Attention is given to some train of thought which deeply interests him. But the table-turners would seldom listen to Common Sense, so completely were they engrossed by their dominant idea. And even when Professor Faraday's "indicator" had supplied the most unequivocal proof that the movement of the table, instead of anticipating and producing that of the hands, is *consequent* upon the pressure which *they* impart, this proof was disposed of by the simple assertion that it had nothing to do with the case; inasmuch as it only showed that Professor Faraday's performers moved the tables with their hands, whereas "*we* KNOW *that we do not.*" Those who make this assertion are (of course) scientifically bound to demonstrate it, by showing that in *their* case the table *does* go round without any deflection of the index by lateral pressure; but they have uniformly refused to apply this test to their own performance, though repeatedly challenged to do so,—in the very spirit of the opponents of Galileo, who would not look through his telescope at the satellites of Jupiter, because they supplied evidence in favour of the Copernican theory.

247. In the investigation of these phenomena, moreover, it was found necessary to treat with complete disregard all the testimony of such as had given themselves up to the "domination" of the table-turning "idea;" for it continually became apparent that – no doubt, quite unintentionally and unconsciously—they would omit from their narrative the point most essential to the elucidation of the mystery :—

Thus, the Writer's scepticism was on one occasion gravely rebuked by a lady, who assured him that, in *her* house, a table had moved round and round, *without being touched.* On inquiring into the circumstances, he found that a hat had been placed upon the table, which was very small and light, and the hands of the performers upon the hat; but the narrator was as sure that the hat could not have carried the table along with it, as she was that the hat moved round without any mechanical force communicated from the hands!—In another case, again, the Writer was seriously informed that a table had been moved round by the *will of a gentleman sitting at a distance from it;* but it came out, upon cross-examination, that a number of hands were laid upon it in the usual way, and that after the performers had sat for some time in silent expectation, the operator called upon the spirit of "Samson" to move the table, which obediently went round:—the rationale being obvious enough to any one who reflects upon the analogy of the whole group with an Electro-biological "operator" and his "subjects" (§ 452).

A long list might be given of similar absurdities; the Writer's experience of which most fully confirmed the conclusions he had previously been led to form, in regard to the want of credibility which attaches to all *testimony* borne by the champions of Mesmerism to the wonders which they declare themselves (doubtless most honestly) to have witnessed; while it prepared him for finding exactly the same sources of fallacy, in the testimony on which the scientific inquirer is called on to accept the marvels of "Spiritualism" (§§ 365, 366).

248. The application of the same principle to the ordinary phenomena of "Table-talking," is no less obvious. There can be no reasonable doubt that these phenomena are manifested in a large number of instances, through the agency of individuals who would not wilfully be parties to deception of any kind; and that the movements which they *involuntarily* and *unconsciously* gave to the tables, are the expressions of the Ideas with which their own Minds are "possessed," as to what the answers should be to the questions propounded. Thus when, in 1853, "Table-talking" first grew out of "Table-turning," several Clergymen, strongly impressed with the belief that it was a manifestation of

Satanic agency, put to the tables a series of what they regarded as test-questions, or performed test-experiments, the responses to which would (as they supposed) afford convincing proof of their hypothesis.

In his *Table-moving tested*, the Rev. N. S. Godfrey began by "tracing the existence of Satanic influence from the time of Moses to the time of Jesus; connecting the 'witch,' the familiar spirit,' the spirit of Python, with the Evil Spirit in its actual and separate existence:" and asserted without the least hesitation, that although "so long as the supernatural gifts of the Spirit remained among men, so long the evil spirits were cast out and their presence detected," yet that when those miraculous powers were withdrawn, they could no longer be discerned, but have continued to exist to the present time, and make themselves known in these "latter times" as the "wandering (seducing) spirits," whose appearance was predicted by St. Paul (1 Tim., iv., 10). That the answers to the "test questions" were exactly contrary to Mr. Godfrey's ideas of truth, was, in his judgment, peculiarly convincing; "for if indeed these tables do become possessed by some of the 'wandering spirits' at the command of the Devil, it would be most impolitic, and quite at variance with the subtlety of his character, to scare people at the very outset." The following answers, therefore, are obviously what Mr. G. expected:—

"I spoke to the table, and said, 'If you move by electricity, stop.' It stopped instantly! I commanded it to go on again, and said, while it was moving, 'If an evil spirit cause you to move, stop.' It moved round without stopping! I again said, 'If there be any evil agency in this, stop.' It went on as before. I was now prepared with an experiment of a far more solemn character. I whispered to the schoolmaster to bring a small Bible, and to lay it on the table when I should tell him. I then caused the table to revolve rapidly, and gave the signal. *The Bible was gently laid on the table, and it instantly stopped.* We were horror-struck. However, I determined to persevere. I had other books in succession laid on the table, to see whether the fact of a book lying upon it altered any of the conditions under which it revolved. It went round with them without making any difference. I then tried with the Bible four different times, and each time with the same result: *it would not move so long as that precious volume lay upon it.* * * I now said, 'If there be a hell, I command you to knock on the floor with this leg (the one next me) twice.' It was motionless. 'If there be *not* a hell, knock twice;' no answer. 'If there be a devil, knock twice;' no motion. 'If there be *not* a devil, knock twice;' *to our horror the leg slowly rose and knocked twice!* I then said, 'In the name of the Lord Jesus Christ, if there be *no* devil, knock twice;' it was motionless. This I tried four several times, and each time with the same result."

249. It is clear that Mr. Godfrey and his associates, if they had not distinctly *anticipated* these results, were fully *prepared* for them. Thus although he assures his readers that, when the Bible was placed on the table, the emotion in the minds of all the parties was *curiosity*, and that, if they *had* a bias, it was *against* the table stopping, the very fact of the experiment being tried by a man imbued with his prepossessions on the subject of Evil Spirits, Witchcraft, &c., sufficiently indicates what his *real* state of mind was, although he may not have been himself aware of it (§ 252 *c*). His *involuntary* muscular actions responded to this, although no *voluntary* movement would have done so, because he had not *consciously* accepted the Idea which had been shaping itself in the under-stratum. The experience of every one must have convinced him that there is often a contrariety between our *beliefs as to our own states of mind*, and the *facts* of that state as they afterwards come to be *self-revealed* to us (§ 439); and it is a very marked peculiarity of these movements, that they often express more truly what is buried (as it were) in the vaults of our storehouse, than what is displayed in the ware-rooms above.—The Rev. E. Gillson, M.A., a Clergyman of Bath, fully partaking of his predecessor's convictions on the subject of Satanic Agency, and also in the excitement prevailing in many circles at that time on the subject of "Papal aggression," gave the following *inter alia* as his experiences (*Table-Talking: Satanic Wonders and Prophetic Signs*, 1853):—

"I placed my hand upon the table, and put a variety of questions, all of which were instantly and correctly answered. Various ages were asked, and all correctly told. In reply to trifling questions, possessing no particular interest, the table answered by quietly lifting up the leg and rapping. But in answer to questions of a more exciting character, it would become violently agitated, and sometimes to such a degree that I can only describe the motion by the word *frantic*. I inquired, 'Are you a departed spirit?' The answer was 'Yes,' indicated by a rap. 'Are you unhappy?' The table answered by a sort of writhing motion (!), which no natural power over it could imitate. It was then asked, 'Shall you be for ever unhappy?' The same kind of writhing motion was returned. 'Do you know Satan?' 'Yes.' 'Is he the Prince of Devils?' 'Yes.' 'Will he be bound?' 'Yes.' 'Will he be cast into the abyss?' 'Yes.' 'Will you be cast in with him?' 'Yes.' 'How long will it be before he is cast out?' He rapped *ten*. 'Will wars and commotions intervene?' The table rocked and reeled backwards and forwards for a length of time, as if it intended a pantomimic acting of the prophet's predictions (Isaiah xxiv., 20). I then asked 'Where are Satan's head-quarters? Are they in England?' There was a slight movement. 'Are they in France?' A violent movement. 'Are they in Spain?' Similar agitation. 'Are they at Rome?' *The table literally seemed frantic.* At the close of these experiments, which occupied about two hours, the invisible agent, in answer to some questions about himself, did not agree with what had been said before. I therefore asked, 'Are you the same spirit that was in the table when we began?' 'No.' 'How many spirits have been in the table this evening?' 'Four.' This spirit informed us that he had been an infidel, and had embraced Popery about five years before his death. Amongst other questions, he was asked, 'Do you know the Pope?' The table was violently agitated. I asked, 'How long will Popery continue?' He rapped ten; exactly coinciding with the other spirits' account of the binding of Satan. Many questions were asked, and experiments tried, in order to ascertain whether the results would agree with Mr. Godfrey's; and on every occasion they did, *especially that of stopping the movement of the table with the Bible.* As we proceeded with our questions, we found an indescribable facility in the conversation, from the *extraordinary intelligence and ingenuity displayed in the table* (!) *E. g.*—I inquired if many devils were posted in Bath. He replied by the most extraordinary and rapid knocking of the three feet in succession, round and round, for some time, as if to intimate that they were innumerable!"

250. A third Clergyman, the Rev. R. W. Dibdin, M.A.,—who communicated to the public the results of his experiences in a Lecture at the Store Street Music Hall, Nov. 8, 1853,—while agreeing with his predecessors in the belief that the movements of the tables are the result of Satanic (or diabolic) agency, differed from them in maintaining "that devils alone (not departed spirits) are the agents in these cases; and being *lying* spirits, it is quite credible that, for purposes of their own, they might *assume* the names of departed men and women." Of course he got the answers he expected on this hypothesis. The following is his set of 'test-questions,' the answers to which—being entirely *opposed* to his own notions of truth—satisfied *him*, and were expected to satisfy his partners in the experiment, of the *diabolical* character of the respondent:—

"'Are we justified by works?' 'Yes.'—'By faith alone?' 'No.'—'Is the whole Bible true?' 'No.'—'Were the miracles of the New Testament wrought by supernatural power?' 'No.'—'By some hidden law of Nature?' 'Yes.'—'Was Oliver Cromwell good?' 'No.'—'Was Charles I. a good man?' 'Yes.'—'Is it right to pray to the Virgin?' 'Yes.'—'Is Christ God?' 'No.'—'Is he a man?' 'No.'—'Is he something between God and man, a sort of angel?' 'Yes.'—'Is he in heaven?' 'No.'—'Where is he?' It spelt slowly H E L L.—As the last letter was indicated, the girl drew her hands quickly off the table, much as a person would do who was drawing them off a hot iron. Her brother-in-law turned very pale, and took his hands off the table also."

251. These phenomena have been cited in fuller detail than may seem requisite; because the character, position, and obvious sincerity of the actors and narrators place them beyond suspicion of intentional deception; and because they afford a singularly

apposite illustration of the principle which the Writer desires to enforce. But that such obvious products of the questioners' own mental states should have been accepted by men of education, occupying the position of religious teachers in the National Church, as the lying responses of *evil* spirits, sent expressly to delude them, can only be deemed—by such, at least, as are prepared to accept a scientific *rationale* of the phenomena—a pitiable instance of the readiness with which minds of a certain type may allow themselves to become "possessed" by dominant ideas.

252. Absurd as their belief may now seem, however, it is in no respect more destitute of foundation than that which is entertained at the present time, by multitudes of persons of high culture and great general intelligence, in the genuineness of messages supposed to be transmitted by *good* "spirits" of departed relatives and friends, to those whom they have left behind them on Earth. These communications always take place through *human agency* of some kind; the individual who is the supposed recipient of them being termed a "medium." The mode of intercourse with "spirits" afforded by the turning and tilting of tables, has now for the most part given place to others of a much simpler and more direct character. Some "mediums" use a small wooden platform, only large enough for the hands to be laid on it, and running easily on castors. This *planchette* was in the first instance furnished with a pointer, which directed itself in succession to the letters or figures of an alphabet-card placed on the table over which it rolled; and thus spelled out words, or indicated numbers. But a simpler process than this has since come into vogue; for if a pencil be attached to the under side of the "planchette," with its point downwards, it will write on a piece of paper placed beneath it, in accordance with the movement of the planchette under the hands of the "medium" laid upon it. In each case, the "mediums" declare that the movements of the planchette are not produced by any manual exertion of their own, but that they are guided by

some agency external to themselves, their hands being simply passive. But other "mediums" take the pencil into their own hands, and write (in the ordinary way) what they conceive to be the messages dictated to them by the "spirits" with which they are in communication. And some, again, carry on supposed conversations with the "spirits;" not only asking questions of them by word of mouth, but giving forth through the same direct channel, the answers which they affirm that they receive.—Now there can be no reasonable doubt that a great many of these phenomena are *genuine* to this extent, that the "mediums" are honest, and believe themselves to be the vehicles of "spiritual" communications. Putting intentional deceit out of the question for the present (§ 254), it is perfectly obvious to such as have had adequate opportunities of studying the *natural* conditions of Reverie and Abstraction (§§ 443—447), and the *artificially-induced* states known as "Electro-biological" and "Hypnotic" (§§ 448, 493), that the condition of the Spiritualistic "mediums" is exactly parallel to that of the "subjects" in these states; and that the supposed communications are nothing else than products of their own automatic mental operations, guided by the principle of suggestion, and expressing themselves in accordance with a certain preformed conception of the mode in which the message is to be made known. Of the influence of such conceptions on the course of thought and action in these curious states, ample evidence will be given hereafter (§ 451, *et seq.*). At present it will be sufficient to cite—as illustrations of the action of dominant ideas of a totally different order from those which brought out the terrifying responses obtained by the clerical seers,—two cases more recently recorded as having occurred within his own experience, by the author of an article on "Spiritualism and its Recent Converts."

a. "Several years ago we were invited, with two medical friends, to a very select *séance*, to witness the performance of a lady, the Hon.

Miss N——, who was described to us as a peculiarly gifted 'medium;' not merely being the vehicle of 'spiritual' revelations of the most elevating character, but being able to convince incredulous philosophers like ourselves of the reality of her 'spiritual' gifts, by 'physical' manifestations of the most unmistakeable kind. Unfortunately, however, the Hon. Miss N—— was not in great force on the occasion of our visit; and nothing would go right It was suggested that she might be exhausted by a most successful performance which had taken place on the previous evening; and that 'the spirits' should be asked whether she stood in need of refreshment. The question was put by our host (a wine-merchant, be it observed), who repeated the alphabet *rapidly* until he came to N, and then went on *slowly;* the table tilted at P. The same process was repeated, until the letters successively indicated were P, O, R, T. But this was not enough. The spirits might prescribe either *port* or *porter;* and the alphabet was then repeated *slowly from the beginning*, a prolonged pause being made at E; as the table did *not* tilt, a bumper of port was administered 'as directed.' It did not, however, produce the expected effect; and with the exception of a 'manifestation' we shall hereafter notice under another head (§ 530), the *séance* was an entire failure.

b. "On another occasion, we happened to be on a visit at a house at which two ladies were staying, who worked the *planchette* on the original method (that of attaching to it a pointer, which indicated letters and figures on a card), and our long previous knowledge of whom placed them beyond all suspicion of anything but *self*-deception. One of them was a firm believer in the reality of her intercourse with the spirit-world; and her 'planchette' was continually at work beneath her hands, its index pointing to successive letters and figures on the card before it, just as if it had been that of a telegraph-dial acted on by galvanic communication. After having watched the operation for some time, and assured ourselves that the answers she obtained to the questions she put to her 'spiritual' visitants were just what her own simple and devout nature would suggest, we addressed her thus:—'*You* believe that your replies are dictated to you by your 'spiritual' friends, and that your hands are the passive vehicles of the 'spiritual' agency by which the planchette is directed in spelling them out. *We* believe, on the other hand, that the answers are the products of your own Brain, and that the planchette is moved by your own Muscles. Now we can test, by a very simple experiment, whether *your* view or *ours* is the correct one. Will you be kind enough to *shut your eyes* when you ask your question, and to let *us* watch what the planchette spells out? If 'the spirits' guide it, there is no reason why they should not do so as well when your eyes are shut, as when they are open. If the table is moved by your own hands, it will not give definite replies except under the guidance of your own vision.' To this appeal our friend replied that she could not think of making such an experiment, as 'it would show a want of faith;' and all our arguments and persuasions could only bring her to the point of *asking the spirits* whether she *might* comply with our request. The reply was, 'No.' She then, at our continued urgency, asked 'Why not?' The reply was, 'Want of faith.' Putting a still stronger pressure upon her, we induced her to ask, 'Faith in what?' The reply was, 'In God.'

c. "Of course, any further appeal in that quarter would have been useless; and we consequently addressed ourselves to our other fair friend, whose high culture and great general intelligence had prepared her for our own rationalistic explanation of marvels which had seriously perplexed her. For having been engaged a short time before in promoting a public movement, which had brought her into contact with a number of persons who had previously been strangers to her, she had asked questions respecting them, which elicited replies that were in many instances such as she declared to be quite unexpected by herself,—specially tending to inculpate some of her coadjutors as influenced by unworthy motives. After a little questioning, however, she admitted to us that she had previously entertained *lurking suspicions on this point*, which she had scarcely even *acknowledged to herself*, far less made known to others; and was much relieved when we pointed out that the planchette merely revealed what was going on in *the under-stratum of her own mind.* Her conversion to our view was complete, when, on her trying the working of the planchette with her eyes shut, its pointer *went astray altogether.*"—(*Quarterly Review*, Oct. 1871, p. 315.)

253. It is often cited as a proof that the performers are *not* expressing by involuntary muscular actions what is passing in

their own minds, that the answers given by the tables are *not known to any of themselves, though known to some other person in the room.* Of this an instance was early recorded by Mr. Godfrey, which corresponds in all essential particulars with cases repeatedly described to the Writer by persons in whose veracity he could place confidence :—

a. "I procured an alphabet on a board, such as is used in a National School; this board I laid down on the floor at some little distance from the table, and I lay down on the ground beside it. I then requested one of the three persons at the table to command it to spell the Christian names of Mr. L——, of B——, by lifting up the leg next him as I pointed to the letters of the alphabet in succession. He did so, and I began to point, keeping the pointer about three seconds on each letter in succession (I must say, that neither of the three persons at the table had ever heard of Mr. L——; and B—— is 150 miles from this place). When I arrived at G, they said, "That's it; the table is lifting its leg." When I came to E, it rose again; and in this way it spelt "George Peter," which was quite correct.

b. So, again, the late Dr. Hare, an American Chemist and Physicist of some reputation, thought that he had obtained *a precise experimental proof of the immortality of the soul* (!) by means of an apparatus by which the answers communicated through the "medium" were spelled out by a hand pointing to an alphabet-dial which was hidden from her eyes. But it is clear from his narrative of the experiment, that her eyes were fixed upon the person to whom the expected answer was known, and that her movements were guided by the indications she received from *his* Involuntary movements.

254. Such "movements of expression" constitute another very curious illustration of the general principle of Ideo-motor action. For the state of expectant attention from which they proceed is almost always mixed up with some degree of *emotional* excitement (§ 265); and there are many persons who *cannot*, by the strongest exercise of Volitional control, refrain from showing what

is the letter or figure they expect, when the pointer comes to it. Still more is this likely to be the case, when the questioner is not on his guard against this source of fallacy; so that, unless a screen be interposed between the "medium" and the person to whom the answer is known, there is no proof whatever of its being derived from any other source than *his* mind. This source of fallacy was very early found out by a sagacious observer, when "spirit rapping" was first introduced into this country by Mrs. Hayden:—

a. Mr. G. H. Lewes, having formed his own conclusions on the matter, from the accounts he had heard of Mrs. Hayden's performances from those who had witnessed them, took an opportunity of personally testing their correctness, with the most satisfactory result. He considered that Mrs. Hayden probably derived her indications as to the times at which to "rap," from some involuntary sign given by the questioner, when his pointer had arrived at the letter which should form the next component of the expected answer; this sign being either an unusual delay in passing to the next letter, or some slight look or gesture which would be perceived by an observer habitually on the watch for such indications. Accordingly, by *purposely* giving such indications, he caused Mrs. Hayden to *rap out* answers of the most absurdly erroneous character, to a series of questions which he had previously written down, and which he had also communicated to another member of the party, for the sake of negativing any subsequent charge of unfairness that might be raised against him; the only *true* reply being the one given to the final question—"Is Mrs. Hayden an impostor?" to which the answer was given by unhesitating raps, as his pointer came upon the letters Y, E, S.

b. The truth of this view of the case was soon confirmed by the results of many similar experiments; and a long series of ludicrous replies could be given, which were spelled-out on various occasions by the direction of waggish questioners. It thus became clear that the raps were made by the "medium" herself (it having been proved that the sounds *can* be produced by a movement in the foot, which shall not be perceptible even to those who are watching it), and that she derived her indications from *the promptings supplied by the questioners themselves, however unintentionally and even unconsciously on their own parts.* And this conclusion was fully borne out by a comparison of the conditions under which Mrs. Hayden was most successful, with those under which her failures (for many failures there were) took place. It was uniformly found that those whose questions had been most accurately and completely answered, were persons of excitable temperament and demonstrative habits, who were accustomed to signify more or less of what was passing in their minds by the automatic movements of gesture, expression, &c. On the other hand, those to whom "the spirits" would give no information, were persons of comparatively imperturbable nature, possessing considerable command over their muscles, and habitually yielding very little to those influences which so strongly manifest themselves in individuals of the opposite temperament. And on one occasion, an eminent man of science, who belongs to the former category, but also possesses a very strong will,—having been at first much surprised at the accuracy of the replies to certain questions which he had put (not being at that time cognizant of the *rationale* of the operation), but having observed that none could be furnished to a gentleman whose temperament was of the opposite kind,—made a second trial, with the strong determination to prevent any indication escaping him of the times at which he expected the "raps;" which trial was as complete a failure on Mrs. Hayden's part, as the first had been a success.

255. The following is a more recent case of the same kind, relating to another American professional "medium," whose gifts (like those of Mrs. Hayden) chiefly lay in playing on the credulity of such as lent themselves to his clever deceptions :—

a. "We were requested by a lady who had known Mr. Foster in America, to accompany her and her son-in-law (an eminent London Physician) on a visit to Mr. Foster, who had arrived in London only a few days previously. We were not introduced to him by name, and we do not think that he could have had any opportunity of knowing our person. Nevertheless, he not only answered, in a variety of modes, the questions we put to him respecting the time and cause of the death of several of our departed friends and relatives, whose names we had written down on slips of paper which had been folded-up and crumpled into pellets before being placed in his hands; but he brought out names and dates correctly, in large red letters, on his bare arm, the redness being produced by the turgescence of the minute vessels of the skin, and passing away after a few minutes, like a blush. We must own to have been strongly impressed at the time by this performance; but on subsequently thinking it over, we could see that Mr. Foster's divining power was probably derived from his having acquired the faculty of interpreting the movements of the *top* of a pen or pencil, though the *point* and what was written by it was hid from his sight, with the aid of an observing power sharpened by practice,* which enabled him to guide his own movements by the indications unconsciously given by ourselves of the answers we expected. For though we were fully armed with the knowledge which had been acquired of the source from which Mrs. Hayden drew her inspiration, and did our utmost to repress every sign of anticipation, we came, on reflection, to an assured conviction that Mr. Foster *had* been keen-sighted enough to detect such signs, notwithstanding our attempt to baffle him. For, having asked him the *month* of the death of a friend, whose name had previously appeared in red letters on his arm, and the *year* of whose death had also been correctly indicated in another way, he desired us to take up the alphabet-card and to point to the successive letters. This we did, *as we believed*, with pendulum-like regularity; nevertheless, distinct raps were heard at the letters J, U. When, however, on the next repetition, we came to L, M, N, Mr. Foster was obviously baffled. He directed us to "try back" two or three times, and at last confessed that he could not certainly tell whether the month was *June* or *July*. The secret of this was, that *we did not ourselves recollect.*

b. "Wishing to clear up the matter further, we called on Mr. Foster, revealed ourselves to him *in propriâ personâ*, and asked him if he would object to meet a few scientific investigators, who should be allowed to subject his powers to fair tests. As he professed his readiness to do so, we brought together such a meeting at our own house; and previously to Mr. Foster's arrival, we explained to our friends the arrangements we proposed. One of these was, that one

* To what a pitch of keenness and rapidity this *discerning power* may be brought by the special education of it has been already shown (§ 185).

of the party should sit outside the "circle," and should devote himself to observing and recording all that passed, without taking any part whatever in the performance. Another was, that instead of writing down names on slips of paper, whilst sitting at the table within Mr. Foster's view, we should write them at a side-table, with our backs turned to him. On explaining these arrangements to Mr. Foster, he immediately said that the first could not be permitted, for that every person present *must* form part of the circle. To the second he made no objection. After handing him our slips of paper carefully folded-up, we took our seats at the table, and waited for the announcement of spiritual visitors. The only one, however, who presented himself during an hour's *séance,* was the spirit of our own old master, whose name Mr. Foster might very readily have learned previously, but about whom he could give no particulars whatever. *Not one of the names written on the papers was revealed.*

c. "The patience of our friends being exhausted, they took their leave; but as Mr. Foster's carriage had been ordered for a later hour, we requested him to sit down again with the members of our own family. 'Now,' we said, 'that these incredulous philosophers are gone, perhaps the spirits will favour us with a visit.' We purposely followed *his* lead, as on our first interview, and everything went on as successfully as on that occasion; until, whilst the name of a relative we had recently lost was being spelled out on our alphabet-card, *the raps suddenly ceased* on the interposition of a large music-book, which was set-up at a preconcerted signal so as to hide the *top* as well as the bottom of our pointer from Mr. Foster's eyes.—Nothing could more conclusively prove that Mr. Foster's knowledge was derived from observation of the movements of the pointer, although he could only see the portion of it not hidden by the card, which was so held as to conceal the lower part of it; and nothing could be a better illustration of the principle of 'unconscious ideo-motor action,' than the fact, that *whilst we were most carefully abstaining from any pause or look from which he might derive guidance, we had enabled him to divine the answer we expected.*—The trick by which the red letters were produced was discovered by the inquiries of our medical friends."—(*Quarterly Review*, October, 1871, p. 332.)

256. It is further asserted, however, that the tables or planchettes

often give true answers to questions proposed to them as to matters of fact, though *none of the parties present* may have any knowledge of what the answers should be; but this, if it be really so, is not only far from being opposed to the Physiological doctrines here advanced, but affords a curious illustration and extension of them. For, as there is no doubt that impressions of which we were once conscious, though we have entirely lost our recollection of them, may direct our trains of thought in Delirium and Dreaming, or may even, as in Somnambulism, govern our actions; so does it seem quite reasonable to attribute the movements by which the table gives its answers, to impressions left by *past* ideas upon the Cerebrum, which may express themselves through the muscular system, without any consciousness of their existence on the part of the operator (§ 425).

257. To this same category are doubtless to be referred a large number of those actions of Mesmeric "subjects," which have been considered by some as most unequivocal indications of the existence of an agency *sui generis*, whilst by others they have been regarded as the results of intentional deception. Many of them are of a kind which the Will *could* not feign, being violent convulsive movements, such as no voluntary effort could produce: but the Mesmeric "subject" being previously "possessed" with the expectation that certain results will follow certain actions (as, for instance, that convulsive movements will be brought on by touching a piece of "mesmerized" metal), and the whole nervous power being concentrated, as it were, upon the performance, the movements follow when the "subject" *believes* the conditions to have been fulfilled, whether they *have* been or *not* (§ 518).—These facts were most completely established by the Commission appointed to investigate the pretensions of Mesmer himself: and whilst they demonstrate the unreality of the supposed "mesmeric" influence (so far, at least, as this class of phenomena is concerned), they also prove the position here contended for; namely, the

sufficiency of the state of *expectant attention,* in those whose Minds can be completely "possessed" by it, to produce effects of the same nature with those which are induced in Hysterical subjects by Emotional excitement (§ 270).

258. Under the same head may be ranked a variety of still more aberrant actions, bordering on Insanity, of which the History of Mankind in successive ages furnishes us with abundant examples :—what is common to all being the entire "possession" of those who perform them by some strongly-excited *dominant idea,* the intensity of which blinds the Common Sense and subjugates the Will, so that it expresses itself in bodily action without the least restraint. The notion may, or may not, be in itself an absurd one. It may be confined to a single individual, or it may spread epidemically through a multitude. It may be one that interests the feelings, or it may be (though seldom) of a nature purely intellectual. The wild but transient vagaries of religious enthusiasm in all ages,—as shown in the Pythonic inspiration of the Delphic priestesses ; the ecstatic revelations of Catholic and Protestant visionaries ; the Flagellant processions of the 13th and 14th centuries ; the Preaching epidemic among the Huguenots in France, and more recently in Lutheran Sweden ; the strange performances of the "convulsionnaires" of St. Médard, which have been since almost paralleled at "revivals" and "camp-meetings ;" the Dancing Mania of the Middle Ages, the Tarentism of southern Italy, the Tigretier of Abyssinia, and the Leaping-ague of Scotland in later times,*—are all, like the "table-turning" and "table-talking" epidemic, which spread through almost the whole civilized world in 1852-3, to be ranged under the same category.

259. The following account given by Dr. Hecker of the prin-

* On the greater number of the foregoing subjects, much curious information will be found in Dr. Hecker's account of the "Dancing Mania," forming part of his Treatise "On the Epidemics of the Middle Ages," translated for the Sydenham Society by Dr. Babington.

cipal features of the Dancing Mania which spread through a large part of Middle Europe in the 14th and 15th centuries, will serve to illustrate those forms of *ideo-motor* action which are intensified by *emotional* excitement :—

a. "In the year 1374, assemblages of men and women were seen at Aix-la-Chapelle, who had come out of Germany, and who, united by one common delusion, exhibited to the public, both in the streets and in the churches, the following strange spectacle:—They formed circles hand in hand, and appearing to have lost all control over their senses, continued dancing, regardless of the bystanders, for hours together, in wild delirium, until at length they fell to the ground in a state of exhaustion. They then complained of extreme oppression, and groaned as if in the agonies of death, until they were swathed in clothes bound tightly round their waists; upon which they again recovered, and remained free from complaint until the next attack. This practice of swathing was resorted to, on account of the tympany which followed these spasmodic ravings; but the bystanders frequently relieved patients in a less artificial manner, by thumping and trampling upon the parts affected. While dancing they neither saw nor heard, being insensible to external impressions through the senses; but were haunted by visions, their fancies conjuring up spirits, whose names they shrieked out; and some of them afterwards asserted that they felt as if they had been immersed in a stream of blood, which obliged them to leap so high. Others, during the paroxysm, saw the heavens open, and the Saviour enthroned with the Virgin Mary; according as the religious notions of the age were strangely and variously reflected in their imaginations.

"Where the disease was completely developed, the attack commenced with epileptic convulsions. Those affected fell to the ground senseless, panting and labouring for breath. They foamed at the mouth, and suddenly springing up began their dance amidst strange contortions.—A few months after this dancing malady had made its appearance at Aix-la-Chapelle, it broke out at Cologne, where the number of those possessed amounted to more than five hundred; and about the same time at Metz, the streets of which place are said to have been filled with eleven hundred dancers. Peasants left their ploughs, mechanics their workshops, housewives their domestic duties, to join the wild revels; and this rich commercial city became the scene of the most ruinous disorder.

"The St. Vitus's dance attacked people of all stations, especially those who led a sedentary life, such as shoemakers and tailors; but even the most robust peasants abandoned their labours in their fields, as if they were possessed by evil spirits; and those affected were seen assembling indiscriminately, from time to time, at certain appointed places, and, unless prevented by the lookers-on, continued to dance without intermission, until their very last breath was expended. Their fury and extravagance of demeanour so completely deprived them of their senses, that many of them dashed their brains out against the walls and corners of buildings, or rushed headlong into rapid rivers, where they found a watery grave. Roaring and foaming as they were, the bystanders could only succeed in restraining them by placing benches and chairs in their way, so that, by the high leaps they were thus tempted to take, their strength might be exhausted. As soon as this was the case, they fell, as it were, lifeless to the ground, and, by very slow degrees, recovered their strength. Many there were, who, even with all this exertion, had not expended the violence of the tempest which raged within them; but awoke with newly revived powers, and again and again mixed with the crowd of dancers; until at length the violent excitement of their disordered nerves was allayed by the great involuntary exertion of their limbs; and the mental disorder was calmed by the exhaustion of the body. The cure effected by these stormy attacks was in many cases so perfect, that some patients returned to the factory or the plough, as if nothing had happened. Others, on the contrary, paid the penalty of their folly by so total a loss of power, that they could not regain their former health, even by the employment of the most strengthening remedies."—(*Epidemics of the Middle Ages*, pp. 87-104.)

In this case we see a notable manifestation of the tendency to *imitation*, which is, in fact, the result of the "hold" taken of the Mind by an idea *suggested* to it (§ 550); that hold being the stronger, in proportion to the want of other sources of healthful activity, as in the two following cases related by Zimmerman :—

b. A Nun, in a very large convent in France, began to mew like a cat; shortly afterwards other nuns also mewed. At last all the nuns began to mew together every day at a certain time and continued mewing for several hours together. This daily cat-concert continued, until the nuns were informed that a company of soldiers was placed by the police before the entrance of the convent, and that the soldiers were provided with rods with which they would whip the nuns until they promised not to mew any more.

c. In the 15th Century, a Nun in a German nunnery fell to biting all her companions. In the course of a short time, all the nuns of this convent began biting each other. The news of this infatuation among the nuns soon spread, and excited the same elsewhere; the biting mania passing from convent to convent through a great part of Germany. It afterwards visited the nunneries of Holland, and even spread as far as Rome.—(*On Solitude*, Vol. ii).

Such "dominant ideas," like emotions, very commonly decline in intensity, when they expend their force in action (§ 265), and the Mind spontaneously returns to its previous condition: and thus it is that we generally find these epidemic delusions passing-away of themselves, without any ostensible cause for their cessation. Sometimes, however, such an Idea may continue to exert a dominant influence over the whole of life; and if the conduct which it dictates should pass the bounds of enthusiasm or eccentricity, we say that the individual is the subject of Monomania.—The nature of this state will be more fully considered hereafter (§ 559).

CHAPTER VII.

OF THE EMOTIONS.

260. Although, as we have seen (§§ 189, 190), there are various forms of Emotional sensibility which are directly called into activity by Sense-perceptions, yet those Emotional states of Mind which directly or indirectly determine a great part of our conduct, belong to the level of the Ideational consciousness; being, in fact, the result of the attachment of the *feelings* of pleasure and pain, and of other forms of emotional sensibility, to certain classes of *ideas*. Thus the Cerebrum and the Sensorium would seem jointly concerned in their production; for whilst the Cerebral hemispheres furnish the *ideational* part of the material, the Sensory ganglia not only give us the consciousness of their result, but invest that result with the peculiar *feeling* which renders it capable of actively influencing our conduct as a *motive* power. This we see clearly, when the Emotional state takes the form of a true *desire*; for when this is felt, even as regards the gratification of a bodily appetite, it involves the existence of an *idea of the object* of desire; but it is only when this idea is associated with the contemplation of enjoyment in the act to which it relates, or of discomfort in the abstinence from that act, that it becomes an impelling force towards the performance of it.—All the higher forms of Emotional consciousness may be decomposed (as it seems to the Writer) in a similar manner. Thus, Benevolence is the pleasurable contemplation of the happiness or welfare of others; and shows itself alike in the habitual entertainment of the abstract or general idea, and in the direction of the conduct with a view to promote this result in any particular instance on which the bene-

volent desire may be fixed. So there is a positive pleasure, in some ill-constituted minds, in the contemplation of the *un*happiness of others; and this (of which Dickens's Quilp is an impersonation) we designate as Malevolence. Combativeness, again, in so far as it is a Psychical attribute, is the pleasurable idea of setting one's self in antagonism with others.

There are individuals who never manifest the least degree of *physical* Combativeness, who yet show a remarkable love of opposition in all their *psychical* relations with others. That objections will be raised by such persons to *any* plan that may be proposed, we can always feel sure, though we may not have the remotest idea as to what the objection may be in each particular case. Persons in whom this tendency exists in a less prominent degree, are apt to see objections and difficulties *first*, although their good sense may subsequently lead them to consider these as of less account, or to be outweighed by the advantages of the scheme. Such was the case with the late Sir Robert Peel. On the other hand, those who are spoken-of as of *sanguine* temperament, are apt to lose sight of the intervening difficulties, in the pleasurable anticipation of the *result*.

So, Pride (or self-esteem) consists in the pleasurable contemplation of our own superior excellencies; whilst the essence of Vanity (or love of approbation) lies in the pleasurable idea of the applause of others. Again, in Conscientiousness we have the love of right, that is, the association of pleasure with the idea of right; Veneration may be defined as the pleasurable contemplation of rank or perfections superior to our own; and the source of Ambition, which is in some degree the antagonistic tendency, lies in the pleasurable idea of self-exaltation. In like manner, Hope is the pleasurable contemplation of future enjoyment; Fear is the painful contemplation of future evil; and Cautiousness is the combination of the desire to avoid anticipated pain, with the pleasurable contemplation (an extremely strong feeling in many individuals) of precautions adapted to ward it off. — The same view may be applied to the love of Order, of Possessions, of Country, of Wit, of

Humour, &c., and to many conditions usually considered as purely Intellectual. And, in fact, the association of any kind of that *emotional sensibility* (§ 157) of which pleasure and pain afford the simplest type, with *any idea*, or *class of ideas*, gives to it an Emotional character; so that emotional states are not by any means limited within the categories under which Psychologists have attempted to range them,—these being, for the most part, *generic terms*, which comprehend certain groups of ideas bearing more or less similarity to each other, but not by any means including all possible combinations.

The truth of this statement must be apparent to all who are familiar with the manifestations of eccentricity and insanity; for we frequently see pleasurable feelings associating themselves with Ideas, which to ordinary minds appear *indifferent*, or are even regarded with pain; and thus are engendered Motives which exert a most powerful influence over the conduct, and which, if not kept in restraint by the Will, render the whole being their slave. Thus one weak-minded youth was driven to commit a murder, by a passion for living where he could see a wind-mill; and another by a passion for possessing himself of every shoe of a particular kind which he chanced to see.—It may be also remarked, in this place, that the impossibility of classing all the Emotional states of mind under a limited number of categories constitutes a most serious and fundamental objection to any system which professes to mark-out in the Cerebrum distinct seats for the animal propensities, moral feelings, &c.

261. By those who regard the Propensities, Moral Feelings, &c., as *simple* states of mind, it is usually said that their indulgence or exercise is attended with pleasure, and the restraint of them with pain. But, if the view here taken be correct, it is the very co-existence of pleasurable or painful *feelings* with the *idea* of a given object, that causes *desire* or *aversion* as regards that object; since the mind instinctively pursues what is pleasurable, and avoids what is painful. And thus, according to the readiness with which these different classes of Ideas are excited in different minds (partly

depending upon original constitution, and partly upon the habitual direction of the thoughts), and to the respective degrees in which they respectively call-forth the different kinds of Emotional sensibility (as to which there is obviously an inherent difference amongst individuals, analogous to that which exists with regard to the feelings of pleasure or pain excited by external sensations,—sights, sounds, tastes, odours, or contacts), will be the tendency of the mind to entertain them, the frequency with which they will present themselves before the mental view, and the influence they will exert in the determination of our conduct.

262. The influence of Emotional conditions, when strongly excited, in directly producing involuntary movements, is readily explained on the idea that the Sensory Ganglia act as the centre of all consciousness, and that the Axial Cord is the real source of all movement. For the excitement of peculiar states of the sensorial centres through the instrumentality of the Cerebrum, will just as readily give rise to automatic movements, as the excitement of similar states by impressions made upon the organs of Sense. And the correspondence is seen to be very close, when the idea distinctly reproduces the sensorial state. Thus, the laughter excited by the act of tickling is a purely consensual movement (§ 79); but, in a very "ticklish" person, the mere *idea* of tickling, suggested by pointing a finger at him, is sufficient to provoke it.—So, again, as laughter may be excited by odd sights or sounds which do not in themselves excite any ideational state, but which act at once upon the "sense of the ludicrous," the same action may be called-forth by the vivid *recollection* of these occurrences; which, being attended with a state of the Sensorium corresponding to that originally produced by the sensation, gives rise to the same involuntary cachinnation. But laughter may also be excited by ideas that are much more removed from actual sensations; as, for example, by those unexpected combinations of ideas of a purely intellectual nature, which we designate as

"witty;" and here, too, we may recognize the very same *modus operandi.* For the mere sound or sight of the *words* excites no feeling of the ludicrous; the *sensation* must develope an *ideational* change; and it is the latter alone, which, reacting downwards upon the Sensorium, and there becoming associated with the Emotional sensibility, excites the impulse to laugh.—The same might be shown to be the case with regard to the act of crying; which may be either purely consensual, being excited by painful sensations; or may be induced by the vivid recollection of past or the anticipation of future sensations; or may be excited by ideas which have no direct relation to sensational states.—Again, the movements which take place under the violent excitement of the passion of anger, are of the same involuntary character; being directly prompted by feelings which may be called-up either by external sensations, or by internal ideas that have a like power of exciting them. Thus the passionate man who receives a blow, instinctively makes another blow in the direction from which it seemed to him to come, without any thought of whether the blow was accidental or intentional; while the idea of an insult, which is a source of mental disturbance, may excite the very same movement, although no bodily suffering had been experienced. There are many of the movements of Expression that are referable in like manner to sensorial states, whether pleasurable or painful, which may arise from ideational as well as from sensational conditions. Thus, as we have seen (§ 156), the cheerful aspect of some individuals is due to a sense of general *physical* well-being, and is altogether discomposed by anything which disturbs this; whilst in others, it may proceed from a happy frame of mind (which may be partly the result of original constitution, and partly of habitual self-direction), disposing them to take the cheerful view of everything that affects themselves or others, notwithstanding (it may be) great bodily discomfort. And the reverse aspect of gloom may in like manner proceed alike from bodily or from mental uneasiness.—All these facts point, therefore, to the conclusion, that whether the elementary states of Emotional sensibility associate themselves with Sensations, with Perceptions, or with Ideas, they are simple *modes of consciousness*, the organic seat of which must be in the Sensorial centres; and this corresponds well with the character of the purely Emotional movements, which, as we have seen, are closely allied to the Sensori-motor in the directness with which they respond to the stimuli that excite them.

263. That the Emotional and Volitional movements differ as to their primal sources, is obvious, not merely from the fact that they are frequently in antagonism with each other,—the Will endeavouring to restrain the Emotional impulse, and either succeeding in doing so, or being vanquished by the superior force of the latter,—but also from the curious fact which Pathological observation has brought to light, that muscles which will still act in obedience to emotional impulses, may be paralysed to volitional, and *vice versâ*. Thus, for example, the arm of a man affected with paralysis, which no effort of his will could move, has been seen to be violently jerked under the influence of the emotional agitation consequent upon the sight of a friend. And in a case of softening of the Spinal cord, the jerking movements which were brought-on by the mere approach of any one to the patient's bed, and still more strongly by putting a question to him, were most violent in the lower limbs, over which he had not the least voluntary power. —It is in the different forms of paralysis of the Facial nerve, however, which is the one most peculiarly subservient to the movements of Expression (§ 195), that we have the best evidence of this distinctness. For it sometimes happens that the muscles supplied by this nerve are paralysed so far as regards the will, and yet are still affected by emotional states of mind, and take their usual part in the automatic actions of respiration, &c.; retaining also their usual tension, so that no distortion is apparent unless voluntary movements be attempted. Thus, to select an action

which may be performed either consensually, emotionally, or voluntarily, a patient affected with this form of paralysis cannot close the eyelid by an act of his Will, although he winks when he feels the uneasy sensation that excites the action, and shuts the lids when the sudden approach of an object to the eye excites the fear of injury to that organ. On the other hand, the paralysed condition may exist in regard to the Automatic and Emotional actions only, so that the muscles lose their tension, the mouth is drawn to one side, the movements of expression are not performed, and there is no involuntary winking: yet the Will may still exert its accustomed control, and may produce that closure of the lids which does not take place in respondence to any other impulse.*

264. The Emotions are concerned in Man, however, in many actions, which are in themselves strictly voluntary. Unless they be so strongly excited as to get the better of the Will, they do not operate downwards upon the Automatic centres, but upwards upon the Cerebral; supplying the *motives* by which the course of thought and of action is habitually determined (§ 331). Thus, of two individuals with differently-constituted minds, one shall judge of everything through the medium of a gloomy morose temper, which, like a darkened glass, represents to his judgment the whole world in league to injure him; and his determinations being all based upon this erroneous view, its indications are exhibited in his actions, which are themselves, nevertheless, of an entirely voluntary character. On the other hand, a person of a cheerful, benevolent disposition, looks at the world around as through a Claude-Lorraine-glass, seeing everything in its brightest and sunniest aspect; and, with intellectual faculties (it may be) precisely similar to those of the former individual, he would come to opposite conclusions; because the materials which form the

* See the detailed accounts of such cases in Sir C. Bell's work on "The Nervous System of the Human Body;" also "Brit. and For. Med. Rev.," vol. iv. p. 500, and vol. xiii. p. 553.

basis of his judgment, are submitted to it in a very different condition.—Various forms of Moral Insanity exhibit the same contrast in a yet more striking light (§ 556); and the distinction between the *sane* and the *insane* is far more difficult to draw in this form of mental disorder, than when Intellectual perversion manifests itself. For we not unfrequently meet with individuals still holding their place in society, who are accustomed to act so much on *impulse*, and to be so little guided by the *rational will*, that they can scarcely be regarded as sane; and a very little exaggeration of such impulses causes the actions to be so injurious to the individual himself or to those around him, that restraint is required, although the intellect is in no way disordered, nor are any of the feelings perverted. We may often observe similar inconsistencies, resulting from the habitual indulgence of one particular feeling, or from a morbid exaggeration of it. The mother who, through weakness of will, yields to her instinctive fondness for her offspring, in allowing it gratifications which she knows to be injurious to it, is placing herself below the level of many less gifted beings. The habit of yielding to a natural infirmity of temper often leads into paroxysms of ungovernable rage, which, in their turn, pass into a state of maniacal excitement. The poor girl who drowns herself after a quarrel with her lover, or the nursemaid who cuts the throat of a child to whom she is tenderly attached, because her mistress has rebuked her for wearing too fine a bonnet, may be really labouring under a "temporary insanity" which drives her irresistibly to a great crime; yet, just as the man who commits a murder in a state of drunken frenzy is *responsible for his irresponsibility* (§ 545), so is the suicide or the murderess, in so far as she has habitually neglected to control the wayward feelings whose strong excitement has impelled her to the commission of her crime.—It not unfrequently occurs, moreover, that a delusion of the *intellect* (constituting what is commonly known as Monomania) has had its source in a disordered state of the *feelings*, which have represented every occurrence in a wrong light to the mind of the individual (§ 559). All such conditions are of extreme interest, when compared with those which are met-with amongst Idiots, and in animals enjoying a much lower degree of intelligence: for the result is much the same in whatever way the balance between the feelings and the rational will (which is so beautifully adjusted in the well-ordered mind of Man) is disturbed; whether by a diminution of the Volitional control, or by an undue exaltation of the Emotions and Passions.

265. This double mode of action of the Emotions—*downwards* through the nerve-trunks upon the Muscular apparatus, and also upon many of the Organic functions (Chap. XIX.),—and *upwards* upon those Cerebral actions which give rise to the higher states of Mental consciousness,—affords a satisfactory explanation of a fact which is practically familiar to most observers of Human nature; namely, that violent excitement of the Feelings most speedily subsides, when these unrestrainedly expend themselves (so to speak) in their natural expressions. Thus it may be commonly noticed that those who are termed *demonstrative* persons are less firm and deep in their attachments, than those who manifest their feelings less: for, without any real insincerity or intentional fickleness, the strongly-excited feelings of the former are rapidly calmed-down by the expenditure of the impulse to bodily action which they have generated; whilst in the latter the very same feelings, acting internally, acquire a permanent place in the Psychical nature, and habitually operate as motives to the conduct. So, again, persons who are "quick-tempered," manifesting great irascibility upon small provocations, real or supposed, are usually soon appeased, and speedily forget the affront; whilst many who make little or no display of anger, are very apt to brood-over and cherish their feelings of indignation, and may visit them upon the unfortunate object of them on some favourable opportunity, long after he had supposed that the occurrence which had given rise to them was

forgotten. There is an instinctive restlessness, or tendency to general bodily movement, in some individuals, when they are suffering under Emotional excitement; the indulgence of which appears to be a sort of safety-valve for the excess of Nerve-force, whilst the attempt at its repression is attended with an increase in the excitement. Most persons are conscious of the difficulty of sitting still when they are labouring under violent agitation, and of the relief which is afforded by active exercise; and this is particularly the case when the movements are such as naturally express the passion that is excited. Thus the combative propensities of the Irish peasant commonly evaporate speedily with the free play of his shillelagh, many irascible persons find great relief in a hearty explosion of oaths, others in a violent slamming of the door, and others (whose excitement is more moderate but less transient) in a prolonged fit of grumbling.

a. This view is most fully confirmed by certain phenomena of Insanity. It is a doctrine now generally received among practical men, that paroxysms of violent Emotional excitement are much more likely to subside, when they are allowed to "work themselves off" freely, without any attempt at mechanical restraint; and maniacal patients are now placed, in all well-managed Asylums, in padded rooms, in which their movements can do no injury to themselves or others.—The following case was related to the Writer by his friend Dr. Howe, of Boston, N.E., the instructor of Laura Bridgman. A half-idiotic youth in the Lunatic Asylum of that place, was the subject (like many in his condition) of frequent and violent paroxysms of anger; and with the view of moderating these, it was suggested that he should be kept for some time every day in rather fatiguing exercise. Accordingly he was employed for two or three hours daily in sawing wood, to which task he made no objection; and the paroxysms of rage never displayed themselves, except on Sundays, when his employment was intermitted. As it was considered, however, to be better for him to spend a part of that day in sawing wood, than to be irascible during the whole of it, his occupation was continued through the entire week, when he became completely tamed-down, and never gave any more trouble by his passionate displays.

So, again, if a ludicrous idea be suggested to our consciousness, occasioning an impulse to laugh, a hearty "guffaw" generally works-off the excitement, and we may be surprised a short time afterwards that such an absurdity should have provoked our risibility; but if we restrain the explosion, the idea continues to "haunt" us, and is continually perturbing our trains of thought until we have given free vent to the expression of it.—Again, it is well known that the depressing emotions are often worked-off by a fit of crying and sobbing; and the "relief of tears" seems manifestly due to the expenditure of the pent-up nerve-force, in the production of an increased secretion. It is noticed in this case, too, that the absence of any such external manifestations of the depressing emotions, gives them a much greater influence upon the course of thought, and upon the bodily state of the individual. Those who really "die of grief," are not those who are loud and vehement in their lamentations, for *their* sorrow is commonly transient, however vehement and sincere while it lasts; but they are those who have either designedly repressed any such manifestations, or who have experienced no tendency to their display; and their deep-seated sorrow seems to exert the same kind of antivital influence upon the Organic functions, that is exercised more violently by "shock" (§ 41); producing their entire cessation without any structural lesion.

b. The Writer once heard the following singular case of this kind:—One of two sisters, orphans, who were strongly attached to each other, became the subject of consumption; she was most tenderly nursed by her sister during a long illness; but on her death, the latter, instead of giving way to grief in the manner that might have been anticipated, appeared perfectly unmoved, and acted almost as if nothing had happened. About a fortnight after her sister's death, however, she was found dead in her bed; yet neither had there been any symptoms during life, nor was there any post-mortem appearance, which in the least degree accounted for this event,—of which no explanation seems admissible, except the depressing influence of her pent-up grief upon her frame generally, through the Nervous system.,

266. The influence of Emotional excitement may operate upon the Muscles, however, not only in giving-rise to the movements which it directly calls forth, but also in affecting the power of the Will over the muscular system, by intensifying or weakening its action. For there can be no doubt that, under the strong influence of one class of feelings, the Will can effect results such as the individual would scarcely even attempt in his calmer moments; whilst the influence of another class of feelings is exercised in precisely the opposite direction, weakening or even paralysing the force which was previously in full activity.

It must be within the knowledge of every one, that, when *first* attempting to perform some new kind of action, the power we feel capable of exerting depends in great measure upon the degree of our *assurance of success.* Of this we have a good example in the process of learning to swim; which is greatly facilitated, as Dr. Franklin pointed-out, by our first taking means to satisfy ourselves of the buoyancy of our bodies in the water, by attempting to pick up an object from the bottom. And every one is aware of the assistance derived from the encouragement of others, when we are ourselves doubtful of our powers; and of the detrimental influence of discouragement or suggested doubt, even when we previously felt a considerable confidence of success. Of the almost superhuman strength and agility with which the body seems endowed, when the whole energy is concentrated upon some Nervo-muscular effort, especially under the influence of an overpowering emotion, the following remarkable example has been communicated to the Writer by a gentleman on whom he can place full reliance, and who was personally cognizant of the fact:—An old cook-maid, tottering with age, having heard an alarm of fire, seized an enormous box containing her whole property, and ran down stairs with it, as easily as she would have carried a dish of meat. After the fire had been extinguished, she could not lift the box a hair's breadth from the ground, and it required two men to convey it upstairs again.

267. But the same Emotion does not always act in the same mode: thus, the fear of danger may nerve one man to the most daring and vigorous efforts to avert it, whilst another is rendered powerless, and gives-way to unavailing lamentations; and the ardent anticipation of success may so unsettle the determinative energy of one aspirant, as to prevent him from attaining his object, whilst another may only be sustained by it in the toilsome struggle of which it is the final reward. Now in order that this variety may be explained, and the *modus operandi* of the Emotions on strictly Volitional actions may be duly comprehended, we must here state two of the essential conditions of the latter: one of which is, that there should be not merely a distinct conception of the purpose *to be* attained, but also a belief that the purpose *will* or at least *may* be attained; whilst the other is, that the attention should be to a great extent withdrawn from other objects, and should be concentrated upon that towards which the Will is directed. The following cases illustrate these principles:—

a. The Writer well remembers being among those, who, forty years ago, tested the validity of the statement put-forth in Sir D. Brewster's "Natural Magic," that four persons can lift a full-sized individual from the ground, high into the air, with the greatest facility, provided they all take-in a full breath previously to the effort, the person lifted doing the same. He could readily understand upon Physiological principles, that a full inspiration on the part of the *lifters* would have a certain degreee of efficacy in augmenting their Nervo-muscular power; but he could not perceive how the performance of the same act by the *person lifted* could have any appreciable effect; and while many of his acquaintances assured him that, when all the conditions were duly observed, the body went up "like a feather," and that they felt satisfied of being able to support it upon the points of their fingers, he found his own experience quite different. Hence he came to the conclusion, after much observation, that the facility afforded by this method entirely depended upon the degree in which it fulfilled the above-mentioned conditions, namely, the *fixation of the attention* upon the effort, and the *conviction of the success* of the method. Whenever the attention was distracted, and confidence weakened by scepticism as to the result, the promised assistance was not experienced.

b. The following little circumstance communicated to the Writer by a friend, is a very characteristic illustration of the same principle. This gentleman related that, having been accustomed in his boyhood to play at bagatelle with other juniors of his family, the party was occasionally joined by a relative who was noted for her success at the game, and who was consequently much dreaded as an opponent; and that, on one occasion, when she was about to take her turn against him, he roguishly exclaimed, "Now, aunty, you will not be able to make a hit;" the effect of which suggestion was, that she missed every stroke,—and not only at that turn, but through the remainder of the evening.

268. Since, then, there is nothing which tends so much to the success of a Volitional effort as a *confident expectation* of its success, whilst nothing is so likely to induce failure as the apprehension of it, and since the tendency of the cheerful or joyous Emotions is to suggest and keep-alive the favourable anticipations, whilst that of the depressing emotions is to bring before the view all the chances of failure, the former will increase the power of the Volitional effort, and the latter will diminish it. And they exert also a direct influence on the Physical powers, through the organs of circulation and respiration; the heart's impulses being more vigorous and regular, and the aeration of the blood being more effectually performed, in the former of these conditions than in the latter.—But an altogether contrary effect may be produced by the operation of these two classes of Emotions on the concentrative power. For the more completely the mental energy can be brought into one focus, and all distracting objects excluded, the more powerful will be the Volitional effort; and the effect of emotional excitement will thus in a great degree depend upon the Intellectual constitution

which the individual may happen to possess. For if he have a considerable power of Abstraction and Concentration, and a full conviction that he has selected the best or the only means to accomplish his end, the intensest fear of the consequences of failure will only *increase* the force of the motive which prompts the effort; and the whole energy of which his nature is capable, will be put forth in the attempt. In a man of this temperament, the most joyous anticipation of success will produce no abatement of his efforts, no distraction of his attention, but will rather tend to keep him steady to his purpose until it shall have been accomplished. But the mind which is deficient in concentrative power is lamentably deranged by any kind of Emotional excitement, in the performance of any Volitional effort. For the fear of failure is constantly suggesting to him new distresses, weakens his confidence in any method suggested for his action, and makes him direct his attention, not to some fixed plan as the best or the only feasible one, but to any and every means that may present a chance of success, or may even serve to avert his thoughts from the dreaded catastrophe; whilst, on the other hand, the joyous anticipation of success leads him to allow his thoughts to direct themselves towards all its agreeable *consequences*, instead of fixing his Intellectual and Volitional energy upon the *means* by which that success is to be attained.

269. We have now to inquire into the influence which the *Will* has upon the Emotions;—a subject of the highest importance in regard to the *direction of the current of thought* and the *determination of our actions.* That the Will *has* such a power, is recognised in those common forms of admonition, "Control your passions," "Govern your temper," and the like. But the success of its exertion will mainly depend upon the judiciousness of the *mode* in which it is attempted; and here, as it seems to the Writer, much assistance is gained from the Physiological method of study

which it has been his aim to develope.—In the *first* place, it may be unhesitatingly affirmed that the Will has no direct power over the emotional Sensibility. We can no more avoid feeling *mental* "hurt," than we can avoid feeling *bodily* "hurt." But we have exactly the same power of *withdrawing the attention* from the mental "hurt," that we have of withdrawing it from bodily pain (§ 124), by *determinately fixing it upon some other object;* and this is the mode in which (as all experience shows) the passions of Children, which are often excited to a degree that is out of all proportion to the exciting cause, and are but little dependent upon Ideational states, are most readily controlled.

The difference between a judicious and an injudicious Mother or Nurse is strikingly shown in the ways in which they respectively deal with the most familiar incident of Child-life. When the little one falls down and hurts itself, and sets up the loud cry of pain and alarm, there are (as Sir Robert Peel used to say), three courses open, —to soothe and 'coddle,' to rebuke and frighten, and to *distract the attention* by the interposition of some new object attractive enough to engage it. Now, the first method, however kindly meant, has the disadvantage of making the Child *attend* to its hurt, and of thus intensifying the feelings connected with it; which, being the very thing to be avoided, should never be had recourse to unless the injury is really serious. The second no doubt gives a motive to self-control; but that motive is inappropriate to the occasion, adding a sense of injustice to the smart of the injury. Whilst the third, by leading the Child to transfer its attention to a more vivid and pleasurable impression, affords time for the smart to die away, and makes the child feel that even when fresh it *can* be disregarded.—As age advances, the judicious Parent no longer trusts to mere sensory impressions for the diversion of the emotional excitement; but calls up in the mind of the Child such ideas and feelings as it is capable of appreciating, and endeavours to keep the attention fixed upon these until its violence has subsided. And recourse is to be had to the same process, whenever it is desired to check any tendency to action which depends upon the *selfish propensities;* appeal being always made to the highest motives which the Child is capable of recognising, and recourse being had to punishment only for the purpose of supplying *an additional set of motives* when all others fail. For a time this process of external Suggestion may need to be continually repeated, especially where there are strong impulses whose unworthy character calls for repression; but if it be judiciously adapted and constantly persevered-in, a very slight suggestion serves to recall the superior motives to the conflict. And in a yet more advanced stage, the Child comes to feel that he has *himself* the power of recalling them, and of controlling his urgent impulses to immediate action.

270. In the *second* place, the Will can exert itself, as already shown, in preventing the *expression* of the excited feelings in action; by determinately bringing the Volitional control over the muscles to antagonize the Emotional impulse,—as in the cases already cited. It seems to the Writer, however, quite a mistake to suppose that the suppression of the muscular expression of an emotion, represses the emotion itself; on the contrary, as he has already shown (§ 265), the emotional state is more likely to last, if it does *not* vent itself in action,—unless, as sometimes happens when persons intentionally "work themselves up into a passion," the muscular movements themselves operate *suggestively* on the mind, so as to augment, instead of relieving, the excitement (§ 494). This reaction is very manifest in persons of the *Hysterical* temperament, who, either from an injudicious system of education, or from habitual want of self-control, or from bodily disorder, give way to the most exaggerated and inconsistent expressions of their feelings,—smiles and tears, laughter and sobbing, being strangely intermingled, and being excited by the most trivial cause. That the deficiency here lies rather in the power of controlling the *thoughts* and *feelings*, rather than in that of directing the action of the *muscles*, appears from the fact that an Hysterical paroxysm may often be kept off by the threat of severe discipline; whilst, by judicious guidance, the patient may be led to cultivate her own power of repressing the first risings of the Emotional excitement,

by the determinate direction of her attention to some other object than her own feelings.

271. But where, *thirdly*, the Emotion is not a mere *passion*, but is a state of *feeling* connected with some definite *idea*, the power of the Will is most effectually exerted in withdrawing the mind from the influence of that idea, by *fixing the attention upon some other*. The power of *self-control*, usually acquired in the first instance in regard to those Emotional *impulses* which directly prompt to action (§ 262), thus gradually extends itself to the habitual *succession of the thoughts;* which, directed to the acquirement of knowledge in the first instance by the dominating Will of the instructor, who uses his pupil's love of praise or fear of punishment as a motive power, comes in time to be so regulated by the Ego himself, as to give him a great *indirect* power over the emotional sensibilities connected with them. For just as, by a determined effort, we restrain ourselves from laughing when laughing would be unseemly, so can a strong Volition keep out of view the idea that excites risibility, by a determined direction of the thoughts into another channel; as when, for example, in a place of worship, it fixes on the prayer or the discourse the attention which had been distracted by some ludicrous interruption. And this *determinate transference of the attention* affords the surest means of escape from the domination of thoughts and feelings which we feel it wrong to entertain. We cannot prevent the rise of these in our minds. As Archbishop Manning has truly said—" The memory of insults or great wrongs *will* arise in the mind, or brain, if you will, at the sight of the person who has outraged us; or by associations of time, place, or any one of endless circumstances; or, again, by the direct suggestion of others. So far, the thoughts [and, it may be added, the feelings prompted by them] may be spontaneous or involuntary on our part. Their presence in the mind is neither good nor evil. Their first impression on the mind, even though it become a fascination or an attraction to

an immoral act, is not immoral, because, as yet, though the Thought has conceived them, the Will has not accepted them."* It is the *acceptance of them by the permission of the Will*, that makes them Voluntary, and brings them within the sphere of Moral control; whilst it is the *intentional direction of the Attention to them* which gives them their Volitional character, and makes the *Ego* fully responsible for whatever he may do at their prompting.—The experience of the Physician here comes in to the aid of the Psychologist, in showing how Volitional control over Emotional states is best maintained :

We will take the case of a Man who has sustained a great shock by the loss of a dearly-loved wife, child, or friend, a disappointed affection, or commercial ruin. His Physical condition is lowered, the power of his Will is weakened, the painful impression seems branded into his innermost nature, he *cannot help* feeling it most acutely, he seems powerless to withdraw himself from it. He may be exhorted to "rouse himself;" every conceivable motive may be suggested to him for doing so; but all in vain. What is needed is the complete *distraction of his attention* from brooding over his misfortune; and the force which the weakened Will cannot of itself exert, must be supplied by the attractive influence of new scenes and persons, and the complete severance from painful associations. He yields himself passively to his advisers; at first "all seems barren, from Dan to Beersheba"; he looks up into the dome of St. Peter's, or down into the crater of Vesuvius, and finds "nothing in it." But gradually his bodily health improves; he begins to show some interest in what he sees and hears; and a judicious companion, like a good nurse, watches for every sign, and encourages every movement in the right direction, noticing what proves most attractive, and secretly planning to bring its attractions into play. At first, the patient seems ashamed of being cheerful, and falls back into his moodiness, as if he felt it a duty to hug the memory of his lost happiness; but these relapses, after a time, become less and less frequent. He begins to find that it is really much pleasanter to

* *Contemporary Review*, Feb. 1871, p 475.

forget himself, and to make himself agreeable to others, than it is to brood morosely over his troubles. With the re-invigoration of his bodily health, his Volitional power gradually returns; and he comes to feel that *he can resist the tendency* to revert to them, by *determinately* giving his attention to the objects around him. The resisting power required becomes less and less, the more frequently it is exerted; and at length the Mental health is completely restored,—the brooding tendency, however, being apt to recur, either when the Will is weakened by physical fatigue, or when old associations are revived with peculiar force and vividness.

272. A valuable lesson may be drawn from these familiar experiences, in regard to the mode of dealing with those unrighteous Thoughts and Feelings, which furnish temptations to immoral action of any kind. The Will may put forth its utmost strength in the way of *direct repression*, and may entirely fail; whilst by exerting the same amount of force in *changing the direction*, complete success may be attained. When the question is not of restraining some sudden impulse of excited passion, but of keeping down an *habitual tendency* to evil thoughts of some particular class, and of preventing them from gaining a dominant influence, it does not answer to be continually repeating to oneself, "I will not allow myself to think of this;" for the repetition, *by fixing the attention* on the very thought or feeling from which we desire to escape, gives it an additional and even overpowering intensity, as many a poor misguided but well-intentioned sufferer has found to his cost. The real remedy is to be found in *the determined effort to think of something else*, and to turn into a wholesome and useful pursuit the energy which, wrongly directed, is injurious to the individual and to society; just as, in "The Caxtons," the Poacher whose love of sport no fear of punishment could restrain, becomes a most valuable Bushman when persuaded to accompany Pisistratus to Australia.—Whilst, then, the Intellectual faculties are exercised in the acquirement of Knowledge and in the pursuit of Truth, by the Volitional direction of their own spontaneous and automatic

activity, the Moral character is formed, and the Conduct mainly determined, by the direction we determinately give to those Motive powers which give *energy* to all our work in life. And there is a strong Physiological probability that the effect of such habitual self-discipline does not end with the Individual, but is exerted upon the Race; the Emotional tendencies having so much of the character of Instincts, that the Hereditary transmission of the form they have *acquired* may be expected in the one case as in the other (§ 84).

CHAPTER VIII.

OF HABIT.

273. There is no part of Man's composite nature, in which the intimate relation between Mind and Body is more obvious, than it is in the formation of *habitual modes of activity*, whether Psychical or Corporeal; the former, like the latter, being entirely conformable to the Laws which express the ordinary course of the Nutritive operations. A general knowledge of these Laws being therefore essential to the student of Mental Physiology, a concise statement of them will be here given.

274. In the *first* place, it is characteristic of every living Organism to *build itself up* according to a certain inherited type or pattern; so that we must attribute to its germ a "formative capacity," in virtue of which it turns to account both the *food* and the *force* which it derives from without,—like the Architect who *directs* the construction of an edifice, which is raised out of the *materials* brought together by one set of workmen, by means of the *labour* furnished by another. But this *constructive* process may undergo considerable modification under the influence of external conditions; so that the reproduction of the parental model is attended with more or less of variation. This influence is peculiarly manifested in the *lower* types both of Animal and Vegetable life; many of which display a remarkable *polymorphism*. Thus the germs of the simpler Fungi, falling upon different kinds of decomposing matter, will develop themselves into forms that differ so strikingly from each other, as to have been accounted not only *specifically* but *generically* distinct. And in the case of the higher types, it is during the *early* stages of development that the modi-

fying influence of external conditions is most strongly exerted; the Constitution of the individual then adapting itself to changes, which, when that constitution has once been fixed, are powerless to alter it. Of this early modifiability no more striking example could be adduced, than the artificial production of Queen-bees from Worker-larvæ (§ 59); the whole course of the development, and the subsequent life-history of the Insect, as well Psychical as Physical, being here determined *for each individual* by the food and nurture it receives. Although among higher animals we know of no case that is comparable to this in *degree*, yet there is ample evidence of the same *kind* of modifiability; of which the following fact mentioned by Sir Charles Lyell affords a good example :—

"Some of our countrymen engaged about the year 1825 in conducting one of the principal mining associations in Mexico, that of Real del Monte, carried out with them some English greyhounds of the best breed, to hunt the hares which abound in that country. The great platform, which is here the scene of sport is at an elevation of about 9,000 feet above the level of the sea, and the mercury in the barometer stands habitually at the height of about 19 inches. It was found that the greyhounds could not support the fatigues of a long chase in this attenuated atmosphere, and before they could come up with their prey, they lay down gasping for breath; but these same animals have produced whelps which have grown up, and are not in the least degree incommoded by the want of density in the air, but run down the hares with as much ease as the fleetest of their race in this country."—(*Principles of Geology*, 11th edit. vol. ii. p. 297.)

275. But, *secondly*, the building-up of the Organism of man or of any one of the higher animals, is not a mere process of addition and extension; for its evolution involves a continual *reconstruction* of every part of the fabric, the intimate substance of each constituent organ being subject to rapid and incessant change. It is this which makes the demand for food so much greater in propor-

tion to the size of the body during the period of growth, than after its completion; the quantity of new material applied to actual *increase*, bearing but a very small proportion to that which is required for the constant *renewal* which that increase involves.—But during the whole period of growth, there is also a *development* of new parts, the original plan not being completed until provision has been made for the perpetuation of the race, as well as for the continued life of the individual. And it is the special characteristic of Man, that the period required for the attainment of complete maturity is much longer in proportion to the entire term of life, than it is among the lower animals. During the whole of this period, each component part of the organism may be said to be undergoing a continual *rejuvenescence;* which is especially shown in the rapidity and completeness with which the effects of injuries are repaired. And thus the modifiability which is characteristic of the organism as a whole during its earliest stages, continues to show itself in each individual organ until its evolution is complete. Thus it is a matter of universal experience, that every kind of training for special aptitudes, is both far more effective, and leaves a more permanent impress, when exerted on the *growing* organism, than when brought to bear on the adult. The effect of such training is shown in the tendency of the organ to "grow to" the mode in which it is habitually exercised; as is evidenced by the increased size and power of particular sets of Muscles, and the extraordinary flexibility of Joints, which are acquired by such as have been early exercised in gymnastic performances.

276. In the *third* place, after the Organism has come to its full maturity, it requires to be *maintained* by the constant exercise of the Nutritive functions; every exertion of its animal Force involving a "waste of tissue," which, unless compensated by a corresponding renovation, would speedily end in death. In fact, it may be said that the whole function of the complicated apparatus

of Organic life (§ 31), when the development of the apparatus of Animal life has been once completed, is to keep it in "working order," and to supply it with the oxygenated blood which is required for the excitement of its activity (§§ 41, 42). But further, while the apparatus of Animal life is subject to "waste" as the very condition of its production of Force, that of Organic life is also subject to "waste" in the course of its constructive operations; so that the entire organism needs a continual *renovation* by Nutritive action; the *rate* of that renovation being mainly determined by the functional activity of each part, but the fundamental plan or law being the same throughout. Thus we everywhere observe a tendency to *the maintenance of the perfect type;* the new materials being substituted for the old in such a manner that, notwithstanding the incessant change in the actual *components* of the Organism, there is no change in its *structure* or *composition.* And this is seen most remarkably in the *reparation of injuries;* which, though not so rapid in the adult, and more limited in its range, is not less complete within that range than it is in the child. There may be said, in fact, to be a continual *rejuvenescence* throughout the whole of vigorous manhood, differing from that of the period of growth only in *degree;* so that each organ has still, to a certain extent, the power of *growing-to* the mode in which it is habitually exercised. And, as Sir James Paget long since pointed-out, the formative power is often exercised, not only in maintaining the *original* type, but also in keeping up some *acquired* peculiarity; as, for example, in the perpetuation of the scar left after the healing of a wound. For the tissue of the cicatrix grows and assimilates nutrient material, exactly as does that of the skin which surrounds it; and thus it not only "grows with the growth" of the child, but is maintained during the whole life of the adult. So, if a child of originally healthy constitution be subjected for a sufficient length of time to such injurious physical conditions as produce a tendency to some particular type of disease, by modifying the normal course of Nutritive action, such tendency (known in Medicine as a *diathesis*), unless early counteracted, not only tends to establish itself during the life of the individual, but to transmit itself hereditarily to the offspring (§ 299). If, on the other hand, a child should inherit such a constitutional tendency, it is during the period of growth and development that appropriate treatment is most efficacious in the removal of it, by restoring the healthy play of the nutritive functions.

277. Familiar as these general facts are, their special application to the Physiology of Habit has not been generally apprehended. There is no part of the Organism of Man, in which the *reconstructive activity* is so great, during the whole period of life, as it is in the ganglionic substance of the Brain. This is indicated alike by the enormous supply of Blood which it receives (for of that large amount which goes to the brain as a whole, § 40, by far the greater proportion is distributed to the "grey matter" of the Cerebral convolutions, of the Sensorial centres, and the Cerebellum); by the evidence furnished by the presence of the products of its oxidation in the excretions, that it undergoes a "waste" proportioned to the demand made upon its functional activity; and by those Microscopic appearances of its tissues, which mark a rapid succession of developmental changes whereby that "waste" is repaired. It is, moreover, a fact of great significance, that the Nerve-substance is specially distinguished by its *reparative* power. For whilst injuries of other tissues (such as the Muscular) which are distinguished by the *speciality* of their structure and endowments, are repaired by substance of a lower or less specialized type, those of Nerve-substance are repaired by a complete reproduction of the normal tissue; as is evidenced in the sensibility of the newly-forming skin which is closing over an open wound, or in the recovery of the sensibility of a piece of "transplanted" skin, which has for a time been rendered

insensible by the complete interruption of the continuity of its nerves. The most remarkable example of this reproduction, however, is afforded by the results of M. Brown-Séquard's experiments upon the gradual restoration of the functional activity of the Spinal Cord after its complete division; which takes place in a way that indicates rather a *reproduction* of the whole of the lower part of the Cord and of the Nerves proceeding from it, than a mere *reunion* of divided surfaces (§ 72). This reproduction is but a special manifestion of the reconstructive change which is *always* taking place in the Nervous system; it being not less obvious to the eye of Reason, that the "waste" occasioned by its functional activity must be constantly repaired by the production of new tissue, than it is to the eye of Sense, that such reparation supplies an actual *loss* of substance by disease or injury.

278. Now in this constant and active reconstruction of the Nervous system, we recognize a most marked conformity to the general plan already shown to be manifested in the Nutrition of the Organism as a whole. For, in the first place, it is obvious that there is a tendency to the production of a *determinate type* of structure; which type is often not merely that of the Species, but some special modification of it which characterized one or both of the progenitors. But this type is peculiarly liable to modification during the early period of life; in which the functional activity of the Nervous system (and particularly of the Brain) is extraordinarily great, and the reconstructive process proportionally active. And this modifiability expresses itself in the formation of the Mechanism by which those *secondarily-automatic* modes of Movement come to be established, which, in Man, take the place of those that are *congenital* in most of the animals beneath him (§§ 191–194); and those modes of Sense-perception come to be *acquired*, which are elsewhere clearly *instinctive* (§§ 160–184). For there can be no reasonable doubt that, in both cases, a Nervous

Mechanism is *developed* in the course of this self-education, corresponding with that which the lower animals inherit from their parents. The *plan* of that *rebuilding* process, which is necessary to maintain the integrity of the organism generally (§ 276), and which goes on with peculiar activity in this portion of it, is thus being incessantly modified; and in this manner all that portion of it which ministers to the *external* life of sense and motion that is shared by Man with the Animal kingdom at large, becomes at adult age the expression of the habits which the individual has acquired during the period of growth and development. Of these Habits, some are common to the race generally, whilst others are peculiar to the individual; those of the former kind (such as walking erect) being universally acquired, save where physical inability prevents; whilst for the latter a special training is needed, which is usually the more effective the earlier it is begun,—as is remarkably seen in the case of such feats of dexterity as require a conjoint education of the perceptive and of the motor powers. And when thus developed during the period of growth, so as to have become a part of the Constitution of the adult, the acquired mechanism is thenceforth maintained in the ordinary course of the Nutritive operations, so as to be ready for use when called-upon, even after long inaction (§ 194).

279. What is so clearly true of the Nervous Apparatus of Animal Life, can scarcely be otherwise than true of that which ministers to the automatic activity of the Mind. For, as already shown, the study of Psychology has evolved no more certain result, than that there are uniformities of mental action, which are so entirely conformable to those of bodily action, as to indicate their intimate relation to a "Mechanism of Thought and Feeling," acting under the like conditions with that of Sense and Motion. The Psychical principles of *association* (§ 218), indeed, and the Physiological principles of *nutrition* (§§ 275–7), simply express—

the former in terms of Mind, the latter in terms of Brain—the universally admitted fact, that any sequence of mental action which has been frequently repeated, tends to perpetuate itself; so that we find ourselves automatically prompted to *think, feel,* or *do* what we have been before accustomed to think, feel, or do, under like circumstances, without any consciously-formed *purpose,* or anticipation of results. For there is no reason to regard the Cerebrum as an exception to the general principle, that, whilst each part of the organism tends to *form itself* in accordance with the mode in which it is habitually exercised, this tendency will be especially strong in the Nervous apparatus, in virtue of that *incessant regeneration* which is the very condition of its functional activity. It scarcely, indeed, admits of doubt, that every state of ideational consciousness which is either *very strong* or is *habitually repeated,* leaves an organic impression on the Cerebrum; in virtue of which that same state may be reproduced at any future time, in respondence to a suggestion fitted to excite it (§ 363). And this *Psychical response,* which is for the Cerebrum what a secondarily automatic movement is for the sensori-motor apparatus, no less certainly depends (in the view of the Physiologist) upon a *reflex action of the Cerebrum* (§ 94), than does an habitual Movement on the *reflex action of the Axial Cord.*

280. The "strength of early associations" is a fact so universally recognized, that the expression of it has become proverbial; and this precisely accords with the Physiological principle, that, during the period of growth and development, the formative activity of the Brain will be most amenable to directing influences. It is in this way that what is early "learned by heart" becomes branded-in (as it were) upon the Cerebrum; so that its "traces" are never lost, even though the conscious memory of it may have completely faded-out (§ 344). For when the organic modification has been once *fixed* in the growing Brain, it becomes a part of the normal fabric, and is regularly *maintained* by nutritive substitu-

tion; so that it may endure to the end of life, like the scar of a wound (§ 276). And there is strong reason for believing that such impressions may *unconsciously* modify our ordinary course of Thought and Feeling (§ 441); while they may themselves be unconsciously modified by it, so as, when awakened to activity after a long period of "latency," to express themselves in a form different from that of the experiences which they originally recorded (§ 365).

281. On the other hand, from the time that the Brain has attained its full maturity, the acquirement of new modes of action, and the discontinuance of those which have become habitual, are alike difficult. Both the Intellectual and the Moral character have become in great degree fixed; so that, although new impressions are being constantly received, they have much less power in directing the course of Psychical action than they had at an earlier period,—that course being henceforth rather determined by the established uniformities, and by the Volitional power of selective attention (§ 131). The readiness with which new Knowledge is now acquired, depends much more upon the degree in which it "fits-in" with those previous habits of thought, which are the expression of the nutritive *maintenance* of the Cerebral mechanism, than it does upon the recording power which expresses a *new formation;* and the record made of it is generally rather that of the *results* of a certain Mental process (involving, it may be, a long sequence of operations), than of the steps of the process itself. Thus nothing is more common than for a man to say, when questioned for the grounds of a confident opinion he may have given upon a particular point,—"I know that I carefully studied the subject before arriving at that opinion; but I have entirely forgotten the reasons which led me to form it." The fixity which certain *modes* of mental activity acquire by habit, is most remarkably shown by the phenomena of Dreaming and Somnambulism; in which states the Cerebrum works *automatically,* without any

Volitional control. For there is ample evidence that just as, in these states, the intense direction of the consciousness to the indications of the "muscular sense" may guide the Movements with extraordinary precision (the distracting influence of vision being for the time suspended, § 192), so the complete engrossment of the consciousness by a particular series of Cerebral changes, without any distracting influences, enables those changes to proceed with more sequential regularity, and thus to evolve a more satisfactory result, than if they were liable to be disturbed either by new sense-impressions, or by volitional interference. But in all such cases, the automatic action follows the course of the *habitual* lines of Thought, and expresses *the result of the whole of that previous training and discipline of the Intellect, which has been carried on under Volitional direction.* The Lawyer could not have written in his sleep a lucid opinion, unravelling the perplexities of a complicated case, if he had not assiduously cultivated the mental habit by which it was elaborated. Nor could the Mathematician in the same state have not merely executed with perfect correctness a lengthened computation, the complexity of which had baffled him in the waking state, but found out a much more direct means of obtaining the result, if his previous training had not been of a kind to develope this particular form of reasoning power. Nor could the Preacher have composed a well-constructed discourse, nor the Poet have evolved an imaginative but consistent and rhythmical series of verses, unless the special faculty required by each had been previously cultivated. (See §§ 231*d*, 487.)

282. During the Decline of life, the influence of early habits and associations often asserts itself in a very remarkable degree; those which have been formed during the middle period being retained with far less of tenacity. And thus it happens that the knowledge *last* acquired is often forgotten *first*; whilst that which was *earliest* learned is retained to the *latest* period. "I am too old to learn" is the continual lament of the aged; the cases in which

the Intellectual activity is retained to such a degree as to enable new forms of thought to be acquired at an advanced period of life, being very rare.—Now this is in precise accordance with the Physiological fact, that Decline essentially consists in the diminution of the formative activity of the organism ; which no longer serves even for the *maintenance* of the Cerebrum, according to the model into which it has gradually shaped itself; so that whilst new modifications of the acquired type are scarcely possible, even those of long standing tend to fade away, the original type being the most enduring.

283. Our best illustrations of the Physiology of Habit are derived from cases in which certain actions, or series of actions, determined by Psychical (cerebral) changes, take place with the least interference from Volitional direction. Such illustrations are afforded by observation of the lower Animals, of Children, and of Idiots. —The whole training of a Horse or a Dog to particular performances, proceeds on the Psychological principle of "contiguous association" ; a set of secondarily-automatic sequences being thus established in the actions of its nervous mechanism, by which a certain response is evoked from it, in accordance with each suggestion to which the animal has been taught to attend. And the tendency of any such series of changes, once established by use, to perpetuate itself organically, is very curiously evidenced by the recurrence of particular actions at fixed intervals of time, without any means of consciously determining its passage, or any incidents that can suggest the return of the period.

Thus a Dog that has been accustomed to receive food at a certain hour and place every day, will come in quest of it with extraordinary punctuality.—The Writer once knew a similar fact in the case of a Swan, which came and tapped with its beak at the door of a cottage at which it received a supply of food, at a certain hour every afternoon.—In like manner, in a family known to the Writer, a Dog, which was accustomed to be washed—to its great dislike—once a fortnight, always kept out of the way on "washing day"; its absence being frequently noticed, and the meaning of it inquired into, before the return of the day had been thought of.—So the Horse of a commercial traveller, after going the same journey a few times, is often known to stop of himself at the houses of the customers, and to stand still for the length of time during which his master's visit usually lasts, becoming restless if it is prolonged; and, if pulled up at a new point on one journey, to stop of his own accord at the same point on the next. (See also § 82 *d.*)

284. The passage of Time would seem to be measured in these cases,—as in another to be presently cited (§ 286),—by that *unconscious* chronometry which determines the duration of Sleep in those who can fix the time of their awaking (§ 481); this chronometry being probably dependent, as was pointed-out by Sir James Paget, on some sequence of organic changes of which the body is the subject. And thus, when the relation of any particular action to some definite lapse of time has once become fixed by association, the action tends to recur automatically, at the intervals determined by habit. There is, however, a considerable difference among individual animals, as to the readiness with which such associations are formed; some having a much more *methodical* tendency than others.

285. No one who observes with intelligence the gradual evolution of the Mind in young Children, can fail to see how constant is the anticipation that what has happened a few times, will recur again under the like circumstances; and how independent is the *sense of dissatisfaction* at the non-recurrence, of any *idea of gratification* to be derived from the occurrence itself. This becomes peculiarly obvious in those "children of a larger growth," who, from want of general culture and of volitional power, have become the slaves of *routine;* their whole course of thought and action being determined rather by what they have been "used to," than by what even ordinary Common Sense would tell them was the best for themselves.

Thus the Writer knew a case in which a family reduced to absolute want during a long depression of trade and manufactures, declined a supply of excellent soup, thickened with barley, merely because "they hadn't been used to barley"; and were only brought to accept it "under protest" by a twenty-four hours' fast.—Again, it was stated by the able Reporter on the Printed Cottons exhibited in the International Exhibition of 1851, that the great obstacle to the improvement of designs lay with the buyers, not with the manufacturers; for that the attempt to introduce the least variation in the patterns of the cheap prints worn by females of the humbler classes had invariably failed; no print being extensively bought by them that departed in the least degree from the pattern which long habit had worked-in (so to speak) upon their brain.—And, quite recently, the Writer has been assured by a large Outfitter, that all his workwomen in a particular department refused to work for a fortnight, merely because they were required to make a slight alteration (not productive of the least additional trouble to themselves) in the pattern of a particular garment.

286. The unreasoning folly shown in these and similar cases of tenacious adhesion to long-established Habits, indicates the strength of the organic tendency which produces the persistence; and this is quite conformable to what the Physiological principles already stated would lead us to anticipate. The tendency is sometimes shown with peculiar force in Idiots, whose whole course of Mental action may be regarded as *automatic*, the power of Volitional control having never been acquired.

The following account was given by Miss Martineau, of a youth under her own observation, who, in consequence of early injury to the brain, never acquired the power of speech, or of understanding the language of others, or of in any way recognizing other minds; but was at the same time strongly affected by sensory impressions. "He could endure nothing out of its position in space or its order in time. *If any new thing was done to him at any minute of the day, the same thing must be done at the same minute every day thenceforward.*" Thus, although he disliked personal interference, his hair and nails having been one day cut at ten minutes past eleven, the next day and every day after, at ten minutes past eleven, he "as if by a fate," brought comb, scissors, and towel; and it was necessary to cut a snip of hair before he would release himself. Yet he had no knowledge whatever of the measurement of time by clocks and watches, and was no less minutely punctual in his observances when placed beyond the reach of these aids. So in regard to form, number, and quantity, his actions were equally methodical. He occupied himself much in making paper-cuttings, which were remarkable for their symmetry. If, when he was out of the room, a brick were taken from the heap with which he amused himself, he would pass his hand over them, spread them a little, and then lament and wander about till the missing one was restored. If seven comfits had once been put into his hand, he would not rest with six; and if nine were given, he would not touch any until he had returned two.—(*Letters on the Laws of Man's Nature and Development*, p. 71. See also *Household Words*, vol. ix. p. 198.)

287. The tendency to Habitual action is so universally recognized as an important part of our Psychical nature, that Man has been said to "be a bundle of habits." Where the Habits have been judiciously formed in the first instance, the tendency is an extremely useful one, prompting us to do that spontaneously, which might otherwise require a powerful effort of the Will. This is especially the case with regard to habits of Intellectual exertion, which are in themselves peculiarly free from any Emotional complication. The Author can speak from long and varied experience, of the immense saving of exertion which arises from the formation of *methodical habits* of mental labour; which cause the ordinary routine to be performed with a far less amount of fatigue, than would be required on a more desultory system (§ 228). Even here, however, care should be taken to avoid allowing oneself to be so much the "slave of habit," that all mental labour, save that which is undertaken at a particular time, or in a particular place, becomes difficult and wearisome. —But, on the other hand, if a bad set of Habits have grown-up

with the growth of the individual, or if a single bad tendency be allowed to become an habitual spring of action, a far stronger effort of Volition will be required to determine the conduct in opposition to them. This is especially the case, when the habitual idea possesses an Emotional character, and becomes the source of *desires* (§ 261); for the more frequently these are yielded-to, the more powerful is the solicitation they exert. And the Ego may at last be so completely subjugated by them, as scarcely to retain any power of resistance; his Will being weakened by the habit of yielding, as the Desire gains strength by the habit of acting upon it.

288. We have thus a definite Physiological *rationale* for that "government of the thoughts," which every Moralist and Religionist teaches to be the basis of the formation of right character, and therefore of right conduct. And the Writer cannot but believe that there are many upon whom the essentiality of Intellectual and Moral *discipline* will be likely to impress itself with greater force, when they are enabled thus to trace out its Physical action, and to see that in the Mental as in the Bodily organism, *the present is the resultant of the past;* so that whatever we learn, think, or do in our Youth, will come again in later life either as a Nemesis or as an Angel's visit.

289. Whilst, then, every one admits the special strength of those *early impressions* which are received when the Mind is most "plastic,"—most fitted to receive and retain them, and to embody them (as it were) into its own Constitution,—the importance of rightly directing the *habits* of thought and feeling during the *whole* stage of Bodily growth, comes to be still more apparent, when we regard those habits as really *shaping* that Mechanism, whose subsequent action mainly determines our Intellectual and Moral character, and, consequently, the whole course of our conscious lives. For large as is the influence which the steady exertion of the *adult* Will, under the guidance of an elevated Moral sense, and of a judgment

matured by experience, can exert in *directing* and *sustaining* the activity of that mechanism, the Ego can only utilize the means which *it* furnishes; the *formation* of the mechanism being greatly dependent upon the *self*-discipline exerted during the stage of adolescence. But the very possession of such self-disciplining power greatly depends on the right exercise, in a yet earlier stage, of the Will, the Moral sense, and the Judgment of the Educator; save where the discipline of "circumstances" has called into activity the best parts of the Child's nature, and has given them the most healthful training. There is nothing, for example, more remarkable than the extraordinary power of *thoughtful self-direction*, or more beautiful than the complete *moral self-abnegation*, which is often shown by a very young girl, upon whom the charge of a still younger family has fallen, through the prolonged ill-health or death of the mother. On the other hand, nothing is more mischievous than the wrongly-directed, though well-intended, discipline often administered by Parents and Teachers who are ignorant of the fundamental facts of child-nature (§ 120); unless it be the education of the Volitional power by "circumstances" that guide it in the *worst* instead of the best direction.

290. It has been from the depth of his conviction as a Physiologist and Psychologist, of the inseparable relation between Corporeal and Mental action, that the Writer has been led, during a life of Educational occupation, to what he may call the Scientific study of that relation, as manifested within the range of his own observation. And the following outline attempts to systematize the general result of the experience thus acquired, on the basis of the *rationale* afforded by the Physiological doctrine of "Unconscious Cerebration," or, in the language of German Psychologists, the "Preconscious Activity of the Soul," hereafter to be expounded (Chap. XIII.). To himself it seems comparatively unimportant whether the doctrine be expressed in terms of Physiology or in terms of Metaphysics; if the *principle* be duly recognized, and the enor-

mous *practical importance* of directing the "preconscious" activity through the Physical nature be admitted and *systematically acted on,*—especially in that *very earliest* stage of Infant Education, which lays the foundation for the Intellectual and Moral Habits of the conscious Life. For as Physiology teaches that external agencies exert their most potent influences on the *bodily* Organism during the period of its development, so Educational experience proves that nothing exerts so great an influence on the *psychical* Organism, as what may be called the *moral atmosphere* which is breathed by it, from the very earliest stage of conscious existence, up to the time of its full maturity. This influence—exerted, on the one hand, through the medium of the body, on the other, through the unconscious action of *example,* in shaping these habits of Feeling which give the *tone* to the character,—is far more potent than is generally supposed; and commencing in the Nursery, it prolongs itself alike in the Home and in the School, through the whole period of Childhood and Youth, and by no means dies out in Adult age. What may be termed the *composition* of that Atmosphere, is, therefore, of fundamental importance; and the following appear to the Writer to be the constituents which science and experience alike recommend :—

(1.) *Order* and *Regularity.*—This should begin even with Infant life, as to times of feeding, repose, &c. The *bodily* habit thus formed, greatly helps to shape the *mental* habit at a later period. On the other hand, nothing tends more to generate a habit of self-indulgence, than to feed a child, or to allow it to remain out of bed, at unseasonable times, merely because it cries. It is wonderful how soon the actions of a young Infant (like those of a young Dog or Horse), come into harmony with systematic "training," judiciously exercised.

The following little incident affords an apposite illustration of the principle just stated, as well as a useful example of the kind of

Nursery-discipline above advocated.—The first child of a young mother was accustomed, before being put into his cradle for his mid-day sleep, to be "hushed off" in the arms of his mother or his nurse. But having been told that this was an undesirable practice, his mother, wishing to break him of the habit, one day laid him down awake in his cradle, and remained behind the head of it, so as to be out of the infant's sight. He screamed so long and so violently, that several times she almost relented, fearing that he would injure himself; but she had firmness to persevere; and after a while the child "cried himself to sleep." Next day the screaming-fit was much shorter, and on the following day shorter still; and in a few days the child ceased to cry when laid down, and never did so again. Now the child was too young to have learned by any process of *reasoning* upon conscious experience, that, as crying was of no use, it might as well keep quiet; but its nature simply adapted itself to the change of circumstances. And in the case of four other children afterwards successively brought up in the same nursery, the habit was established from the beginning.

Again, while an Infant of a few months old, being allowed to sleep with its mother, and suckled by her every time that it wakes and cries, contracts not only a habit of wakefulness, but an organic longing (so to speak) for the accustomed solace, another that is kept away during the night from the smell of the milk (which seems to be the immediate provocative), will soon accustom itself to be fed at stated intervals only, and will only wake at these.

And so, in those later stages in which the *conscious* influence of Order and Regularity comes to exert itself, there is still an *unconscious* persistence of the habits early shaped upon them, which tends to repress desire in that earliest phase in which it has not shaped itself into a definite idea. Every one who has been practically engaged in Education, well knows the Moral value of orderly discipline; but few are fully aware of the enormous influence which it unconsciously exercises over the formation of that "Mechanism of Thought and Feeling," which determines a large part of our conscious life.

The Writer has been thus led to regard Military Drill as an

important part of Education; not merely as promoting a healthy physical development, and as preparing every youth, if occasion should require, to serve in the ranks of National defenders; but even more for the Moral value of the enforcement of strict order and discipline, and prompt obedience to the "word of command." This is especially seen in those Industrial Schools and Reformatories, the boys in which have been previously accustomed to a life of uncontrolled freedom. That they have a power of *self*-control, seems to them an entirely new idea; and there is no better way of bringing it to their apprehension, than this systematic "playing at soldiers," in which they usually feel a great enjoyment.—It may be further remarked that the whole system of "making a soldier" out of a raw recruit, proceeds upon the same principle of the *unconscious* formation of acquired Instincts, through the combined action of bodily and mental Habit. The "pluck" by which the British Soldier is especially distinguished, is clearly as much Physical as Psychical; and much of it seems to depend on the *unfelt* sense of mutual support, which is derived from the maintenance of order. Every officer of experience knows full well that if the "formation" of the British ranks is once "broken," a panic *sauve qui peut* is almost certain to spread; and that the success of a "storm" mainly depends upon whether the "forlorn hope" by which it is led, can "form" within the fortress. The determined maintenance of the admirable "formation" in squares, directed by Wellington, was what enabled the inferior British force to hold its ground at Waterloo until the Prussians came up; whilst the failure of the first attack upon the Redan at Sebastopol, was attributed by the officers concerned in it to the impossibility of "forming" the men who had gained an entrance into the fortress.—This Moral influence of discipline was never more grandly exhibited, than in the manner in which the soldiers on board the "Birkenhead" went to their death; while the want of it was as sadly shown in the behaviour of the "navvies" on board the "Northfleet." Yet the "raw material" of the latter was just as good as that of the former; and a few months' drill would probably have turned out as good a product. The Writer does not agree with the "Spectator" (Jan 25, 1873) in regarding this influence as exercised through the ideational and moral consciousness. For it seems to him to operate much more directly, through a channel which has been formed by habit, without any distinct consciousness of the *how* and the *why*. Herodotus tells us that the Scythian warriors, returning victorious from a campaign, found that their slaves had risen in their absence, and had fortified themselves in a stronghold from which it would have been very difficult to dislodge them by force; but that they lost all their courage at the cracking of their masters' whips, which terrified them into submission. In this case—as also, after the battle of Gravelotte, when the riderless horses assembled at the bugle-call,—there was surely nothing higher than an immediate *automatic* response to the sound which had habitually called it forth; that response being Emotional in the one case, Motorial in the other.

But, with the advance of years, and the development of the power of *self*-control, the aim should be rather to foster *its* independence by relaxing external coercion, so far as may prove safe, than systematically to restrain the healthy spontaneity of the individual within trammels that tend to become formal and mechanical; the consequence of such prolonged restraint too often being, that when the individual *is* freed from it, he runs altogether wild, through not having been trained in the habit of *self*-discipline.

(II.) With the preceding, there naturally connects itself the principle of *Duty* or *Obligation;* which, indeed, grows up as the *internal* correlative to the *external* coercion. The child very early comes to feel what it *ought* to do; and this sense of obligation, which at first has its Mother or its Nurse as its object, extends itself after a while to its Father and its Schoolmaster; the feeling of Duty becoming *religious*, when the notion of obligation to an unseen Power is first comprehended. Now the child is *unconsciously* impressed, long before it distinctly shapes out an idea of the fact, by the manifestation of this sense of duty on the part of those around it; and is very quick to perceive what it supposes (perhaps wrongly) to be a departure from it. And while every such example tends *unconsciously* to weaken the feeling on the part of the child, the unconscious influence of steady persistence, especially when the child is made aware that such persistence is disagreeable, is enormous. Particularly is this the case with respect to Punishment; for when the child can be brought to feel that its infliction is not vindictive, but is prompted only by a sense of duty, and is guided by love, so that it is as painful to the Parent as it is to the Child, the basis is unconsciously laid for the habit of trustful submission to the life-discipline of an All-wise and All-merciful Father. But here, again, it is essential to healthful development, that with the advance of years this submission, even to Duty, should not be maintained in its *slavish* form; but that the growth of that highest form of submission should be fostered and encouraged, which consists in the *identification* of our own Will with the Divine (§ 338).

(III.) In order that the feeling of Duty may be led to develop itself into notions of *Right* and *Justice*, it is essential that the Moral atmosphere should be pervaded by those attributes. There is in most Children a very strong instinctive sense of *their own* rights, or what they consider to be such; the action of which is so admirably delineated in the "Rejected Addresses," that the description, in "The Baby's Debut," of the nursery quarrel it provoked, is almost worthy of the great Poet whom it parodied:—

> "This made him cry with rage and spite:
> Well, let him cry, it serves him right.
> A pretty thing, forsooth!
> If he's to melt, all scalding hot,
> Half my doll's nose, and I am not
> To draw his peg-top's tooth."

And respect for the *rights of others* is one of the earliest lessons which the child has to be taught through his consciousness. But that lesson is far more effectual, when the preparation for it has been made, and the illustration of it afforded, by the habitual example of the *elders* of the family; that of older brothers and sisters being in this respect (as in many others) even more *impressive* than that of parents, whom the child is apt to regard *unconsciously* as in a distinct category from itself. Hence arises the vast importance of giving a right direction to the character of the *first* child in a family; and those who have had the opportunity of observing the immense influence, either for good or for evil, which that example insensibly as well as sensibly exerts, will fully confirm the Writer's estimate of its importance.—It is wonderful how powerfully what may be called the *current* of daily life carries along with it, *without any consciousness of its influence,* those who are subject to it. Thus, in a Household, or in a School, where order and regularity prevail, and where the sense of duty is predominant, the preference of duty to pleasure comes to be the general habit; the lessons are prepared as a "matter of course," before mere amusement (proper intervals of *recreation* being presupposed) is had recourse to; and no written rule, or system of punishments, to enforce it, is found requisite. The Child's knowledge (when it comes to *think* on the matter) that its neglect or carelessness as to the preparation of lessons will give pain to a Parent whom it loves, *unconsciously* determines its habit of application; and the Schoolboy's feeling towards a Master who has acquired his respect and confidence, instinctively governs his whole conduct towards him. "It's a shame to tell Arnold a lie, because he always believes one," was the *resultant* of a very large number of impressions made by the character of that noble type of an educator upon the mind of the lad who gave utterance to this oft-quoted phrase; and if these impressions could be separately reviewed, it would assuredly be found that the operation of many of them had never distinctly come before the consciousness of their recipient.

Of one who, within a narrower sphere, exercised a power probably not inferior to Arnold's "of commanding the reverence and reconstituting the wills, of the least manageable class of human beings," it was said by one of his most distinguished pupils, whose own character rendered him peculiarly amenable to that influence, "I have often reflected on this singular power, and tried to make out where the secret of it lay. Though there were doubtless cases which it could not reach, it daily achieved triumphs which most teachers would believe impossible."* He traces it mainly to the combination of great and varied Intellectual power, acting through well-formed habits of thought, with profound sense of right pervading the whole life and conversation, and with the insight derived from a thorough and affectionate sympathy with boy-nature, quickened by a temperament of peculiar vivacity. "The spirit of duty in his house was no withered ghost of custom, but such a living reality as it befalls few to witness. Though the machinery of rules and habits devised for the maintenance of punctuality and order was more complicated and extensive than I have ever seen in operation elsewhere, never was there less indolent trust in mere routine. *The mechanism served, and never ruled;* and at the remotest point, felt the thrill of some high purpose as its moving power." . . . "The earnestness with which he insisted on the smallest things being done *well*, was an indication of the same kind; manifesting his watchfulness against the least slovenliness of conscience, his resolution to close the most trivial aperture through which looseness and disorder could find entrance into life." . . . "Boyhood, which detests, as they deserve, all kinds of sham and pretence, easily places itself at the disposal of a sincerity so profound as this; owns as a true guide *one who lives under an authority of the rules he imposes, and whose administration of command is in itself an exercise of obedience.*"—It may be doubted whether this was *consciously* owned, save by those whose own Intelligence was sufficiently developed to analyse the sources of the power thus exerted; to the Writer (whose privilege it was to experience it from his earliest years) it seems to have acted in the first instance through the deeper and more direct channels of Unconscious Influence. It must have been as the *resultant* of a multitude of separate impressions, many of them not rising to the level of conscious apprehension, that "his perceptive eye *looked into order* a thousand things for which there was no audible guidance or command."

* See the Rev. James Martineau's admirable delineation of the Character of Dr. Lant Carpenter as a Schoolmaster, in the Memoirs of his Life, pp. 342-352.

(IV.) Last, but not least, among the components of a healthy Moral atmosphere, is *love* towards others; showing itself in that habitual *kindness*, which springs from consideration for their *feelings* as well as for their *rights*. The fostering of this principle by her own example, is one of the most important functions of the Mother or Nurse; upon whom falls the whole of the earliest direction of the Infant's nature, and the formation of those dispositions which *colour* the whole future life. It is a beautiful fact in Human Nature, that the Maternal instinct, which manifests itself among the lower animals in the supply of the mere Physical wants of the offspring, extends itself to the sphere of Mental and Moral development; so that the first *affections* of the infant spring out of its earliest association between its own simple sense of *well-being*, gradually quickened into *happiness*, and the personality of the Mother; who in the first instance imparts sustenance and comfort to its body, and who, when this duty has been discharged, delights to call into exercise its powers of observation, to cherish its awakening intelligence, and to minister in every suitable way to its pleasures. This association is formed, and shows itself in the Infant's actions, long before it can have risen into the sphere of distinct consciousness; but the foundation is thus laid for that abiding affection, which, if sustained by judicious culture, as infancy passes into childhood, and childhood into youth and adolescence, makes the link between Mother and Offspring more enduring than any other.

It is the experience of those who have endeavoured to awaken any dormant sense of good, which may not have been crushed-out from the nature of the most hardened Criminals by a life of brutal indulgence in every bad passion, that if they retain any recollection of a Mother's love, this affords a loop-hole through which what remains of their better nature may be reached, though the wall of sullenness and obstinacy within which it was shut up may at first seem utterly impenetrable.—The Writer well remembers, at a distance of forty years, the impression produced on his mind by hearing a case of this kind narrated by Dr. Tuckermann, of Boston, N.E.; a man probably unequalled for his success in the line of action above indicated, which was entirely prompted and guided by his implicit confidence in the principle of Love.

But whilst the unconscious influence of Love is essential to the due formation of the affectional part of Child-nature, it is no less important that it should be rightly directed. Nothing is more common among Mothers whose own characters have not been judiciously trained, either by their own experience or by the discipline of others, than that *indulgence* of their children's desires, which, when it grows into a habit, forms both the Physical and the Psychical constitution in accordance with it; so that a "Mechanism of Thought and Feeling" comes to be established, which re-acts in direct response to the impressions made upon it from without.

291. Every observant Educator must have seen instances in which the same Child has almost seemed two different individuals, according to the "atmosphere" in which it happens to be. Where a nature originally good and docile has been "spoiled" for a time by injudicious indulgence and by irregularity in the habits of the family, and where the child has early been transplanted into a household in which strict rule, judiciously tempered by kindness, has created a more healthy atmosphere, it is wonderful how the whole character shapes itself insensibly on an improved model, so long as that atmosphere is breathed; while it is no less wonderful how rapidly and completely the child falls back into the old ways, when it is again subjected to the old influences, before the improved habits have become ingrained, and it has come to the stage of conscious recognition of their superiority.—It is the great advantage of the early establishment of a judicious *system* of Nursery-management, that the habits formed under its influence make the Child directly connect the well-being of which it is conscious, not merely with the love that *gives*, but with the love that *withholds;* so that by the time the Mother can impress its consciousness with the idea that

the *truest kindness* may consist in the *refusal* of a gratification, the foundation for the apprehension and recognition of that idea has already been unconsciously laid. And when, with advancing years, this assurance comes to be strengthened, at the same time by Paternal authority, and by the child's own enlarging experience, —when a firm *trust* has grown up in the loving wisdom and justice of the Father, and in the judicious love of the Mother,—this constitutes the surest basis for the recognition of the like attributes in the dispensations (whether joyous or grievous) of that Infinite Being, whom Theodore Parker used thus to address, "O God, our Father and our Mother too!"

292. While the early Habits are thus in a great degree determined for each individual by the *family* influences under which he is brought up, he soon comes under those *social* influences which in a great degree shape the future course of his Mental life,—constituting that *aggregate* which was designated by the Greeks as the Νόμος. This term (sometimes translated "custom" and sometimes "law") may be considered as expressing the *custom which has the force of law*, and which is often far less easily changed than any written law; becoming so completely ingrained in the Constitution of a People or a Class, as to constitute a "second nature," which only a long course of the "discipline of circumstances" can alter. The following admirable analysis by Mr. Grote of the Greek conception of the Νόμος, is as applicable to the present time as to the Classical epoch :—

"This aggregate of beliefs and predispositions to believe, Ethical, Religious, Æsthetical, Social, respecting what is true or false, probable or improbable, just or unjust, holy or unholy, honourable or base, respectable or contemptible, pure or impure, beautiful or ugly, decent or indecent, obligatory to do or obligatory to avoid, respecting the status and relations of each individual in the society, respecting even the admissible fashions of amusement and recreation,—this is an established fact and condition of things, the real origin of which is for the most part unknown, but which each new member of the society is born to and finds subsisting. It is transmitted by tradition from parents to children, and is imbibed by the latter almost unconsciously from what they see and hear around, without any special season of teaching or special persons to teach. It becomes a part of each person's nature,—a standing habit of mind, or fixed set of mental tendencies, according to which particular experience is interpreted and particular persons appreciated. It is not set forth in systematic proclamations, nor impugned nor defended: it is enforced by a sanction of its own,—the same real sanction or force, in all countries,—by fear of displeasure from the Gods, and by certainty of evil from neighbours or fellow-citizens. The community hate, despise, or deride any individual member who proclaims his dissent from their social creed, or even openly calls it in question. Their hatred manifests itself in different ways, at different times and occasions; sometimes by burning or excommunication, sometimes by banishment or interdiction of fire and water; at the very least, by exclusion from that amount of forbearance, goodwill, and estimation, without which the life of an individual becomes insupportable; for society, though its power to make an individual happy is but limited, has complete power, easily exercised, to make him miserable. The orthodox public do not recognize in any individual citizen a right to scrutinize their creed, and to reject it if not approved by his own rational judgment. They expect that he will embrace it in the natural course of things, by the mere force of authority and contagion, as they have adopted it themselves; as they have adopted also the current language, weights, measures, divisions of time, &c. 'Nomos (Law and Custom), King of all' (to borrow the phrase which Herodotus cites from Pindar), exercises plenary power, spiritual as well as temporal, over individual minds; moulding the emotions as well as the intellect, according to the local type,—determining the sentiments, the beliefs, and the predisposition in regard to new matters tendered for belief, of every one,—fashioning thought, speech, points of view, no less than action,—and reigning under the appearance of habitual, self-suggested tendencies."—(*Plato and the other Companions of Sokrates*, vol. i. p. 249.)

293. It thus becomes extremely difficult—in fact, in most cases,

impossible—to discriminate clearly between the tendencies to Thought and Feeling which are *shaped* by the Νόμος, and those which *grow out* of the congenital Constitution,—that constitution having itself (it seems probable) been moulded under the pressure of the like Νόμος on the ancestral Race. Some information may doubtless be drawn from the comparison of the diverse mental habits of the several children of the same family, brought up, as nearly as may be, under the same influences. But we find a better evidence of congenital peculiarity in the distinctive characteristics of those who have been very early transplanted from one Νόμος into another of an entirely different character. True it is that during that earlier period when the *general* characters of Humanity are evolving themselves, when the observing powers are most active, and the mental development is chiefly sustained by the direct assimilation of the *pabulum* they afford, the child of one Race readily adapts itself to the habits of another; and, as a general rule, the children of the less civilized races show a remarkable *quickness* of perception, which sometimes engenders the belief that their capacity is not below our own. But their Intellectual development very commonly stops at a point which leaves them "great children" all their lives; and the *special* characters of the *type* then begin to manifest themselves unmistakeably. This is particularly noticeable where the ancestral habits have been *nomadic*; the craving for a return to them being often so strong, as to render the restraints of a settled life irksome or even insupportable.

Thus the Writer has been informed of several cases in which boys of the native races of Australia, who have been brought up in schools either in the Colonial capitals or in Britain, and who have seemed to profit very fairly by the education they have received, and to form attachments to those among whom they have lived, have seized the very earliest opportunity of going off to the "bush" and of rejoining their countrymen.—And the same tendency shows itself generally in our own "street Arabs," who can only be trained to steady habits by an elastic system of discipline skilfully adapted to their nature; * the restraints of the ordinary school-system, which are well suited to the children of the *settled* members of our labouring population, being quite intolerable to those who have been *wanderers* for generations back.

Here, then, we have another very striking instance of that Hereditary Transmission of *acquired tendencies*, which would be probably found by careful investigation to be scarcely less clearly manifested among the different Races of Men, than among different Breeds of Dogs (§§ 93, 301).

294. Again, the Psychical tendencies of every individual undergo a consecutive change during the progress of his life, which is quite independent of external influences. Infancy, Childhood, Youth, Adolescence, Adult age, the period of Decline, and Senility, have all their characteristic phases of Mental as of Corporeal development and decline: this being shown, not merely in the general advance of the Intellectual powers up to the period of middle life, and in their subsequent decay; but in a gradual change in the balance of those springs of action which are furnished by the Emotional states, the pleasures and pains of each period being (to a certain extent) of a different order from those of every other. This diversity may be in great part attributed to changes in the Physical constitution. Thus, the Sexual feeling which has a most powerful influence on the direction of the thoughts in adolescence, adult age, and middle life, has comparatively little effect at the earlier and later periods.

* The existence of a *distinct Class* of this kind, a good deal resembling that of the Bushmen of the Hottentot country and the Pariahs of India, is well known to those who have investigated the social condition of our Great Cities. It is continually receiving additions from the degraded members of the class above, into which, on the other hand, some of its own are occasionally elevated; but, if left to itself, it maintains its ground with remarkable persistence, and furnishes a large proportion of our *professional* Criminals.

So, again, the thirst for novelty, and the pleasure in mental activity, which so remarkably characterize the young, when contrasted with the obtuseness to new impressions and the pleasure in tranquil occupations, which mark the decline of life, seem related to the greater activity of the changes of which the Nervous system is the subject during the earlier part of life, and to its diminished activity as years advance. There are other changes, however, which cannot be so distinctly traced to any Physical source, but which yet are sufficiently constant in their occurrence to justify their being regarded as a part of the developmental history of the Psychical nature; so that each of the "Seven Ages of Man" has its own character, which may be with difficulty defined in words, but which is recognized by the apprehension, as it forces itself upon the experience, of every one.

295. But whilst recognizing, as a fact of observation, the large share which congenital Constitution and external influences have in the formation of those tendencies of Thought and Feeling, which make up the Character of each individual, we must equally rely on the assurance of our own Consciousness, that we have within us a Power, which, if we use it aright, can in great measure control the excesses and supplement the deficiencies of these tendencies, and can direct them to good and useful instead of evil and injurious ends. The strength of this Power, which we term *Will*, mainly depends upon the *constancy with which it is exercised;* the ascendancy of *principles* of action determinately adopted by the Reason, over the strong *impulses* of passion or desire, being only possible (as will be shown hereafter, § 333), when that ascendancy has been *habitually* maintained; and this not only in the ordinary course of conduct, but in the government of the thoughts. There is, in fact, a continual action and reaction between Habit and Will, in our *mental*, just as there is in our *bodily* life. For as, when we train ourselves to execute a parti-

cular movement, the Bodily mechanism in the first instance *shapes itself* into accordance with our Will, and then comes to execute the same movement without or even (it may be) *against* our intention, so does our Mental mechanism in the first instance *grow-to* the mode in which the Will exercises it, and then, having taken on a particular method of activity, tends, by the automatic power it has acquired, either to supply a strong *motive* to the determination of the Will, or to act altogether independently of it (§ 331).

296. It is, moreover, by an action and reaction between Habit and Will, similar to that which regulates our conduct, that our *Beliefs* are determined. For although it is often maintained that, as Belief is in itself involuntary, we are not responsible for what we believe, yet a careful examination of the conditions under which Beliefs are generated, will show that in so far as we volitionally regulate those conditions, we make ourselves to that extent *indirectly* responsible for the Beliefs which automatically spring from them (§§ 321-326). And in the same proportion we become responsible for our share in the direction of that Νόμος, which exerts so powerful an influence on the formation of the Beliefs of the rising generation.

297. But, further, there is strong reason to believe that our power of benefiting the future of our Race by the formation of right Habits of Thought, is not limited to those results which we give to the world, whether by the publication of our opinions, or by our conduct in life; being even more directly exerted in that Hereditary transmission, which the Physiologist cannot but regard as no less evidently manifested in tendencies to Thought, than it is in tendencies to Action. And it will be appropriate, therefore, here briefly to consider this subject; which is one, however, that especially needs all the elucidation that close, prolonged, and extensive study of facts can alone afford.

298. *Hereditary Transmission.*—The view of the relation of Mental Habits to peculiarities of Bodily organization, whether

congenital or *acquired*, which it has been the object of this chapter to develope, must be extended to that remarkable *hereditary transmission* of Psychical character, which presents itself under circumstances that entirely forbid our attributing it to any agency that can operate subsequently to birth; and which it would seem impossible to account for on any other hypothesis, than that the "formative capacity" of the Germ (§ 274) in great degree determines the subsequent development of the Brain, as of other parts of the body, and (through this) its mode of activity. And this formative capacity, which is the Physiological expression of what is commonly spoken of as the "original constitution" of each individual, is essentially determined by the conditions, dynamical and material, of the Parent-organisms. That "like produces like" is certainly the *general* rule. For not only do *Species* maintain their fixity, so long as the continuance of the same external conditions tends to keep up the mode of Life which belongs to each; but *Races*, whose origination in *varietal modifications* of the specific type can be distinctly traced historically, often display a scarcely inferior permanence. It is, therefore, by a *limitation* of this general rule, that Family or Individual peculiarities do *not* tend to reproduce themselves in anything like the same degree; and the explanation of that limitation seems to lie in the fact, that while the characters of Species or of Races express the most *general* and *constant* of those influences which tend to produce varietal modifications, those of Families or Individuals express influences of far inferior *range*, and of far less duration. In particular, it is to be noted that while the *more general* characters are shared by *both* parents, the *more special* commonly belong to *one* only.—A great deal of discussion has taken place, as to whether the Male or the Female parent exerts the greater influence over the character of the offspring; and while experience does not yet justify any definite conclusion on this point, the question seems to have been entirely ignored, whether the union of two different natures may not produce—as in the combination of an acid and a base—a *resultant* essentially dissimilar to either of them.

299. This much, however, may be confidently affirmed, that where *general Constitutional taints*, that is to say, *abnormal habitudes* of Nutrition, have been acquired, these tend to propagate themselves hereditarily; and that they do so with the most certainty, when *both* parents partake of them. It may also be affirmed that every repetition of such transmission tends to increase the mischief; so that by "breeding in and in," the injurious external conditions remaining the same, a very slight original departure from healthy nutrition may become intensified in successive generations into a most serious abnormality. Of this we have an example in the production of Cretinism, which may be characterized as Idiocy combined with bodily deformity. (This is not essentially connected, as some have supposed, with Goitre; for it presents itself in some localities where goitre does not prevail, whilst elsewhere goitre prevails without developing itself into cretinism. But in the Vallais of the Alps, the conditions of the two appear concurrent; and the result of their conjoint action through a succession of generations becomes most distressingly apparent.)—There is one class of cases, moreover, in which a particular abnormal form of Nutrition that is distinctly *acquired* by the *individual*, exerts a most injurious influence upon the offspring;—that, namely, which is the result of such habitual Alcoholic excess, as *modifies the nutrition* of the Nervous system (§§ 138, 155).

a. We have a far larger experience of the results of habitual Alcoholic excess, than we have in regard to any other "Nervine Stimulant;" and all such experience is decidedly in favour of the *hereditary transmission* of that acquired perversion of the normal Nutrition which it has engendered in the individual. That this manifests itself sometimes in congenital Idiocy, sometimes in a predisposition to

Insanity, which requires but a very slight exciting cause to develop it, and sometimes in a strong craving for Alcoholic drinks, which the unhappy subject of it strives in vain to resist, is the concurrent testimony of all who have directed their attention to the inquiry. Thus Dr. Howe, in his Report on the Statistics of Idiocy in Massachusetts, states that the habits of the parents of 300 idiots having been learned, 145, or nearly one half, were found to be habitual drunkards. In one instance, in which both parents were drunkards, seven idiotic children were born to them. Dr. Down, whose experience of Idiocy is greater than that of any other man in this country, has assured the Writer that he does not consider Dr. Howe's statement as at all exaggerated. Dr. W. A. F. Browne, the first Medical Lunacy Commissioner for Scotland, thus wrote, when himself in charge of a large Asylum:—"The drunkard not only injures and enfeebles his own nervous system, but entails mental disease upon his family. His daughters are nervous and hysterical; his sons are weak, wayward, eccentric, and sink under the pressure of excitement of some unforeseen exigency, or the ordinary calls of duty." Dr. Howe remarks that the children of drunkards are deficient in bodily and vital energy, and are predisposed by their very organization to have cravings for Alcoholic stimulants. If they pursue the course of their fathers, which they have more temptation to follow, and less power to avoid, than the children of the temperate, they add to their hereditary weakness, and increase the tendency to Idiocy or Insanity in their constitution; and this they leave to their children after them. The experiences of those who, like Hartley Coleridge, have inherited the craving for Alcoholic excitement, together with the weakness of Will which makes them powerless to resist it, whilst all their better nature prompts the struggle, must satisfy any one who carefully weighs them, how closely connected their Psychical state is with the Physical constitution which they inherit, and how small is their own moral responsibility for errors which are mainly attributable to the faults of their progenitors. As Robert Collyer (of Chicago) has well said in an admirable sermon on "The thorn in the flesh:"—"In the far-reaching influences that go to every life, and away backward as certainly as forward, children are sometimes born with appetites fatally strong in their nature. As they grow up, the appetite grows with them, and speedily becomes a master, the master a tyrant, and by the time he arrives at his manhood the man is a slave. I heard a man say that for eight-and-twenty years the soul within him had had to stand, like an unsleeping sentinel, guarding his appetite for strong drink. To be a man at last under such a disadvantage, not to mention a saint, is as fine a piece of grace as can well be seen. There is no doctrine that demands a larger vision than this of the depravity of human nature. Old Dr. Mason used to say that as much grace as would make John a saint, would barely keep Peter from knocking a man down."

300. With such evidence of the Hereditary Transmission of general *diatheses*, or modes of Nutrition, of which we can distinctly trace the acquirement in the history of the progenitor, we seem fully justified in applying the same doctrine to such particular *habits* as may be regarded, from the Physiologist's point of view, in the light of *expressions* of special modifications of Nervous organization.

A very curious example of the transmission of tendencies to special Automatic movements, the secondary *acquirement* of which tendencies is altogether beyond doubt, is afforded by the following curious fact established by the researches of M. Brown-Sequard:—In the course of his masterly experimental investigations on the functions of the Nervous system, he discovered that, after a particular lesion of the Spinal cord of Guinea-pigs, a slight pinching of the skin of the face would throw the animals into a kind of Epileptic convulsion. That this artificial Epilepsy should be constantly producible in Guinea-pigs, and not in any other animals experimented on, was in itself sufficiently singular; and it was not less surprising that the tendency to it persisted, after the lesion of the Spinal cord seemed to have been entirely recovered from. But it was far more wonderful that when these epileptic Guinea-pigs bred together, their offspring showed the same predisposition, without having been themselves subjected to any lesion whatever; whilst no such tendency showed itself in any of the large number of young, which were bred by the same accurate observer from parents that had not thus been operated on.

301. The same view may be extended to that higher class of *secondarily automatic* actions, which can only be performed under the guidance of Sensation, and which therefore involve some Psychical change in each case, as one of the links in the sequence. For as it is impossible not to recognize the influence of Habit,—that is to say of the Voluntary repetition of similar acts under similar circumstances,—in establishing a condition of the Nervous apparatus which leads to the Automatic performance of such acts, it is conformable to all Physiological probability that a *tendency* to them should be hereditarily transmitted.

a. "On what a curious combination of corporeal structure, mental character, and training," says Mr. Darwin, "must *handwriting* depend! Yet every one must have noted the occasional close similarity of the handwriting in father and son, although the father had not taught his son. A great collector of franks assured me that in his collection there were several franks of father and son hardly distinguishable except by their dates. Hofacker, in Germany, remarks on the inheritance of handwriting; and it has been even asserted that English boys, when taught to write in France, naturally cling to their English manner of writing."

b. The Writer has been assured by Miss Cobbe, that in her family a very characteristic type of handwriting is traceable through *five generations.*

c. The following case which occurred in the Writer's own family, and which (he is assured) can be exactly paralleled elsewhere, indicates the "constitutional" character of handwriting:—A gentleman who emigrated to the United States, and settled in the backwoods, before the end of last century, was accustomed from time to time to write long letters to his sister in England, giving an account of his family affairs. Having lost his right arm by an accident, the correspondence was temporarily kept up by one or other of his children; but in the course of a few months he learned to write with his left hand; and, before long, the handwriting of the letters thus written came to be indistinguishable from that of his former letters.

d. Again, there are some "nervous" men who always seem to require to do *something* with their hands when they are speaking earnestly; and what particular "trick" each individual may learn, depends very much upon accident. Thus, in the old times of dependent watch-chains and massive bunches of seals, these were the readiest playthings; and now that watches are commonly worn in the waistcoat pocket, the hands of such persons may often be seen unconsciously stealing upwards to "twiddle" with their watch-keys.—There is a well-known story of a barrister who acquired the "trick" of winding and unwinding a piece of string on his fingers when addressing the court, and who was thrown into confusion when the opposing counsel stole "the thread of his discourse."—Not long since, when listening to a very interesting extempore sermon, the Writer observed that the preacher was continually opening and shutting his Bible, and shifting it from side to side of the pulpit; this being probably a mere "trick" of which the preacher was quite unconscious, and the Bible merely supplying the place of the bunch of seals, the watch-key, or the bit of string, in giving his "idle hands" something "to do."—Such "tricks," like particular "gestures" indicative of Emotional excitement, are often repeated in successive generations, under circumstances that forbid the idea of their having been learned by imitation.

302. The case seems to be even stronger in regard to *drawing*, and to *playing* on Musical instruments; for it seems quite certain that the power of attaining Artistic proficiency in either, and the readiness with which it is acquired, depends in great degree upon congenital Temperament. No doubt every child may be *taught* to draw, or to play a musical instrument, after a certain fashion: but there are some whom no teaching or self-effort will ever carry beyond a certain mechanical exactness; whilst there are others who "take to" drawing or to musical performance as their natural language, and who, with very little instruction, learn to express themselves in either with singular force and beauty.

The Writer knows one family in which this Artistic temperament is widely diffused; the several members of it almost always "taking" either to Drawing or to Music, and sometimes to both. On the other hand, he knows other cases, in which, with a considerable acquired interest (rather Intellectual, however, than Artistic) both in Pictorial and Musical art, and with considerable manual dexterity (as shown in other ways), there is a greatly-regretted inability to acquire anything more than a stiff formal execution, either in Drawing or in Music. To the *first*, "free-hand" Drawing, to the *second*, "mechanical" Drawing, comes most naturally.

303. Hence it seems clear that there is a different Hereditary tendency to the performance of certain classes of Movements; just as, in different Nations, there is a different hereditary tendency to the production of certain Vocal sounds. As in the case of Handwriting, it is impossible to say how much of this is due to what we are accustomed to call "spontaneous" variation (this being itself the expression of *organic* influences on the Parental constitution), and how much to intentional "culture;" but it may be fairly affirmed as probable that *both* are concerned in it; and that the manual dexterity with which a Mozart or a Caracci could express his conceptions, was as much an inherited gift, as the Genius from which those conceptions emanated.

304. From these the transition is direct to those special *Mental* aptitudes, which we can scarcely do otherwise than regard as dependent on the conformation of the Physical mechanism (§ 51*b*), and of whose original *acquirement* and subsequent *transmission* we have evidence that can scarcely be gainsaid (§§ 93, 201—207). And if this be the case with *special* tendencies to Thought, much more is it likely to be true of such as are common to Mankind in *general*. For these, being parts of Man's ordinary nature, are in great degree susceptible of modification by early training; and in proportion as the modifications so acquired tend to become *constitutional* in the individual, will the probability of their hereditary transmission be increased. And thus we are justified in believing, that in so far as we improve our own Intellectual powers and elevate our own Moral nature by watchful Self-Discipline, we are not merely benefiting ourselves and those to

whom our personal influence extends, but are improving the Intellectual and Moral Constitution which our children and our children's children will inherit from us. It is when we regard not merely the accumulation of Knowledge, but the development of the thinking power of the Race—the "universal Human Reason" —as *progressive*, that we feel the strongest call to exert ourselves to foster and direct that development. For every man who leaves behind him the expressions of great Thoughts, the record of noble Deeds, or the creations of a lofty Imagination, not merely helps to educate each successive generation, as it comes, in the use and enjoyment of them, but contributes to *enlarge its capacity for such use and enjoyment, and this in an ever-increasing degree.*—What motives to the highest exertion of our powers can be more inspiring or more disinterested? And yet they spring directly from a Philosophy which is stigmatized by many as "material" and "degrading."

CHAPTER IX.

OF THE WILL.

SECTION 1.—*Influence of the Will on Bodily Movement.*

305. "I AM, I OUGHT, I CAN, I WILL," are (as has been recently well said) the only firm foundation-stones on which we can base our attempt to climb into a higher sphere of existence. The *first* implies that we have a faculty of *Introspection*, which converts a simple state of consciousness into *self*-consciousness, and thus makes it the object of our own contemplation:—the *second*, that we have submitted that state of consciousness (whether Thought or Feeling) to our *moral judgment*, which has pronounced its verdict upon it:—the *third*, that we are conscious of a *freedom* and a *power* to act in accordance with that judgment, though drawn by cogent motives in some different direction;—and the *fourth*, that we *determinately exercise* that power. Hence we may define *Volition* or *Will* as *a determinate effort to carry out a purpose previously conceived;* and this effort may be directed to the performance of either the Mental or the Bodily acts which are adapted to carry that purpose into execution.—The manner in which this Volitional power is exerted in either case, and the conditions of its exercise, constitute our present subject of enquiry.

306. In our examination of the different forms of Nervous activity presented to us in the ascending series of Animal life (Chap. II.), we have found, as we approach Man, blind unreasoning Instinct gradually giving place as a spring of action to rational Intelligence. But neither the performance of Reasoning processes, nor the execution of their results, necessarily involves the exercise

of Will—at least in the sense in which it is here defined. For we have seen that, even in Man, intellectual operations of a high order may go on *automatically*,—one state of consciousness calling forth another in strict accordance with the "laws of thought," without any Volitional interference (§ 216); and also that *ideational* as well as *emotional* states may express themselves in Muscular action, not only without any exertion of the Will, but even in opposition to it (§ 240). And this will hereafter become still more obvious, when we investigate the phenomena of those abnormal states in which the Will is in more or less complete abeyance (Chaps. XIV. XV.).—Now if, under the light afforded by this principle, we carefully study the actions of even those among brutes whose nature has been most completely shaped into accordance with that of Man by habitual association with him, we see that they afford no indication of the existence of any other spring, than the Idea or Feeling with which the mind of the animal may be at the moment possessed; in this respect corresponding closely with those of the young Child, in whom the power of *self*-control has not yet come into exercise, and whose conduct is entirely determined by the "preponderance of motives." *

a. Thus a Dog which is fondly attached to its master, and awaits his return home with pleasurable anticipation, at once runs to meet

* Exception has been taken to this phrase, on the ground that Motives are of such different orders as not to be commensurable in force; and that we have no other ground for the estimation of their relative strength, than the actual preponderance of one aggregate when weighed against the other. But it seems to be forgotten that this is our only criterion of the relative weights (*i. e.* of the forces of downward attraction) of any different *material* substances—*e. g.* lead and feathers; and that no one disputes its applicability in the case of moral *evidence*, in which the considerations *for* or *against* any particular proposition are at least as diverse in kind as the motives inclining us *to* or *from* any particular action. We say that our judgment is determined *pro* or *con.* by the way in which the "balance of evidence" inclines; and when, as sometimes happens, the verdict of a jury is given *against* what an impartial judge considers to be the "preponderance of evidence," it is because some motive has swayed their decision, which ought not to have been admitted into the scale.

him on hearing his voice or his step, and manifests its delight in seeing him by expressive movements. But if the dog is conscious of having done something at which its master will be displeased, it slinks away and tries to keep out of his sight.

b. The whole system of "training" a dog or a horse, like the early education of a child, consists in bringing such "motives" to bear upon it, as are adapted to its "nature." A creature that has no capacity for loving right or hating wrong for its own sake, can only be made to comprehend that certain actions will bring reward, certain others punishment. And the direction of its conduct is clearly determined by the preponderance of such motives. Thus if a dog or a child which has a propensity to thieve, be punished in the first instance by a slight castigation, the deterrent influence of the prospective repetition of that punishment may be outweighed by the attractive influence of a tempting *bonne bouche*, and the offence may be repeated; the infliction of a sufficiently severe punishment, however, will then serve as a deterrent; and so long as the memory of the punishment remains vivid, this will continue effective against a temptation of like strength, although it may be overborne by a yet stronger attraction.

c. The well-known lines of Dr. Watts, commencing "Let dogs delight to bark and bite," embody, as it seems to the Writer, the true idea of the relation of the self-determining power to the automatic tendencies. When "bears and lions growl and fight," it is because "it is their nature to;" they cannot *make themselves* other than they are; and we can only mould them to *our* wills, by bringing to bear upon them a motive strong enough to restrain their own automatic impulses. But when we urge our children not to "let their angry passions rise," it is because we believe that they *can* acquire, if they can be led to *try* hard enough, the independent *self*-regulating power which we are ourselves conscious of exercising.

307. Now the Man in full possession of his Volitional power can use it (1) in giving bodily effect to his mental decision, by either putting in action the Muscles which will execute the movement he has determined-on, or by restraining them from the action to which they are prompted by some other impulse; and (2) in controlling and directing that succession of Mental operations, by

which the determination is arrived at.—In the prosecution of our enquiry as to the mechanism of Mental self-direction, we shall find ourselves greatly aided by the indications we may draw from the study of the mode in which the Will operates on the Bodily organism.—The distinction between voluntary and involuntary *movements,* is recognized by every Physiologist; but it has been customary to assign these characters to the *muscles* by which certain of these two classes of movements are respectively performed. Thus the Heart, the Muscular coat of the Stomach and Intestines, the Iris, &c., are said to be "involuntary" muscles, because no *intentional effort* of the conscious Ego can either excite or check their contractions, although some of them may be acted on by Emotional states. On the other hand, the muscles of our limbs are termed "voluntary," because we can use them to carry out the purposive determinations of the Will. But the muscles which are concerned in the act of Respiration, are both "voluntary" and "involuntary;" for while the ordinary movements of breathing are as "automatic" as are those of the Heart, we yet have a certain measure of Volitional control over them, by which we can regulate their actions in subservience to the purposes of Speech. And, further, there is not a single one of the so-called "voluntary" muscles, which may not be automatically thrown into violent contraction (as in cramp or tetanus), which the Will vainly attempts to restrain; whilst a large part of their ordinary sequential actions are performed "mechanically," without anything more than an *initiation* by the Will, which, though it can *check* them at any time, is *not* exercised in constantly sustaining them (§§ 16, 17).—Hence we see that the distinction between voluntary and involuntary Muscles is good only to this extent, that there are certain muscles which are *entirely* removed from the control of the Will, their contractions being altogether involuntary; while the actions of all others may be either voluntary or involuntary, according as they are called into exercise by the Will, or by an automatic prompting of which we may or may not be conscious. And we must, therefore, look *higher*—that is, to the *sources* and the *channels* of the Nerve-force that excites the Muscles to contraction,—for the real distinction between their several modes of activity.

308. It has been shown that the development of Nerve-force, whatever may be the mode in which it manifests itself, depends upon a reaction between the Nervous substance (especially its vesicular form) and the circulating Blood (§§ 40—42). And this Blood has a double function ; for, on the one hand, it supplies the material at the expense of which the Nerve-substance is formed, or rather forms itself, by Nutritive action, and so, as it were, lays up a store of *potential energy ;* whilst, on the other, it supplies the Oxygen, by the action of which upon the Nerve-substance (as in the oxidation of the zinc-element of the galvanic battery) this *potential* energy is converted into *actual* energy. Of this conversion we have the most conspicuous example in the production of the Nerve-force which calls forth Muscular movement, and its transmission along the motor nerves ; this production and transmission being extremely analogous to that generation of an Electric current, which takes place in a Galvanic apparatus of battery and wires, immediately that the circuit is closed. Now the ordinary state of activity of the Nervous system would seem to correspond closely with that of a moderately-charged Electric battery, which can be *dis*charged by the completion of the circuit ; this discharge relieving the tension for the moment, until it is restored by the chemical reaction between the blood and the ganglionic substance. And as there is strong reason to believe that the amount of the Nerve-force generated stands in no less direct a relation than the strength of the Electric current to the activity of that reaction, it is obvious that this, in its turn, will essentially depend on the amount of oxygenated blood which is allowed to pass through the capillaries of the ganglionic substance. It will be hereafter shown

that a *reduced* supply of Blood, which may be attributed with great probability to contraction of the Arteries supplying the Brain, called forth by the Vaso-motor system of nerves, is the essential condition of Sleep, as of other states of inactivity of the Sensorial centres (§§ 472, 473) in which the ordinary stimuli call forth no response. On the other hand, we have a yet stronger assurance in the phenomena of morbid excess of activity in the Brain or Spinal cord, that an enlargement of the Arterial trunks, and an *increase* in the quantity of Blood which passes through the Capillaries—constituting the state termed *hyperœmia*—becomes the cause of an *augmented tension* of the Nerve-centres; so that they are much more easily called into action by slight stimuli, and discharge themselves with greater force; whilst the tension may increase to such a degree, that a spontaneous Centric discharge takes place, analogous to that of an overcharged Leyden jar.—Under the guidance of this clue, it seems possible to arrive at a tolerably distinct conception of the nature of the Physical antecedence of every kind of Nervo-muscular action, from the simple Excito-motor up to the Volitional.

309. Starting, then, with the act of Coughing, as one which may be either *reflex*, *centric*, or *volitional*, it is to be remarked that the very same Muscles are employed in executing it, and that they are co-ordinated in the very same manner (presumably, therefore, by the same mechanism), whether the impulse to cough originates in an irritation of the lining membrane of the air-passages, or arises from a hyperæmia of the Respiratory ganglion, or is produced by an act of Will (§ 17). In the former case, the gradual increase of the feeling of irritation excited by the impression conveyed to the ganglion by the afferent nerve, may be considered to represent a progressively augmenting tension of that centre, which, when it arrives at a certain strength, discharges itself in motor nerve-force; just as a Leyden jar that is being gradually charged by an electrical machine, discharges itself,

if fitted with a "discharging electrometer," so soon as its electric tension has risen to the degree to which the electrometer is adjusted. This discharge of nerve-force will take place in spite of the strongest Volitional effort to restrain it; and immediately that it has occurred, the feeling of irritation subsides for a time, to return again by gradual accumulation, and again to relieve itself in the same manner. But sometimes (as in Hooping-cough) it is clear that there is, combined with the local irritation, an unduly excitable state of the Respiratory nerve-centres, so that a very small stimulus shall call forth a most violent paroxysm of coughing; and this, it can scarcely be doubted, is the expression of a *hyperæmic* state of their ganglionic substance. In the later or "spasmodic" stage of that malady, after the local irritation has subsided altogether, the paroxysmal cough is still kept up by the hyperæmic state of the centres, producing spontaneous discharges of nerve-force, each relieving its tension for a time.

310. Ascending to those centres whose action is more intimately related to Mental states, we seem able to recognize so intimate a connection between the Nervous *tension* which expresses itself in muscular movement, and the state of Mental *attention* (Chap. III.), that the two may be fairly regarded as dependent on the same Physical antecedent,—a *hyperæmia* of the particular centre whose activity is thus exalted, resulting from the regulative action of the Vaso-motor system of nerves on the Muscular walls of the Arteries. Thus, when we are intently *listening* for some sound, we not only hear distinctly what might be otherwise inaudible to us, but we start when it occurs; and, if any strong Emotion be connected with the sound, not only is our sensorial perception quickened, but the involuntary start is more violent.—This, again, points to the same locally-augmented afflux of blood to the centres of *emotional* action, as an essential condition of Emotional excitement; a view which corresponds well with the known effects of Emotion upon the local Circulation, as is shown in the act of

blushing, and, still more remarkably, in the modification which Emotional excitement produces in the Secretions (§§ 565, 566).—The same view, again, will apply to *Ideational* states; for, as has been shown, these states, when excited with sufficient intensity, express themselves in movements which the Will vainly endeavours to restrain (§§ 235, *et seq.*),—these movements being particularly violent and uncontrollable, when the Ideas in which they originate are accompanied by strongly excited Feelings (§§ 258, 259). It may, therefore, be fairly inferred that the intensity of any *ideational* state is the expression of the hyperæmic condition of some particular part of the Cerebrum, as that of a *sensational* state is of a hyperæmic condition of some part of the Sensorium.—An Ideational *hyperæmia*, or determination of blood to some part of the Cerebral convolutions, may discharge itself either (1) in calling-forth a directly respondent Movement, by stimulating the motor centres through the descending fibres of the medullary substance, or (2) by exciting a like Ideational hyperæmia in some other part of the Cerebral convolutions, through the commissural fibres, or (3) by calling up Sensorial states similar to those primarily excited by sense-impressions. Of the first of these modes of discharge, we have an example in such ideo-motor actions as the yawning that is producible in a predisposed subject by the mere verbal suggestion of the idea; the second is that which may be presumed to operate in the suggestive maintenance of every "train of thought;" whilst the third is the source of those truly "subjective sensations" which are generated by ideational states (§§ 141—150).—Now as we have it in our own power, by the Volitional act of Attention, to intensify any of these states, it seems probable that the physiological condition of that intensification must be the increase of the *local hyperæmia* in the nervous centre or part of a nervous centre which is its instrument, through the agency of the Vaso-motor system of nerves.

311. Carrying back our inquiry, now, to the nature of the

Cerebral change which initiates a Volitional action, we find reason for attributing this also to a local *hyperæmia* of some part of that cortical layer which constitutes the instrument of Ideation. For all Volitional action, it will be remembered, is based on an *idea* of what is to be done (§ 305), whether this have reference to Bodily movements, or to Mental exertion. And it seems clear that the same Vaso-motor action which is the condition of the state of *attention* to that idea, will, if exerted to produce a still greater local *hyperæmia,* give effect to it in a spontaneous motorial discharge. And thus we are led to regard the immediate source of *ideo-motor* and of *volitional* movements as the same; and the Volitional effort as really exerted in augmenting the nervous tension of the part of the cortical substance of the Cerebrum, which is concerned in the formation of the Idea of the thing to be done. This doctrine finds a remarkable confirmation in two orders of facts;—(1) that there is practically every gradation between those, *voluntary* actions, which (under *permission* of the Will) simply express dominant ideas, and those actions which proceed from distinct and cogent *volitional* determinations;—and (2) that *emotional* states have a most powerful influence either in augmenting or in diminishing the motor force which the Will can call forth (§§ 266—268). For the known influence of the Emotions on the Vaso-motor system of nerves, and the manner in which they intensify those Ideational states which express themselves in movement, afford a strong indication that they exert their effect on Volitional action by increasing the local hyperæmia of the cortical substance. And this conclusion will be shown to derive yet stronger confirmation from the remarkable result of Dr. Ferrier's recent experiments. (See Appendix.) —The *restraining* influence of the Will on bodily movement (as when we make an effort to stifle a cough, to resist a yawn, to repress laughter, or to keep down the expression of some passionate impulse) seems really to consist in putting the antagonist Muscles into action; and we experience just the same sense of effort in doing this, that we do in trying to stop a horse that is running away, or to check the rotation of a wheel.

312. Now the strongest Volitional effort may be inoperative, through some defect of the apparatus by which the Nerve-force is transmitted to the muscles which are to execute the behests of the Will; as happens in paralysis. But there are states of absolute incapacity for such effort; the mental *desire* existing, while the energy needed to carry it into effect is deficient. That this incapacity arises from deficient supply of blood to the ideational (Cerebral) nerve-centre, appears probable from the familiar fact, that a general deficiency of Volitional power over the muscles is a marked feature of the physical depression which betokens feebleness of the circulation, being especially noticeable in sea-sickness; while a defect in the distributive action of the Vaso-motor system of nerves (such as that of which we have evidence in many local congestions) might very well account for such cases as the two following, which are recorded by Professor J. H. Bennett (*Mesmeric Mania* of 1851) on the authority of Sir Robert Christison:—

a. "The first was that of a gentleman who frequently could not carry out what he *wished* to perform. Often, on endeavouring to undress, he was two hours before he could get off his coat, all his mental faculties, Volition excepted, being perfect. On one occasion, having ordered a glass of water, it was presented to him on a tray, but he could not take it, though anxious to do so; and he kept the servant standing before him half an hour, when the obstruction was overcome."

b. "In the other case the peculiarity was limited. If, when walking in the street, this individual came to a gap in the line of houses, his will suddenly became inoperative, and he could not proceed. An unbuilt-on space in the street was sure to stop him. Crossing a street also was very difficult; and on going in or out of a door, he was always arrested for some minutes. Both these gentlemen graphically described their feelings to be 'As if another person had taken possession of their will.'"—(*The Mesmeric Mania of* 1851, p. 16.)

This state seems akin to that form of *hysterical paralysis*, in which the defect lies not so much in the *want of power*, as in the want of that *belief in the possession of the power*, which is essential to its exercise (§ 267). A strong motive will here sometimes take the place of Volition; and no motive is so efficacious, as that *confident expectation* of cure, which is awakened either by Religious faith, or by the belief in the occult powers of Mesmerism, Spiritualism, &c. Thus it has been that numberless pseudo-miracles have been worked on patients of this class by Religious Enthusiasts; whilst they furnish to Mesmerists and Spiritualists the subjects of "wonderful cures," effected by the agencies which they profess to wield.—Such cases are peculiarly interesting to the Psychologist, from their parallelism to those in which there is a like suspension of volitional control over the course of Thought (§ 454).

SECTION 2.—*Influence of the Will on Mental Action.*

313. Now since, according to the view which it has been the special purpose of this Treatise to develope, the relation of the Will to *mental* is essentially the same as that which it has to *bodily* action, the measure of its exertion will be the *sense of effort* which we experience, in *intentionally* exciting, directing, or restraining any particular form of mental activity. It has been already pointed out (Chap. III.) that the Attention may be *involuntarily* fixed upon certain states of consciousness, through the attraction they exert upon the individual Mind, in virtue either of its original constitution, or of its acquired habitudes; and further that this attraction determines much of the automatic action of our faculties (§ 228). When most strongly exerted, it causes the consciousness to be so completely engrossed by one train of ideas, that the mind is, for the time, incapable of any other ideational change: external impressions on the Sensorium, either not being perceived at all (the individual being as insensible to

them as if he were in a profound sleep), or not giving rise to any Cerebral action (§ 124). — But these automatic tendencies of the Mind may be to a certain extent antagonized by the Will, which keeps them in check (just as it restrains many of the automatic impulses to bodily movement) by the special power which it exerts over the Attention. This it can detach from subjects which have at the time the greatest attractiveness for it, and can forcibly direct it to others from which their attraction would otherwise divert it. And in its most complete and powerful exercise (which is not within the capacity of every one), it can so entirely limit the Mind to one train of thought, that the state of Abstraction induced by the Will may be as complete as that which in some individuals is of *spontaneous* occurrence (§ 446).

314. Now when our current of Thought is flowing-on smoothly and uninterruptedly, we are no more conscious of effort than we are in the act of breathing; in fact, an effort may be required either to check the current, or to turn it into another channel. But so soon as a difficulty or obstruction arises,—as when the Mathematician finds a "hitch" in his computations, the Poet wants a phrase to complete his verse, or the Lawyer does not see his way through some intricacy in his "case,"—the Will is called into play to overcome it, by determinately projecting the Mind in search (as it were) of the desiderated idea. So, when the Attention is distracted, either by some sense-impression, or by the intrusion of some inappropriate idea or feeling, we have to "make an effort" to keep it fixed upon our "train of thought;" the degree of that conscious *effort* being the measure of our Volitional exertion. And the same is the case when Cerebral fatigue weakens the hold of even an attractive subject; and we have to *force ourselves* to keep our attention fixed upon it, so as to complete the task we may have set ourselves to perform.

315. In proportion as we are able thus to concentrate our Attention on the subject proper to the time, and to exclude all

distracting considerations whilst pursuing the trains of thought which the contemplation of it suggests, will be our power of advantageously employing our Intellectual Faculties in the acquirement of Knowledge and in the pursuit of Truth; and all men who have been distinguished by their Intellectual achievements, have possessed this faculty in a considerable degree. It is one which is "eminently capable of cultivation by steady intention of mind and habitual exercise;" and the more frequently it is put in practice, the easier the exercise becomes. In fact, when a man has once brought his Intellectual faculties under the mastery of his Will, to such an extent as to induce the state of Abstraction whenever he pleases, this state becomes secondarily automatic; and the fixed direction of the thoughts, which at first required a *constant volitional effort* for its maintenance, comes to be continued *without any consciousness of exertion*, so long as the Will may permit.—We have in our own *consciousness of effort*, and in our experience of *subsequent fatigue*, a very strong indication that the power which thus controls and directs the current of thought, is of the same *kind* with that which calls forth Volitional contraction in the muscles, though exerted in a different mode. And just as the strongest exertion of Will is required to produce or sustain muscular contraction, either when the sense of muscular fatigue is already strongly experienced, or when we are antagonizing a powerful automatic impulse, so in the determination of Mental effort in a particular direction, we find ourselves necessitated to make the greatest Volitional exertion, when we are already labouring under the sense of Cerebral fatigue, or when the attention is powerfully solicited by some other attractive object. And it is after any such contest with our natural tendencies, that we experience the greatest degree of exhaustion; the merely automatic action of the Mind, which is attended with no effort, being followed by comparatively little fatigue.

The Writer is satisfied from his own experience, that a most valuable indication may be hence drawn, in regard to the regulation of the habits of Intellectual labour. To individuals of ordinary mental activity, who have been trained in the habit of methodical and connected thinking, a very considerable amount of *work* is quite natural; and when such persons are in good bodily health, and the subject of their labour is congenial to them,—especially if it be one that has been chosen by themselves, as furnishing a centre of attraction around which their thoughts spontaneously tend to range themselves,—their intellectual operations require but little of the controlling or directing power of the will, and may be continued for long periods together without fatigue. But from the moment when an indisposition is experienced to keep the attention fixed upon the subject, and the thoughts wander from it unless coerced by the Will, the mental activity loses its spontaneous or automatic character; and (as in the act of walking, § 16) more *effort* is required to maintain it volitionally during a brief period, and more *fatigue* is subsequently experienced from such exertion, than would be involved in the continuance of an automatic operation through a period many times as long. Hence he has found it practically the greatest economy of Mental labour, to work vigorously when he feels disposed to do so, and to refrain from exertion, so far as possible, *when it is felt to be an exertion*.—Of course this rule is by no means universally applicable; for there are many individuals who would pass their whole time in listless inactivity, if not actually spurred-on by the feeling of necessity. But it holds good for those who are sufficiently *attracted* by objects of interest before them, or who have in their worldly position a sufficiently strong motive to exertion, to make them feel that they *must* work; the question with them being, *how* they can attain their desired results with the least expenditure of mental effort (§ 228).

316. In the foregoing instances we are *distinctly conscious* of the *Volitional* effort, because there is a *struggle* between opposing tendencies. But there are many other cases in which the *guidance* of the Will is exercised so gently, that we are only aware of its exercise when our attention is drawn to its effects. Thus, as already pointed out, whilst the movements of Respiration are essentially

automatic (being probably altogether so in most of the lower animals), they are in Man so far under the control of his Will, that he can utilize them for the purposes of speech; and yet, unless an unusually severe or long-continued strain is put upon the vocal organs, requiring an actual *effort* for its sustentation, we are not aware that during all the time we have been speaking, we have been interfering by an act of Will with the automatic uniformity of our respiratory movements.—Now this has its precise counterpart in Mental action; for the determining power of Volition is employed, in however slight a degree, whenever the succession of thought is not *perfectly spontaneous,*—whenever, in fact, we *purposely guide its course in any particular direction,* even for the apprehension of ideas most familiar to us. Thus, as will be shown hereafter (§ 372), *all determinate Recollection involves the exercise of volitional control over the direction of the thoughts;* and hence, if this control be suspended, and the mind be left to its own automatic activity, the power of recalling even the most familiar ideas is completely annihilated. So, again, *the determinate exercise of the Judgment, which involves the comparison of ideas, can only take place under the guidance of the Will;* which selects those which are appropriate, and brings them into collocation with each other (§ 227). And it is the readiness with which this process is usually performed, which constitutes the source of that *Common Sense,* whereon we rely in the ordinary conduct of life (Chap. XI.). But we cannot use this test, even in the simplest case, without a Volitional selection from among the records of our experience, of that which may be brought into comparison with the idea whose validity is to be tried. The simple credulity of the Child depends upon his having no stock of experience upon which to fall back, for the correction of the erroneous notions which he may himself form, or which may be imparted to him by others. But the deficiency of Common Sense, which we occasionally meet-with in grown-up Men and Women, depends, not so much upon the want of experience, as on the want of power to profit by it; their minds not having been duly trained in that volitional exercise, which, when it once comes to be habitual, is performed with so slight an effort that it is scarcely perceptible even to ourselves. Slight as this effort may be, however, it is the one thing needful; and it may be unhesitatingly laid down, that, *if the directing power of the Will be entirely suspended, the capability of correcting the most illusory ideas by an appeal to Common Sense is for the time annihilated.*

317. Of this we have a typical example, familiar to every one, in the state of *Dreaming* (§ 482): which is a condition of Automatic mental activity, usually of an irregular kind; the combinations and successions of Ideas being often of the most extraordinary character, and inconsistent not merely with our most familiar experience, but also with each other. Yet, as has been most truly remarked, *nothing surprises us in dreams.* We are never struck with the impossibility of the events which we seem to witness; but we accept as genuine, with child-like simplicity, all the wonderful combinations which successively rise to our consciousness.—The same must be the case in *any* state of mental activity in which there is a similar abrogation of Volitional control: and the records of "absence of mind" (§ 445) afford abundant examples of the absurd incongruities which occur, when the Will is *temporarily* prevented by the mental preoccupation from summoning Common Sense to check the ideas which external impressions suggest; while those of Insanity, in which there is a *persistent* deficiency in the power of self-direction of the thoughts, show that no belief is too absurd to be accepted, however inconsistent with the most direct and most constant experience (§ 559).—Hence we see that if the Mind should lose for a time all power of Volitional self-direction, it cannot recall any fact, even the most familiar, that is beyond its immediate grasp;—its attention being engrossed with the idea that may be before it for the moment, no incongruity prevents that idea from presenting itself with all the

vividness of reality;—it cannot bring any notion with which it may be possessed to the test of "Common Sense," but *must* accept it as a belief, if it be impressed on the consciousness with adequate force;—it cannot shake off the yoke of any "dominant idea" however tyrannical, but *must* execute its behests.

318. We have, now, however, to consider a much more obscure question,—namely, the *nature* of that self-determining agency to which we give the name of Will. Is it, as some think, the mere *resultant* of the general (spontaneous or automatic) activity of the Mind, and dependent, like it, upon Physical antecedents? Or is it a Power, which, being completely independent of these conditions, is capable of acting *against* the preponderance of motives?—as if, when one scale of a balance is inclining downwards, a hand placed on the beam from which the other scale is suspended, were to cause that lighter scale to go down.

319. Now that the Will *is* something essentially different from the general resultant of the automatic activity of the Mind, appears to the Writer to be proved, notmerely by the evidence of our own consciousness of the possession of a *self*-determining power (Chap. I.), but by observation of the striking contrasts which are continually presented in abnormal states of Mind, between the automatic activity and the power of volitional control. For, in the first place, it is the special attribute of all "nervine stimulants," such as Alcohol, Opium, and Hachisch, as well as of those morbid poisons which induce Delirium, to *exalt* the automatic activity of the Mind, while *diminishing* the power of volitional control; and this not only *relatively* but *absolutely*. A most instructive example of this general fact is furnished by the description given by Dr. Moreau of his own experience in regard to the Hachisch (§ 537); and the "Confessions of an English Opium-eater" exhibit the same characteristic phenomena (§ 542). Moreover, the continual use of these stimulants has a manifest tendency to produce a *permanent* weakening of the Volitional power (§ 543), which often

shows itself hereditarily even where the offspring have not themselves given way to the habit (§ 299 *a*). We have seen that the whole mental life of Coleridge was one of singular automatic activity (§ 231), whilst there was a no less marked deficiency in the power of volitional self-direction; and there can be little doubt that this deficiency, probably constitutional in the first instance, was aggravated by the habitual use of the nervine stimulants which augmented the automatic activity of his Psychical nature.—But, further, the complete *suspension* of the power of volitionally directing the current of thought and feeling, will be shown to be the essential feature, not merely of the states of Dreaming and Delirium, but also of natural and induced Reverie, and of natural and induced Somnambulism; while the *weakening* of that power, usually in concurrence with an exaltation of some Emotional tendency, is the special characteristic of Insanity.

The variety of phases which these different states present, is chiefly dependent upon the following conditions:—(1) The relative degree in which the Mind is in a state of receptivity for *external* impressions, or is attending only to what passes *within itself*; (2) the degree in which the coherence of the successive states is maintained by preformed Associations; and (3) the degree in which the normal operation of the Intellectual faculties is disturbed by Emotional excitement, either general, or limited to one class of feelings.—The influence of the *first* of these elements is remarkably seen in the contrast between *natural* and *artificial* Reverie (§§ 443, 448), also between some forms of *natural* and *artificial* Somnambulism (§§ 488, 492): and not less between different forms of Insanity, in which last condition we find some patients constantly brooding over particular trains of thought, and almost incapable of being turned from the contemplation of these by external suggestions; whilst others are no less remarkable for the instability of their mental states, and for the readiness with which a new direction may be given to the thoughts by sensory impressions.—The influence of the *second* element is strikingly manifested in the difference between the various phases of the state of Dreaming, and in the contrast between the incoherence of the commoner forms of this (§ 482), and that consistency in the trains of thought which generally characterizes the state of Somnambulism; this last again being strongly contrasted with the states of Delirium and Mania (§ 548, 553), which are especially characterized by the complete *confusion* of the Intellectual powers, all previous states of consciousness being (as it were) jumbled together, and the order of their recurrence and the nature of the new combinations which may arise out of them, being irreducible to any principle of orderly sequence.—The influence of the *third* element is well seen in those forms of artificial Reverie and Somnambulism, in which the *feelings* as well as the *ideas* admit of being played upon by external influences; for it is easy to bring the mind of the "subject" under the domination of any particular Emotion, by taking the appropriate means to excite it; and, so long as this may continue, the language and actions most obviously display its impress. Thus it is often sufficient to ask the Biologized subject (§ 451), "Why are you so angry?" "Why are you so sad?" &c.,—to induce these conditions respectively, the suggestions being here conveyed verbally: whilst in the Hypnotic state there is often a very curious Emotional susceptibility to the influence of Muscular associations (§ 494). But it is in Insanity (Chap. XVIII.) that we best see the influence of Emotional states upon the course of thought and action. For here we find them supplying impulses to bodily movements, which the weakened Will cannot resist, although the Intellect distinctly apprehends the evil consequences of such actions; or, on the other hand, we find them directing the whole course of mental activity, giving a wrong colour to all the ideas which call them into exercise, and so attracting the attention to the trains of thought founded upon these, that they come to attain a complete domination over the mind, and hence over the conduct, to which they supply *motives* of such potency that the weakened Will can neither resist them, nor withdraw the mind from attending to them.

320. It will aid us in the examination of the mode in which the Will determines our *actions*, if we first examine the influence it has on the formation of our *opinions*,—a subject on which, in the Writer's judgment, a grave misapprehension is prevalent. For it is very frequently asserted that it does not rest with any Man to

determine what he shall *believe* or what he shall *disbelieve;* that he cannot help giving or refusing his assent, according to the *preponderance of evidence;* and that he is, therefore, "no more responsible for his opinions, than he is for the colour of his skin." Now whilst fully recognizing it as a fundamental fact of consciousness, that Assent is an *automatic* action, over which the Will can exert no *direct* influence, the Writer has now to show that the Will has an immense *indirect* power of a twofold nature;—(1) through the *habitual discipline* by which it gives shape to the Intellectual fabric; and (2) through its power of modifying the *relative force* of different evidentiary considerations, by the degree of Attention given to each.

SECTION 3.—*Influence of the Will on the Formation of Beliefs.*

321. It may be freely admitted that there *are* certain Propositions which claim our *immediate* and *entire* assent; but between these and the Beliefs to which we give our assent as *on the whole preferable,* after mentally balancing a variety of considerations bearing upon them, there is every gradation. The cogency of the propositions of the first kind depends upon the fact that they are consistent with our previous convictions, and that nothing can be said against them; whilst the uncertainty we feel as to the second, results from the fact that there is "much to be said on both sides." And further, whilst those of the first kind are equally accordant with the Mental Constitution (whether *original* or *acquired*) of every one (§ 377), those of the second are very differently estimated by different individuals, all equally desirous of arriving at the truth, according to their conformity or disaccordance with that *aggregate of preformed opinion* which has grown up in the Mind of each. For just as—if so rude a comparison be permitted—we try whether a new piece of furniture which is offered us *does* or *does not fit* into a certain recess in our apartment, and accept or decline it accord-

ingly, so we try a new Proposition which is offered to our Mental acceptance. If it either at once *fits-in*, or can by argument or discussion *be brought to fit-in*, to some recess in our fabric of Thought, we give our assent to it, by admitting it to its appropriate place. But if it neither *fits* in the first instance, nor can by any means be brought to fit, the Mind *automatically* rejects it.

a. Of this we have a marked illustration in the opposite receptions given to the asserted wonders of Spiritualism. To those, for example, who have been trained in Scientific habits of thought, the statement of a dozen persons that a lady was transported two miles through the air, in a state of trance, and came down upon the table of a darkened room, the doors and windows of which were securely closed,—or that Mr. Home, in a like condition, floated out of the open window of one room, and into that of another, at a height of seventy feet above the ground,—seems a simple absurdity, to which no ordinary testimony would induce their assent. And, further, to any one who has so far studied the constitution of the Human Mind, as to be aware of the influence of "dominant ideas" in producing false perceptions (§ 186), it becomes obvious that no amount of testimony given by witnesses who are "possessed" by such ideas has the least evidentiary value.—On the other hand, to those whose previous training utterly incapacitates them for the appreciation of Scientific truth, who scornfully repudiate the objections urged by men of science as those of prejudiced or interested opponents, who have a craving for "spiritual manifestations" as proofs of the continued existence of their departed friends, who are utterly ignorant of the nature of the "subjective" as distinguished from the "objective," who implicitly believe what they term the "evidence of their senses," and who are consequently quite prepared to mistake the creations of their own imaginations for external realities, such statements appear not only *credible* but *probable*; in fact, the more inconsistent the asserted phenomena are with every-day experience, the more readily do they give their entire assent to them, as *fitting-in* with their previous conceptions of the supernatural powers of "the Spirits."

Again, since the preformed Mental habits thus determine not merely the primary acceptance or rejection of the Proposition, but the issue of the further attempts to *make it fit*, not only will different individuals draw very different conclusions from the same data, but the same Ego will form different judgments as to the very same matter at different stages of his Mental life; without any change in the external materials of his judgment, but solely from changes in his own fabric of Thought,—one recess (to revert to our former simile) having *grown large enough* to admit what it formerly refused, and another having *contracted* so as no longer to give place to what it originally admitted.

b. Every one who has gone through a sufficiently long course of Intellectual experiences, and has been accustomed to reflect upon them, must be conscious that this has often occurred to himself. The Writer, in common (he doubts not) with many Men of Science, has often been surprised, on turning over the records of his earlier beliefs, to find how many of them he would now absolutely reject; not because they have been disproved by additional evidence, but because he has himself *grown out of them;* either from no longer attaching the same value to evidence on which he formerly relied, or from looking at the whole subject from an entirely different stand-point. These purely *Intellectual* diversities of judgment are closely related to those modifications in our Memory of actual occurrences, which are unconsciously produced by our habits of thought in relation to the subject of them (§ 365). How much our conclusions on any matter into which *Emotional* considerations enter, are swayed by the state of feeling in which we may be at the time, is a matter of familiar experience; a night's rest often completely reversing our judgment, by altering our estimation of the data on which that judgment was based.

322. Thus, then, while no one, *constituted as he is at the time*, may be able to help *giving* his assent to certain propositions, and *refusing* his assent to others, every one who has learned to direct his own Intellectual activity is responsible for the use he has made of his power, in the *construction of that Mental fabric*, the aptitude or inaptitude of which for the reception of a new proposition determines his acceptance or rejection of it.—There are numerous pro-

positions which are scarcely less "self-evident" to the minds of such as have given special attention to the subjects to which they relate, than are the Axioms of Geometry to the Common-Sense of Mankind generally (§ 200); the admission of them into the Mental fabric of such persons being *immediate*, and their *fit* into appropriate places being *exact*, in virtue of its special preparedness for their reception. And thus the *unhesitating assent* which one man gives to a proposition or set of propositions, the *nescience* of another who avowedly forms no opinion about them (pronouncing the matter "unknowable"), and the *positive denial* of them by a third who denounces them as altogether monstrous and absurd, are all the expressions of antecedent states of mind, which partly arise out of the *original* constitution of each individual, but partly depend on the self-discipline he has habitually exercised in his search for Truth.

Every one, for example, who has been trained in Scientific habits of thought, recognises the cogency of the evidence afforded by Spectroscopic observation, in regard to the Chemistry and Physics of the Celestial luminaries. For if he has not himself observed the phenomena, he accepts the testimony of those who have; the concurrence of independent observers, and the accordance of their statements with the antecedent probabilities established by other investigations, giving the fullest validity to that testimony. And the deductions from those phenomena are so simple and direct, requiring neither the elaborate computations of Astronomy, nor the combinations of probabilities which Geological reasonings involve, that no *special* education is required for the recognition of their claim on his assent. Hence it may be fairly said that any man who *refuses* to accept them, is responsible for the state of mind which dictates that refusal; unless his mind is so deficient (either by original constitution, or through want of appropriate training) in the power of *apprehending* scientific Truth, that he *cannot* yield his assent to that which becomes perfectly obvious to every man of ordinary intelligence who bestows sufficient attention on the matter.

323. But a large part of the Propositions offered to our accept-

ance, relate to matters as to which the evidence is far less cogent and conclusive; and the question is not so much whether they do or do not *fit at once*, as whether they can be *brought to fit* by argument and discussion. Now here, again, a sincere desire to arrive at *Truth*, without the least *wish* to come to any particular conclusion, being presupposed, the balance of judgment will be entirely determined by the individual's previous habits of thought; as on these will depend the relative weight he attaches to the several arguments *pro* and *con*. (This is so well understood at the Bar, that a skilful advocate, in pleading before a Judge, will shape his argument according to his knowledge of that judge's "turn of mind.") And there are very few persons who are so entirely devoid of Intellectual prejudices, or inclinations to particular modes of thought, as to be altogether free from their influence; that influence being especially dangerous, when we are altogether unconscious of its existence (§ 389). Those in whom it is chiefly "conspicuous by its absence" are said to be distinguished by their "judicial habit of mind."

324. As soon, however, as *any other motive* than the desire to arrive at Truth enters into the formation of our beliefs, the Will comes to have a far more powerful influence. That "we easily believe what we wish" is a proverb which Experience shows to be so often true, that Science is called on to give the *rationale* of a fact which seems opposed to what has been said of the *automatic* nature of our Intellectual decisions. The opposition, however, is more apparent than real. In the discussion of a question of Intellectual Truth, as in debating with one's self a question of Morals (§ 210), the Will has the power of keeping some considerations out of view, and thereby *diminishing* their force, whilst it fixes the attention upon others, and thereby *increases* their force. And in this manner the Will can *indirectly* determine the inclination of the balance of *evidence* which commands *belief*, as it can the balance of *motives* which determines *conduct*

Its action may be compared to that (now happily, in our own country, a matter of history) of a partisan Judge presiding over a Political trial, in which the Prisoner's life or death depends on the verdict of an impartial Jury. For a Judge who is determined to procure an adverse verdict, has various means of influencing the decision. In the first place he refuses even to consider the objections which the prisoner's counsel may be justified in urging against the indictment; and accepts the reply of the crown lawyer as all-sufficient, when it does not really meet one of the points raised for the defence. Again, he treats the witnesses for the crown with the utmost consideration, assumes the truth of every statement they make, and not only asks no inconvenient questions himself, but places every obstacle he can in the way of the cross-examination which tends to expose the inconsistencies of their testimony, or to convict them of interested motives. On the other hand, he treats the witnesses for the defence as if they were utterly unworthy of credit; and allows the utmost licence to the crown-counsel who endeavours to lower the value of their testimony by unjustifiable insinuations or bullying assumptions. And in his "summing-up," he so forcibly presents to the jury both the law and the evidence on one side, and so determinately keeps down the force of law and evidence on the other, that the Jury, however honest their intentions, may be *forced* into giving a most iniquitous verdict, for the injustice of which it is the Judge who is really responsible.

325. This influence of the Will is all the more powerful, when we do not discuss the question with *others*, but only with *ourselves.* For we can far more easily withdraw our attention from the suggestions which occur to our own minds, than we can from the very same considerations forcibly urged as arguments by others. And, further, if we only look at the matter from *our own* point of sight, we are almost sure to take but a limited view of it. Every one, therefore, who really desires to arrive at Truth upon a subject which is open to question, will seek to acquaint himself with the view that may be taken of it by others;—as when a Judge says "I should like to hear that point argued," knowing that the Counsel on the two sides will bring forwards all that is to be said on each.

Upon his candid readiness to listen to all that they can fairly urge, and his trained ability to estimate their arguments at their just weight, will depend the worth of his final decision. This procedure is especially important with Minds which have been habituated to the worship of *Idols* of any kind whatever; for to these they are perpetually, without being aware of it, "sacrificing their intellectual and moral independence."

a. Thus Dr. Channing was led by the representations of Samuel J. May, to perceive that he had been thus sacrificing to the idol of "unanimity" in holding his peace upon the Slavery question. The conduct of those two great and good men, as recorded in the Memoir of the former (vol. iii., pp. 156-159), is a noble moral as well as intellectual lesson, which should teach charity, as well as fidelity to principle. "At first," said Mrs. L. M. Child, who seems to have made the earliest attempt to draw Dr. Channing's attention to the subject, "I thought him timid, and even slightly time-serving; but I soon found that I formed that estimate from ignorance of his character. I learned that it was justice to *all*, not popularity for *himself*, which made him so cautious. He constantly grew upon my respect, until I came to regard him as the wisest, as well as the gentlest, apostle of humanity."—When subsequently appealed-to by Mr. May, he did not raise any objections to the fundamental doctrines of the Abolitionists; but excused himself from participating in their agitation, on account of the severity of their denunciations, and the vehemence, heat, and excitement caused by their meetings. This called forth an indignant protest from the Abolitionist advocate; who urged that any imprudences of this kind were due to the silence of such men as Dr. Channing, who, acknowledging the awful injustice of Slavery, had not raised their voices in remonstrance against it. "It is not our fault," he said, "that those who might have pleaded for the enslaved so much more wisely and eloquently, both with the pen and the living voice, than we can, have been silent. Why, Sir, have *you* not spoken?"—This appeal went "home."—"Brother May," replied Channing, after some minutes' consideration, "I acknowledge the justice of your reproof: I have been silent too long." And he forthwith prepared himself to "speak out."—No one who knew Dr. Channing could suspect for a moment that he would have refused to this argument, if it had suggested itself to his own mind, the same weight that it had with him when brought forwards by another. His unconscious bias in the opposite direction had prevented it from ever occurring to him.

b. The case of the Rev. Blanco White affords another illustration of the same principle. Born in Spain, and brought up as a Roman Catholic Priest, but rebelling against the principle of Authority, he found refuge in the Church of England, specially attaching himself to the Liberals of Oxford, by whom he was highly esteemed; so that when Dr. Whately was appointed to the Archbishopric of Dublin, he named Blanco White his domestic chaplain. Whilst holding this position, he published a controversial work entitled "Second Travels of an Irish Gentleman in search of a Religion," which fell under the notice of the Rev. George Armstrong, who had already, on conscientious grounds, resigned his preferment in the Irish Church; and a correspondence took place between them. Mr. Armstrong urged upon Blanco White, that, upon the principles he had avowedly adopted, it was not possible for him, as a matter of logical consistency, to continue in the position he was then holding; and being brought to admit this, he at once relinquished it, though by so doing, he severed ties of the closest nature, personal as well as professional.—The whole life of Blanco White showed such a thorough fidelity to principle, such a readiness to make any and every sacrifice which his conscience demanded, that it cannot for a moment be imputed to him that he had intentionally kept out of his view the logical result of his own train of reasoning. He simply did not see it, until it was pointed out to him.

On the other hand, it not unfrequently happens in a controversy between honest opponents, that the effect of an adverse argument is in the first instance to stir up an antagonistic spirit, which prevents its weight from being duly appreciated; and that it is only when the argument is calmly and quietly reviewed at a subsequent time, that its real cogency makes itself felt. With most persons indeed, the first effect of an assertion which runs counter to their ettled beliefs, is to make them think what can be said *against* it; so that the most candid and truth-seeking of men generally require *time* for the digestion and assimilation (so to speak) of any such proposition.

326. But another Proverb, or concrete expression of a vast body of familiar experience,—tells us that "there are none so blind as those that *won't* see;" and it is in wilfully turning away the eyes of their minds from inconvenient truths, whether facts of Nature or results of Thought, that the moral responsibility of such persons for their opinions really consists. As the opponents of the Copernican system refused to look at the satellites of Jupiter through the telescope of Galileo, so there are too many who refuse to admit even a gleam of reason into the dark chambers of their Intellects; where they hide as sacred treasures the antiquated beliefs of past ages, the worthlessness of which would be at once apparent if the full light of day were permitted to shine in upon them.—On the other hand, as Nelson at Copenhagen turned his blind eye to the signal for his recall, which he did not think it for the honour of his country to obey, so may we rightly keep from our Mental vision, not merely the direct promptings of self-interest, but such arguments as we instinctively *feel* to be sophistical, though we may not be able logically to expose their fallacy; and it is in cultivating and quickening this instinct, that the habitual desire to act on the highest principle of *right* most powerfully operates on the Intellect (§ 389).

327. There is no subject as to which the influence of the Νόμος (§ 292) shows itself more strongly, than it does in regard to Religion and Morals; none as to which it is more difficult for a man to free himself from the influence of those Habits of thought and feeling, which, impressed upon him at the earliest dawn of his Intelligence, have "grown with his growth, and strengthened with his strength." But as there are many signs which it is impossible to disregard, of the awakening of a general spirit of inquiry into the foundation of our Beliefs on these subjects, it may not be inappropriate here to consider the three principal *tendencies to*

thought which are most potent in the direction of that enquiry, and the three modes in which we make ourselves responsible for its results, by the deliberate adoption of one or other of them.

1. The first of these is *implicit reliance on Authority*, which is the fundamental tenet of the Roman Catholic Church. This tendency, when originally implanted in the mind, and fostered by a system of training most skilfully devised to fix and develope it, acquires a most powerful hold over Intellects which are free and independent upon all other matters. Taught from his earliest years that Faith and Morals are beyond Human ken, impressed with the claims of the successors to the Apostles as the conservators of Divine Truth, warned that to doubt is sinful, alarmed by the dangers which he sees to be inseparable from unrestrained freedom, and feeling comfort in being relieved from all responsibility as to the formation of his opinions, the sincere Catholic submits himself unreservedly to the dictation of his Church, and gives his unhesitating assent to every dogma she imposes upon him. And it is only when, in endeavouring to strengthen her grasp upon the Intellect of her members, she asserts a control over matters on which they *cannot help* thinking for themselves, that she incurs any danger of a general revolt against her authority.—That there is a certain type of mind, to the Constitution of which (whether original or acquired) the system of implicit reliance on external support is most congenial, is shown by the thorough acceptance of it by men of rare acquirements and great ability, who have been brought up in Protestant Churches, and have made great sacrifices in quitting them. Upon such, however, lies a much heavier responsibility for the *adoption* of this system, than upon the former for their simple *acceptance* of it. For they determinately *surrender*, in regard to the most momentous of all subjects, that freedom to form their own opinions, which they would regard it as not only their *right*, but their *duty*, to exercise in all other matters; and submit

to such limitations upon that freedom, in every Philosophical or Scientific investigation which can have even the remotest bearing upon Religion, as the Church may at any time see fit to impose. While holding themselves free from all responsibility, therefore, they do in effect *make themselves* responsible for whatever they may think, say, or do, at the bidding of the Authority to which they have deliberately chosen to submit themselves.

II. The truly *independent* thinker, on the other hand, who upholds the duty of individual judgment on *all* subjects, is responsible, not directly for the conclusions he arrives at, but for the right use of his reason in the search for Truth. It is not the least among the evil results of the tyranny of the Νόμος, that it tends to drive into *antagonism* every one who feels called on to resist it; and thus to engender a *defiant* attitude where *firmness* alone is needed, an iconoclastic and controversial disposition where the *judicial* habit of Mind is specially required. Those who put themselves forwards to attack the cherished beliefs of the world at large, are bound to master the whole of each question they bring under discussion, and not to content themselves with a one-sided or imperfect view of it; and they have no more right to put aside an evidentiary fact or deduction merely because it *looks* old and worn out, than to adopt another without due examination because it is new and specious. In particular it behoves those who rest on experience as the basis of all knowledge, to beware of excluding all experience save their own. As a man who has no "musical ear" may deny the soul-stirring power of a Handel or a Beethoven, or as one who is "colour-blind" cannot recognize either the glorious hues of Nature herself, or the reflection of them in the picture of a consummate artist, so the man whose Mental constitution leads him to fix his attention too exclusively on experiences of one kind, is too prone to deny the reality of those in which he does not himself share, and to regard as "unknowable" what *other* Minds assert to be within *their* apprehension (§ 328). So, again, there is a tendency on the part of independent thinkers to an excessive confidence in their own conclusions, which a due sense of Human frailty would restrain; and to a corresponding intolerance of the different conclusions of others, who, while honestly exercising the like independence of thought, may have acquired, by the greater comprehensiveness of their survey of the subject, a better title to the acceptance of their judgment upon it. The very fact that a man has emancipated himself from the tyranny of the Νόμος, is evidence that he has a power of Will which enables himself to exercise *self*-discipline; and it is all the more incumbent on him, therefore, to take heed that he does not, in abjuring the worship of one Idol, set up another—his own Individuality—in the place of it. A due respect for the "common consciousness of Mankind," though he may not himself share it, will engender a wholesome distrust of any belief that directly opposes it; and this will lead him, before finally adopting that belief, to subject its basis to the most careful scrutiny. And while rejoicing in his own freedom, and doing what in him lies to stir up in others a desire for the same "glorious liberty," it especially behoves him not to think too hardly of those whose Mental constitution and habits of thought are different from his own, for their assent to propositions which to him appear not only untrue, but irrational, perhaps even immoral (§ 441).

III. Between these two tendencies, there is a third which is far more widely prevalent than either,—that, namely, of *passive acquiescence* in the forms of thought in which the Ego has been brought-up. For one man who determinately sets himself to seek a definite basis for the opinions he professes,—who, after making the best use of his faculties and opportunities, finds that basis either in external Authority or in the authority of his own Reason,—and who, having found what his Intellect approves as Truth, acts upon his convictions to his personal detriment (or, what is far more trying, to the injury of those most dear to him),—there are multitudes who do not feel called upon to enquire for themselves, but consider themselves justified not only in accepting a body of doctrine which they regard as on the whole beneficial, but in recommending it to the acceptance of others. Such persons cannot be truly said to *believe* a set of propositions, the evidence of which they have never studied, and the very language of which (framed as the expression of ideas that have long since passed away) they do not understand. But *dis*belief in them is equally out of their thoughts. Their Mental fabric has been built up under the direction of a Νόμος which has shaped it into accordance with the furniture it is to receive. And unless something occurs to make them question the validity of their position, they continue to hold it without any suspicion of its possible untenability.—Now such persons are responsible for their acquiescence, in so far as this has been induced either by passive indolence, or by a timorous apprehension of the possible results of inquiry, whether in unsettling and disturbing their own minds, or in injuriously affecting their worldly interests. The results of such enquiry, honestly pursued, may be so far satisfactory to their reason, as fully to justify them in resting where they are. Some, again, may be led to the conclusion that the adoption, to however small an extent, of the principle of Authority, leaves them no other logical basis than the Authority of the Church. While others, in whom the spirit of independence is more pronounced, find themselves driven by it in the opposite direction; and are led by the application of the very same logical tests, to pure Individualism,—that is, to the implicit adoption of those opinions, and those only, which express the experiences of the individual's own mind.

328. It is of great importance, in our search for Truth, that we should set out with clear ideas respecting the object of that search and our means of attaining it. From the Psychological point of view, what is accepted as *truth* by each individual is "that which he troweth;" in other words, that which is consistent with the

constitution of his own Mind, and with his previously-acquired convictions (§ 321). But the Truths which claim our acceptance may be ranged under two distinct orders :—the first of which includes all such propositions as are purely *subjective*, the evidence of their truth being *internal*, and consisting in their conformity to ideas which are essentially the creations of our own minds ;—whilst the second includes all such propositions as represent *objective* realities, the evidence of the truth of which is *external*, consisting in the conformity of these ideal representations to actual facts. There is a third order, consisting of propositions, which are supported by evidence of both kinds ; but these need not be separately considered.—Now to the first of these orders belong all the so-called "necessary truths ;" their necessity *to us* arising out of the existing constitution of our own minds, whether original or acquired (§ 201), or out of their exact conformity with some other ideas, which we have already either accepted as "self-evident," or assume as the foundation of our reasoning.

a. Thus Geometry, which consists in the study of the relations of Space, is founded upon two sets of propositions ; one set being the self-evident axioms or "first truths" we have already considered (§ 200) ; whilst the other consists of definitions, which, while professing to represent objective realities, are really, for the most part, intellectual abstractions, with which nothing external to the mind actually corresponds. Now many simple propositions often cited as examples of "necessary truths," carry their own conviction to our minds, simply because either the contrary or anything else would be obviously inconsistent with some one of these fundamental ideas. Thus we at once see that the proposition that "two straight lines can enclose a space" is absolutely contradictory to that "common-sense" conception of a straight line, which is clearer than any definition yet framed ; as is also the proposition that "any two sides of a triangle can be either equal to or less than the third side."—It is by a succession of such steps, each securely cemented to the one on which it rests, that we are led to the higher propositions of Geometry, every one of which is as "necessary" a truth to him who has thoroughly mastered it, as are those just cited to the tyro; its "self-evidence" consisting in the mutual and indissoluble cohesion of the entire succession of ideas, which cohesion they derive from nothing else than the constitution of our own minds.

b. The higher Mathematics, again, rest on a new set of ideas, which carry us still further from the range of objective experience; these ideas, in some form or other, involving the notions of the infinitely-great and the infinitely-small, and of never-ending approximation to one or to the other. By the student whose mind has attained a certain stage of development, these ideas are as readily apprehended as are the ideas of a geometrical point or line by the tyro; and the propositions which he builds upon them are no less "necessary truths" to him, than is the 47th of the first Book of Euclid to the youth who has thoroughly mastered the train of reasoning which leads up to it.

Hence it seems clear that our capacity for apprehending Truths of the first order, entirely depends on the Constitution of our own Minds; and must necessarily be, like it, *progressive.* And for this view we have ample historic confirmation in the fact, that many propositions formerly accepted universally as "necessary truths," are now no less universally abandoned as untenable or even absurd; whilst, on the other hand, *we* now accept as "necessary" many propositions which our ancestors would have scouted as preposterous.

c. Thus the dogma of the Aristotelian philosophy, that, the Circle being the most perfect of figures, the celestial bodies *must* therefore move in circles, continued to hold its sway until the time of Kepler. And in like manner the proposition that Celestial motions continue without diminution because "natural," whilst Terrestrial motions *must* come to an end because "unnatural," was part of the unquestioned philosophical creed until the time of Newton; whose first Law of Motion is based on the idea that motion, whether celestial or terrestrial, is just as "natural" as rest.

How far what *we* now regard as "necessary truths" may require modification in the future, it is impossible for us to judge;

simply because we can no more conceive of anything beyond the range of mental development we have ourselves attained than a man born blind can picture visual objects. But foreshadowings of such a requirement are not wanting :—

d. From certain recondite investigations which have been recently prosecuted by Mathematicians of distinguished ability, the unexpected conclusion has been drawn that more than three dimensions in Space are *ideally* possible. The antagonism of this proposition, however, not only to our actual experience, but to any conceivable extension of it, leads to the suspicion that some fallacy lurks in the primary mathematical expression of "dimensions of space," on which the whole train of reasoning is founded; and that although the result may be perfectly true as regards its conformity to that fundamental idea, it may not be true as representing any possible objective reality,—being, in fact, an ingenious mathematical quibble, not a real extension of our knowledge. That we have no right, however, to tie ourselves down to the limits of our own experience in such a matter, has been ably urged by Prof. Helmholtz; who has worked out the case of an Insect living on a plane surface, which could only know *two* dimensions of space by experience, and to which the notion of *three* would be probably as "unthinkable" as that of *four* or more is to ourselves.

329. Proceeding now to propositions of the *second* order, the Truth of which depends on the conformity of the *ideal* statement to the objective *reality*, it is obvious that their basis is entirely *experiential*; and that it must, therefore, be subject to continual modification from the enlargement of our range of observation, and the increasing precision of our methods. The first question in regard to any *particular* proposition, is whether it accurately represents the facts of the case; and the first question in regard to any *general* proposition, is whether it accurately represents the facts of all the particular cases to which it applies. The conviction we feel as to any of the Truths of this order, rests on a basis very different from the preceding. There are fallacies of observation, fallacies of testimony, and fallacies of reasoning, against each of

which it is necessary to guard; and it is rather on the cumulative value of the evidence in their favour, on the mutual confirmation afforded by different methods of investigation, and on the absence of opposing considerations, that our convictions are based. The strength of such convictions may be such as to make them scarcely inferior in practical reliableness to the "necessary truths" of the purely subjective order. But even those which command our most unhesitating assent, will be accepted by the real Philosophér with a certain "reserve of possibility;" as the nearest approximations to objective reality that the present state of knowledge may justify, but as liable to modification by the extension of that knowledge.

a. Thus, while no one doubts that there is an *actual* distance between the Earth and the Sun, no Astronomer expects that we shall ever be able to obtain more than an approximate determination of it; and yet it is upon the basis of this determination, that the estimate of the distances of the other Planets entirely rests.

b. So, while no Chemist doubts that the different Elements have precise "combining equivalents" or "atomic weights," no one would venture to affirm that these are at present exactly known. And while our present knowledge of these numerical relations is sufficiently precise for our existing requirements, it cannot be accounted improbable that new methods of research may modify our present estimates, perhaps by opening-up altogether new views of these relations.

Thus, then, all the Truths of our second order *must* be progressive; that is, as our ideas can only *approach* to precise conformity with the objective realities they represent, a nearer approach will be for ever possible; and this not merely from increased exactitude of observation, but from the augmented capacity of our minds to utilize its results. For, as has been well remarked, what *we* look upon as a straight line, the prolongation of which to infinity would only increase the distance between its two extremities, may really be seen by beings of a wider range of vision as part of a circle returning into itself.—It is this *progressive* character which imparts to objective Science one of its greatest elements of value as an intellectual discipline, and one of its greatest attractions as a pursuit. For what can be more conducive to a noble but self-restrained independence of thought, than the conviction that whatever we may accept as *authoritative* in the teachings of those whom we regard as our best guides in any department of investigation, must be accepted *provisionally*, to be tested by the results of further inquiry, as our own conclusions will be in their turn? What, again, can be a better lesson of humility, than the remembrance that our own work will in its turn be reviewed by those who shall come after us; and that however complete and satisfactory it may appear to ourselves, our successors will find much to add, if not to correct? And what can be a stronger stimulus to the zealous exercise of our best powers, than the conviction that though we may never be able to attain to "absolute" truth, yet we can be for ever approximating to it; ever striving upwards, so as either ourselves to reach, or to help our successors to reach, a still loftier elevation, whence a yet more comprehensive view may be obtained? "Tendre à la perfection, sans jamais y pretendre," will ever be the animating spirit of the genuine Philosopher; as the "forgetting the things behind, and reaching forth unto the things before" of the greatest of Christian Apostles, will continue to the end of time to nerve the efforts of every true aspirant after Moral excellence. And if we sedulously cultivate this spirit, our Habits of Thought will shape themselves in accordance with it; provided that we set before us an end which is not only worthy in itself, but is also suited to our capacity. "Let every man *find* his work," Carlyle has somewhere wisely said, "and *do* it." The conformity between the *objects* of Human Knowledge and the *faculties* of the Human Mind, is such—however we may account for it—as to provide fitting work for every one; and in proportion as each *does* "with his might" whatsoever he "findeth to do," will be the value of his own share in that *progress* in which true Vitality consists, and of his contribution to the progress of others,—not merely by the additions he may make to the general stock of Knowledge, but through the influence he exerts by his mode of seeking for it. For, as was admirably said by Dr. Thomas Brown,—

"There is a Philosophic Spirit which is far more valuable than any limited attainments in Philosophy; and the cultivation of which, therefore, is the most precious advantage that can be derived from the lessons and studies of many Academic years:—a spirit which is quick to pursue whatever is within the reach of human intellect; but which is not less quick to discern the bounds that limit every human inquiry, and which, therefore, in seeking much, seeks only what man may learn:—which knows how to distinguish what is just in itself from what is merely accredited by illustrious names; adopting a truth which no one has sanctioned, and rejecting an error of which all approve, with the same calmness as if no judgment were opposed to its own:—but which, at the same time, alive with congenial feeling to every intellectual excellence, and candid to the weakness from which no excellence is wholly privileged, can dissent and confute without triumph, as it admires without envy; applauding gladly whatever is worthy of applause in a rival system, and venerating the very genius which it demonstrates to have erred."

SECTION 4.—*Influence of the Will on the Direction of the Conduct.*

330. While the actions of the Biologized or Hypnotized subject are entirely determined, as will be shown hereafter (Chaps. XIV., XV.), by the motive power of Ideas and Feelings, the man in full possession of his Volitional faculty has the power (1) of *refraining from bodily action* under the immediate pressure of motives; and (2) of so far *modifying the relative force of motives* by the mode in which he mentally contemplates them, that their preponderance may be completely reversed. Hence his ultimate determination, whilst still governed by the *preponderance*

of motives, may be entirely different from that on which he would have acted if he had given way to his first impulse. For just as we may direct our intellectual operations by an exercise of Volition, so as to fix upon certain *ideas* only, out of the many which present themselves to our consciousness, and to limit our attention to certain peculiar aspects of these (§ 324), so may we fix our attention upon any one or more among the *motives* which tend to determine our action, and keep these (as it were) in a strong light before the mental eye; whilst, by withdrawing our attention from others, we virtually throw them into the back-ground, as we can do with regard to objects of sensation (§ 123). And further, by calling the Reasoning powers into operation, and bringing them to bear upon the questions at issue, so as to follow-out each of the modes of action that are before the mind to its probable consequences, the Will indirectly brings a set of *new* motives, arising out of these consequences, before the judgment; and these, at first overlooked, may become important elements in the decision. On the other hand, by thus reasoning-out the probable consequences of an action, motives which at first presented themselves in great strength, may lose more or less of their force, and even become altogether futile.

331. Now if we examine into the different kinds of *motive powers*, which, under the *permission* or the *intentional direction* of the Will, are the sources of Human action, we shall find that they may be ranged under the following heads:—(1). *Previously acquired habits*, which automatically incite us to do as we have been before accustomed to do under the like circumstances, without the idea of prospective pleasure or pain, or of right or wrong, being at all present to our minds (Chap. VIII). Where the habits have been judiciously formed in the first instance, this tendency is an extremely useful one, prompting us to do that spontaneously, which might otherwise require a powerful effort of the Will: but if, on the other hand, a bad set of habits have grown-up with the

growth of the individual, or if a single bad tendency be allowed to become an habitual spring of action, a far stronger effort of Volition will be required to determine the conduct in opposition to them. This is especially the case, when the habitual idea possesses an Emotional character, and thus becomes the source of *desires;* for the more frequently these are yielded-to, the more powerful is the solicitation they exert.—(2). *Emotional states*, which incite us to particular actions, by the expectation of gratification, either in the acts themselves, or in some consequences which our reason leads us to anticipate from them; or by the expectation of pain, if the act be not performed. All those *desires* and *aversions* which have so large a share in determining our conduct, come under this category: and to it must likewise be referred all those considerations which are simply *prudential;* these usually having reference to the *remoter* effects which our actions are likely to have upon our own welfare or upon that of others, and thus bringing before the mind, as elements in its determination, certain additional objects of desire or aversion.—(3). *Notions of Right and of Duty*, which, so far as they attach themselves to our actions, give them a *moral* and *religious* character. These may act simply as Ideas, whose coercive power depends upon the intensity with which they are brought before the mind; but they obtain a much stronger influence, when they acquire an Emotional character, from the association of the feeling of *desire* with the idea of *obligation*,—that is, when we feel a *wish* to do that which we are conscious we *ought* to do,—an association which it is peculiarly within the capability of the Will to cherish and strengthen. And still more potent is the operation of these combined motives, when a constant *habit* of acting upon them has been formed, so as to give them the force of *fixed principles;* for if the question be always looked-at *first* in its Moral aspect, and a clear perception is attained of its *right* and its *wrong* side, the strongest desires and the strongest aversions are repressed in their nascent stage, without exerting any influence.

The difference between the *habitual*, the *prudential*, and the *moral* aspects of the very same action, may be made apparent by a very simple illustration:—We will suppose that a man has been accustomed to take a ride every day at a particular hour; his whole nature so accommodates itself to the *habit*, that he feels both mentally and physically uncomfortable at any interruption to the usual rhythm. But suppose that, just as the appointed hour comes round, the sky becomes overcast, threatening the rider with a drenching if he perseveres in his intention; his decision will then be founded on a *prudential* consideration of the relative probabilities of his escaping or of his being exposed to the shower, and of how far the enjoyment he may derive from his ride is likely to be replaced by the discomfort of a thorough wetting. But suppose, further, that instead of taking a mere pleasure-ride, a Medical man is about to set-forth on a professional visit to a patient whose condition requires his aid; a new motive is thus introduced, which alters the condition of the whole question, making it no longer one of prudence only, but one of *morality*. Another motive which should give the question a Moral aspect, would be consideration for himself, and the risk of life or health he might run: this should be decisive, where the motive which impels him to the act in question is merely that of self-gratification; but if it bring into antagonism his duty to his patient and his desire to benefit him, and on the other hand his duty to himself and his regard for the ulterior welfare of those who may be immediately dependent upon him, the question has its right and its wrong aspect on both sides (§ 210), and the right may only be determinable after a careful balance of the considerations involved.

332. In connection with the foregoing, it will not be inappropriate to notice the manner in which the principle of *Love*, early fostered by judicious Nursery training (§ 290, IV.), comes to modify the strictness of Volitional action on the principles of Right and Justice in the subsequent intercourses of life. For it is *genuine consideration for the feelings of others*, which con-

stitutes the distinction between the *courtesy* of the *true* "gentleman" or "lady" (in whatever rank of life), and that mere external *politeness* which is nothing more than a social habit, and which may veil the meanest and most selfish dispositions. There is nowhere, perhaps, a more beautiful instance of complementary adjustment between the Male and the Female character, than that which consists in the predominance of the Intellect and Will, which is required to make a *man* successful in the "battle of life," and of the lively Sensibility, the quick Sympathy, the unselfish Kindliness, which give to *woman* the power of making the happiness of the home, and of promoting the purest pleasures of social existence. When we analyse the nature of that *tact* which is usually so much more strongly displayed by Women than by Men, we find it (strange as this may seem) to rest in part on the same basis as Man's ordinary *common sense* (§ 374); being, like it (as was suggested to the Writer by Mr. J. S. Mill, p. 486), the resultant of the unconscious co-ordination of "a long succession of small experiences, mostly forgotten, or perhaps never brought out into distinct consciousness;" these experiences, however, being of a kind to impress the Sensibilities rather than the Reason, to be perceived by Sympathy rather than by Ideation. Like the higher form of "common sense" (§ 383), Tact is often so strongly manifested at a very early period of life, that we can scarcely refuse to it the character of an *original* intuition; whilst it is also eminently capable of being *acquired*, or at any rate *improved*, by a Volitional culture which directs the attention to the impressions fitted to develope it: and it is in this way that a Woman comes to possess a *direct insight* as to what is due to consideration for others, which the duller and more rationalistic apprehension of Man can seldom attain. This, unchecked by a disciplined Moral sense, is apt to run to excess; tending to "make things pleasant," at the expense of honest consistency. But, when so restrained, it is an endowment so trustworthy, that every wise man will trust to the guidance of womanly Tact, wherever the question is one which it can fittingly resolve.

333. If, now, taking our stand upon the foregoing Physiological and Psychological facts, and leaving out of view all embarrassing questions about "freedom" and "necessity," we apply ourselves to the practical inquiry as to the *mode* in which the Will regulates the ordinary course of our daily life, we shall be led to the following conclusions :—

I. The Will is constantly *initiating* movement (as in walking or writing), or *directing* movement (as in speaking, § 307), without any present consciousness of motives; this initiation or direction being, in fact, the expression of a *remotely-formed determination* deliberately made and systematically acted on. Thus when a man chooses a certain profession, or undertakes a certain office, and does so with the fixed purpose of faithfully discharging its responsibilities, the habits he forms become a "second nature" to him; he does not "stop to think" whether he shall or shall not perform any action which clearly forms part of his duties; but his Will says to his body, "Do this," and the body does it accordingly :—

Thus, no right-minded Medical man ever "allows himself to think" of his own personal risk, when called to attend a case of malignant Scarlatina: he determinately puts either himself or his horse in motion, to obey a summons which comes to him in the ordinary course of his professional duty; and if he does "allow himself to think" of the risk of conveying the infection to his family, or to other patients, it is only as a motive to taking all possible precautions against doing so.—Again, cases every now and then occur, in which a Medical man may feel sorely tempted by feelings of pure humanity, as well for the sufferer as for those around him, to put an early termination to the hopeless agony of his patient, just as he would put a dog or a horse "out of its pain;" and his Moral *right*, or even *duty*, to practise such a Euthanasia, has been seriously advanced and supported by arguments of no little force. But here he will fall back on that general rule of the Profession, which binds every member of it to do his utmost to prolong life and to mitigate suffering; this being clearly beneficent on the whole, while the least infraction of it would lead to the gravest dangers in practice.

In such circumstances, the Will carries out *fixed principles of action;* which, having been adopted by the Reason, under the guidance of the Moral sense, habitually rule the Conduct.

II. But suppose, in the next place, that, as is constantly happening, these fixed principles of action come into collision with other motives, which strongly appeal to the "likes" or "dislikes" of the individual:—

The Official, for example, may be sorely tempted to desert his post for a day, by some prospect of pleasure or advantage to himself; or the ill-paid Union-doctor to "shirk" attendance on some tedious and uninteresting "case," which will bring him neither credit nor remuneration. Now, a man who is determined to make *duty* his first consideration, will not allow himself to dwell upon his personal preferences, but will say to himself, "*I ought,*" or "*I ought not,*" as the case may be; thus *fixing his attention* on the principle of action which he has deliberately adopted, and thereby strengthening his determination to adhere to it; whilst, in the same measure, he weakens the force of the temptation by *withdrawing his attention* from it.

Here, again, the Will (though less immediately) carries out a fixed principle of action; its power being secondarily exerted in intensifying the sense of obligation, and in keeping out of view the advantages and pleasures to be derived from an infraction of the strict rule of duty. And this secondary action of the Will comes to be the principal mode of its operation, when the subject of the temptation so far hesitates, as to *discuss* the question with himself; the *ruling principle* being abandoned, and the question coming to be decided by motive influences of a lower class.

Thus, the Official may say to himself, "What harm will come of my absence?" or, "Would not a day's holiday do me a great deal of good?"—And the Union-doctor may find excuses for himself, in his hopelessness of doing any effectual service, in the unthankful spirit with which his best endeavours have been treated, and in the exhaustion of his own power which seems likely to result from the continuance of his efforts.

But, in such a discussion with one's self, the Will may still take an important part. For it can *select*, among the motives which present themselves, those which the Moral Sense approves as the most worthy, and can *intensify the force* of these by *fixing the attention* upon them; whilst it can, in like manner, keep to a great extent out of sight those which it feels ought not to be admitted, and can thus *diminish their force.* And thus at last, while the decision is really formed by the "preponderance of motives," it is the action of the Will in modifying the force of those motives, that really determines *which* shall preponderate.—The Will is here, therefore, the expression of the higher Reason, controlling the operation of the selfish Propensities.

III. But, further, the Will can *put a check* upon the bodily action to which some strong internal impulse would *directly* prompt; so that *time is gained* for consideration, by which the further course is guided. This exertion of the Will may proceed from a fixed determination "not to give way" to such impulses; and this determination, originally formed on a deliberate Moral judgment, becomes strengthened by every exercise of it. Here, again, our Volitional action is the expression of a Habit, which has become a part of our "second nature," overcoming the promptings of our original disposition.—If, after restraining the *immediate* impulse to action, we deliberate upon further steps, the Will has exactly the same power of modifying the decision as in the preceding instance:—

Thus, to take a not uncommon case, a man considers himself to have received an affront or injury, which his first impulse is to resent. He restrains himself, however, by a strong effort, from immediate action; that effort being the determinate expression of the general conclusion he may have long ago arrived at, that such immediate action is undesirable. Still he thinks that the matter requires *some* notice; and his judgment tells him that whilst he is still labouring under an excitement of feeling, he is not in a fit condition to decide what is best to be done.* And when, after a further prolongation of this Volitional restraint, he comes at last calmly to consider the matter, his action is at last determined by the preponderance of the motives which his Will has selected as most fitting to be admitted into the discussion.

IV. These experiences of Self-regulation have their parallel in the experience of our endeavours to influence the conduct of *others.* For suppose, in the first place, that we are appealing to a man whom we know to act habitually and determinately upon his "sense of duty;" — our whole aim is to convince his Reason that his Duty points in a certain direction, and we feel assured that, if once satisfied of this, he will carry out his determination to the best of his ability (§ 325 *a*, *b*).

V. Again, we have to do with a less "resolute" man,—one who may admit that he *ought* to pursue a certain course, but who distrusts his own power to follow it out. We then endeavour to strengthen both his sense of Moral obligation, and his Volitional power of acting upon it; for here the encouraging assurance that he *can* do so if he will only *try*, gives the same kind of added force to the mental as it does to the bodily effort (§ 266). It is by giving a fixed basis of principle, or *point d'appui*, to this effort, that a

* It is within the experience of many, that nothing so much relieves the mind under such circumstances, or forms such a good basis for subsequent *action* or *inaction*, as *writing a letter*, in which adequate expression is given to the disturbed feeling. The act itself discharges the mind of much of its Emotional excitement, on the principle formerly stated (§ 265); the fact that everything which has *to be* said *has been* said, relieves the thoughts from the recurring tendency to seek for the modes of expressing it; and when, after the lapse of a day or two, the letter (not having been sent) is reconsidered, the judgment can be calmly exercised in either toning it down, or putting it aside altogether.

definite "resolution" or "pledge" is often valuable. Thus it is proved by ample experience, that many a man who has not enough strength of Will to keep him from yielding to Alcoholic seduction, has enough to make him "keep the pledge" he has taken against it: the mere repetition to himself of a determination to do so, having the good effect of augmenting the force of that determination, and of helping him to keep out of the way of temptation. As it has been said that "a woman who deliberates is lost,"—the mere entertainment of the idea of a violation of chastity showing how strong a hold the temptation to it has already gained upon her,—so may it be said of the man who is strongly tempted to "break his pledge," that if he once allows himself to "think" about it, the force of that principle is grievously weakened. But we may strengthen his determination by directing his attention to the various collateral motives which we may deem most likely to influence him; the operation of these, however, being most advantageous, when they give steadiness and fixity to the principle of action.

VI. But, lastly, when we are dealing with a man of brutal nature, who is callous to all appeals to his sense of duty, and whose attitude is one of dogged defiance, we have to search out the most impressible part of his nature, and endeavour to work upon this by an appeal to some feeling that may be roused into motive force. Thus, on one side there may be some lingering affection for mother or sister, wife or child, which may be vivified by a skilful touch (§ 290 IV.). On the other, the dread of consequences may be wrought-upon; the grief and shame that will be brought upon those for whom there is still a regard, being often more potent deterrents than the prospect of punishment to the individual. It is in the *direction of the attention* towards all the *deterrent* motives which are found to have any potency, in the *withdrawal* of it from all those which *attract* to ill-doing, and in the prospect of reward for every exertion of *self-control*, that the work of Criminal Reformation essentially consists; the efficacy of that work being largely increased by the patient sympathy that has a softening influence on even the most brutal natures, and by the encouragement to "try again" after repeated lapses. And so, in time, the tendency to act on the "impulse of the moment" comes to be amenable to the control of the Moral Will; while *principles* grow up under judicious discipline, which, approving themselves to the reason, may ultimately acquire fixity and steadiness sufficient to determine the conduct, without any recurrence of the conflict between opposing motives.—This is, of course, more feasible with juvenile than with adult Criminals; since bad habits, once constitutionally established, are not easily changed; whilst during the period of growth, not only may *bad* habits be more easily and completely eradicated, but *good* habits may be planted and fostered, growing with the growth, and strengthening with the strength (§ 289).

334. Thus we see that, the less the potency of Volitional control, the more completely is the Conduct of the individual governed by the *direct* "preponderance of motives;" whilst the interposition of the Will operates for the most part in one of two ways; either (1) by the determinate adherence to some fixed principle of action,—just-as a man who is falling over a precipice tenaciously holds-on to any ledge to which he can cling;—or (2) by that modification of the relative force of opposing motives, which is effected by the determinate attention to some, and determinate *in*attention to others. It is in that important period of each life, when the Youth is first left to his own direction, and has to make his own choice of the principles which are henceforth to be the guides of his life, that the value of a *resolute determination* to "turn to the right, and keep straight on," is the greatest; and however potent may have been the influence of judicious training and discipline *in giving a right direction* to the thoughts and feelings, and in repressing or diverting the violence of

passion, it is only the established habit of *self*-direction and *self*-control that can give real steadiness to a resolution, real force to a determination. And thus we may truly say that the power which the Will of any individual can exert in a great crisis, is the "resultant" of his whole previous Mental life; being proportional to the degree in which he has *habituated* himself to keep the spontaneous or automatic activity of his mind under Volitional control, instead of allowing himself to be the sport of his intellectual vagaries, the slave of his passionate impulses.

As Dr. J. D. Morell has well said (*Introduction to Mental Philosophy*, p. 375), "The education of the Will is really of far greater importance, as shaping the destiny of the individual, than that of the Intellect; and it should never be lost sight of by the practical Educator, that it is only by the amassing and consolidating our volitional *residua* in certain given directions, that this end can be secured. Theory and doctrine, and inculcation of laws and propositions, will never of themselves lead to the uniform habit of right action. It is by doing, that we learn to do; by overcoming, that we learn to overcome; by obeying reason and conscience, that we learn to obey; and every *right act* which we cause to spring out of pure principles, whether by authority, precept, or example, will have a greater weight in the formation of character than all the theory in the world."

335. But to carry into *action* the Volitional determination, to give to the I WILL its practical effect, something more is usually needed than the mere "preponderance of motives." The Idea of *the thing to be done* (which we have seen to be the necessary antecedent of all Volitional action, § 305), may, indeed, be so decided and forcible, when once fully adopted, as of itself to produce a degree of Nervous tension that serves to call forth respondent Muscular movements,—as in the purely Ideo-motor form of action (§ 235). Thus, cases are not uncommon, in which persons who have had some difficulty in "making up their minds" to a particular course, find themselves borne along, as by the rush of a stream that has

been let out by the opening of a flood-gate, when once they have committed themselves to it. But in general, a distinct exertion of the Will is needed to give to the Ideational state the energy requisite to call forth the action that expresses it ; and this is especially the case, where either some powerfully opposing motive diminishes the force of the preponderance, or a state of fatigue causes the bodily mechanism to be less easily called into action. Reasons have been already adduced (§§ 308, 309) for the belief, that the Volitional exertion really consists in an intensification of the hyperæmic state of the Ideational centre ; which will produce an augmented tension of its nerve-force, whose discharge through the *motor* centres calls forth the muscular movement. And this may take place without a corresponding intensification of the *idea* itself ; if, according to the doctrine previously advanced (§ 100), we only become *conscious* of Cerebral changes as Ideas, when their influence has been reflected downwards to the Sensorium.

336. It must not be forgotten, however, that the Volitional power may be turned to a *bad* as well as to a good account ; and that the value of its results will entirely depend upon the *direction* in which it is employed. The thoughts may be so determinately drawn away from the higher class of motives, the suggestions of Conscience, of Affection, or of Benevolence, so habitually disregarded, and the whole attention so completely fixed upon the gratification of the selfish or malevolent propensities, that the Human nature acquires far more of the *Satanic* than of the Divine character ; the highest development of this type being displayed by those who use their power of self-control for the purposes of hypocrisy and dissimulation, and cover the most malignant designs under the veil of friendship. Such men (whose portraiture is presented by our great Dramatist in the character of Iago) show us to what evil account the highest Intellect and the most powerful Will may be turned, when directed by the baser class of motives

and we cannot but feel that they are far lower in the Moral scale, than those who have never known the meaning of Love and Truth, Kindness and Honesty,

337. Of this latter class there are some, who, from original constitution and early influences of the most degrading kind, are altogether destitute of anything but a *brutal* nature: these ought to be treated as irresponsible beings, and, as such, restrained by external coercion from doing injury to society. But this class is small in proportion to that of individuals who *act* viciously, simply because they have never been led to *know* that any other course is open to them, or to *feel* any motives that might give them a different impulse. The experience of those who have undertaken the noble work of Juvenile Reformation, has satisfied them that the cases are few, if any, in which there is not "a holy spot in the child's heart," on which an impression may be made by appropriate suggestions; and that by following the method of the good nurse (§ 269), the power of self-control, which seems in the first instance altogether absent, may be awakened and cherished, the lower propensities repressed by a judicious mixture of restraint and distraction, and the higher tendencies called by the genial warmth of sympathy into full activity, so that the little reprobate most truly becomes "a new creature."—If it be assumed as a fundamental principle, that *every* part of our Nature has its *use* as well as its *abuse*,—our propensities and passions not being evil in themselves, but evil only in their excess and misdirection,—it is wonderful what effects may be produced by the judicious *guidance* of their energy towards innocent or worthy objects. A latent nobleness and vigour of character not unfrequently shows itself under such treatment, in youths who have been (in a manner) forced into antagonism by the ill-judged sternness of parents, and who, when left to themselves, have committed extravagances of conduct that have caused them to be stigmatized as hopeless outcasts; while many a naughty girl who has been driven

by mismanagement into rebellion at home, has been moulded into an admirable woman by the skilful discipline of a wise schoolmistress (§ 272).

338. There is a *negative* type of character again, on which the wisest Educator finds it difficult to make any permanent impression, through constitutional want of self-determining power. The intellectual capacity may not be below, or may even be above, the average, and generally the disposition may be amiable; and yet there may be little power of resisting a seductive temptation, or of holding fast to any fixed principle of action. Individuals of this type are, to a great degree, the "creatures of circumstances." Under conditions favourable to the operation of the better part of their Nature, they may not only lead blameless and useful lives, but be credited with Moral excellencies which they do not really possess. For let the same individuals be subjected to the insidious influence of attractive but immoral companionship, or come to a rugged and thorny place in a path of Duty that had previously been smooth and pleasant to them, having no stability of purpose, they fall away; and when they have once entered on the downward course, they can only be checked in it by voluntarily submitting themselves to renewed control.

It is an old and just observation, that youths who have been "brought up at their mothers' apron-strings," are the most likely to "go wrong" when first thrown upon their own guidance; and that when such once begin to go astray, they soonest run into wild excesses. The *rationale* of this seems to be, that the tendency of such an education is usually to repress, instead of fostering, habits of independence and self-regulation; and too frequently to weaken, instead of strengthening, the force of Moral obligation, by attaching to *small* things the same importance as to *great*. If a lad is constantly watched and never trusted, he is almost sure to abuse his liberty when he first acquires it. And if he is taken to task as severely for spilling ink on a table-cloth or for tearing his clothes, as for telling a lie or appropriating what does not belong to him, it is not to be wondered-at that he should come to regard the graver offences in the same light as those which he feels to be venial.

With a character of this type, the object of the judicious Educator will be to *invigorate the whole nature*, corporeal as well as physical; to find out what worthy objects of pursuit have the most attraction for his pupil, and to aid and encourage his steady pursuit of them, not by removing difficulties from his path, but by helping him to surmount them; and in this manner to foster habits of self-reliance, which, when once formed, whether in regard to manly exercises, or to the work of the intellect, may be looked to as available for the Moral direction of the conduct.

339. The highest exercise of the Will is shown in those who are endowed with vigorous Intellectual powers, and whose strong Emotional nature gives force to all their tendencies to action; but who determinately fix their attention on the *divine ideal*, and steadily endeavour to shape their character and direct their conduct in accordance with it. This is *not* to be effected by dwelling exclusively on any one set of motives, or by endeavouring to repress the energy which is in itself healthful. Even the idea of Duty, *operating alone*, tends to reduce the individual to the subservience of a slave doing his master's bidding, rather than to make him master of himself; but it gives most powerful *aid* in the acquirement of that power of fixing the thoughts and affections on "things on high," which most effectively detaches them from what is earthly and debasing. It is by the *assimilation*, rather than by the *subjugation*, of the Human Will to the Divine, that Man is really lifted towards God; and in proportion as this assimilation has been effected, does it manifest itself in the life and conduct; so that even the lowliest actions become holy ministrations in a temple consecrated by the felt presence of the Divinity. Such was the Life of the Saviour; towards that standard it is for the Christian disciple to aspire.

BOOK II.

SPECIAL PHYSIOLOGY.

CHAPTER X.

OF MEMORY.

340. There is no part of our purely Psychical activity, the relation of which to Physical conditions is more obvious and more intimate, than that *reproduction of past states of consciousness*, which, when supplemented by the *recognition* of them as having been formerly experienced, we call *Memory.* It is now very generally accepted by Psychologists as (to say the least) a probable doctrine, that any Idea which has once passed through the Mind may be thus reproduced, at however long an interval, through the instrumentality of suggestive action; the recurrence of any other state of consciousness with which that idea was originally linked by Association, being adequate to awaken it also from its dormant or "latent" condition, and to bring it within the "sphere of consciousness." And as our ideas are thus linked in "trains" or "series," which further inosculate with each other like the branch lines of a railway or the ramifications of an artery, so, it is considered, an idea which has been "hidden in the obscure recesses of the mind" for years—perhaps for a lifetime,—and which seems to have completely faded out of the *conscious* Memory (having never either recurred spontaneously, or been found

capable of recall by volitional Recollection), may be reproduced, as by the touching of a spring, through a *nexus* of suggestions, which we can sometimes trace-out continuously, but of which it does not seem necessary that all the intermediate steps should fall within our cognizance.* Such a reproduction not unfrequently takes place, when persons revisiting certain scenes of their childhood, have found the renewal of the sensorial impressions of *places* bring vividly back to their minds the remembrance of *events* which had occurred in connection with them; and which had not only been long forgotten by themselves, but, if narrated to them by others, would not have been recognised by them as having ever formed part of their own experience. And it is not a little significant that the basis of such memories appears capable of being laid at a very early period of life; as in the two following cases, of which the first is recorded by Dr. Abercrombie, whilst the second was mentioned to the Writer by the subject of it:—

a. "A lady, in the last stage of chronic disease, was carried from London to a lodging in the country:—there her infant daughter was taken to visit her, and, after a short interview, carried back to town. The lady died a few days after, and the daughter grew up without any recollection of her mother, till she was of mature age. At this time she happened to be taken into the room in which her mother died, without knowing it to have been so:—she started on entering it, and, when a friend who was with her asked the cause of her agitation, replied, 'I have a distinct impression of having been in this room before, and that a lady who lay in that corner and seemed very ill, leaned over me and wept.'"—(*Intellectual Powers*, 5th Ed. p. 120.)

b. Several years ago, the Rev. S. Hansard, now Rector of Bethnal Green, was doing clerical duty for a time at Hurstmonceaux in Sussex; and while there, he one day went over with a party of

* This disappearance of some of the links from Consciousness, "as completely as if they had never formed part of the series," is a fact admitted by Psychologists of all schools, whatever may be their *rationale* of it.—See Mr. John Mill's Note to his Edition of James Mill's "Analysis," vol. i. p. 106.

friends to Pevensey Castle, which he did not remember to have ever previously visited. As he approached the gateway, he became conscious of a very vivid impression of having seen it before; and he "seemed to himself to see" not only the gateway itself, but donkeys beneath the arch, and people on the top of it. His conviction that he *must* have visited the Castle on some former occasion,—although he had neither the slightest remembrance of such a visit, nor any knowledge of having ever been in the neighbourhood previously to his residence at Hurstmonceaux,—made him enquire from his mother if she could throw any light on the matter. She at once informed him that being in that part of the country when he was about *eighteen months* old, she had gone over with a large party, and had taken him in the pannier of a donkey; that the elders of the party, having brought lunch with them, had eaten it on the roof of the gateway where they would have been seen from below, whilst he had been left on the ground with the attendants and donkeys.—This case is remarkable for the vividness of the Sensorial impression (it may be worth mentioning that Mr. Hansard has a decidedly Artistic temperament), and for the reproduction of details which were not likely to have been brought up in conversation, even if he had happened to hear the visit itself mentioned as an event of his childhood, and of such mention he has no remembrance whatever.

c. "A remarkable case is mentioned by a writer (Miss H. Martineau?) of a congenital idiot who had lost his mother when he was under two years old, and who could not have subsequently been made cognizant of anything relating to her; and who yet, when dying at the age of thirty, "suddenly turned his head, looked bright and sensible, and exclaimed in a tone never heard from him before, 'Oh my mother! how beautiful!' and sunk round again—dead.'"—(*Household Words*, vol. ix. p. 200.)

341. Although it is commonly stated that Memory consists in the renewal of past Sensations and of the Ideas they have excited, it may be questioned whether impressions are really left on our minds by anything else than *Ideas;* and whether the reproduction of *Sensations*, independently of the presence of the object of them, is not a secondary change, dependent upon the reaction

of Ideational (Cerebral) changes upon the Sensorium. It is certain that the most vivid reproduction of sensations is often consequent upon the recurrence of the ideational states with which they were originally associated. Thus a Roman Catholic friend of the Writer, who, when a boy, had gone to Confession for the first time with his mouth full of the taste of a sweet cake, which he had been eating just before, and his digestion of which had been emotionally disturbed, never went on the same errand for some years, without the distinct recurrence of the same flavour. Again, it is by no means uncommon for those who suffer acutely from Sea-sickness, to experience nausea at the mere sight of an agitated ocean, especially if a wave-tossed vessel be within view; and a like feeling, it is said, has been excited by the sight of a toy, in which (by a peculiar combination of levers) the motion of a ship was imitated with peculiar fidelity. The Writer, indeed, was once assured by a lady that she had herself been affected with an actual paroxysm of sea-sickness, through having witnessed the departure of a friend by sea on a stormy day.—Such facts, indeed, are so familiar as to have become proverbial; for the common expression "it makes me sick to think of it" is nothing else than the expression of a Sensorial feeling excited by an Ideational state.

342. This Sensorial feeling may, indeed, be so intense, as to reproduce any bodily action that originally supervened on its first excitement. Thus Van Swieten relates of himself, that, having chanced to pass a spot where the bursting of the dead body of a dog produced such a stench as made him vomit, on passing the same spot *some years afterwards* he was so vividly affected by the recollection, that the sickness and even vomiting recurred. So it must be within the experience of every one, that tears rise at some painful or tender reminiscence; that the mental reproduction of circumstances which originally produced a blush of shame or self-consciousness, will call forth not merely the same emotion, but the same expression of it; and that laughter is as often provoked by the remem-

brance of some ludicrous incident, as it is by its actual occurrence. These facts, indeed, are so familiar, that they may seem too trivial to deserve notice; but they have just the same significance as the equally familiar fact, that Coughing may be produced either by the stimulus of an irritation in the throat, or by the stimulus of a Volitional, *i. e.*, Cerebral determination (§ 17). For they all tend to show that *the immediate instrumentality of our Sensational consciousness is always the same*, whether its remote Physical antecedent be an impression on the organs of the *external* senses, transmitted to the Sensorium through their afferent nerves, or be a change in the cortical substance of the Cerebrum—the instrument of the *internal* senses—transmitted downwards to the Sensorium by the nerve-fibres which constitute its medullary substance (§ 100).

342. It seems a strong confirmation of this doctrine, that if we *wish* to reproduce any Sensational state,—whether Visual, Auditory, Olfactive, Gustative, or Tactile,—we first recall by *Recollection* (§ 370) the notion of some object by which that state was formerly produced; and it is only by giving our attention strongly to that notion, that we can bring ourselves to see, hear, smell, taste, or feel that which we desire to experience. Indeed it is not every one who can thus reproduce Sensational states, the general notion being most commonly all that is arrived-at; of this we have a good illustration in the conception we form of the face of an absent friend,—the number of persons who are able to reproduce the Visual image with sufficient distinctness to serve as a model for delineation, being comparatively small, although a much larger number would be able to say how far such a delineation realized their own conception of the countenance, and to point-out in what it might depart from this.—A further confirmation of this view is to be found in the familiar fact, that the *expression* of a countenance, which directly appeals to our Ideational consciousness, is much more distinctly recollected by most persons than

the *features*, the recognition of which is more dependent upon the recall of antecedent Sensational experiences.—What is true of the act of Recollection in this particular, is probably true also in great degree of *spontaneous* Memory; for although such a case as that to be presently related (§ 344 *d*) might seem to indicate that there may be a mere Sensorial memory—words and phrases being reproduced by their sound *alone*, without the attachment of any distinct *meaning* to them,—yet it can scarcely be doubted by those who have carefully studied the phenomena of Dreaming and Delirium, that what was really reproduced in the the first instance was the patient's *idea* of her old master reciting as he walked up and down his passage; and that it was this idea which prompted her utterance of the words and phrases whose sounds had come to be habitually associated with it.

343. Now, it is obviously upon this recording of impressions, so that they are reproduced as Ideas when the appropriate *suggesting strings* are pulled, that all our accumulated knowledge depends. For when we say that we "know" a language, or an author, or a department of science, we do not mean that the whole or even any part of that knowledge is present to our minds at the time; since, as Sir William Hamilton has justly remarked, "the infinitely greater part of our spiritual treasures lies always beyond the sphere of our consciousness."* The perfection of our knowledge consists, in fact, in the readiness and precision with which the appropriate words or ideas *spontaneously* present themselves, whenever we *desire* to bring them *within* the sphere of our consciousness; and this action depends upon the strength of the association previously formed between the word or idea actually before the mind at each moment, and that which furnishes the response to it. Thus, in speaking a foreign language with which we are thoroughly conversant, the *automatic* play of suggestion calls up the successive words or phrases that express the equivalents of those in which our thoughts have shaped themselves. In quoting a book with which we are

familiar, the sequence of a long passage may be suggested by the mention of its first words, or by the starting of the idea that forms the subject of it. And when the man of science is called upon to "explain" a fact, his mind goes forth, as it were, in the direction most likely to lead to the recall of similar facts which he has previously learned, and to that of some principle common to them all. —On the other hand, we say that we have "forgotten" a word or an idea, when we are conscious of the fact that we *must* have once known it, but cannot reproduce it at the moment. Thus, as a recent writer has remarked :—

a. "If we have *ever* known a thing, the question whether we can be said to know it at any particular time, is simply whether we can *readily reproduce it* from the storehouse of our memory.—There are some ideas, which, if we may use so material an illustration, are systematically arranged in cupboards to which we have immediate access, so that we generally know exactly where to find what we want; this is the case with the knowledge that we have in constant daily use. And yet to whom has it not occurred to be unable to recollect, on the spur of the moment, a name or a phrase that is generally most familiar to him; just as he often fails to remember where he laid his spectacles, or his pencil-case, only five minutes before?—There are other ideas, again, which we know we have got put away *somewhere*, but cannot find without *looking for them;* as when we meet an acquaintance whom we have not seen for a long time, and recognise his face without being able to recall his name; or when we go to a foreign country, the language of which we have once thoroughly mastered, and find ourselves in the first instance unable either to speak or to understand it. In these cases, the lost ideas are pretty certain either to be found, if we look for them, by putting in action that associative train of thought which we term recollection; or to turn up, spontaneously and unexpectedly, when the *effort* to recollect has proved a failure, and we have abandoned the search as hopeless.—There is other knowledge, again, which we are not conscious either of possessing, or of ever having possessed; as in the conjugal experience familiar to most of us, in which a husband assures the wife of his bosom (the converse case being perhaps hardly less frequent) that she *never did* tell him of some occurrence which *he should most certainly have remembered* if she had: and yet he may be brought to recollect, days or weeks afterwards, by the accidental shining-in of a light upon some dark corner of his 'chamber of imagery,' that the communication was really made, but was put away without any account being taken of it at the time."—(*Quarterly Review*, Oct. 1871, p. 318.)

A distinguished Equity Judge has recently favoured the Writer with the following experience:—

b. It has frequently occurred to him that "further proceedings" having been taken in a "cause" which he had "heard" some years previously, and had dismissed altogether from his mind, he has found himself in the first instance to have totally forgotten the whole of the *former* proceedings, not being even able to recollect that the "cause" had been previously before him. But in the course of the argument, some word, phrase, or incident has furnished a suggestion, that has served *at once* to bring *the whole case* vividly into his recollection; as if a curtain had been drawn away, and a complete picture presented to his view.—The entireness of his previous forgetfulness was probably due to the habit common to Barristers, of "getting up" their cases only to forget them as soon as possible (§ 362).

344. Now there is very strong Physiological reason to believe that this "storing-up of ideas" in the Memory is the psychological expression of physical changes in the Cerebrum, by which ideational states are permanently registered or recorded; so that any "trace" left by them, although remaining so long outside the "sphere of consciousness" as to have *seemed* non-existent, may be revived again in full vividness under certain special conditions,—just as the invisible impression left upon the sensitive paper of the Photographer, is developed into a picture by the application of particular chemical re-agents. For in no other way does it seem possible to account for the fact of very frequent occurrence, that the presence of a fever-poison in the blood,—perverting the

normal activity of the Cerebrum, so as to produce *Delirium* (§ 548) —brings within the "sphere of consciousness" the "traces" of mental experiences long since past, of which, in the ordinary condition, there was no remembrance whatever. Thus, the revival, in the delirium of fever, of the remembrance of a Language once familiarly known, but long forgotten, has been often noticed. The following case was mentioned to the Writer many years ago by a Medical friend, as having fallen under his own observation :—

a. "An old Welch man-servant, who had left Wales at a very early age, and had lived with one branch or another of this gentleman's family for fifty years, had so entirely forgotten his native language, that when any of his Welch relatives came to see him, and spoke in the tongue most familiar to *them*, he was quite unable to understand it ; but having an attack of fever when he was past seventy, he talked Welch fluently in his Delirium."

The following cases, recorded by Dr. Rush of Philadelphia, have points of interest peculiar to each :—

b. "An Italian gentleman, who died of yellow fever in New York, in the beginning of his illness spoke English, in the middle of it French, but on the day of his death only Italian."

c. "A Lutheran clergyman of Philadelphia informed Dr. R. that Germans and Swedes, of whom he had a considerable number in his congregation, when near death always prayed in their native languages ; though some of them, he was confident, had not spoken these languages for fifty or sixty years."

The following case, mentioned by Coleridge, is one of the most remarkable on record : its distinguishing feature being that the patient could never have known anything of the meaning of the sentences she uttered :—

d. "In a Roman Catholic town in Germany, a young woman, who could neither read nor write, was seized with a fever, and was said by the priests to be possessed of a devil, because she was heard talking Latin, Greek, and Hebrew. Whole sheets of her ravings were written out, and found to consist of sentences intelligible in themselves, but having slight connection with each other. Of her Hebrew sayings, only a few could be traced to the Bible, and most seemed to be in the Rabbinical dialect. All trick was out of the question; the woman was a simple creature; there was no doubt as to the fever. It was long before any explanation, save that of demoniacal possession, could be obtained. At last the mystery was unveiled by a physician, who determined to trace back the girl's history, and who, after much trouble, discovered that at the age of nine she had been charitably taken by an old Protestant pastor, a great Hebrew scholar, in whose house she lived till his death. On further inquiry it appeared to have been the old man's custom for years to walk up and down a passage of his house into which the kitchen opened, and to read to himself with a loud voice out of his books. The books were ransacked, and among them were found several of the Greek and Latin Fathers, together with a collection of Rabbinical writings. In these works so many of the passages taken down at the young woman's bed-side were identified, that there could be no reasonable doubt as to their source."—*Biographia Literaria*, edit. 1847, vol. i. p. 117.

345. The same occurrence has been noticed as a consequence of accidental blows on the head; though these more commonly occasion the *loss* than the *recovery* of a language. The following case of this kind is mentioned by Dr. Abercrombie, as having occurred in St. Thomas's Hospital :—

"A man who had been in a state of stupor consequent upon an injury of the head, on his partial recovery spoke a language which nobody in the hospital understood, but which was soon ascertained to be Welch. It was then discovered that he had been thirty years absent from Wales, and that, before the accident, he had entirely forgotten his native language. On his perfect recovery *he completely forgot his Welch again*, and recovered the English language."—*Op. cit.* p. 148.

346. If the following case, given by Dr. Abercrombie as having been related to him, be correctly recorded, the "traces" may be

registered under conditions in which the Mind *seems* altogeth[illegible] dormant :—

"A boy at the age of four suffered fracture of the skull, for which he underwent the operation of the trepan. He was at the time in a state of perfect stupor, and, after his recovery, retained no recollection either of the accident, or of the operation. At the age of fifteen, however, during the delirium of fever, he gave his mother an account of the operation, and the persons who were present at it, with a correct description of their dress, and other minute particulars. He had never been observed to allude to it before; and no means were known by which he could have acquired the circumstances which he mentioned."—*Op. cit.*, p. 149.*

347. It seems perfectly clear, then, that under what we cannot but term purely *physical* conditions, strictly *mental* phenomena present themselves. It is common to the whole series of cases, that the automatic action of the "Mechanism of Thought" does that which Volition is unable to effect. Whether it be the *toxic* condition of the blood, or the simple excitement of the cerebral circulation generally, or the special direction of blood to a particular part of the brain, it is beyond our present power to tell; but as *all* Brain-change is (like the action of any other mechanism) the expression of Force, the production of these *unusual* mental phenomena by the instrumentality of an unusual reaction between the blood and the brain-substance is no more difficult of comprehension, than that of those *ordinary* forms of Psychical activity, which we have seen reason to regard as the results of the translation (so to speak) of one form of Force into another (§ 42).

348. It must be freely admitted that we have at present no certain knowledge of the mode in which the recording process is effected :

* A variety of interesting cases, illustrating the general principles above stated, will be found in Dr. Abercrombie's little volume, and in Dr. Prichard's valuable Treatise on Insanity.

but looking to the considerations already adduced, as to the manner in which the Sensori-motor apparatus—the instrument of our *bodily* activity—appears to *grow to* the mode in which it is habitually exercised (§ 278), we seem justified in assuming that the same thing is true of the Cerebrum, which is the instrument of our *mental* activity; and the following may be suggested as a Physiological *rationale* for the phenomena under consideration:—The record of each of those states of consciousness, of the aggregate of which the acquirement of a Language is made up, must consist in some change in the nutrition of the brain; say, for example, the development of a certain group of nerve cells and nerve-fibres, constituting one connected system. The material particles constituting this system are continually changing; but, according to the laws of Nutrition (§ 276), the structure itself is kept up by *re*-position of *new* matter in the precise form of the *old*. So long as this structure remains in acting connection with other parts of the Brain habitually called into play, the *conscious* memory of the Language is retained; that is, the individual *wishing* to recall the word or phrase that expresses the idea present to his mind, can do so. But by disuse this becomes more and more difficult. Thus it happens to the Writer, as doubtless to many others, that if an unusually long interval elapses without his having occasion to speak French, he finds himself unable to call to mind French words and phrases, which, if spoken to him, or seen in writing, he at once understands; and yet, after being a week or two in France, and in the daily habit of speaking the language, he finds his ideas shaping themselves in French *in the first instance*, without the process of translation. The Physiologist would say that the nerve-tracks which disuse has rendered imperfect, have restored themselves by use; so that the part of the brain which has recorded the Language, has been brought back into ready connection with that which ministers to the current play of ordinary Thought. But a more prolonged disuse gradually produces such a disseverance, that no Volitional effort can bring about the recall of equivalents in a language once even more familiar than that of later years; and yet the Mechanism of the earlier thought is still preserved in working order, ready to be called into action by some unwonted stimulus.

349. The intimacy of the relation between the Psychical phenomena of Memory and the Physical condition of the Brain, is further shown by the effect of fatigue and of the impaired nutrition of old age in *weakening* the Memory; and of disease and injury of the brain in *impairing* it or *destroying* it.

350. Every one is conscious of the difference in the activity of the reproductive faculty on which Memory depends, according as his mind is fresh, or his head feels tired. The latter state, in which the automatic activity and the directing power of the Will are alike reduced, is clearly dependent, like the feeling of Muscular fatigue, on the deterioration either of the organ, or of the blood, or of both combined, which results from the prolonged exercise of it (§ 474); and it is especially in our inability to *recollect* something which we *wish* to call to mind, that the failure of power shows itself. An interval of repose completely restores the powers, obviously (to the mind of the Physiologist) by the renovation of the worn-out brain-tissue, and by the purification of the blood that has become charged with the products of its "waste."—So, *transient* lapses of Memory are often traceable to a *general* lowering of the circulation produced by exhaustion; the memory returning with the recovery of general power.

Thus Sir H. Holland tells us:—"I descended on the same day two very deep mines in the Hartz Mountains, remaining some hours under ground in each. While in the second mine, and exhausted both from fatigue and inanition, I felt the utter impossibility of talking longer with the German Inspector who accompanied me. Every German word and phrase deserted my recollection; and it was not until I had taken food and wine, and been some time at rest, that I regained them again."—*Op. cit.*, p. 160.

351. The impairment of the Memory in Old Age commonly shows itself in regard to *new* impressions; those of the earlier period of life not only remaining in full distinctness, but even, it would seem, *increasing* in vividness, from the fact that the Ego is not distracted from attending to them by the continual influx of impressions produced by passing events. The extraordinary persistence of early impressions, when the mind seems almost to have ceased to register new ones, is in remarkable accordance with the law of Nutrition already referred to (§ 282). It is when the Brain is *growing*, that a definite *direction* can be most strongly and persistently given to its structure. Thus the habits of thought come to be formed, and those nerve-tracks laid-down which (as the Physiologist believes) constitute the mechanism of association, by the time that the brain has reached it maturity; and the nutrition of the organ continues to keep up the same mechanism in accordance with the demands upon its activity, so long as it is being called into use. Further, during the entire period of vigorous manhood, the Brain, like the Muscles, may be taking-on some additional growth, either as a whole or in special parts; new tissue being developed and kept up by the nutritive process, in accordance with the modes of action to which the organ is trained. And in this manner a store of "impressions" or "traces" is accumulated, which may be brought within the sphere of consciousness, whenever the right suggesting-strings are touched. But as the nutritive activity diminishes, the "waste" becomes more active than the renovation; and it would seem that while (to use a commercial analogy) the "old-established houses" keep their ground, those later firms whose basis is less secure, are the first to crumble away,—the nutritive activity, which yet suffices to maintain the *original* structure, not being capable of keeping the subsequent additions to it in working order. This *earlier* degeneration of *later*-formed structures is a general fact perfectly familiar to the Physiologist.

352. The effects of Disease and Injury on the Memory are so marvellous and diverse, that only a very general indication of them can be here given. Cases are very common, in which the form of impairment just spoken of as characteristic of old age, shows itself to a yet greater extent; the Brain being so disordered by attacks of apoplexy or epilepsy (for example), as to be altogether incapable of registering any *new* impressions; so that the patient does not remember anything that passes from day to day, whilst the impressions of events which happened *long before* the commencement of his malady, recur with greater vividness than ever. On the other hand, the Memory of the long-since-past is sometimes entirely destroyed; whilst that of events which have happened subsequently to the malady is but little weakened. The Memory of *particular classes of ideas* is frequently destroyed; that (for example) of a certain Language or some other branch of knowlenge, or of the patient's domestic or social relations. Thus, a case was recorded by Dr. Beattie, of a gentleman, who, after a blow on the head, found that he had lost his knowledge of Greek, but did not appear to have suffered in any other way. A similar case has been recently communicated to the Writer, in which a lad, who lay for three days insensible in consequence of a severe blow on the head, found himself on recovering to have lost all the Music he had learned, though nothing else had been thus "knocked out of him.' Similar losses of acquired Languages have been noted as results of Fevers.—Dr. Abercrombie relates a curious case, on the authority of an eminent medical friend, in which a surgeon who suffered an injury of the head by a fall from his horse, on recovering from insensibility, gave minute directions in regard to his own treatment, but was found to have lost all remembrance of having a wife and children; and this did not return until the third day. (*Op. cit.* p. 156). One of the most curious examples of this *limited* loss of Memory occurred in the case of Sir Walter Scott, who, having produced one of his best works under the

pressure of severe illness (§ 124 *e*), was afterwards found to have entirely forgotten what he had thus constructed.

a. "The book (says James Ballantyne) was not only written but published, before Mr. Scott was able to rise from his bed; and he assured me, that when it was first put into his hands in a complete shape, *he did not recollect one single incident, character, or conversation it contained!* He did not desire me to understand, nor did I understand, that his illness had erased from his memory the *original incidents* of the story, with which he had been acquainted from his boyhood. These remained rooted where they had ever been; or, to speak more explicitly, he remembered the general facts of the existence of the father and mother, of the son and daughter, of the rival lovers, of the compulsory marriage, and the attack made by the bride upon the hapless bridegroom, with the general catastrophe of the whole. *All these things he recollected*, just as he did before he took to his bed; *but he literally recollected nothing else*,—not a single character woven by the romancer, not one of the many scenes and points of humour, nor *anything with which he was himself connected*, as the writer of the work."—(*Life of Walter Scott*, chap. xliv.)

b. A case has been lately mentioned to the Writer by his friend Dr. J. R. Reynolds, in which a Dissenting minister, apparently in perfectly sound health, went through an entire pulpit service on a certain Sunday morning with the most perfect consistency,—his choice of hymns and lessons, and his *extempore* prayer, being all related to the subject of his sermon. On the following Sunday morning, he went through the introductory part of the service in precisely the same manner,—giving out the same hymns, reading the same lessons, and directing his *extempore* prayer in the same channel. He then gave out the same text, and preached the very same sermon as he had done on the previous Sunday. When he came down from the pulpit, it was found that he had not the smallest remembrance of having gone through precisely the same service on the previous Sunday; and when he was assured of it, he felt considerable uneasiness lest his lapse of memory should indicate some impending attack of brain-disease. None such, however, supervened; and no *rationale* can be given of this curious occurrence, the subject of it not being liable to fits of "absence of mind," and not having had his thoughts engrossed at the time by any other special pre-occupation.

353. Sometimes, again, the Memory of *persons* is unimpaired, whilst that of *places* remains vigorous; so that *persons* are only recognized when seen in the accustomed localities.

This was remarkably shown in the case of a gentleman of considerable Scientific ability, with whom the Writer had been in close intimacy from boyhood, and who had been accustomed frequently to spend an evening at his house. When he had passed the age of seventy, but while retaining an unusual degree of bodily vigour, he was observed by his friends to be forgetful of circumstances which had happened not long previously, and occasionally to show a want of comprehension of any unusual words. For example, when the Writer happened to speak of the beautiful Comet then visible (1855), his friend said, "What do you mean?", and could only be made to understand by being shown the comet from a window. Again, though continually at the British Museum, the Royal Society, and the Geological Society, he would be unable to refer to either by name, but would speak of "that public place." He still continued his visits to his friends, and recognized them in their own homes, or in other places (as the Scientific Societies) where he had been accustomed to meet them; but the Writer, on meeting him at the house of one of the oldest friends of both, usually residing in London, but then staying at Brighton, found that he was not recognized; and the same want of recognition showed itself when the meeting took place out of doors. The want of memory of words then showed itself more conspicuously; one word being substituted for another, sometimes in a manner that showed the chain of association to be (as it were) bent or distorted, but sometimes without any recognizable relation. Thus on calling one day at the Writer's residence, and finding neither him nor Mrs. C. at home, he asked his son (then quite a lad) "how his *wife* was," meaning, of course, his *mother*. But about the same time, he told a friend that "he had had his umbrella washed," the meaning of which was gradually discovered to be, that he had had his hair cut. This impairment progressively increased to such a degree that he could with difficulty make himself understood, though he was generally able to recognise the meaning of what was said to him, and would assent if the right words were supplied. His general health continued vigorous for some time, but his Memory progressively failed; and at last he became obviously unfit to take care of himself. It was curious that after he had lost the ordinary power of expressing himself intelligibly, and even that of comprehending what was said to him, he would *swear* most tremendously when opposed in anything he tried to do, although he had not been accustomed to use such language in his previous life; and the same has been noticed in other cases of a like kind. His life was terminated by an attack of apoplexy; which confirmed the judgment that had been previously formed as to the progressive impairment of the nutrition of the Brain (§ 355).

354. The loss of the Memory of Words, exemplified in the preceding case, is a special disorder which not unfrequently presents itself: the patient understanding perfectly well what is said, but not being able to reply in any other terms than *yes* or *no*, or by affirmative or negative gestures; not from any paralysis of the muscles of articulation, but from incapability of expressing the ideas in language. To this condition, the term *Aphasia* has been recently applied. Sometimes the memory of words is impaired merely, so that the patient mistakes the proper terms. And in some instances there is an obvious association, though an irrelevant one, between the word used and the word that ought to have been used; thus the case of a Clergyman has been lately mentioned to the Writer, who continually confuses "father" and "son," "brother" and "sister," "gospel" and "epistle," or the like. But sometimes there is no recognizable relation, so that the patient speaks a most curious jargon. Again, the Memory of only *a particular class of words*, such as Nouns or Verbs, may be lost; or the patient may remember the letters of which a word is composed, and may be able to *spell* his wants, though he cannot speak the *word* itself; asking for *bread* (for example) by the separate letters b, r, e, a, d.—A very curious affection of the Memory is that in which the *sound* of spoken words does not convey any idea to the mind; yet the individual may recognize in a written or printed list of words, those which have been uttered by the speaker, the *sight* of them enabling

him to understand their meaning. Conversely, the sound of the word may be remembered, and the idea it conveys fully appreciated; but the visual memory of its written form may be altogether lost, although the component letters may be recognized.—For this class of phenomena, in which there is rather a severance of the Associative connections that have been formed between distinct states of consciousness, than an actual annihilation of the impression left by any of the latter, the term "Dislocation of Memory" has been proposed by Sir H. Holland; but, as he justly remarks, "no single term can express the various effects of accident, disease, or decay, upon this faculty, so strangely partial in their aspect, and so abrupt in the changes they undergo, that the attempt to classify them is almost as vain as the research into their cause." (*Chapters on Mental Physiology*, p. 146.)

355. It has been recently affirmed that this class of cases affords distinct evidence of the dependence of a particular affection of the Mind upon a particular disorder of the Brain; for a large number of cases of *Aphasia* have been collected, in which a degeneration of structure was found after death in a certain part of the anterior lobe of one side (usually the *left*) of the Cerebrum. This degeneration seems most commonly due to an impairment of Nutrition, consequent upon deficient supply of blood; the arterial trunk which supplies this part being plugged up either by a fibrinous clot* (constituting the condition termed "embolism"), or by a morbid deposit upon its lining membrane; the latter being the condition found in a case recently communicated to the Writer by a Medical friend, who was a near relative of the patient.—But it would certainly be premature to speak of this relation as an established

* The fibrinous clot is brought from the Heart, where it is produced as a consequence of valvular disease; and the acute observation of Dr. J. Hughlings Jackson, who first pointed out the frequent dependence of this brain-lesion upon "embolism" of the middle meningeal artery, has supplied the *rationale* of the frequent occurrence of Apoplexy in connection with Heart-disease, which had long since been recognised as a fact of observation.

fact; for the association between the Mental state and the Cerebral lesion has been shown to be by no means constant *; and there is, moreover, good reason to think that several states essentially different have been grouped together under the general designation Aphasia.

356. It is not a little remarkable that even in some of those cases in which the impairment of Memory has resulted from a fever or other disease, or from an accident, *the lost power should suddenly return.* Thus Dr. Rush, of Philadelphia, was acquainted with a person of considerable attainments, who, on recovering from a fever, was found to have lost all his acquired knowledge. When his health was restored, he began to apply himself to the Latin Grammar; and while, one day, making a strong effort to recollect a part of his lesson, *the whole of his lost impressions suddenly returned to his remembrance*, so that he found himself at once in possession of all his former acquirements. The like sudden restoration, after an equally sudden loss for a much longer period, is a remarkable feature of a very interesting case presently to be detailed (§ 360 *a*). And the Physiologist who accepts the doctrine of the original dependence of Memory on a physical registration, and the dependence of the reproductive power upon the activity of the Blood-circulation, will be disposed to look for the causation of such *limited* and *temporary* lapses, in modifications in the Circulation produced by *local* and *transient* alterations in the calibre of the arteries, under the influence of the Vaso-motor system of nerves (§ 113). And it seems a confirmation of this view, that in both these cases the recovery took place under Emotional excitement, which exerts a peculiar power over that portion of the Nervous system.

357. There is another class of familiar phenomena, which affords strong evidence of the dependence of the *recording process* upon Nutritive changes in the brain. Every one is aware that what is

* See especially Dr. Bateman's Treatise on Aphasia, London, 1870.

rapidly learned—that is, merely "committed to Memory,"—is very commonly forgotten as quickly, "one set of ideas driving out another." That thorough apprehension of what is learned, on the other hand, by which it is made (as it were) part of the Mental fabric, is a much slower process. The difference between the two is expressed by the colloquial term "cramming," as distinguished from "learning;" the analogy being obvious to the overloading the stomach with a mass of food too great to be digested and assimilated within a given time, so that a large part of it passes *out of* the body without having been applied to any good purpose *in* it. A part of this difference obviously consists in the formation of *Mental Associations* between the newly acquired knowledge and that previously possessed; so that the new ideas become linked on with the old by suggesting chains. Such is especially the case when we are applying ourselves to the study of any branch of knowledge, with the view of permanently mastering it; and here the element of *time* is found practically to be very important.

Thus it is recorded of Lord St. Leonards, that having (as Sir Edward Sugden) been asked by Sir T. F. Buxton what was the secret of his success, his answer was,—"I resolved, when beginning to read Law, to make everything I acquired *perfectly my own*, and never to go to a second thing till I had entirely accomplished the first. Many of my competitors read as much in a day as I read in a week; but at the end of twelve months, my knowledge was as fresh as on the day it was acquired, whilst theirs had glided away from their recollection."—(*Memoirs of Sir T. F. Buxton*, chap. xxiv.)

358. In this Assimilating process it is obvious that the new knowledge is (as it were) turned over and over in the Mind, and viewed in all its aspects; so that by coming to be not merely an *addition* to the old, but to *interpenetrate* it, the old can scarcely be brought into the "sphere of consciousness" without bringing the new with it. And from the considerations already adduced, it seems almost beyond doubt that the formation of this Asso-

ciative *nexus* expresses itself in the Physical structure of the Brain, so as to create a mechanism whereby it is perpetuated so long as the Nutrition of the organ is normally maintained.

359. Still more direct and cogent evidence of the dependence of Memory upon a *registering* process that consists in some Nutritive modification of the Brain-tissue, is afforded by the fact well known to Medical men, that when a person has sustained a severe injury to the head which has rendered him temporarily insensible, he generally finds himself, on recovering from his insensibility, unable to retrace the events which had *immediately* preceded, though his remembrance of what had gone before is not at all impaired.

The following example of this frequent occurrence has been communicated to the Writer by his friend the Rev. S. Hansard, whose remarkable reminiscence of a very early impression has been already noticed (§ 339 *b*). He was driving his wife and child in a phaeton, when the horse took fright and ran away; and all attempts to pull him in being unsuccessful, the phaeton was at last violently dashed against a wall, and Mr. H. was thrown out, sustaining a severe concussion of the brain. On recovering, he found that he had forgotten the *immediate* antecedents of the accident; the last thing he remembered being that he had met an acquaintance on the road about two miles from the scene of it. Of the efforts he had made, and the terror of his wife and child, he has not, to this day, any recollection whatever.

360. The same indication that *time* is needed for the effectual performance of the registration, may be drawn from another class of phenomena familiar to all. In what we call "learning by heart,"—which should be rather called learning by Sense, instead of by Mind,—we try to imprint on our memory a certain sequence of words, numbers, musical notes, or the like; the reproduction of these being mainly dependent upon the association of each *item* with that which follows it, so that the utterance of the former, or the picture of it in the mind's eye, *suggests* the next. We see this plainly enough, when children are set to learn a piece of

poetry of which their Minds do not take-in the meaning; for the *rhythm* affords here a great help to the suggestive action; and nothing is more common than to hear words or clauses—transferred, perhaps, from some other part of the poem—substituted for the right ones, though not only inappropriate, but absolutely absurd, in the lines as uttered. So, again, if the child is at fault, he does not think of the meaning of the sentence, and of what is wanted to complete it, but "tries back" over the preceding words, that their *sound* may suggest that of the word he has forgotten. This form of Memory sometimes endures through life, even in individuals of great acquirements; and where it exists in unusual strength, it seems rather to *impede* than to aid the formation of that *nexus* of associations, which makes the acquired knowledge a part of the Mind itself.

Thus it is stated by Dr. Abercrombie, that Dr. Leyden, who was distinguished for his extraordinary power of learning languages, could repeat correctly a long Act of Parliament, or any similar document, after having once read it. Being congratulated by a friend on his remarkable gift, he replied that instead of being an advantage, it was often a source of great inconvenience to him. This he explained by saying that when he wished to recollect a particular point in anything which he had read, he could only do it by repeating to himself the whole from the commencement, till he had reached the point he wished to recall.—*Op. cit.*, p. 101.

Here the "recording process" would seem to be very little higher in character than that by which the brain of the German servant-maid became impressed by the sounds of the Hebrew sentences recited by her master (§ 344 *d*). For it must have been the sequence of words, rather than the relations of the ideas conveyed by them, that constituted the chain of association. And we seem able to trace the Physiological working of this process, in the fact known to every school-boy who has to commit to memory fifty lines of Virgil, that if he can "say them to himself," even slowly and bunglingly, just before going to sleep, he will be able to recite them much more fluently in the morning. For we have here an obvious indication that the renovation of the brain-substance which takes place during sleep, going on without interruption by new impressions on the Sensorium, gives *time* for the *fixation* of the *last* impressions by nutritive change.—We have, moreover, a remarkable *converse* phenomenon in the rapid fading-away of a Dream, which, at the moment of waking, we can reproduce with extraordinary vividness; for the "trace" left by its details is soon obliterated by the new and stronger impressions made on our waking consciousness: so that, a few hours afterwards, we are often unable to revive more than the general outline of the dream, and perhaps not even that, unless we have told it to another when it was fresh in our minds, of which act a "trace" would be left.

361. There are two classes of persons who are professionally called upon for great *temporary* exercises of Memory, viz., Dramatic performers and Barristers. An Actor, when about to perform a new "part," not only commits it to memory, but "studies" it, so as to make it part of himself; and all really *great* actors identify themselves for the time with the characters they are performing (§ 463). When a "part" has once been thoroughly mastered, the performer is usually able to go through it, even after a long interval, with very little previous preparation. But an Actor is sometimes called upon to take a new "part" at a very short notice; he then simply learns it by heart, and speedily forgets it.

A case of this kind is cited by Dr. Abercrombie, as having been the experience of a distinguished Actor, on being called on to prepare himself in a long and difficult part, at a few hours' notice, in consequence of the illness of another performer. He acquired it in a very short time, and went through it with perfect accuracy; but forgot it to such a degree, immediately after the performance, that although he performed the character for several days in succession, he was obliged every day to prepare it anew, not having time to go through the process of "studying" it. When questioned respecting the mental process which he employed the first time he performed the part, he said that he entirely lost sight of the audience, and seemed to have nothing before him but the pages of the book from which he had learned it; and that, if anything had occurred to interrupt this illusion, he should have instantly stopped.—*Op. cit.*, p. 103.

362. In the case of Barristers, who are called upon to "get up" the "briefs" which are supplied to them, to master the facts, to apply to them the principles of Law, and to present them in the Court in the form which they deem most advantageous to the cause they have undertaken to plead, the very highest faculties of Mind are called into active exercise: but in consequence, it would seem, of the want of previous connection with the case (of which they know nothing but what is set down in their brief), and of the complete cessation of that connection so soon as the decision has been given, they very commonly "forget all about it," by the time that they have transferred their attention to their next brief.

A curious case of this kind was mentioned to the Writer a few years ago by an eminent Barrister (since elevated to the Judicial Bench), whose great scientific attainments led to his being frequently employed in Patent-cases. A heavy case of this kind being placed in his hands, he was reminded of having been engaged by the same parties in the same case, when it had been first brought to trial about a year previously. He had not the slightest remembrance of its having ever been before him; none of the particulars of it seemed familiar to him; and he was only convinced that he really *had* taken part in the previous trial, by finding the record of his engagement in his fee-book. Even when he came to "get up" the case again, no remembrance of his former attention to it came within his "sphere of consciousness."—(See also § 343 *b*).

363. It seems, then, to admit of question, whether *everything* that passes through our Minds thus leaves its impression on their

material instrument; and whether a somewhat too extensive generalization has not been erected on a rather limited basis. For the doctrine of the indelibility of Memory rests on the spontaneous revival, under circumstances indicative of some change in the physical condition of the Brain, of the long dormant "traces" left by such former impressions as are referable to one or other of the three following categories:—(1) States of Consciousness as to places, persons, languages, &c., which are *habitual* in early life, and which are, therefore, likely to have directed the *growth* of the Brain; (2) Modes of Thought in which the formation of Associations largely participates, and which are likely to have modified the course of its *maintenance* by Nutrition after the attainment of maturity; or (3) Single experiences of peculiar force and vividness, such as are likely to have left very decided "traces," although the circumstances of their formation were so unusual as to keep them out of ordinary Associational remembrance. Thus, in the remarkable case mentioned in § 346, it would seem that all the Mind the patient then had must have been concentrated upon the impressions made upon his Sensorium, which were thus indelibly branded, as it were, upon his Organism; but that these "traces," being soon covered up by those resulting from the new experiences of restored activity, remained outside the "sphere of consciousness," until revived by a Physical change, which reproduced the images of the objects that had left them.

364. This Reproduction, however, is not all that constitutes Memory; for there must be, in addition, a *recognition* of the reproduced state of Consciousness as one which has been formerly experienced; and this involves a distinct Mental state, which has been termed the "consciousness of agreement." Without this recognition, we should live in the present alone; for the reproduction of past states of Consciousness would affect us only like the succession of fantasies presented to us in the play of the

Imagination. We should only be conscious of them as *present* to us at the time of their recurrence, and should not in any way connect them with the past. Hence this Consciousness of Agreement between our present and our past Mental experiences, constitutes the basis of our feeling of *personal identity*; for, if it were entirely extinguished, there would be nothing to carry on that feeling of identity from one moment to another. I am satisfied that *I* am the person to whom such and such experiences happened yesterday, or a month, or a year, or twenty years ago; because I am not only conscious at this moment of the ideas which represent those experiences, but because I recognize them as the revived representations of my past experiences. But I may be told by others that things have happened to me in the past, of which I can call up no remembrance whatever, even when they mention circumstances likely to revive their traces by Association. And in *this* case, I cannot recognize my own identity with the subject of these experiences, save as I do so *indirectly* by reliance on the testimony of those who relate them. Sometimes, indeed, we come so completely to *realize* such forgotten experiences, by repeatedly *picturing* them to ourselves, that the ideas of them attain a force and vividness which equals or even exceeds that which the actual memory of them would afford. In like manner, when the Imagination has been exercised in a sustained and determinate manner,—as in the composition of a work of fiction,—its ideal creations may be reproduced with the force of actual experiences; and the sense of personal identity may be projected backwards (so to speak) into the characters which the Author has "evolved out of the depths of his own consciousness,"—as Dickens states to have been continually the case with himself. And something of the same kind has happened to most persons, however unimaginative they may be, in the reproduction of ideas which have previously only passed through the mind in Dreams; for almost every one has had occasion, at some time or other, to say "Did this really happen to me, or did I dream it?"—the past *mental* experience having been as complete in the one case as in the other.

a. A remarkable case of this kind has lately been related to the Writer.—A Lady of advanced age, who retains a remarkable degree of *general* Mental activity, continually dreams about passing events, and seems entirely unable to distinguish between her dreaming and her waking experiences, narrating the former with implicit belief in them, and giving directions based upon them, until corrected. Though at first impatient of such corrections, she now readily accepts them, having become quite aware of her infirmity.

365. Though we are accustomed to speak of Memory as if it consisted in an *exact* reproduction of past states of Consciousness, yet experience is continually showing us that this reproduction is very often *inexact*, through the modification which the "trace" has undergone in the interval. Sometimes the trace has been partially obliterated; and what remains may serve to give a very erroneous (because imperfect) view of the occurrence. And where it is one in which our own Feelings are interested, we are extremely apt to lose sight of what goes against them, so that the representation given by Memory is altogether one-sided. This is continually demonstrated by the entire dissimilarity of the accounts of the same occurrence or conversation, which shall be given by two or more parties concerned in it, even when the matter is fresh in their minds, and they are honestly desirous of telling the truth. And this diversity will usually become still more pronounced with the lapse of time: the trace becoming gradually but unconsciously modified by the habitual course of thought and feeling; so that when it is so acted on after a lengthened interval as to bring up a reminiscence of the original occurrence, that reminiscence really represents, *not* the actual occurrence, but the modified trace of it. And this is the source of an enormous number of "fallacies of testimony," which recent experiences of Mesmerism and

Spiritualism have brought into strong light. For the very prevalent disposition to believe in the marvellous is found so to change the form of the original record, without the least intention of doing so, that the most truthful narrator may come to believe implicitly in a version of an occurrence, which differs in some most essential point from the facts of the case as known to himself (or herself) at the time.—The following example of this change is given by Miss Cobbe, who has specially directed attention to these "Fallacies of Memory :—

"It happened once to the Writer to hear a most scrupulously conscientious friend narrate an incident of Table-turning, to which she appended an assurance that the table rapped when *nobody was within a yard of it.* The writer being confounded by this latter fact, the lady, though fully satisfied of the accuracy of her statement, promised to look at the note she had made ten years previously of the transaction. The note was examined, and was found to contain the distinct statement that the table rapped when *the hands of six persons rested on it!* The lady's memory as to all other points proved to be strictly correct; and in this point she had erred in entire good faith."—(*Hours of Work and Play*, p. 100.)

366. The old story of "the three black crows" is, in fact, continually being repeated; so that incidents in which there is really nothing wonderful, come to be magnified into most surprising and perplexing phenomena. If some sagacious inquirer, however, would perseveringly take up the clue, and carry a searching investigation backwards to the *fons et origo* of every such narrative, he would probably succeed as well as Dr. Noble did in the following case, which he has kindly communicated to the Writer for publication :—

"It was, I think, in 1845, that Mesmerism was exciting much interest in Manchester, as in many other parts of the country; and a great deal was made, by enthusiastic advocates, of Miss Martineau's adhesion to the cause. This lady had a maid, whom, in her record of mesmeric experiences, she called J. This person was said to be a wonderful clairvoyant; and besides what was published of her doings, unauthenticated representations of the same became the frequent topic in discussions on the subject. It was stated to me again and again, as to a sceptic, that Miss Martineau's J. could converse, when in her mesmeric state, in languages that she had never learnt, and of which she knew nothing when in her ordinary condition. Of course, I could not admit the verity of such an astonishing fact upon mere hearsay. But it was confidently averred that Lord Morpeth (the late Earl of Carlisle) had tested J. as to this power, and had found it real. It happened about this time, soon after I had heard these things, that being in Liverpool I met at dinner a brother-in-law of Miss Martineau; and him I questioned as to what he knew of the thaumaturgic doings of which I had heard. He told me that what I had heard was not quite accurate; but that the fact was, J. had replied in the vernacular to questions proposed by Lord Morpeth in the foreign tongue. This statement, I thought, though wonderful enough, diminished the marvellous character of the incident materially; for I reasoned that, probably, the nature of the queries might approximately be inferred from the intonation used in putting them. However, a few weeks later, Lord Morpeth had occasion to come to Manchester, presiding at an Athenæum Soirée; and the late Mr. Braid, during this visit, invited Lord Morpeth to come and witness his own particular experiments, suggestive of the true interpretation to be affixed to Mesmeric phenomena, so far as they were real. I was present during the exhibition, whereat Lord Morpeth assisted, actively and with great interest. I remembered the story about J., and took advantage of the opportunity to ask his Lordship what really happened. Lord Morpeth said, "No, it was not so; I certainly spoke to J. in a foreign language, and, by an unmeaning articulation of sound, she imitated my speech after a fashion; that was all!" And thus disappeared the last shred of this marvel.—I have told this anecdote so frequently since the period in question, that I am absolutely sure of the literal accuracy of my recollection in this matter."

367. This modification of the "traces" left by the original events or narrations, in accordance with the preconceptions of those through whom they are successively transmitted, has doubtless been, in all ages, a fertile source of those Religious Myths, which are accepted by some as representations of actual events, and repudiated by

others as mendacious inventions; whilst others again regard them, with more justice, as having grown up by a process of gradual accretion and modification, and as representing the Ideas current among the disciples of any particular Creed, between the time of the occurrences on which they are based, and that of their full development as a body of doctrine. Modern criticism has done no greater service, than in showing that disbelief in the Myths themselves may be perfectly consistent with a full recognition of the honesty and good intentions of those with whom they originated, their memory alone being in fault.

368. In that remarkable abnormality known as *Double Consciousness* (§ 489), there is a *break*, more or less complete, between the two states, A and B, as regards the conscious Memory of the occurrences of each; for the experiences of state A constitute one phase of existence, and the experiences of state B an entirely distinct phase. Still, although the memory of actual occurrences may be wanting, there is often recognizable a persistence of *habitual modes of thought or feeling,* such as in their aggregate constitute the Character. Thus in a very interesting case elsewhere cited (§ 448 *a*) as having been shown to the Writer by Mr. Braid, the subject of it, who was a firm teetotaller, having been repeatedly assured by Mr. Braid that he was drunk, and his own feelings of unsteadiness (induced by Muscular suggestion) falling in with this assurance, it was most amusing to witness the conflict in his mind between the idea thus *forced* into it, and the idea derived from his habitual practice that he "*could* not have had any gin and water." And so in other cases (§§ 453, 490) in which considerable excitement of feeling had been induced during the abnormal state, the "trace" of it obviously lasted into the ordinary state, without any *ideational* remembrance, either of the excitement itself, or of the occurrences that had called it forth.—Hence, then, we may conclude, that if it were possible completely to obliterate all memory of the past, so that the conscious life of the individual

would begin altogether *de novo*, there would be this degree of connection between the new and the old life,—that the *modes* of thought and feeling established by Habit would tend to reproduce themselves; in so far, that is, as the similarity of external conditions should tend to call them into corresponding action. Considering, however, the extraordinary complexity of the associations which make up that record of past experiences whereon nearly all our Mental operations are based, it could not be expected that the new experiences should act on the mechanism altogether in the same manner as they would do if the old could be recalled. And it will suffice to confirm the view here advocated, if it can be shown that there are *definite indications* of such action. Such, the Writer believes, may be clearly recognized in the following case; which is here cited at length as probably being the most remarkable instance of the *temporary extinction of all memory of the past*, without the stupor which would prevent the reception of new impressions, that has ever been recorded by an intelligent observer.*

a. The subject of this case was a young woman of robust constitution and good health, who accidentally fell into a river and was nearly drowned. She remained insensible for six hours after the immersion; but recovered so far as to be able to give some account of the accident and of her subsequent feelings, though she continued far from well. Ten days subsequently, however, she was seized with a fit of complete stupor, which lasted for four hours; at the end of which time she opened her eyes, but did not seem to recognize any of her friends around her: and she appeared to be utterly deprived of the senses of hearing, taste, and smell, as well as of the power of speech. Her Mental faculties seemed to be entirely suspended; her only medium of communication with the external world being through the senses of Sight and Touch, neither of which appeared to arouse *ideas* in her mind, though respondent *movements* of various kinds were excited through them. Her vision at short distances was quick; and so great was the exaltation of the general sensibility upon the surface of the

* See Mr. Dunn's Narrative in the *Lancet*, November 15 and 29, 1845.

body, that the slightest touch would startle her; still, unless she was touched, or an object or a person was so placed that she could not help seeing the one or the other, she appeared to be quite lost to everything that was passing around her. She had no notion that she was at home, not the least knowledge of anything about her; she did not even know her own mother, who attended upon her with the most unwearied assiduity and kindness. Wherever she was placed, there she remained during the day. . . . . Her appetite was good; but having neither taste nor smell, she ate alike indifferently whatever she was fed with, and took nauseous medicines as readily as delicious viands. All the automatic movements unconnected with sensation, of which the Spinal Cord is the instrument, seemed to go on without interference; as did also those dependent upon the sensations of sight and touch; whilst the functions of the other Sensory ganglia, together with those of the Cerebral hemispheres, appeared to be in complete abeyance. The analysis of the facts stated regarding her ingestion of food seems to make this clear. She swallowed food when it was put into her mouth; this was a purely automatic action, the reception by the lips being excited by tactile sensation, whilst the act of deglutition, when the food has been carried within reach of the pharyngeal muscles, is excited without the necessary concurrence of sensation. But she made no spontaneous effort to feed herself with the spoon, showing that she had not even that simple idea of helping herself, which infants so early acquire; though after her mother had conveyed the spoon a few times to her mouth, so as to renew the association between the muscular action and the sensorial stimulus, the patient continued the operation. It appears, however, to have been necessary to repeat this lesson on every occasion; showing the complete absence of memory for any idea, even one so simple and so immediately connected with the supply of the bodily wants. The difference between an *instinct* and a *desire* or *propensity* (§§ 57, 261) is here most strikingly manifested. This patient had an *instinctive* tendency to ingest food, as is shown by her performance of the actions already alluded to; but these actions required the stimulus of the present sensation, and do not seem to have been connected with any notion of the character of the object *as food;* at any rate, there was no manifestation of the existence of any such notion or idea, for she displayed no *desire for food* or drink in the absence of the objects, even when she must have been conscious of the uneasy sensations of hunger and thirst.—The very limited nature of her faculties, and the *automatic* life she was leading, appear further evident from the following particulars. One of her first acts on recovering from the fit, had been to busy herself in picking the bed-clothes; and as soon as she was able to sit up and be dressed, she continued the habit by incessantly picking some portion of her dress. She seemed to want an occupation for her fingers, and accordingly part of an old straw bonnet was given to her, which she pulled into pieces of great minuteness: she was afterwards bountifully supplied with roses; she picked off the leaves, and then tore them into the smallest particles imaginable. A few days subsequently, she began forming upon the table, out of these minute particles, rude figures of roses and other common garden-flowers; she had never received any instructions in drawing. Roses not being so plentiful in London, waste paper and a pair of scissors were put into her hands; and for some days she found an occupation in cutting the paper into shreds; after a time these cuttings assumed rude figures and shapes, and more particularly the shapes used in patchwork. At length she was supplied with proper materials for patchwork; and after some initiatory instruction, she took to her needle and to this employment in good earnest. She now laboured incessantly at patchwork from morning till night, and on Sundays and week-days, for she knew no difference of days; nor could she be made to comprehend the difference. She had no remembrance from day to day of what she had been doing on the previous day, and so every morning commenced *de novo.* Whatever she began, that she continued to work at while daylight lasted; manifesting no uneasiness for anything to eat or drink, taking not the slightest heed of anything which was going on around her, but intent only on her patchwork. She gradually began, like a child, to register ideas and acquire experience. This was first shown in connexion with her manual occupation. From patchwork, after having exhausted all the materials within her reach, she was led to the higher art of worsted-work, by which her attention was soon engrossed as constantly as it had before been by her humbler employment. She was delighted with the colours and the flowers upon the patterns that were brought to her, and seemed to derive special enjoyment from the harmony of colours; nor did she conceal her want of respect towards any specimen of work that was placed before her, but immediately threw it aside if the arrangement displeased her. She still had no recollection from day to day of what she had done, and every morning began something new, unless her unfinished work was placed before her; and after imitating the patterns of others, she began devising some of her own.

The first *ideas* derived from her former experience, that seemed to be awakened within her, were connected with two subjects which had naturally made a strong impression upon her; namely, her fall into the river, and a love-affair. It will be obvious that her pleasure in the symmetrical arrangement of patterns, the harmony of colours, &c., was at first simply *sensorial;* but she gradually took an interest in looking at pictures or prints, more especially of flowers, trees, and animals. When, however, she was shown a landscape in which there was a river, or the view of a troubled sea, she became intensely excited and violently agitated; and one of her fits of spasmodic rigidity and insensibility immediately followed. If the picture were removed before the paroxysm had subsided, she manifested no recollection of what had taken place; but so great was her feeling of dread or fright associated with water, that the mere sight of it in motion, its mere running from one vessel to another, made her shudder and tremble; and in the act of washing her hands, they were merely placed in water. From this it may be inferred that simple *ideas* were now being formed; for whilst the actual sight or contact of moving water excited them by the direct Sensorial channel, the sight of a picture containing a river or water in movement could only do so by giving rise to the *notion* of water.—From an early stage of her illness, she had derived obvious pleasure from the proximity of a young man to whom she had been attached; he was evidently an object of interest when nothing else would rouse her; and nothing seemed to give her so much pleasure as his presence. He came regularly every evening to see her, and she as regularly looked for his coming. At a time when she did not remember from one hour to another what she was doing, she would look anxiously for the opening of the door about the time he was accustomed to pay her a visit; and if he came not, she was fidgetty and fretful throughout the evening. When by her removal into the country she lost sight of him for some time, she became unhappy and irritable, manifested no pleasure in anything, and suffered very frequently from fits of spasmodic rigidity and insensibility. When, on the other hand, he remained constantly near her, she improved in bodily health, early associations were gradually awakened, and her Intellectual powers and memory of words progressively returned.—We here see very clearly the composite nature of the Emotion of Affection. At first, there was simple pleasure in the presence of her lover, excited by the gratification which the impress of former associations had connected with the *sensation*. Afterwards, however, it was evident that the pleasure became connected with the *idea;* she *thought* of him when absent, expected his return (even showing a power of measuring time, when she had no memory for anything else), and manifested discomfort if he did not make his appearance. Here we see the true *Emotion*, namely, the association of pleasure with the *idea;* and the manner in which the *desire* would spring out of it. The desire, in her then condition, would be inoperative in causing voluntary movement for its gratification; simply because there was no Intellect for it to act upon.

Her Mental powers, however, were gradually returning. She took greater heed of the objects by which she was surrounded; and on one occasion, seeing her mother in a state of excessive agitation and grief, she became excited herself, and in the emotional excitement of the moment suddenly ejaculated, with some hesitation, "What's the matter?" From this time she began to articulate a few words; but she neither called persons nor things by their right names. The pronoun "this" was her favourite word; and it was applied alike to every individual object, animate and inanimate. The first objects which she called by their right names were wild flowers, for which she had shown quite a passion when a child; and it is remarkable, that her interest in these and her recollection of their names should have manifested itself, at a time when she exhibited not the least recollection of the "old familiar friends and places" of her childhood. As her Intellect gradually expanded, and her *ideas* became more numerous and definite, they manifested themselves chiefly in the form of *emotions;* that is, the chief indications of them were through the signs of Emotional excitement. These last were frequently exhibited in the attacks of insensibility and spasmodic rigidity, which came on at the slightest alarm. It is worth remarking that similar attacks, throughout this period, were apt to recur three or four times a day, when her eyes had been long directed intently upon her work; which affords another proof how closely the Emotional cause of them must have been akin to the influence of Sensory impressions, the effects of the two being precisely the same.

The mode of recovery of this patient was quite as remarkable as anything in her history. Her health and bodily strength seemed completely re-established, her vocabulary was being extended, and her mental capacity was improving; when she became aware that her lover was paying attention to another woman. This idea immediately and very naturally excited the Emotion of jealousy; which, if we analyse it, will appear to be nothing else than a painful *feeling* connected with the *idea* of the faithlessness of the object beloved. On one occasion this feeling was so strongly excited, that she fell down in a fit of insensibility, which resembled her first attack in duration and severity. This, however, proved sanatory. When the insensibility passed off, she was no longer spell-bound. The veil of oblivion was withdrawn; and, as if awakening from a sleep of twelve months' duration, she found herself surrounded by her grandfather, grandmother, and their familiar friends and acquaintances, in the old house at Shoreham. *She awoke in the possession of her natural faculties and former knowledge; but without the slightest remembrance of anything which had taken place in the year's interval, from the invasion of the first fit up to the present time.* She spoke, but she heard not; she was still deaf, but being able to read and write as formerly, she was no longer cut off from communication with others. From this time she rapidly improved, but for some time continued deaf. She soon perfectly understood by the motion of the lips what her mother said; they conversed with facility and quickness together, but she did not understand the language of the lips of a stranger. She was completely unaware of the change in her lover's affections, which had taken place in her state of "second consciousness;" and a painful explanation was necessary. This, however, she bore very well; and she has since recovered her previous bodily and mental health.

369. It is clear from what has preceded, that Memory is essentially an *automatic* form of Mental activity. By far the larger part of our

Psychical operations depend on the mechanism by which past states of consciousness *spontaneously* reproduce themselves: and while the Metaphysician accounts for this reproduction on the principle of "association of ideas," the Physiologist holds that in the formation of such associations, certain modifications took place in the organization of the Brain, which determine its mode of responding to subsequent suggestions; so that, under the stimulus of new impressions either from without or from within, the long-dormant "traces" of former mental states are caused to reproduce themselves as Ideas and Feelings.—But while this faculty is *essentially* automatic, it is one which is peculiarly capable of being guided and disciplined by the Will; and as there is no part of our Mental action, in which what the Will *can* and what it *cannot* do is more clearly distinguishable, it will be worth while to dwell somewhat particularly on the subject.

370. The act of *Recollection* is the Volitional exercise of the Automatic power of "reproduction," which we put in practice whenever we *try to remember* something that does not spontaneously present itself when wanted; our knowledge of *what* is wanted, which directs the action of the Will (§ 324), extending only to the conviction that we have *once* known it, and that the desiderated Idea would be found hid away somewhere in the recesses of our minds, if we could only tell in what direction to look for it. This conviction may be derived from the very circumstances of the case:—as when we meet a man whose *face* is perfectly familiar to us, and with whom we feel sure that we have been in personal communication on some previous occasion, but whose *name* we forget; or when, in speaking a foreign language, the word or phrase that would express our thought does not occur to us, although we know that we *must* have not only have once *learned* it, but also *retained* it in our minds. Or it may be impressed on us by the assurance of others that a certain communication has been made to us, of which we have entirely lost the remembrance.

371. In the first place, it may be positively affirmed that we cannot call up any Idea by a direct exertion of Volition; for it is as necessary a condition of the operation of the Will on the Brain, as it is of its action on the Muscles, that we should have before our consciousness the *idea of what is willed;* and if we have got this already, it cannot be what we *want.* How, then, do we "call to mind" an idea, of which all that we *know* is, that it represents a *something* which has at some former time been an object of consciousness?

372. The process really consists in the fixation of the Attention upon one or more of the ideas *already present to the mind,* which may directly recall, by suggestion, that which is desiderated; the very act of thus *attending to* a particular idea serving not only to intensify the idea itself, but also to strengthen the associations by which it is connected with others. There are certain ideas so familiar to us, that they seem necessarily to recur upon the slightest prompting of suggestion; yet even with regard to these, the volitional Recollection at any particular time involves the process just described. Thus if a man be asked his own name, he usually finds no difficulty in giving the proper answer; because it only requires that his attention should be directed to the idea involved in the words "my name," to suggest the words of which that name may consist. But if the individual should be in that state of "absence of mind," which really consists in the fixation of the attention upon some *internal* train of thought (§ 445), he may not be able on a sudden to transfer his attention to the new idea that is forced upon his consciousness *ab externo;* and may thus hesitate and bungle, before he is able to answer the question with positiveness.—So, again, it sometimes happens in old age, that men fail to recollect their own names, or the names of persons most familiar to them, in consequence of the weakening of the bond of direct association; and they then only recall it by the operation to be presently described. And in those abnormal states

of mind, in which the power of Volitionally directing the current of thought is for a time suspended, the individual cannot recall either his own name or any other most familiar idea, if "possessed" with the conviction that such recollection is impossible (§ 462).

373. But supposing that—the Mind being in full possession of its ordinary powers—the desiderated idea does not at once recur suggestively, on the direction of the attention to some idea already present to our consciousness: we then apply the same process to other ideas which successively come before us, selecting those which we recognize as most likely to suggest that which we require, and following out one train of thought after another, in the directions which we deem most suitable; until we either succeed in finding the idea of which we are in search, or give-up the pursuit as hopeless. Thus a man who is making up his accounts, and finds that he has expended a sum in a mode which he cannot recollect, sets himself to consider what business he has done, where he has recently been, what shops he may have entered, and so on. Or when a man meets another whom he recognizes as an acquaintance without remembering his name, he runs over a number of names (one being suggested by another, when the attention is directed to them), in hopes that some one of these may prove to be that which will furnish the clue to the one he has forgotten; or he thinks of the place in which he may have previously seen him, this being recalled by fixing the attention on the association suggested by the sight of his face and figure, or by the sound of his voice, or by his personality altogether; or he endeavours to retrace the time which has elapsed since he last met with him, the persons amongst whom he then was, or the actions in which he was engaged; that some one or other of these various associations may suggest the desiderated name. Or, when a man tries to retrace some "train of thought" which has formerly passed through his mind, but of which he only remembers that the *subject* of it had been before

him, he may often recover it by following it out (as it were) from the original starting-point; when the whole, with its conclusion, will often flash into the mind at once.

Thus, the Writer well recollects that, when going to register the birth of one of his own children, he found, when approaching the Office, that he had entirely forgotten the intended name, which had been decided on after a considerable amount of domestic discussion; and only brought it to his remembrance by "trying back" over the reasons which had determined the one finally selected.

374. Nothing can more clearly prove the *essentially automatic* nature of the process, than the fact familiar to every one, that when we have been striving by all the means at our command to recover a lost idea—as, for example, to remember where we have put away an important paper—and have abandoned the effort as hopeless, it "comes into our heads" some time afterwards, under circumstances which seem to justify the conclusion that we have started, by our volitional effort, a train of Cerebral mechanism, which *unconsciously* evolves a result that is then brought to our consciousness by being transmitted to the Sensorium. This subject, however, will be better discussed hereafter, in connection with other phenomena of "Unconscious Cerebration" (§§ 419—423). —The *automatic* action of Memory, as contrasted with its *volitional* exercise, is further shown by its activity in the states of Dreaming, Somnambulism, Intoxication, Delirium, and Insanity; in which the directing power of the Will is suspended, while the Automatic power of the Brain is in full play. (See Chaps. XV.—XVII.)

375. But while the faculty of Memory immediately depends upon a mechanism over the working of which the Will has only an indirect control, the *culture* and *discipline* by which that mechanism is shaped and directed is essentially Volitional. And since all acquirement of knowledge depends, in the first place, on our *recording* power, and, in the second, on our power of *finding*

what has been stored-away (so to speak) in our "Record-office," the culture of an *exact* and *ready* Memory is one of the most important parts of Intellectual Education.*—Now the recording power mainly depends upon the degree of Attention we give, whether automatically or volitionally (§ 118), to the idea to be remembered; and this will depend, for the most part, upon the degree of attraction it has for us. Thus a person who is fond of any particular study, will generally have a better memory for the ideas which come before him in the pursuit of it, than for those belonging to any other subject; whilst, on the other hand, however determinately a student may labour to acquire some branch of knowledge for which he has no natural aptitude, he finds that it is not only much more difficult to keep his Attention fixed upon it, but that a more close and prolonged fixation is required to imprint it on his Memory.—The reproducing power, again, altogether depends upon the nature of the Associations by which the new idea has been linked-on to other ideas which have been previously recorded, and which enter into our habitual current of thought. Some associations are *local* or *accidental*, forming themselves, instead of being formed at any bidding of our own; over these we have very little control. But those which are most useful to us in the acquirement of Knowledge, and over the formation of which we have the most power, may be distinguished as *rational;* being based on the fundamental relations of the ideas themselves, the perception of which gives to the new idea a definite place in our fabric of thought (§ 321). This is the kind of Memory which it is most desirable to cultivate to volitional attention; since it tends to bring-up the ideas we have previously acquired, whenever we have special occasion to reproduce them. A merely *verbal* memory can scarcely be said to give us *knowledge;* it merely supplies us with the symbols by which knowledge can be acquired (§ 198).

* This subject is treated with great practical ability in Dr. Abercrombie's "Inquiries concerning the Intellectual Powers."

CHAPTER XI.

OF COMMON SENSE.

376. The term "Common Sense" has been used in a vast variety of acceptations, of which a most learned collection will be found in Sir William Hamilton's supplemental note to Dr. Reid's Essay; but it will be convenient here to use that of Dr. Reid himself, who says that the office of Common Sense, or the "first degree of reason," is to "judge of things self-evident," as contrasted with the office of "the second degree of reason" (or Ratiocination), which is "to draw conclusions that are not self-evident from those that are."—The distinction between "Common Sense," and "Ratiocination" or the "discursive power," is regarded by Sir William Hamilton as equivalent to that which the Greek philosophers meant to indicate by the terms νοῦς and διάνοια; and our colloquial use of the former, as corresponding to that cultivated Common Sense which is often distinguished as "good sense," is thereby justified.—There are, however, two principal forms of this capacity, which it is desirable clearly to distinguish.

377. The *first* is what the *philosopher* means by Common Sense, when he attributes to it the formation of those original convictions or ultimate beliefs, which cannot be resolved into simpler elements, and which are accepted by every normally-constituted human being as *direct cognitions* of his own mental states (§ 199). It might, indeed, be maintained that this "necessary" acceptance of propositions which only need to be intelligibly stated to command unhesitating and universal assent, cannot be rightly termed an act of *judgment*. But just as Sense-perceptions, which are *intuitive* in the lower Animals, have to be *acquired* in Man by a process of

self-education in the earliest stages, in which acts of judgment are continually called-for (§§ 160—184), so may we regard the autocratic deliverances of the *universal* Common Sense of Mankind as really having, in the first instance, the characters of true judgments, each expressing the *general resultant* of uniform experience,—which may be partly that of the Individual, and partly that of the Race embodied in the Constitution of each member of it (§§ 201, 202).

378. The *second* or *popular* acceptation of the term Common Sense, on the other hand, is that of an attribute which judges of things whose self-evidence is *not* equally apparent to every individual, but presents itself to different individuals in very different degrees, according in part to the original constitution of each, and in part to the range of his experience and the degree in which he has profited by it. This is the form of Common Sense by which we are mainly guided in the ordinary affairs of life: but inasmuch as we no longer find its deliverances in uniform accordance, but encounter continual divergences of judgment as to what things *are* "self-evident,"—some being so to A whilst they are not so to B, and others being self-evident to B which are not so to A,—it cannot be trusted as an autocratic or infallible authority. And yet, as Dr. Reid truly says, "disputes very often terminate in an appeal to common sense;" this being especially the case, when to doubt its judgment would be ridiculous.

379. If the view here taken be correct, these two forms—which may be designated respectively as *elementary* and as *ordinary* Common Sense—have fundamentally the same basis; and we may further connect with them as having a similar *genesis*, those *special* forms of Common Sense, which are the attribute of such as have applied themselves in a Scientific spirit to any particular course of inquiry, —things coming to be perfectly "self-evident" to men of such special culture, which ordinary men, or men whose special culture has lain in a different direction, do not apprehend as such.

380. The judgment of Common Sense as to any "self-evident" truth, may be defined as *the immediate or instinctive response* that is given (in Psychological language) by the automatic action of the mind, or (in Physiological language), by the reflex action of the brain, to any question which can be answered by such a direct appeal. The nature and value of that response will depend upon the *acquired condition* of the mind, or of the brain, at the time it is given; that condition being the *general resultant* of the whole Psychical activity of the individual. The particular form of that activity is determined, as we have seen, in the *first* place, by his original constitution; *secondly*, by the influences which have been early brought to bear upon it from without; and *thirdly*, by his own power of self-direction. And it may be said that while the *elementary* form of Common Sense depends mainly upon the first of these factors, its *ordinary* form chiefly arises out of the first and second, and its *special* forms almost exclusively out of the third;—the response being given, in each case, by a nervous mechanism, in the organization of which the generalized results of the past experiences of consciousness (whether of the race or of the individual) have become embodied.—This doctrine may be made more intelligible to those who have not been accustomed to look at questions of this kind from the Physiological point of view, by reverting to what has been shown as to the Sense-perceptions and the Instinctive movements dependent upon them (Chap. V.); the capacity for each of these forms of activity, like that for "judging of things self-evident," being *acquired* by the action of *experience* on the original constitution.

381. The parallel between the Cerebral action which furnishes the mechanism of thought now under consideration, and the action of the Sensori-motor apparatus which furnishes the mechanism of sense and motion, is extremely close. We have seen that there are certain Sense-perceptions, which, although *not* absolutely intuitive, very early come to possess—in every normally

constituted Human being—the immediateness and perfection of those corresponding perceptions which *are* intuitive in the lower animals (§ 161); and that with these are associated certain respondent motions, which, though acquired by practice in the first instance, ultimately come to be performed as by a "second nature." Certain of these motions, such as *walking-erect*, are *universally* acquired; and thus obviously come to be the expressions of the *original* endowments of the mechanism, trained by an *experience* very similar in the uniformity of its character to that which educates the elementary form of Common Sense. For it must be clear to any one who compares the erect progression of a child who has just learned to walk, with that of a "dancing dog" or even of a chimpanzee, that while experience makes its acquirement possible in each case, only an organism which is at the same time structurally adapted for erect progression, and possessed of a special *co-ordinating* faculty, can turn such experience to full account (§ 192). The balancing the body in the erect position at starting, the maintenance of that balance by a new adjustment of the centre of gravity as the base of support is shifted from side to side and from behind forwards, and the alternate lifting and advance of the legs, involve the harmonious co-operation of almost all the muscles in the body. Although this co-operation is brought about in the first instance by the purposive direction of our efforts towards a given end, under the guidance of our visual and muscular sensations, yet when we have once learned to walk erect, we find ourselves able to maintain our balance without any exertion of which we are conscious; all that is necessary for the performance of this movement being that a certain *stimulus* (volitional, or some other) shall call the mechanism into activity.—But further, we have seen that *special* powers of Sense-perception can be acquired by the habitual direction of the attention to particular classes of objects; and that special movements come to be the *secondarily automatic* expressions of them (§ 193). How nearly related these are to the preceding, we may assure ourselves by attending to the process by which an adult learns to walk on a narrow base, such as a rope or the edge of a plank. For the co-ordinating action has here to be gone through afresh under altered and more special conditions, so as to give a greater development to the balancing power; yet when this has been fully acquired, it is exerted automatically with such an immediateness and perfection, that a Blondin can cross Niagara on his rope with no more danger of falling into the torrent beneath, than any ordinary man would experience if walking without side-rails along the broad platform of the suspension-bridge which spans it. Now since in those cases in which Man *acquires* powers that are *original* or *intuitive* in the lower animals, there is the strongest reason for believing that a mechanism forms itself in *him* which is equivalent to that congenitally possessed by *them*, we seem fully justified in the belief that in those more special forms of activity which are the result of prolonged "training," the Sensori-motor apparatus *grows-to* the mode in which it is habitually exercised, so as to become fit for the immediate execution of the mandate it receives (§ 194): it being often found to act not only without intelligent direction, but without any consciousness of exertion, in immediate response to some particular kind of stimulus,—just as an Automaton that executes one motion when a certain spring is touched, will execute a very different one when set going in some other way.

382. There is strong analogical ground, then, for the belief that the higher part of the Nervous mechanism which is concerned in Psychical action, will follow the same law; embodying the generalized result of its experiences, so as to become able to evolve, by a direct response, a result of which the attainment originally required the intervention of the conscious mind at several intermediate stages of the process. What there is strong ground for believing in regard to the *perceptional* consciousness, may fairly be extended to the *ideational*, which is so intimately related to it,

that it may be said to be only a higher development of the same form of Psychical activity. And thus our Intellectual conviction of the existence of the external world (for example) would be derived from the effect produced upon our *original* constitution by the *automatic generalization of a multitude of separate experiences;* the *resultant* of this generalization having probably been embodied in the Nervous mechanism, long before the Intelligence is sufficiently developed to cognosce the *idea* which mentally represents it. The conviction, however, that those other "first truths" of a purely Intellectual character (§ 199), the aggregate of which constitutes the *elementary* Common Sense of the Philosopher, are not only true within the range of actual experience, but *must be universally true,* is one which requires not only a more advanced stage of Intellectual development for its formation, but a Mental fabric specially prepared for its reception. And the unhesitating adoption of any proposition as "self-evident," which thus distinctly transcends the experience of the Individual, implies a congenital tendency to that mode of thought, which belongs to the constitution of the Race (§ 201).

383. The like view may be extended to that acquired aptitude "for judging of things self-evident," here designated as *ordinary* Common Sense; the deliverances of which may be regarded as based on the aggregate of our past experiences, which have ranged themselves in the unconscious depths of our Intellectual nature by a process of automatic co-ordination, and have become embodied in our Cerebral organization. We often find it strongly manifested by persons of very limited acquirements, who are said to have a "fund of native good sense." On the other hand, we often meet with a singular want of it in persons of great learning, whose judgments about things that are "self-evident" to men of ordinary capacity are obviously untrustworthy. And if we examine into the nature of this difference, we shall find it to lie partly in the original Constitution of the individual, and partly in the range of

the *unconscious co-ordinating action*, which in the former case brings the whole experience to bear upon the question, whilst the decisions of the latter are based upon a limited, and therefore one-sided, view of it,—the defect of judgment being due either to an original want of the co-ordinating power, or to disuse of the exercise of it through the limitation of the attention to special fields of study. It may often be noticed that Children display a power of bringing "common sense" to bear upon the ordinary affairs of life, which seems much beyond that of their elders; and yet a very sensible child will often grow into a much less sensible man. Now the reason of this seems to be, that the Child perceives the application of "self-evident" considerations to the case at issue, without being embarrassed by a number of other considerations (perhaps of a trivial or conventional nature) which distract the attention and unduly influence the judgment of the adult. And the deliverances of a child's "common sense" thus often resemble those of the old "Court Fools" or "Jesters," whose function seems to have been to speak out "home truths" which timid courtiers would not venture to utter. Moreover, as has been well remarked, "it is quite possible for minds of limited power to manage a small range of experience much better than a large, to get confused (as it were) with resources on too great a scale, and therefore to show far more Common Sense within the comparatively limited field of childish experience, than in the greater world of society or public life. This is probably the explanation of a thing often seen,—how very sagacious people instinctively shrink from a field which their tact tells them is too large for them to manage, and keep to one where they are really supreme. (*Spectator*, Feb. 3, 1872.)

384. Now, in so far as our *conscious* Mental activity is under the direction of our Will, we can improve this form of Common Sense, as to both its range and the trustworthiness of its judgments, by appropriate training. Such training, as regards the purely Intellectual aspect of Common Sense, will consist in the determinate

culture of the habit of honestly seeking for Truth,—dismissing prejudice, setting aside self-interest, searching out all that can be urged on each side of the question at issue, endeavouring to assign to every fact and argument its real value, and then weighing the two aggregates against each other with judicial impartiality. For in proportion to the steadiness with which this course is *volitionally* pursued, must be its effectiveness in shaping the mechanism whose *automatic* action constitutes the "unconscious thinking," of which the results express themselves in our Common-Sense judgments.

Such was eminently the habit of mind of Joseph Hume; a man whom it was the fashion to abuse and ridicule, simply because his honest and consistent advocacy of great principles now universally accepted, placed him in advance of his time; but who in private life, as the Writer has been informed by a member of his family, was so noted for the excellence of his judgment, that he was continually resorted to by his friends for advice. This was readily and explicitly given, and was almost invariably justified by the event; but he could seldom assign reasons for his conclusions. All he would say was, "Such is my opinion, but I cannot tell you how I have arrived at it." And thus his judgments were obviously the deliverances of his originally strong "Common Sense," improved by the discipline of the determinate and systematic direction of his conscious thinking to the attainment of Truth, the reaction of which on his automatic mechanism imparted to its operations the like tendency.

385. The *ordinary* Common Sense of mankind, disciplined and enlarged by appropriate culture, becomes one of the most valuable instruments of Scientific inquiry; affording in many instances the best, and sometimes the only, basis for a rational conclusion. A typical case, in which no special knowledge is required, is afforded by the "flint implements" of the Abbeville and Amiens gravel beds. No logical proof can be adduced that the peculiar shapes of these flints were given to them by Human hands; but no unprejudiced person who has examined them now doubts it. The evidence of *design*, to which, after an examination of one or two

such specimens, we should only be justified in attaching a probable value, derives an irresistible cogency from accumulation. On the other hand, the *im*probability that these flints acquired their peculiar shape by *accident*, becomes to our minds greater and greater as more and more such specimens are found; until at last this hypothesis, although it cannot be directly disproved, is felt to be almost inconceivable, except by minds previously "possessed" by the "dominant idea" of the modern origin of Man. And thus what was in the first instance a matter of discussion, has now become one of those "self-evident" propositions, which claim the unhesitating assent of all whose opinion on the subject is entitled to the least weight.

386. We proceed upwards, however, from such questions as the Common Sense of mankind generally is competent to decide, to those in which *special* knowledge is required to give value to the judgment: and here we must distinguish between those departments of inquiry in which Scientific conclusions are arrived-at by a process of strict reasoning, and those in which they partake of the nature of Common-Sense judgments. Of the former class we have a typical example in Mathematics, and in those "exact sciences" which make use of mathematics as their instrument of proof; but even in these, it is "common-sense", which affords not only the *basis*, but the *materials* of the fabric. For while the Axioms of Geometry are "self-evident" truths which not only do not require proof, but are not capable of being proved in all their universality (§ 200), every step of a "demonstration" is an assertion of which our acceptance depends on our incapability of conceiving either the *contrary* or *anything else* than the thing asserted. And thus the certain assurance of the Q. E. D. felt by every person capable of understanding a Mathematical demonstration, depends upon the conclusive "self-evidence" of each step of it. But we not unfrequently meet with individuals, not deficient in *ordinary* Common Sense, who cannot be brought to see this "self-evidence;" whilst,

on the other hand, the advanced Mathematician, when adventuring into new paths of inquiry, is able to take a great deal for granted as "self-evident," which at an earlier stage of his researches would not have so presented itself to his mind. The deliverances of this acquired intuition can in most cases be readily justified by the reasoning process which they have anticipated. But the *genius* of a Mathematician—that is, his special aptitude developed by special culture—will occasionally enable him to *divine* a truth, of which, though he may be able to prove it experientially, neither he nor any other can at the time furnish a logical demonstration (§ 205). In this *divining* power we have clear evidence of the existence of a capacity which cannot be accounted for by the *mere* "co-ordination" of antecedent experiences, whether of the Individual or of the Race; and yet, as already shown, such co-ordination has furnished the stimulus to its development (§§ 202—207).

387. Of those departments of Science, on the other hand, in which our conclusions rest (like those of ordinary Common Sense) not on any one set of experiences, but upon *our unconscious co-ordination of the whole aggregate of our experiences,*—not on the conclusiveness of any one train of reasoning, but on *the convergence of all our lines of thought towards one centre,*—Geology may be taken as a typical example. For this inquiry brings (as it were) into one focus, the light afforded by a great variety of studies,—Physical and Chemical, Geographical and Biological; and throws it on the pages of that great Stone Book in which the past history of our globe is recorded. And its real progress dates from the time when that "common sense" method of interpretation came to be generally adopted, which consists in seeking the explanation of past changes in the forces at present in operation, instead of invoking (as the older Geologists were wont to do) the aid of extraordinary and mysterious agencies.

Of the adequacy of common sense to arrive at a decisive judgment under the guidance of the *convergence* just indicated, we have a good example in the following occurrence:—A man having had his pocket picked of a purse, and the suspected thief having been taken with a purse upon him, the loser was asked if he could swear to it as his property. This he could not do; but as he was able to name not only the precise sum which the purse contained, but also the pieces of money of which that sum consisted, the jury unhesitatingly assigned to him the ownership of the purse and its contents.—A mathematician could have calculated, from the number of the coins, what were the chances against the correctness of a mere guess; but no such calculation could have added to the assurance afforded by common sense, that the man who could tell not only the number of the coins in the purse, but the value of each one of them, must have been its possessor.

388. Familiar instances of the like formation of a basis of Judgment by the unconscious co-ordination of experiences, will be found in many occurrences of daily life; in which the effect of special training manifests itself in the formation of decisions, that are not the less to be trusted because they do not rest on assignable reasons:—

a. Thus, a Literary man, who has acquired by culture the art of writing correctly and forcibly, without having ever formally studied either grammar, the logical analysis of sentences, or the artifices of rhetoric, will continually feel, in criticizing his own writings or those of others, that there is something faulty in style or construction, and may be able to furnish the required correction, whilst altogether unable to say *in what* the passage is wrong, or *why* his amendment sets it right.

b. Or, to pass into an entirely different sphere, a practised Detective will often arrive, by a sort of divination, at a conviction of the guilt or innocence of a suspected person, which ultimately turns out to be correct; and yet he could not convey to another any adequate reasons for his assurance, which depends upon the impression made upon his mind by *minutiæ* of look, tone, gesture, or manner, which have little or no significance to ordinary observers, but which his specially-cultured Common Sense instinctively apprehends (§ 423).

389. But, in the ordinary affairs of life, our Common-Sense

judgments are so largely influenced by the Emotional part of our nature—our individual likes and dislikes, the predominance of our selfish or of our benevolent affections, and so on,—that their value will still more essentially depend upon the earnestness and persistency of our self-direction towards the Right. The more faithfully, strictly, and perseveringly we try to disentangle ourselves from all selfish aims, all *conscious* prejudices, the more shall we find ourselves progressively emancipated from those *unconscious* prejudices, which cling around us as results of early misdirection and habits of thought, and which (having become embodied in our organization) are more dangerous than those against which we knowingly put ourselves on our guard. And so, in proportion to the degree in which we habituate ourselves to try every question by first principles, rather than by the supposed dictates of a temporary expediency, will the Mechanism of our "unconscious thinking" form itself in accordance with those principles, so as often to evolve results which satisfy both ourselves and others with their "self-evident" truthfulness and rectitude.—It has been well remarked by a man of large experience of Human nature and action, that the habitual determination to do the *right* thing, marvellously clears the Judgment as to matters purely intellectual or prudential, having in themselves no moral bearing.

Of this we have a good illustration in the advice which an eminent and experienced Judge (the story is told of Lord Mansfield) is said to have given to a younger friend, newly appointed to a Colonial judgeship:—"Never give reasons for your decisions; your judgments will very probably be right, but your reasons will almost certainly be wrong." The meaning of this may be taken to be:—"Your legal instinct, or specially-trained Common Sense, based on your general knowledge of Law, guided by your honesty of intention, will very probably lead you to correct conclusions; but your knowledge of the technicalities of law is not sufficient to enable you to give reasons for those conclusions, which shall bear the test of hostile scrutiny."

390. But further, in any of those complicated questions that are

pretty sure to come before us all at some time or other in our lives,—as to which there is "a great deal to be said on both sides;" in which it is difficult to say what is prudent and even what is right; in which it is not duty and inclination that are at issue, but one set of duties and inclinations at issue with another, —Experience justifies the conclusion to which Science seems to point, that the habitually well-regulated mind forms its surest judgment by trusting to the automatic guidance of its Common Sense; just as a rider who has lost his road is more likely to find his way home by dropping the reins on his horse's neck, than by continuing to jerk them to this side or that in the vain search for it. For continued argument and discussion, in which the feelings are excited on one side, provoke antagonistic feelings on the other; and no true balance can be struck until all these adventitious influences have ceased to operate. When all the considerations which ought to be taken into the account have been once brought fully before the mind, it is far better to leave them to *arrange themselves*, by turning the conscious activity of the mind into some other direction, or by giving it a complete repose. If adequate time be given for this unconscious co-ordination, which is especially necessary when the Feelings have been strongly and deeply moved, we find, when we bring the question again under consideration, that *the direction in which the mind gravitates* is a safer guide than any judgment formed when we are fresh from its discussion (§ 431).

391. Not only may the range and value of such Common-Sense judgments be increased by appropriate culture in the individual, for, of all parts of our higher nature, the aptitude for forming them is probably that which is most capable of being transmitted hereditarily; so that the descendant of a well-educated ancestry constitutionally possesses it in much higher measure than the progeny of any savage race. And it seems to be in virtue of this automatic co-ordination of the elements of judgment, rather than

of any process of conscious ratiocination—by the exercise of the νοῦς rather than of the διάνοια—that the Race, like the Individual, emancipates itself from early prejudices, gets rid of worn-out beliefs, and learns to look at things as they are, rather than as they have been traditionally represented. This is what is really expressed by the "Progress of Rationalism." For although that progress undoubtedly depends in great part upon the more general diffusion of knowledge, and the higher culture of those intellectual powers which are exercised in the acquirement of it, yet this alone would be of little avail, if the self-discipline thus exerted did not act *downwards* in improving the mechanism that evolves the "self-evident" material of our reasoning processes, as well as *upwards* in more highly elaborating their product. If we examine, for instance, the history of the decline of the belief in Witchcraft, we find that it was not killed by discussion, but perished of neglect. The Common Sense of the best part of mankind has come to be ashamed of ever having put any faith in things whose absurdity now appears "self-evident;" no discussion of evidence once regarded as convincing is any longer needed; and it is only among those of our hereditarily-uneducated population, whose general intelligence is about upon a par with that of a Hottentot or an Esquimaux, that we any longer find such faith entertained. —In the Writer's belief, the "Spiritualistic" doctrines of the present day will be looked upon by the Intelligence of future generations, with the same pitying wonder that *we* extend to the old belief in Witchcraft.

392. There is, in fact, a sort of under-current, not of actually formed Opinion, but of tendency to the formation of opinions, in certain directions, which bursts up every now and then to the surface; exhibiting a latent preparedness in the public mind to look at great questions in a new point of view, which leads to most striking results when adequately guided. That "the hour is come—and the man," is what History continually reproduces;

neither can do anything effectively without the other. But a great idea thrown out by a mind in advance of its age, takes root and germinates in secret, shapes the "unconscious thought" of a few individuals of the next generation, is by them diffused still more widely, and thus silently matures itself in the "womb of time," until it comes forth, like Minerva, in full panoply of power.

393. Those who are able to look back with intelligent retrospect over the Political history of the last half-century, and who witness the now general pervasion of the public mind by truths which it accepts as "self-evident," and by Moral principles which it regards as beyond dispute, can scarcely realise to themselves the fact that within their own recollection the fearless assertors of those truths and principles were scoffed at as visionaries or reviled as destructives. And those whose experience is limited to a more recent period, must see, in the rapid development of public opinion on subjects of the highest importance, the evidence of a previous unconscious preparedness, which may be believed to consist mainly in the higher development and more general diffusion of that automatic co-ordinating power, which constitutes the essence of Reason as distinct from Reasoning.

394. Thus, then, every course of Intellectual and Moral self-discipline, steadily and honestly pursued, tends not merely to clear the mental vision of the Individual, but to ennoble the Race; by helping to develope that Intuitive power, which arises in the first instance from the embodiment in the Human Constitution of the *general resultants* of antecedent experience, but which, in its highest form, far transcends the experience that has furnished the materials for its evolution,—just as the *creative* power of Imagination shapes-out conceptions which no merely *constructive* skill could devise (§ 408).

The following extract from a letter addressed to the Writer by the late Mr. J. S. Mill, Jan. 29, 1872, will be interesting as an expression of his matured views on several questions discussed in the preceding pages :—

"I have long recognized as a fact that judgments really grounded on a long succession of small experiences mostly forgotten, or perhaps never brought out into very distinct consciousness, often grow into the likeness of intuitive perceptions. I believe this to be the explanation of the intuitive insight thought to be characteristic of women ; and of that which is often found in experienced practical persons who have not attended much to theory, nor been often called on to explain the grounds of their judgments. I explain in the same manner whatever truth there is in presentiments. And I should agree with you that a mind which is fitted by constitution and habits to receive truly and retain well the impressions made by its passing experiences, will often be safer in relying on its intuitive judgments, representative of the aggregate of its past experience, than on the inferences that can be drawn from such facts or reasonings as can be distinctly called to mind at the moment.—Now you seem to think that judgment by what is called Common Sense is a faculty of this kind ; and, so far as the genesis of it is concerned, I think you are right ; but it seems to me that there is a great practical difference. [The difference which Mr. Mill goes on to point out, is that which the Writer has above defined as the distinction between the *ordinary* and the *special* forms of Common Sense.]

"On the Physiological side of Psychology, your paper raises questions of great and increasing interest.—When states of Mind in no respect innate or instinctive have been frequently repeated, the Mind acquires, as is proved by the power of Habit, a greatly increased facility of passing into those states ; and this increased facility must be owing to some change of a physical character, in the organic action of the Brain. There is also considerable evidence that such acquired facilities of passing into certain modes of Cerebral action can in many cases be transmitted, more or less completely, by Inheritance. The limits of this transmission, and the conditions on which it depends, are a subject now fairly before the scientific world ; and we shall doubtless in time know much more about them than we do now. But so far as my imperfect knowledge of the subject extends, I take much the same view of it that you do, at least in principle."

CHAPTER XII.

OF IMAGINATION.

395. THE form of Ideational activity which we distinguish as *Imagination*, has the same basis as that which exerts itself in Reasoning processes; but works in a very different manner. In its lowest and simplest exercise, Imagination, or the image-making power, consists in that reproduction of the mental "idea" or representation of an object formerly perceived through the senses, which is more generally understood by the term *Conception.* In strict language, every such reproduction of an image, however distinctly traceable to the laws of Association, is an act of Imagination. This, however, is not the generally understood meaning of the term; which is usually applied to the faculty by which, in the *first* place, the materials supplied by experience or direct apprehension are recombined in such forms as to gratify the sense of beauty or fitness, rather than to satisfy the reason; and by which, in the *second* place, some higher form of Beauty than experience has ever presented, or some more profound Truth than reason could bring within our grasp, is discerned by direct "insight,"—the "vision and the faculty divine." Thus we may distinguish between the *constructive* and the *creative* Imagination; and each may be exercised in every department of Human knowledge, whether relating to external Nature or to Human nature. It is a very limited view of the Imagination to regard its operation as restricted to works of Art, of which the object is to gratify the Æsthetic sense; for, as will be presently shown, its highest exercise is put forth in the discovery of great fundamental Truths, as well in Science as in Art, in the Universe around as in the Mind of Man.

396. The exercise of the *constructive* Imagination requires a mind not only richly stored with materials, but endowed with the power of readily reproducing them; together with a judgment guided by correct taste and artistic sensibility, whereby the most appropriate images are selected and combined into the forms most effective for the production of the desired impression. In fact, the mental process by which the constructive Imagination works, is only an ideal rendering of that which Praxiteles is said have employed when commissioned to produce a statue of Venus for the people of Cos,—namely the selection of the most beautiful parts of the most beautiful female figures he could obtain as models, and the combination of these into one harmonious whole. It is that by which, for the most part, the Architects of the present time essay to produce their designs. And it differs but little in kind from that which the Mechanician employs, when devising a piece of Machinery for a given purpose, which only requires some new combination of levers, wheels, &c., involving no new *principle* of action. The following description seems accurately to represent the method in which the constructive Imagination ordinarily works:—

"The general idea, or the subject in its *outlines*, must be supposed to be already present to the mind of the writer. He accordingly commences the task before him with the expectation and the desire of developing the subject more or less fully, of giving to it not only a greater continuity and a better arrangement, but an increased interest in every respect. As he feels interested in the topic which he proposes to write upon, he can, of course, by a mere act of the Will—although he might not have been able, in the first instance, to have originated it by such an act—detain it before him for any length of time. Various conceptions continue, in the meanwhile, to arise in the mind, on the common principles of Association; but, as the general outline of the subject remains the same, they have all a greater or less relation to it. And partaking, in some measure, of the permanency of the outline to which they have relation, the writer has an opportunity to approve some and to reject others, according as they impress him as being suitable or unsuitable to the nature of the subject. Those which affect him with emotions of pleasure, on account of their perceived fitness for the subject, are retained and committed to writing; while others, which do not thus affect and interest him, soon fade away altogether."—(*Upham's Elements of Mental Philosophy*, Vol. i. p. 385.)

The process of *selection* among the variety of Conceptions that spontaneously occur to the mind, and then rapidly give place to others if not Volitionally retained, has been thus graphically described by Dr. Reid :—

"We seem to treat the thoughts that present themselves to the fancy in crowds, as a great man treats the courtiers who attend at his levee. They are all ambitious of his attention. He goes round the circle, bestowing a bow upon one, a smile upon another; asks a short question of a third, while a fourth is honoured with a particular conference; and the greater part have no particular mark of attention, but go as they came. It is true, he can give no mark of his attention to those who were *not there;* but he has a sufficient number for making a choice and distinction."—(*Essay* iv. chap. 4.)

397. The faculty of Imagination, therefore, works within the same limits as that of *Recollection* (§ 373). The recurrence of images is essentially *automatic;* but the mind can determinately place itself in the condition most favourable to their reproduction, and can project itself, as it were, in search of them. While some persons are obliged to wait until the memory supplies them with the image they desiderate, there are others who are distinguished by the exuberance of this reproductive power; so that they have only to ask themselves for an appropriate simile or metaphor, and it immediately occurs to them. When apologising for the "lavish imagery" of Scott's historical style (which caused the French to designate his Life of Napoleon as "Walter Scott's last novel"), his biographer says of him :—"Metaphorical illustrations, which men born with prose in their souls hunt for painfully, and find only to murder, were, to him, the natural and necessary offspring and playthings of ever-teeming fancy. He could not write a note

to his printer, he could not speak to himself in his diary, without introducing them." Some, again, use images merely as illustrations, dismissing each as it has served its purpose, and availing themselves of new figures as they pass before the mental vision. But the "constructive" Imagination is most characteristically exercised in the elaboration of some work which consists of a methodical aggregation of separate images, that are grouped round what may be termed a central idea. Thus Walter Scott, whom we may regard as a typical instance of this form of imagination (without by any means denying his possession of the higher attribute), generally took as the basis of his novels (as Shakspere did of his plays) some story or stories previously existing; and meditated on this fundamental conception, until it had taken a distinct form in his own mind. When it had reached this stage, the further development of it seems to have been generally very easy to him. Waking early, he used to lie "simmering" for an hour or two; and then, rising and seating himself at his writing-table, he would throw off chapter after chapter as fast as his pen could trace the words. How much he was aided by Memory, is curiously shown in the use which he made of casual suggestions or occurrences. Thus we are told by Mr. Lockhart, with reference to the composition of "Ivanhoe" (of which the earlier part was written towards the close of the severe illness in which the "Bride of Lammermoor" was produced, § 124 *e*), that—

a. "The introduction of the charming Jewess and her father originated in a conversation Scott held with his friend Skene during the severest season of his bodily sufferings. 'Mr. Skene,' says that gentleman's wife, 'sitting by his bedside, and trying to amuse him as well as he could in the intervals of pain, happened to get upon the subject of the Jews, as he had observed them when he spent some time in Germany in his youth. Their situation had naturally made a strong impression; for in those days they retained their own dress and manners entire, and were treated with considerable austerity by their Christian neighbours, being still locked up at night in their own quarter by great gates; and Mr. Skene, partly in seriousness, but partly from the mere wish to turn his mind at the moment upon something that might occupy and divert it, suggested that a group of Jews would be an interesting feature if he could contrive to bring them into his next novel.' Upon the appearance of 'Ivanhoe,' he reminded Mr. Skene of this conversation, and said, 'You will find this book owes not a little to your German reminiscences.' Mrs. Skene adds—'Dining with us one day, not long before 'Ivanhoe' was begun, something that was mentioned led him to describe the sudden death of an Advocate of his acquaintance—a Mr. Elphinstone—which occurred in the *Outer-house* soon after he was called to the bar. It was, he said, no wonder that it had left a vivid impression on his mind, for it was the first sudden death he had ever witnessed; and he now related it so as to make us all feel as if the scene was passing before our eyes. In the death of the Templar in 'Ivanhoe,' I recognised the very picture—I believe I may safely say the very words.'

"Before 'Ivanhoe' made its appearance, I had myself been formally admitted to the author's secret; but had he favoured me with no such confidence, it would have been impossible for me to doubt that I had been present some months before at the conversation which suggested, and indeed supplied, the materials of one of its most amusing chapters. I allude to that in which our Saxon terms for animals in the field, and our Norman equivalents for them as they appear on the table, and so on, are explained and commented on. All this Scott owed to the after-dinner talk one day, in Castle Street, of his old friend, Mr. William Clerk."—(*Life of Scott*, chap. xlvi.)

The following particulars supplied by Mr. Morritt of Rokeby, respecting Scott's mode of obtaining materials for the romantic poem he was composing on the traditions connected with that picturesque seat, are interesting from the additional light they throw on the method in which Scott's "constructive" imagination made use of the materials supplied by fact and legend; sometimes fitting the scene to the legend, and sometimes the legend to the scene:—

b. "I had of course," says Mr. Morritt, "had many previous opportunities of testing the almost conscientious fidelity of his local descriptions; but I could not help being singularly struck with the light which this visit threw on that characteristic of his compositions. The morning after he arrived, he said, 'You have often given me materials for romance: now I want a good robber's cave, and an old church of the right sort.' We rode out, and he found what he wanted in the ancient slate quarries of Brignal and the ruined Abbey of Eggleston. I observed him noting down even the peculiar little wild flowers and herbs that accidentally grew round and on the side of a bold crag near his intended cave of Guy Denzil; and could not help saying that, as he was not to be upon oath in his work, daisies, violets, and primroses would be as poetical as any of the humbler plants he was examining. I laughed, in short, at his scrupulousness; but I understood him when he replied, that in Nature herself no two scenes were exactly alike, and that whoever copied truly what was before his eyes, would possess the same variety in his descriptions, and exhibit apparently an imagination as boundless as the range of nature in the scenes he recorded; whereas whoever trusted to Imagination would soon find his own mind circumscribed, and contracted to a few favourite images, and the repetition of these would sooner or later produce that very monotony and barrenness which had always haunted descriptive poetry in the hands of any but the patient worshippers of truth. Besides which, he said, 'local names and peculiarities make a fictitious story look so much better in the face.' In fact, from his boyish habits he was but half satisfied with the most beautiful scenery, when he could not connect with it some local legend; and when I was forced sometimes to confess with the Knife-grinder, 'Story! God bless you! I have none to tell, sir!'—he would laugh and say, 'Then let us make one: nothing so easy as to make a tradition.' "—(*Op. cit.*, chap. xxv.)

398. The practice of the Italian *improvisatori*, who, without preparation, compose and utter a long series of verses upon any given subject, affords a characteristic example of the readiness with which the Memory supplies the requisite materials to such as have a special gift for versification, and have improved that gift by culture.

The practice is of frequent occurrence in Italy; and the facilities which the structure of the Italian language affords to versification and rhyme, are of great assistance towards it. The *improvisatore* delivers his verse, generally accompanied by a guitar, and with a sort of chanting cadence; and he spins out hundreds, nay at times thousands, of lines with apparent ease; whole dramas have indeed been thus delivered. It must not be imagined, however, that this kind of extempore poetry is of the best kind; in reality, very few of these compositions can stand the test of publication. Still they have the merit of the flow of language, and the quick adaptation of accessory ideas and images to the main subject, which rivet the attention and excite the surprise of the listeners.

Some examples there are, indeed, of *improvisatori*, who, uniting high culture to a great original gift of extemporaneous versification, have acquired, by continual practice, a facility in producing unpremeditated verses which would not only bear perusal, but even pass the ordeal of severe criticism.—Another species of *impromptu* peculiar to Italy is the extempore Comedy; in which only the outline of the plot, with an indication of the characters, is given to be filled up at the spur of the moment by the performers. It is said that the great painter, Salvator Rosa, had a remarkable gift for this kind of performance; in which, in more recent times, Louis Riccoboni and his wife Plamenia showed a proficiency so extraordinary, that it was suspected that they did not really act *all' improvista*, but that they were imposing on the public by preconcerted schemes. This, however, was clearly proved not to be the case; and Riccoboni gave the following account of the peculiar combination of endowments required for success in this *rôle*:—"An actor of this description, always supposing an actor of genius, is more vividly affected than one who has coldly got his part by rote. But figure, memory, voice, and even sensibility, are not sufficient for the actor *all' improviste*; he must be in the habit of cultivating the imagination, pouring forth the flow of expression, and prompt in those flashes which instantaneously vibrate in the plaudits of the audience."

399. The most remarkable example of Improvisation which our own country has supplied, was one of an entirely different character; Theodore Hook's improvised versification being rather witty and humorous, than what is ordinarily understood as imaginative.

a. "Thus, very early in his career, he displayed his extraordinary gift of extemporaneous singing, at a dinner given by the Drury Lane Company to Sheridan. The company was numerous, and generally strangers to Mr. Hook; but, without a moment's premeditation, he composed a verse upon every person in the room, full of the most pointed wit and with the truest rhymes, unhesitatingly gathering into his subject, as he rapidly proceeded, in addition to what had passed during the dinner, every trivial incident of the moment. Every action was turned to account; every circumstance, the look, the gesture, or any other accidental effects, served as occasion for more wit; and even the singer's ignorance of the names and condition of many of the party seemed to give greater facility to his brilliant hits, than even acquaintance with them might have furnished. Mr. Sheridan was astonished at his extraordinary facility. No description, he said, could have convinced him of so peculiar an instance of genius; and he protested that he should not have believed it to be an unstudied effort, had he not seen proof that no anticipation could have been formed of what might arise to furnish matter and opportunities for the exercise of this rare talent."—(*Life of Mathews*, vol. ii., p. 59.)

We are told by the writer of a notice of Theodore Hook in the "Quarterly Review" (vol. lxxii.), that, either early in the evening or very late, his improvisation was indifferent; the reason apparently being that his gift required the *stimulus* of wine and society, but that too much of the former would prevent him from keeping up the necessary continuity of idea.

b. Another friend describes a meeting between Coleridge and Hook, at which the latter "burst into a bacchanal of egregious luxury, every line of which had reference to the author of the 'Lay Sermons' and 'Aids to Reflection'" An absurd scene followed, every individual's share in which "had apt, in many cases exquisitely witty commemoration." In walking home, Coleridge gave the narrator "a most excellent lecture on the distinction between talent and genius; and declared that Hook was as true a genius as Dante."

But Hook further enjoyed, in a very remarkable degree, that faculty of *dramatic imagination*, which enables its possessor to identify himself for a time with the personality of another, so as not merely to *act the part* of that other, but (ideally) to *be* that other (§ 364). Thus, when staying in a country house with men distinguished in Parliamentary or Forensic oratory, he would give, when "in the vein," a whole series of imitations; representing not merely the *manner* but the *matter* of each speaker so closely, that the audience could almost believe that they were listening to the man himself. It was only when he was "primed" for such a performance by alcoholic stimulants, and encouraged by the admiration of his audience, that his gift would manifest itself in full force; and its *automatic* rather than its *volitional* character is obvious from the fact, that its greatest displays took place just at that stage of alcoholic excitement, in which the Automatic activity of the mind is increased, whilst the controlling power of Volition is weakened (§ 544).

400. The power of Improvisation for which certain Musicians have been especially distinguished, requires for its highest exercise a combination of the *constructive* power acquired by training, with the *creative* power which nothing but Genius can supply. Any Musician who practises the art, can extemporise a set of "variations" upon an air, or work up a "subject" into a fugue, according to certain set forms; and the product can be in a great degree anticipated by any one who is familiar with the "manner" of the player. To this extent, therefore, the power is under Volitional control; whilst, in so far as the improvisation is *creative*, it is purely Automatic; and all that the Musician can himself do,

is to place himself in conditions favourable to the spontaneous flow of his ideas.

a. Of all Musicians, Mozart seems to have had *inventive* power most completely *at call*, so that he could never draw upon it, without (in commercial phrase) his draft being instantly "honoured." Thus, it is recorded of him that, "in the performance of his own Concertos, he never confined himself to the precise melody before him, but varied it from time to time with singular grace and beauty, according to the inspiration of the moment." This feat seems the more remarkable, as, in such a performance, he could not give free scope to his invention, as in a solo fantasia, but had always to keep himself in due relation to his Orchestra.—In the following incident, we have an example of his unsurpassed inventive power when working entirely unfettered, and under the stimulus of the warmest sympathy. It should be premised that the Concert at which it occurred was one given by Mozart at Prague, the inhabitants of which ancient city had shown themselves singularly appreciative of "Le Nozze di Figaro," which had been produced there a short time before. The concert, every piece in which was of Mozart's own composition, was to end with an improvisation on the pianoforte. "Having preluded and played a fantasia, which lasted a good half-hour, Mozart rose; but the stormy and outrageous applause of his Bohemian audience was not to be appeased, and he again sat down. His second fantasia, which was of *an entirely different character*, met with the same success; the applause was without end; and, long after he had retired to the withdrawing room, he heard the people in the theatre thundering for his reappearance. Inwardly delighted, he presented himself for the third time. Just as he was about to begin, when every noise was hushed, and the stillness of death reigned throughout the theatre, a voice in the pit cried, '*From Figaro.*' He took the hint, and ended this triumphal display of skill by extemporising a dozen of the most interesting and scientific variations upon the air *Non piu andrai.* It is needless to mention the uproar that followed."—(*Holmes's Life of Mozart*, p. 278.)

401. The mental processes which are concerned in the exercise of constructive Imagination, are also in operation in the production of other embellishments of ordinary prosaic language to which we give

distinctive names. That which comes nearest to Imagination—so near, indeed, that no distinct line of demarcation can be drawn between them—is *Fancy*. Notwithstanding that much has been written on the relative characters of these two attributes,* it seems impossible to define them otherwise than by the states of Mind to which their products are respectively addressed. For whilst the *Imagination* works on the sense of beauty, of grandeur, of sublimity, or, by way of contrast, on that of aversion, awe, or even terror,—Fancy plays with the grotesque, the whimsical, the ridiculous, in short, with the "fanciful;" its function being rather to amuse by its superficial "caprices" and "strange conceits," than to appeal to the deeper nature.

"Fancy," says Wordsworth, "is given to quicken and beguile the temporal part of our nature; Imagination, to incite and support the eternal." And he gives, as an example, the use of the same image by two writers; in the one case as a mere "conceit," in the other as a grand stroke of imaginative genius. The effect of the following couplet, ascribed to Lord Chesterfield—

> "The dews of the evening most carefully shun,
> They are tears of the sky for the loss of the sun,"

"is a flash of surprise, and nothing more; for the nature of things does not sustain the combination." But when Milton marks the sympathy of Nature with the transgression of Adam, in the grand lines—

> "Sky lowered, and muttering thunder, some sad drops
> Wept at completion of the mortal sin."

"the effects from the act, of which there is this immediate consequence and visible sign, are so momentous, that the mind acknowledges

* Dr. Trench, in his "Study of Words," assigns to Wordsworth the credit of having, in his original preface to his "Lyrical Ballads" (reprinted separately in his collected works), distinctly limited the respective provinces of Fancy and Imagination. But, however clear may be the boundary lines which Wordsworth marked out in the domain of Poetry, we fail to trace them when we follow the Imagination into other departments of Art, still more into Science and Human Nature.

the justice and reasonableness of the sky weeping drops of water as if with human eyes."

The pages of Shakspere, and many of those of Jeremy Taylor, are full of the products both of Fancy and Imagination; but while in George Wither and Charles Lamb the Fancy predominates, Milton is the pre-eminent type of Imagination (as distinguished from fancy) in Poetry, as Walter Scott may be considered to be in Prose.

402. Closely akin to Fancy, but differing from it in the kind of feeling which it excites, is Wit; the distinctive character of which (as now understood) is "a felicitous association of objects not usually connected, so as to produce a pleasant surprise." It is this *intellectual surprise* which is the essence of pure Wit; but Humour is often mixed up with it. The lowest form of wit is that mere play on *words*, which is known as "punning;" this being, according to the definition of Addison, "a conceit arising from the use of two words that agree in sound but differ in the sense," or, as might be said in certain cases, use of the same word in two different senses (§ 403 *b.*). Thus a man having undertaken to pun on any subject that might be given him, and the word "King" having been named, replied that "the King was no subject." Here the wit lay in the double meaning attached to the word "subject," or, as Addison would say, to the unexpectedness of the relation brought out between what are really two distinct words agreeing only in sound. The effect is heightened when this *double entendre* presents itself more than once in the same sentence:—

a. Thus Bishop Mountain, being asked by George II., to whom he should give the Archbishopric of York then vacant, replied, "If thou hadst faith as a grain of mustard seed, thou wouldst say to this mountain" (pointing to himself), "'Be thou removed, and be thou cast into the sea (see).'"

b. The Writer once heard another felicitous application of the same idea, at a meeting of the Geological Society, at which Dr. Buckland, then Dean of Westminster, had been speaking on a paper by Sir Roderick Murchison, in which much stress was laid on the effects of "waves of translation," as carrying large blocks of stone onwards until their force was spent, when the blocks sank to the bottom. As the Dean sat down, a facetious member expressed the hope "that the Right Reverend gentleman might be speedily carried along by a wave of translation, and deposited in a sea (see) as good as that of his predecessor," the Bishop of Oxford, who was sitting by his side.

403. But in the higher form of Wit, the play is rather upon *ideas* than upon *words*, the effect consisting in the peculiar felicity of the illustration; as when Douglas Jerrold characterized *dog*matism as being *puppy*ism full grown,—a remark most true and apposite in itself, while deriving a peculiar "point" from the unexpectedness of the ideal relation developed between the two words, and from the ludicrous contrast between the *real* derivation of the first of them and the *humorous* one assigned to it. In *repartees*, the effect very often depends on the unexpectedness of the contrast between the idea conveyed by the original remark and that of the reply; or by the happy turn given to the former, by which it is made to convey something very different from, perhaps quite contrary to, what the speaker intended.

a. Of the first kind, the matter-of-factness of the Scotch mind furnishes many admirable illustrations, of which the following is perhaps one of the most characteristic:—During an agitation to put down Sunday trains between Edinburgh and Glasgow, a Minister in the latter city was accustomed to go down to the railway-station in the morning, and endeavour to dissuade any persons whom he knew from setting-off by the train then starting. Happening to encounter a lady of his congregation, who was going to spend the day with a relative at Falkirk, and having learned from her where she was going, he said, "You are not going to Falkirk, you are going to hell;" to which she replied very composedly, "Well, anyhow, I have got a return-ticket."

b. Of the second kind, a good example is offered by the ready answer given to a debater, who, addressing the adversary who was to reply to him, somewhat magniloquently enounced, "I want common sense;" to which the quiet rejoinder was, "Exactly." The turn thus given was to the meaning of the word "want;" which the speaker used as expressing his sense of its deficiency in his adversary, who neatly directed it against himself.

404. These two examples, again, serve to illustrate the connection between Wit and Humour. Whilst the former, in its purity, is purely *intellectual*, the latter, in its purity, has reference to the varieties of *human character* and the actions which proceed from them. But in a large number of cases, the two are commingled; and the amusement we derive from their exercise often arises from the concomitant circumstances, or from the manner of utterance; so that a joke which was felt to be admirable at the time, falls comparatively flat when repeated under other conditions.

a. Thus, in the well-known reply made by Charles Lamb to Coleridge's question, "Lamb, did you ever hear me preach?" "Why, I never heard you do anything else,"—the effect is described by one who was present as having been most ludicrously heightened by the 'grand manner" in which the question was asked by Coleridge, and the stammering utterance, ending in a kind of explosion, of his friend's witticism.

405. It may be affirmed, without hesitation, that the power of presenting the peculiar combinations and contrasts in which Wit consists, is purely *automatic;* being one of which many persons, highly distinguished for general ability, and possessing a keen enjoyment of the good things said by others, are themselves entirely destitute. And all that the Will can do, as in other operations of the constructive Imagination, is to place the mind in the conditions most favourable for its exercise. Of such conditions, sympathetic appreciation undoubtedly ranks among the first. A brilliant talker, who, in congenial society, can give forth an unceasing succession of coruscations of Wit, like meteor-streams, feels the dulness of a set of matter-of-fact listeners as a

complete extinguisher of the internal glow from which the flashes emanate.—The power is doubtless capable of improvement by culture and exercise; but it cannot be *acquired* by any one who has not a native capacity for it.—The following example, which recently presented itself within the Writer's own experience, is so singularly illustrative of the contrast between the Volitional and the Automatic forms of Intellectual operation, that those who recognize the case to which he alludes will excuse him for making it public:—

a. "A gentleman distinguished alike for his great and varied ability, for his faculty of scientific exposition, and for his conversational gifts, found himself the subject of a sudden failure of Intellectual power, the result of continued overstrain; which manifested itself especially in the inability to keep up a continuous train of thought, such as was required in lecturing. Having determined at once to suspend his work and go abroad, he was dining with a few of his most intimate friends a few days before his departure; and surprised them by the undiminished flow of his wit, and the readiness of his repartee. As one of them afterwards said to the writer, 'He never was more brilliant.'"

406. It would not be correct to say that *Humour* is itself dependent upon the Imagination; since what we term *humorous* may derive its character entirely from the oddity of the mode of expression or action *as it presents itself to ourselves.* Dean Ramsay truly says of one of his good stories:—"The humour of the narrative is unquestionable, and yet no one has tried to be humorous; the same idea differently expressed might have no point at all;" as in the following case:—

a. "The Laird of Balnamoon, having dined at the house of a friend who by mistake had given him cherry brandy, instead of his usual beverage, port-wine, found himself more than ordinarily affected by his liquor; and while being driven home by his servant Harry in an open carriage, over an intervening moor, his unsteadiness of head caused his hat and wig to fall on the ground. Harry having got down to pick them up, and having restored them to his master, the laird was satisfied with the hat, but demurred to the wig. 'It's no my wig, Harry, lad; it's no my wig,' and refused to have anything to do with it. Harry lost his patience; and, anxious to get home, remonstrated with his master, 'Ye'd better tak it, sir, for there's nae waile (choice) o' wigs on Munrimmon Moor.' This was a simple matter-of-fact remark on the part of the servant; and its humour consists in its contrast with the circumstances under which it was made, heightened by its quaint form of expression."

407. But in all works of Imagination which aim to present varieties of Human character, a power of ideally developing and combining such varieties is essential; and without the sense of Humour, the portraiture wants warmth and reality. Of all great masters in the art of construction, Milton is the one who shows the least of this sense; in fact he scarcely seems to have possessed it at all. In Shakspere, on the other hand, Humour is redundant; and every one of his humorous characters has an individuality of his (or her) own. Thus Falstaff will occur to every one as the very incarnation of Humour, with no small addition of Wit; while Dogberry and Verges, Bottom and Snug, with the whole series of Clowns and Jesters, no one of whom is a repetition of another, bespeak the wonderful acuteness of his discriminative observation, the readiness with which his memory reproduced its stores, and the sympathy which his own many-sided nature possessed with every phase of Humour delineated. In those of Sir Walter Scott's novels in which the scenes and characters are those of his own country, we find abundant examples of the richest Humour; every one of which seems either to have had an individual prototype, or to have been a generalised form derived from a variety of sources. Thus the Baron of Bradwardine, the Antiquary, Paulus Pleydell, Dominie Sampson, Mause and Cuddie Headrigg, Meg Dods, and a number of others, are felt by those who are well acquainted with the Scottish life of the past to be so intensely *natural,* that every one of them *might* have been a real character. And the same is true

of the best of Dickens's and of Thackeray's imaginary constructions; in which these great Humourists have so completely identified themselves, as it were, with the several types they delineated, as to make each of them speak and act as he (or she) would have done in actual life. It is certain, indeed, that most of these (as in Walter Scott's case) are developments of actual types; while those which are purely *ideal*—the work of the *creative* rather than of the *constructive* Imagination—lack "flesh-and-blood" reality.

408. The exercise of the *creative* Imagination involves that peculiar quality of mind which we distinguish as *Genius*;—a special gift of which no definite account can be given, the possessor of it not being himself able either to trace its origin, or to describe (save as regards its external conditions) its mode of working. Although this term is often applied to superior Intellectual power of *any* kind, yet it may certainly be most appropriately limited to that of which *invention*, *origination*, or *insight* is the distinctive characteristic, and of which the products bear the well-marked stamp of *individuality*. Working upon the materials it derives from observation and reflection, Creative genius not merely developes and recombines these, but evolves products *altogether new*, beyond the scope of the reasoning power, and deriving their value from their expression of a *higher truth* than is at the time attainable in any other way. When Dr. Channing, in discoursing of Poetry, said that "Genius is not a creator, in the sense of fancying or feigning what does not exist: *its distinction is, to discern more of Truth than ordinary minds*," he himself uttered a truth of which the profundity and generality bear the stamp of true genius; for it is equally applicable to every exercise of the highest form of creative imagination. Its *lower* form may be distinguished as *Ingenuity*;* the products of which may either be altogether worthless, or may approach those of true genius.

* Although this word is now commonly used with reference chiefly to Mechanical invention, yet it was often applied by the older writers to Poets and Advocates.

a. A typical example of the difference between *Ingenuity* and true *Genius* is afforded by the contrast between Kepler and Newton. Strongly impressed with the belief that some "harmonic" relation must exist among the distances of the several Planets from the Sun, and also among the respective times of their revolution, Kepler passed a large part of his early life in working out a series of *guesses* at this relation; some of which now strike us as not merely most improbable, but positively ridiculous. His single-minded devotion to Truth, however, led him to abandon each of these hypotheses in its turn, so soon as he had proved its fallacy by submitting it to the test of its conformity to observed facts; while his fertile Ingenuity furnished him with a continual supply of new guesses, which presented themselves in turn as creations of his Imagination, to be successively dismissed when they proved to be nothing else than imaginary. But he was at last rewarded by the discovery of that relation between the times and the distances of the Planetary revolutions, which, with the discovery of the ellipticity of the orbits, and of the passage of the "radius vector" over *equal areas in equal times*, has given him immortality as an Astronomical discoverer. But these discoveries cannot be regarded as based on any higher Mental attribute than *persevering ingenuity*; for so far was he from divining the true *rationale* of the planetary revolutions, that we learn from his own honest confessions that he was led to the discovery of the elliptic orbit of Mars by a series of happy accidents, which turned his erroneous guesses into a right direction, and to that of the "equal areas" by the notion of whirling force emanating from the Sun; whilst his discovery of the relation between the times and distances was the *fortunate* guess which closed a long series of *un*fortunate ones, many of which were no less "ingenious."—Now it was by a grand exertion of Newton's *constructive* imagination, based on his wonderful mastery of Geometrical reasoning, that, starting with the conception of two *forces*, one of them tending to produce continuous uniform motion in a straight line, the other tending to produce a uniformly-accelerated motion towards a fixed point, he was able to show that if these

We speak even now of an "ingenious argument," when we have in view rather the skill with which it is conducted, than the truth it is to support; in fact, our admiration is sometimes most called forth by the Ingenuity which is exerted to sustain a position we regard as untenable.

dynamical assumptions be granted, Kepler's *phenomenal* "laws," being necessary consequences of them, must be *universally* true. And while that demonstration would have been alone sufficient to give him an imperishable renown, it was his still greater glory to divine the profound truth, that the fall of the Moon towards the Earth—that is, the deflection of her path from a tangential line to an ellipse—*is a phenomenon of the same order* as the fall of a stone to the ground; and thus to show that the mutual attraction of all masses of matter which we call Gravitation, pervades the whole Universe, and everywhere follows the same Law.

The grand scientific Truth thus divined by the "insight" of Newton, was at once proved to be such, not only by its conformity with all the phenonema of Nature then known, but by the power which it furnished of *predicting* phenomena as yet undiscovered—such as the "perturbations" produced by the mutual attractions of the Planets.

409. In the provinces of Æsthetics and Morals, however, the test of the Truths brought to light by real Genius lies not in the confirmation of them by reasoning processes, but in that conformity to *our own highest Nature*, which enables us to recognise them by "direct apprehension." In this way, it has been well said, a great Moral or Religious Teacher, who first proclaims truths which we feel to be of universal applicability, "reveals us to ourselves." And the Poet who, as Dr. Channing says of Wordsworth, "sees under disguises and humble forms everlasting beauty," and makes us conscious of the loveliness of the *primitive feelings* which constitute the universal affections of Humanity, strikes chords to which our own hearts instinctively respond. It is the undoubted attribute of real Genius, that the Intellectual Truth, the Moral Power, or the Artistic Beauty of its conceptions *ultimately* gains the recognition of such as seek the highest good in either pursuit. That recognition may not be immediate, because the advance upon the previous culture may have been too sudden. But all really great Thought tends to elevate Human nature to its own level; and

the Philosophers, the Prophets, and the Poets, whom we now venerate as the noblest benefactors of our Race, have earned their claim to that distinction, not by bringing to us messages from other spheres which they alone were privileged to visit, but by enunciating Truths which our expanded Intellect accepts as self-evident, by proclaiming great Principles which our deepened insight perceives to constitute the basis of all Morality, by creating forms of Beauty to which our heightened and purified sense looks up as standards of Ideal perfection. And this could not be, unless the intuitions of Genius call forth echoes from the depths of our own souls; awaking dormant faculties, which can apprehend if they cannot create, which can respond if they cannot originate. The "Principia" of Newton, unintelligible to the great mass of his most learned contemporaries, are now the A B C of the student of the higher Mathematics. The dramas of Shakspere, appreciated by the theatre-goers of his day only for the pleasure to be derived from their poetry and their action, are now read and pondered by every student of Human Nature as the embodiment of the profoundest and most universal knowledge. And the grand Symphony of Beethoven, which was laid aside as incomprehensible by the most cultivated Musicians of his time, is now the delight, not alone of the select few, but of the many whom the more advanced culture of the present generation has made capable of appreciating a great work of Musical art.

410. It cannot be questioned by any one who carefully considers the subject under the light of adequate knowledge,* that the *creative*

* To represent Science and Poetry as antagonistic, and to maintain (as Mr. Ruskin continually does) that the Poet sees into Nature both *more deeply* and *more truly* than the Scientific inquirer, simply shows this able writer's ignorance of what science really is. The Truths which the man of Science and the Poet respectively discover, are of different orders, but cannot be antagonistic. And if Mr. Ruskin would acquaint himself with the methods and results of Spectrum Analysis, he would find that the strictest scientific reasoning has led to the discovery of truths which the boldest imagination of the Poet would scarcely have conceived, and which his profoundest insight could not possibly have discerned.

Imagination is exercised in at least as high a degree in Science, as it is in Art or in Poetry. Even in the strictest of Sciences—*Mathematics*—it can be easily shown that no really great advance, such as the invention of Fluxions by Newton, and of the Differential Calculus by Leibnitz—*can* be made, without the exercise of the Imagination; and it is interesting to observe that whilst the fundamental idea of both systems had been long pondered over by previous thinkers, it was reserved for the genius of Newton and Leibnitz to divine the two modes of realising and applying it, the introduction of which marks the most important era in the modern history of the science.* —Not less conspicuous evidence of the exercise of this faculty is presented in the highest forms of Mechanical invention. Some of the greatest triumphs of modern ingenuity, however, have been the products rather of the *constructive* than of the *creative* imagination. Thus the Steam-engine of Newcomen was developed into the Steam-engine of Watt by a succession of steps, each of which was to a certain degree suggestive of the next; and it would perhaps not be incorrect to affirm that the *genius* of Watt was more displayed in his invention of the "governor," than in any other part of the wonderful mechanism he devised.

When the complete Double-action Steam-engine, with its heavy fly-wheel, was employed for the execution of any *work* which was variable in regard to the *resistance* to be overcome, it became apparent that some means must be found of adjusting the *force* of the engine to that variation; since, if the engine were suddenly relieved of any large part of that resistance, and were still driven by the same quantity and pressure of steam, its velocity would immediately be accelerated to an injurious and perhaps dangerous extent.

* An eloquent defence of Mathematical Science from the charge of *cramping* the exercise of the Imagination, was made a few years since by Professor Sylvester (whose own contributions to it have been eminently marked by the exhibition of this faculty), in his address as President of the Mathematical Section of the British Association. See "Nature," vol. I., p. 238.

Watt, therefore, set himself to devise for his engine some means of *self-regulation;* and sought it in an entirely new direction—the application of *centrifugal force.* The "governor" consists of a pair of heavy balls, which are so fixed to a vertical spindle as to revolve with it, and to diverge from one another in proportion to its rate of revolution, which depends on that of the fly-wheel. This divergence, by a system of levers, acts on a throttle-valve in the steam-pipe; and thus, whenever a *diminution* in the work of the engine causes the fly-wheel to revolve more rapidly, the *divergence* of the balls *reduces* the quantity of steam admitted to the cylinder, so as to keep the *power* proportional to the *resistance;* whilst, on the other hand, when *more* work is thrown on the engine, the diminished rate of the fly-wheel causes the balls to *approach* each other, so as to open the throttle-valve and admit *more* steam to do it. The great *beauty* of this invention, beyond its perfect adaptation to the required purpose, consists in its utilisation of the very Force which constitutes the source of danger. For it has several times happened that where the "governor" has been accidentally out of gear, the fly-wheel of the engine has revolved so rapidly, when work was taken off, as to break in pieces by its own centrifugal force; the pieces flying apart, and causing loss of life and destruction of property.

The Steam-Hammer of Nasmyth may be cited as a more recent product of *creative* Imagination; and it is worthy of note that its inventor is the son of a distinguished Artist, and himself strongly partakes of the Artistic temperament.

411. In the domain of Art, the *creative* Imagination shows itself in the production of *ideal* representations, whose grandeur and beauty transcend all actual experience, and which appeal to our most refined sensibilities, our most elevated emotions. As typical examples of such creations, no one who is capable of what is truly great in Art would hesitate in naming the finest Madonnas of Raffaelle, and the best Landscapes of Turner, the Venus of Milo, and the Theseus or the Apollo Belvedere. In each of these glorious works, the *mens divinior* most clearly displays itself to all whose own nature is capable of discerning it; and, as already remarked, it is their

noble prerogative to *raise* the nature of such as cultivate the appreciation of them.—Lastly, in the domain of Poetry, few would hesitate in ascribing the highest measure of *creative* power to Shakspere; who possessed, in an unrivalled degree, the power of *idealising* every type of humanity which he brought before our mental vision; and in so doing displayed to us the working of every variety of character, with a fundamental truth which makes his delineations represent, *not* the men or women of any particular time or place, but those of *all* time and *every* place;—his marvellous insight enabling us to see them in his pages more truly than if they were actually performing their several parts in the world's drama before our own eyes.

412. If, now, we inquire into the mode in which Genius works, we find ourselves baffled at the outset by the slightness of our materials; since no one who is unpossessed of the creative Imagination can study its mode of operation in himself, while those who do possess it are seldom given to self-analysis. This much, however, is very clear:—that from whatever source the creative power of Genius is derived, it is capable of being *improved* by culture; and that in its highest exercise it is directed by *knowledge* acquired by study, and disciplined by *judgment* based on extended experience. As was well said by Sir Joshua Reynolds in one of his Discourses, "it is by being conversant with the inventions of others, that we learn to invent, as by reading the thoughts of others we learn to think. It is in vain for Painters or Poets to endeavour to invent, without materials on which the mind may work, and from which inventions must originate. Nothing can come of nothing. Homer is supposed to have been possessed of all the learning of his time; and we are certain that Michael Angelo and Raffaelle were equally possessed of all the knowledge in the art, which had been discovered in the works of their predecessors." The same is undoubtedly true of Handel, Mozart, Beethoven, and all other great Musical inventors.—Another condition, essential to the exercise of

Genius, is the power of *fixing the attention* on the object to be attained; and this power may be exerted, as already shown (Chap. III.), either *automatically* or *volitionally*. With some inventors, the interest of the subject on which their minds may be engaged is so absorbing, that they can with difficulty be drawn off from it; with others, the fixation of the attention can only be sustained by a determined effort. When Newton was asked how he attained his grand results, he replied, "By always thinking about them." But even "always thinking about them" might not have enabled any other mind than Newton's to arrive at them. A somewhat parallel case is afforded by Wordsworth; who has himself given us such an insight into his mode of working, as distinctly marks the dependence of his creative power upon continued meditation on the great themes he had set before himself, and, at the same time, upon the influences surrounding him, alike in his daily life, and during his meditative moods. Still, neither reflection nor the influences of Nature could have *made* Wordsworth a poet; and we can only attribute the faculty he possessed to an inborn gift, the working of which is essentially *automatic*. And it will be shown in the next chapter that there is strong reason to regard this automatic operation as often carried on *beneath the consciousness*, by a mechanism which the Will may bring into action, but of which it thenceforth loses the control. This automatic action of Genius, originally prompted volitionally, is thus distinctly set forth by Wordsworth himself:—

The poems in these volumes (the "Lyrical Ballads") will be found distinguished by this—that each of them has a worthy *purpose*. Not that I always began to write with a distinct purpose formally conceived; but habits of meditation have, I trust, so prompted and regulated my feelings, that my descriptions of such objects as strongly excite those feelings will be found to carry along with them a *purpose*. If this opinion be erroneous, I can have little right to the name of a Poet. For *all good poetry is the spontaneous overflow of powerful feelings*; and though this be true, poems to which any value can be attached were never produced on any variety of subjects but by a man who, being possessed of more than usual organic sensibility, has also thought long and deeply. For our continued influxes of feeling are modified and directed by our thoughts, which are, indeed, the representatives of all our past feelings; and as, by contemplating the relation of these general representatives to each other, we discover what is really important to men, so by the repetition and continuance of this act, our feelings will be connected with important subjects; till at length, if we be originally possessed of much sensibility, such habits of mind will be produced, that, *by obeying blindly and mechanically the impulses of those habits*, we shall describe objects, and utter sentiments, of such a nature, and in such connection with each other, that the understanding of the reader must necessarily be in some degree enlightened, and his affections strengthened and purified.

That power which was thus *acquired* by Wordsworth, as the result of self-discipline, may be almost certainly said to have been *original* with Shakspere, the type of the Poet "of imagination all compact," whose instinctive genius seems to have been alone sufficient, with the aid of his large general culture, to enable him to find the solution of those problems in Human Nature, which Wordsworth set himself purposively and patiently to unravel.—A somewhat parallel contrast has been already drawn (§§ 232, 234) between the pure *spontaneity* of Mozart's creative genius, originally cultured by rigorous training, and the *painstaking* manner in which Haydn, throughout his whole life, applied himself to Musical Composition.

413. Thus, then, we are able to arrive at a pretty clear understanding of what the Will *can*, and what it *cannot* do, in regard to the exercise of the Imagination. As already pointed out, the Will can no more reproduce images which are not actually before the consciousness, than it can recall a word or a fact which has faded from the Memory. It can only *select* from the ideas which *spontaneously* present themselves, those which are most suitable to the object in view; and, *intensifying* these by fixing the attention upon them, it can use them (so to speak) as "feelers" for others,—just as when we

try to recollect something that has been forgotten, by dwelling on the idea most likely to suggest it (§§ 370-374). Then, again, the Will can exercise a most important *control* over the Imagination, by *withdrawing* the Attention from ideas which the Judgment, guided and chastened by the *habitual direction of the thoughts to pure and noble objects*, rejects as degrading and unseemly. And it is for want of such control, that some of those Poets and Prose writers who show the strongest evidences of genius, let their imagination run riot in scenes of impurity, and use their gifts to deprave the Taste and confuse the Moral sense of their readers.*

414. Thus, although the work of the Imagination is itself purely automatic, yet the Will can both *set it going* and *keep it going* by the fixation of the attention; it can in great degree *guide* its activity; and it has (in the well-balanced mind) an entire *control* over its results, so far as regards the *expression* of them. In fact, there is no part of our Intellectual nature, to which the simile of the relation of the horse and his rider (§ 24) is more exactly applicable.—But *indirectly* the Will may do a great deal to improve the action of the Imagination, by the culture and discipline to which it subjects the *general activity* of the Mind. For there can be no doubt that the automatic succession of our Ideas and Feelings is greatly influenced by the direction we voluntarily give to them; so that the imaginative creations of the Poet or the Artist show the impress of the habitual tone of his thoughts; while the inventive power of the Scientific discoverer mainly depends upon the thoroughness of the grasp of his subject, which he has attained by his previous study of it. Further, the development given to the primary Idea will in great degree depend upon the constructive

* The disciples of what has been recently termed the "fleshly school" of Poetry, adopting the theory that all passions, however gross, are fitting subjects for it, because they are Human, push that theory to the extent of endeavouring to glorify the merely animal lust by vivid descriptions of its unrestrained exercise; instead of, like Shakspere, pourtraying it in such a manner as to bring into strong relief the superior beauty of pure love.

power which has been acquired by systematic training: thus, the Painter or Sculptor must have mastered both the principles and practice of his art, before he can produce any really great work; and the Musician must have diligently applied himself to the study of counterpoint and instrumentation, in order to be able to give the fullest expression to his Ideas. In these modes, then, the Imaginative faculty may be directed and invigorated, cultivated and chastened, by Volitional effort; while its *productiveness* depends essentially on its own inherent fertility, and on the energy of its automatic action (§§ 232, 397).

415. Before quitting the subject of the Imagination, it may be well to remark, that although no one can acquire (if he has it not in his original constitution) the creative power of Genius, yet every one can train himself to appreciate its products; his capacity for such appreciation growing and intensifying in proportion as it is exercised aright. The more we fix our attention on the highest ideals of Art, and withdraw ourselves from the influence of those lower forms of it which in any way connect themselves with the grosser parts of our nature (§ 413, *note*), the stronger will be the intuitive preference we shall acquire for what is noble and elevating, the more thorough our intuitive distaste for all that is mean and degrading. And a truly great work of Art not only commands the admiration of such as are already capable of appreciating its excellence, and contributes to form the taste of that much larger number whose dormant capacity for the enjoyment of it needs only to be called forth and rightly directed; but tends to develope in successive generations an ever-increasing capacity for such enjoyment, which will thus add largely to the happiness of the Race.

Thus there can be no doubt that the enjoyment which all persons of cultivated taste now feel in the Picturesque, is a growth of modern times; its germ having been furnished (as the very word implies) by that perception of the beautiful in landscape-scenery, which first began to show itself in the works of Painters scarcely three centuries ago. Not many generations back, Switzerland, now "the playground of Europe," was regarded but as a rugged and toilsome pass, that had to be surmounted by the traveller who desired to find his way to "sunny Italy;" and the reply given to the Writer by a domestic of average intelligence, who had been visiting Tintern Abbey, "Yes, it's very pretty; but what a pity it's in ruins!" would have expressed the general sentiment of the educated public.

That even the least cultivated, however, may have real appreciation for Pictures which express a high ideal of Humanity, appears from the marked preference shown for the best works of this class in the collection of Sir Richard Wallace, during its exhibition at the Bethnal Green Museum.—"Who would not try to be a good woman, who had such a child as that?" was the spontaneous utterance of a female of the artizan class, who had been gazing intently at one of the beautiful representations by Murillo of the Infant Jesus in the arms of his Mother.

CHAPTER XIII.

OF UNCONSCIOUS CEREBRATION.

416. Having thus found reason to conclude that a large part of our Intellectual activity—whether it consist in Reasoning processes, or in the exercise of the Imagination—is essentially *automatic*, and may be described in Physiological language as the *reflex action of the Cerebrum*, we have next to consider whether this action may not take place *unconsciously*. To affirm that the Cerebrum may act upon impressions transmitted to it, and may elaborate Intellectual results, such as we might have attained by the intentional direction of our minds to the subject, *without any consciousness* on our own parts, is held by many Metaphysicians, more especially in Britain, to be an altogether untenable and even a most objectionable doctrine. But this affirmation is only the Physiological expression of a doctrine which has been current among the Metaphysicians of Germany, from the time of Leibnitz to the present date, and which was systematically expounded by Sir William Hamilton,—that the Mind may undergo modifications, sometimes of very considerable importance, without being itself conscious of the process, until its *results* present themselves to the consciousness, in the new ideas, or new combinations of ideas, which the process has evolved. This "Unconscious Cerebration," or "Latent Mental modification," is the precise parallel, in the higher sphere of Cerebral or Mental activity, to the movements of our limbs, and to the direction of those movements through our visual sense, which we *put in train* volitionally when we set out on some habitually-repeated walk, but which then proceed not only *automatically*, but *unconsciously*, so long as our attention continues to be uninterruptedly diverted from them.

(§ 16). And it was by reflection on this parallelism, and on the peculiar structural relation of the Cerebrum to the Ganglionic tract which seems to constitute the *Sensorium* or centre of consciousness, alike for the *external* and for the *internal* senses (§ 100), that the Writer was led (in ignorance of the teachings both of German Metaphysicians, and of Sir William Hamilton, whose lectures had not at that time been published) to the idea that Cerebral changes may take place *unconsciously*, if the Sensorium be either in a state of absolute torpor, or be for a time non-receptive as regards those changes, its activity being exerted in some other direction; or, to express the same fact Psychologically, that mental (?) changes, of whose *results we subsequently* become conscious, may go on *below the plane* of consciousness, either during profound sleep, or while the attention is wholly engrossed by some entirely different train of thought.*

417. To the Writer it seems a matter of no practical consequence, whether the doctrine be stated in terms of Metaphysics or in terms of Physiology—in terms of *mind*, or in terms of *brain*,—provided it be recognised as having a positive scientific basis. But since, in the systems of Philosophy long prevalent in this country, *consciousness* has been almost uniformly taken as the basis of all strictly *mental* activity, it seems convenient to designate as Functions of the Nervous System all those operations which lie

* Subsequently to the first publication of his views on this subject (in the fourth edition of his "Human Physiology," 1852) the Author learned from his friend, Dr. Laycock, to whose Essay on the "Reflex Action of the Brain" he has already referred (§ 94 *note*) as a most important contribution to Mental Physiology, that he had fully intended to convey the idea that such reflex action might be *unconscious*. As no distinct statement was made to that effect, and as all Dr. Laycock's illustrative examples were of a kind in which consciousness was involved, the Writer may be excused for having—in common with others who were following the same line of enquiry—failed to apprehend Dr. Laycock's meaning on this point. But he willingly accepts Dr. Laycock's statement of it: and now re-states the grounds on which he himself independently arrived at the same conclusion, simply as justifying the claim which the question has to a thorough re-consideration on the part of British Metaphysicians.

below that level. And there is this advantage in approaching the subject from the Physiological side,—that the study of the automatic actions of other parts of the Nervous System furnishes a clue, by the guidance of which we may be led to the scientific elucidation of many phenomena that would otherwise remain obscure and meaningless. For, as we have seen, each of the Nervous centres has an independent "reflex" activity of its own, sometimes "primary" or "original," sometimes "secondary" or "acquired;" while our *consciousness* of its exercise depends upon the impression which it makes upon the Sensorium, which is the instrument alike of the external and of the internal senses. Looking, therefore, at all the automatic operations of the Mind in the light of "reflex actions" of the Cerebrum, there is no more difficulty in comprehending that such reflex actions may proceed without our cognizance,—their results being evolved as *intellectual products*, when we become conscious of the impressions transmitted along the "nerves of the internal senses" from the Cerebrum to the Sensorium,—than there is in understanding that impressions may excite muscular movements through the "reflex" power of the Spinal Cord without the necessary intervention of sensation (§ 68). In both cases, the condition of this mode of unconscious operation is, that the *receptivity* of the Sensorium shall be suspended *quoad* the changes in question; either by its own functional inactivity, or through its temporary engrossment by other impressions. —It is difficult to find an appropriate term for this class of operations. They can scarcely be designated as Reasoning processes, since "unconscious reasoning" seems a contradiction in terms. The designation *unconscious cerebration* is perhaps as unobjectionable as any other, and has been found readily intelligible. (See § 428.)

418. The following passages from Sir William Hamilton's Lectures will show the aspect in which this subject presented itself to a pure Metaphysician :—

"Are there, in ordinary, Mental modifications—*i.e.* Mental activities and passivities—of which we are unconscious, but which manifest their existence by effects of which we are conscious?

"I do not hesitate to affirm that what we are conscious of is constructed out of what we are not conscious of;—that our whole knowledge, in fact, is made up of the unknown and the incognisable. . . There are many things which we neither know nor can know in themselves, but which manifest their existence indirectly through the medium of their effects. This is the case with the Mental modifications in question: they are not in themselves revealed to Consciousness; but as certain facts of Consciousness necessarily suppose them to exist, and to exert an influence on the mental processes, we are thus constrained to admit, as Modifications of Mind, what are not phenomena of Consciousness.

"Consciousness cannot exist independently of some peculiar modification of Mind. We are only conscious, as we are conscious of a determinate state or Mental modification, the existence of which supposes a change or transition from some other state or modification. But as the modification must be present before we have a consciousness of it, we can have no consciousness of its rise or awakening, for this is also the rise or awakening of consciousness.—(*Lectures*, Vol. i., pp. 348, 349.)

"To Leibnitz belongs the honour of originating this opinion, and of having supplied some of the strongest arguments in its support. He was, however, unfortunate in the terms which he employed to propound his doctrine. The latent modifications—the unconscious activities of mind—he denominated 'obscure ideas,' 'obscure representations,' 'perceptions without apperception or consciousness,' 'insensible perceptions,' &c. In this he violated the universal use of language. For perception, and idea, and representation all properly involve the notion of consciousness;—it being, in fact, contradictory to speak of a representation not actually represented—a perception not really perceived—an actual idea of whose presence we are not aware. It has been in consequence of this misuse of terms, that the Leibnitzian doctrine has not been more generally adopted; and that, in France and in Britain, succeeding philosophers have admitted as an almost self-evident truth, that there can be no modification of mind devoid of consciousness. As to any refutation of the Leibnitzian doctrine, I know of none."—(*Op. cit.* p. 262.)

Mr. J. S. Mill, however, while fully accepting Sir William Hamilton's statement as to the facts on which his doctrine of "Mental Latency" is based, objects to his mode of expressing them as no less erroneous than that of Leibnitz: considering "unconscious *mental* modification" as a contradiction in terms; and attributing the phenomena to unrecognized changes in the substance of the Brain, which he regards as the constant physical antecedents of mental modifications,—thus explicitly accepting the doctrine of "unconscious cerebration." *

419. A very apposite example of this form of activity is afforded by a phenomenon, which, although familiar to every one who takes note of the workings of his own mind, has been scarcely recognised by Metaphysical inquirers; namely, that when we have been *trying to recollect* (§ 373) some name, phrase, occurrence, &c.,—and, after vainly employing all the expedients we can think-of for bringing the desiderated idea to our minds, have abandoned the attempt as useless,—it will often occur *spontaneously* a little while afterwards, suddenly flashing (as it were) into our consciousness, either when we are thinking of something altogether different, or on awaking out of profound sleep.—Now it is important to note, in the *first* case, that the Mind may have been entirely engrossed in the mean time by some entirely different subject of contemplation, and that we cannot detect any link of association whereby the result has been obtained, notwithstanding that the whole "train of thought" which has passed through the mind in the interval may be most distinctly remembered; and, in the *second*, that the missing idea seems more likely to present itself when the sleep has been profound, than when it has been disturbed. The first form of the phenomenon has been thus admirably described by Miss Cobbe:—

"It is an every-day occurrence to most of us to forget a particular

* See his *Examination of Sir William Hamilton's Philosophy*, chap. xv.

word, or a line of poetry, and to remember it some hours later, when we have ceased consciously to seek for it. We try, perhaps anxiously, at first to recover it, well aware that it lies somewhere hidden in our memory, but unable to seize it. As the saying is, we 'ransack our brains for it,' but, failing to find it, we at last turn our attention to other matters. By-and-bye, when, so far as consciousness goes, our whole minds are absorbed in a different topic, we exclaim, 'Eureka! the word or verse is so-and-so.' So familiar is this phenomenon, that we are accustomed in similar straits to say, 'Never mind; I shall think of the missing word by-and-bye, when I am attending to something else'; and we deliberately turn away, not intending finally to abandon the pursuit, but precisely as if we were possessed of an obedient secretary or librarian, whom we could order to hunt up a missing document, or turn out a word in a dictionary, while we amused ourselves with something else. The more this common phenomenon is studied, the more I think the observer of his own mental processes will be obliged to concede, that, so far as his own conscious self is concerned, the research is made absolutely *without him*. He has neither pain, nor pleasure, nor sense of labour in the task, any more than if it were performed by somebody else; and his conscious self is all the time suffering, enjoying, or labouring on totally different grounds." —(*Macmillan's Magazine*, November, 1870, p. 25.)

So says also Oliver Wendell Holmes :—

"We wish to remember something in the course of conversation. No effort of the will can reach it; but we say, 'Wait a minute, and it will come to me,' and go on talking. Presently, perhaps, some minutes later, the idea we are in search of comes all at once into the mind, delivered like a prepaid parcel laid at the door of consciousness, like a foundling in a basket. How it came there, we know not. The mind must have been at work, groping and feeling for it in the dark; it cannot have come of itself. Yet, all the while, our consciousness, *so far as we are conscious of our consciousness*, was busy with other thoughts."—(*Mechanism in Thought and Morals*, p. 43.)

420. So frequently has this occurred within the Writer's experience, that he is now in the habit of trusting to this method of recol-

lection, where he has reason to feel sure that the desired idea is not far off, if the mind can only find its track ;—as when it relates to some occurrence (such as a payment of money, or meeting with a person whose face is familiar to him, but whose name he cannot recall) which he knows to have taken place within a few days previously. For he has found himself much more certain of recovering it, by withdrawing his mind from the search when it is not speedily successful, and by giving himself up to the occupation appropriate to the time, than by inducing fatigue by unsuccessful efforts. And this is not his own experience only, but that of many others. The fact has been noticed by Sir H. Holland ;* from whom he has learned that the above plan has been put into successful action by many to whom he had recommended it. By Sir Henry, however, the success was regarded as due simply to the refreshment which the mind has received by change of thought : but the considerations to be presently adduced seem to justify the belief, that the train of action, which we volitionally set going in the Cerebrum in the first instance, continues to work by itself after our attention has been fixed upon some other object of thought ; so that it goes on to the evolution of its result, not only without any continued exertion on our own parts, but also without our consciousness of any continued activity. The advantage of thus detaching our attention from it seems to be, that it runs on *undisturbed* by our fruitless and distracting attempts ; just as, to use Miss Cobbe's happy illustration, our "obedient secretary" is more likely to find what we want, when we leave him to search for it *in his own way*, instead of worrying him with continual directions to look in this, that, or the other place.

421. The following circumstance, mentioned by Mr. Macgregor in his "Thousand Miles in the Rob Roy Canoe," is a good example of this automatic reproduction of a lost idea, after the will had searched for it in vain :—

* "*Chapters on Mental Physiology*," p. 66.

When on the Meurthe, he saw three women on the bank of the river, in great alarm, who searched in vain for two boys supposed to have gone away to fish, but missing for many hours. They eagerly asked Mr. Macgregor to tell them whether he had seen them, and implored him with tears to advise them what to do. "I tried," he says, "all I could to recollect; but no, I had not seen the boys; and so the women went away distracted, and left me sorrowful. But suddenly, *when toiling in the middle of a very difficult piece of rock-work*, lowering the boat (and therefore *no longer trying to remember*), I remembered having seen those boys, so I ran over the fields after the anxious mamma, and soon assured her that the children had been safe an hour ago."

422. Another instance, which has been kindly communicated to the Writer (with others to be cited hereafter, § 435) by a Graduate of the University of London, as having occurred to himself, is interesting from the evidence it affords that the automatic reproduction of the lost impression is a matter of *time;* the process having apparently not been completed, so as to produce the recognition by the *mind* of the desiderated name, when it was first presented to the *eye* :—

"One day I was summoned to a Town at some distance, to see a friend lying dangerously ill at a Physician's house. While in the railway-train, I found that I could not remember either the name of the Physician or his address. I vainly endeavoured to recall them. I became much excited; but bethought me that if I consulted a Post Office Directory, I should see and recognize the name. I consulted the Directory on reaching the hotel, but the name seemed not to be there. Soon after, while I was ordering some refreshment, the name flashed on my consciousness. I left the astounded waiter, rushed to the Directory, and there saw the name; and what is more, *I am sure that I had noticed it on my first inspection, without recognizing it as the name I sought.*"

423. Of the recovery of a lost idea after profound sleep, the following incident (which was communicated to the Writer as of recent occurrence, after a Lecture he had been giving on this

subject, in 1868, at the Royal Institution) affords a remarkable example :—

The Manager of a Bank in a certain large town in Yorkshire could not find a duplicate key, which gave access to all the safes and desks in the office, and which ought to have been in a place accessible only to himself and to the Assistant-manager. The latter being absent on a holiday in Wales, the Manager's first impression was that the key had probably been taken away by his Assistant in mistake; but on writing to him, he learned to his great surprise and distress that he had not got the key, and knew nothing of it. Of course, the idea that this key, which gave access to every valuable in the bank, was in the hands of any wrong person, having been taken with a felonious intention, was to him most distressing. He made search everywhere, tried to think of every place in which the key might possibly be, and could not find it. The Assistant-manager was recalled, both he and every person in the bank were questioned, but no one could give any idea of where the key could be. Of course, although no robbery had taken place up to this point, there was the apprehension that a robbery might be committed after the storm had blown over, when a better opportunity might be afforded by the absence of the same degree of watchfulness. A first-class detective was then brought down from London, and this man had the fullest opportunity given him of making inquiries; every person in the bank was brought up before him; he applied all those means of investigation which a very able man of this class knows how to employ; and at last he came to the Manager and said, "I am perfectly satisfied that no one in the bank knows anything about this lost key. You may rest assured that you have put it away somewhere yourself; but you have been worrying yourself so much about it, that you have forgotten *where* you put it away. As long as you worry yourself in this manner, you will not remember it; but go to bed to-night with the assurance that it will be all right; get a good night's sleep; and in the morning you will most likely remember where you have put the key." This turned out exactly as was predicted. The key was found the next morning in some extraordinarily secure place, which the Manager had not previously thought of, but in which he then felt sure he must himself have put it.

424. It is a most remarkable confirmation of this view, that Ideas which have passed out of the *conscious* memory, sometimes express themselves in *involuntary muscular movements,* to the great surprise of the individuals executing them. Thus, while the answers given by the "talking-tables," or by the pointing or writing of the "planchette," in most instances express the ideas consciously present to the minds of the operators (§ 252), true answers are often given to questions as to matters of fact, notwithstanding that there may be either entire *ignorance* (proceeding from complete forgetfulness) of those facts, or absolute *disbelief* in the statement of them. The following examples of this remarkable occurrence (which were attributed by their narrators to "spiritual" agency) are selected from many that might be cited to the same effect :—

a. The Writer was assured that a "planchette," made in Bath, which had been on a visit in various families for several months, having been asked where it was made, replied "Bath;" although the questioners all thought it came from *London*, and disbelieved its statement, which was afterwards verified.

b. The Rev. Mr. Dibdin, M.A. (in his *Lecture on Table-Turning* published in 1853), states that he and a friend having directed the table to say, "How many years is it since her Majesty came to the throne?" the table struck *sixteen*, though no one present knew the date of her accession; and having directed it to "give the age of the Prince of Wales," which was not known either to Mr. Dibdin or his friend, the table struck *eleven*, and then raised the foot a little way. On referring to an Almanack, both these numbers were found to be correct.—Further, the question being put (in the house of a tailor), "How many men are at work in the shop below?" the table replied by striking *three*, and giving *two* gentle rises; on which the employer, who was one of the party, said, "There are *four* men and *two* boys, so *three* is a mistake;" but *he afterwards remembered* that one of the young men was out of town.

425. This last fact affords a simple and direct clue to much that seems mysterious in these phenomena : for it is obviously as unphilosophical to attribute to an assumed "spiritual" agency the

motions of the hands which give written expression to ideas formerly—though not at the time—present to the consciousness, as it would be to attribute to the like "possession" the utterance of words and sentences in a language that has been entirely forgotten, and was perhaps never even understood (§ 344 *d*). In the one case, as in the other, the records of the old impressions, left in the deeper stratum of unconsciousness, disclose their existence through the automatic motor apparatus. And that this is the true account of the phenomena in question, seems evident from other examples, in which (as in the last of the cases just now cited) the subsequent reminiscence proved that the idea, which was contrary to the belief of the questioner at the time, was the correct reproduction of one which had been formerly recorded, but which had passed out of the conscious memory:—

a. The author of an article on "Spiritualism and its Recent Converts," states himself to have been informed by an eminent literary man, in whose veracity he had the fullest confidence, that "the spirit of a friend, whose decease had taken place some months previously, having announced itself in the usual way, and the question having been put, 'When did I last see you in life?' the answer given was inconsistent with the recollection of the interrogator. But, on his subsequently talking over the matter with his family, it was brought to his remembrance that he *had* seen his deceased friend on the occasion mentioned, and had spoken of it to them at the time, although he had afterwards quite forgotten the circumstance."—(*Quarterly Review*, October, 1871, p. 319.)

b. Another instance, supplied by Mr. Dibdin (*op. cit.*), affords yet more remarkable evidence to the same effect; especially as being related by a firm believer in the "diabolical" origin of Table-talking:—A gentleman, who was at the time a believer in the "spiritual" agency of his table, assured Mr. Dibdin that he had raised a *good* spirit instead of *evil* ones—that, namely, of Edward Young, the poet. The "spirit" having been desired to prove his identity by citing a line of his poetry, the table spelled out, "Man was not made to question, but adore." "Is that in your 'Night Thoughts'?" was then asked. "No."

"Where is it, then?" The reply was, "J O B." Not being familiar with Young's Poems, the questioner did not know what this meant; but the next day he bought a copy of them; and at the end of the "Night Thoughts" he found a paraphrase of the Book of Job, the last line of which is, "Man was not made to question, but adore." Of course he was very much astonished; but not long afterwards he came to Mr. Dibdin, and assured him that he had satisfied himself that the whole thing was a delusion,—numerous answers he had obtained being obviously the results of an influence unconsciously exerted on the table by those who had their hands upon it; and when asked by Mr. Dibdin how he accounted for the dictation of the line by the spirit of Young, he very honestly confessed, "Well, the fact is, I must tell you, that I had the book in my house all the time, although I bought another copy; and *I found that I had read it before.* My opinion is that it was *a latent idea,* and that the table brought it out."

426. There are other cases, again, in which two distinct trains of Mental action are carried on simultaneously,—one *consciously,* the other *unconsciously;* the latter guiding the movements, which may express something quite unrelated to the subject that *entirely* and *continuously* engrosses the attention. This is only a higher form of the automatic movements already referred to (§ 194) as executed under the like circumstances; those now in question being the expressions of *mental,* not of mere *bodily* habit,—that is to say, the resultants of a previous training, which has left its impress on the organization. Here, again, we profit by Miss Cobbe's graphic sketches (*loc. cit.*, p. 26):—

"We read aloud, taking-in the appearance and proper sound of each word, and the punctuation of each sentence; and all the time we are not thinking of these matters, but of the argument of the author; or picturing the scene he describes; or, possibly, following a wholly different train of thought.

"Similarly, in writing with 'the pen of a ready writer,' it would almost seem as if the pen itself took the business of forming the letters and dipping itself in the ink at proper intervals, so engrossed are we in the thoughts which we are trying to express. We unconsciously cerebrate—while we are all the time consciously buried in our subject—that it will not answer to begin two consecutive sentences in the same way; that we must introduce a query here, or an ejaculation there, and close our paragraphs with a sonorous word and not with a preposition. All this we do, not of *malice prepense*, but because the well-tutored sprite, whose business it is to look after our *p*'s and *q*'s, settles it for us, as a clerk does the formal part of a merchant's correspondence.

"Music-playing is of all others the most extraordinary manifestation of the powers of unconscious cerebration. Here we seem not to have one slave, but a dozen. Two different lines of hieroglyphics have to be read at once, and the right hand has to be guided to attend to one of them, the left to another. All the ten fingers have their work assigned as quickly as they can move. The mind, or something which does duty as mind, interprets scores of A sharps and B flats and C naturals into black ivory keys and white ones, crotchets and quavers and demi-semiquavers, rests, and all the mysteries of music. The feet are not idle, but have something to do with the pedals; and, if the instrument be a double-action harp [or an organ], a task of pushings and pullings more difficult than that of the hands. And all this time the performer, the *conscious* performer, is in a seventh heaven of artistic rapture at the results of all this tremendous business, or perchance lost in a flirtation with the individual who turns the leaves of the music-book, and is justly persuaded she is giving him the whole of her soul."

427. Another example of "latent" mental or rather Cerebral action is afforded, as Sir William Hamilton has pointed out, by the process (first noticed by Hartley) whereby one idea, A, comes *directly* to suggest another idea, C, to which it is unrelated; the link of connection being supplied by a former *intermediate* idea, B, which has passed altogether out of the consciousness. A careful analysis of the sequence of our *conscious* mental action would show that this is an extremely common occurrence; in fact, it is the basis of the whole doctrine of *resultants*, which has been already advanced as the *rationale* of our Common-sense judgments (§ 382).

The following circumstance, which happened to the Writer within a few hours after penning the above sentences, might seem almost too trivial to be recorded, if it were not so exactly "to the point." His friend Dr. Sharpey having for years acted as one of the Secretaries to the Royal Society, the writer had been in the constant habit of communicating with him on Royal Society business. But a few months previously, Dr. S. had resigned this post; a fact of which the writer was most fully cognizant at the time, and the recollection of which would have prevented him from applying to Dr. S. on any secretarial matter. Having wished to obtain some information which it would have been the function of the Secretary to give, and meeting Dr. S. at the "Athenæum," he at once asked him for it, as he had been wont to do; the mere *recognition* of Dr. S. prompting this application (as a reflex action of the Cerebrum), *without the conscious excitement of the idea* of Dr. Sharpey's Secretariat, which had originally been the connecting link. Had this idea been brought up, the writer is sure that he should have at once *remembered* Dr. Sharpey's resignation; more especially since he had been speaking to Professor Huxley, his successor in the Secretariat, only a few minutes before, in the same room, with reference to the duties of his office.

428. The explanation of this fact usually adopted by British Metaphysicians, is that the intermediate idea, B, is in reality *momentarily* present to the consciousness, but that its presence is not remembered. This is, however, a mere assumption, resting entirely on a foregone conclusion, and incapable of any kind of proof. For, as Mr. J. S. Mill has well remarked, when this obliteration of the intermediate ideas is complete "it is, to our subsequent consciousness, exactly as if we did not have them at all; we are incapable, by any self-examination, of being aware of them." Mr. Mill entirely agrees with Sir William Hamilton as to the facts; but prefers the Physiological mode of expressing them: "If we admit (what physiology is rendering more and more probable) that our mental feelings, as well as our sensations, have for their physical antecedents particular states of the nerves; it may well be believed

that the apparently suppressed links in a chain of association really are so; that they are not, even momentarily, felt; the chain of causation being continued only physically by one organic state of the nerves succeeding another so rapidly that the state of mental consciousness appropriate to each is not produced."—(*Examination of Sir William Hamilton's Philosophy*, p. 285.)

429. It is by a process of *direct suggestion* of ideas through sensations which have originally acted through an intermediate succession of Mental states, that, as already pointed out, we derive our notion of *solid form*, by a process of *mental construction*, from the two dissimilar perspectives projected on our retinæ (§ 168). And a like case of *abbreviation*, by the omission of the intermediate processes by which a composite idea was originally formed, is supplied by the two operations to be next adverted to; the first of them familiar to every one, the second well known among experts:—When we apply ourselves to the perusal of a book, for the purpose of making ourselves acquainted with the author's meaning, if its subject be one into which we readily enter, if the writer's flow of thought be in a course which we easily follow, and if his language be appropriate to express his ideas, we acquire the meaning of one *sentence* after another, without any conscious recognition of the meaning of each of its component *words*; it being only when the language is ill chosen, or when we do not readily follow the author's train of thought, that we direct our attention to the signification of the individual words, and become conscious of their separate meaning. Yet it is certain that a particular impression must have been made by each of these words upon the Cerebrum, before we can comprehend the notion which they were collectively intended to convey; and when a child is first learning to read, that impression gives rise to a distinct idea of the meaning of each word separately.—So an expert calculator, who may have originally had no more than an ordinary facility in apprehending the relations of numbers, casting his eye rapidly from the bottom to the top of

a column of figures, will name the total without any conscious appreciation of the value of each individual figure; having acquired by practice somewhat of that *immediate insight*, which is so remarkable a form of intuition in certain rare cases (§ 205). It is certain that a distinct ideational state must have been *originally* called up by the sight of each individual figure; and yet an impression made by it upon the Cerebrum, which does not produce any *conscious* recognition of its numerical value, comes to be adequate for the evolution of the result.

430. But whilst, in the preceding instances, no higher act of Mind is required, than the production of one ideational *resultant* from the combination of simpler elements, there are cases in which processes of a far more elaborate nature are carried on, without necessarily affecting our consciousness. Most persons who attend to their own Mental operations are aware, that when they have been occupied for some time about a particular subject, and have then transferred their attention to some other, the first, when they return to the consideration of it, may be found to present an aspect very different from that which it possessed before it was put aside; notwithstanding that the mind has since been so completely engrossed with the second subject, as not to have been consciously directed towards the first in the interval. Now a part of this change may depend upon the altered condition of the Mind itself; such as we experience when we take up a subject in the morning, with all the vigour which we derive from the refreshment of sleep, and find no difficulty in overcoming obstacles and disentangling perplexities, which checked our further progress the night before, when we were too weary to give more than a languid attention to the points to be made out, and could use no exertion in the search for their solutions. But this by no means accounts for the *entirely new development* which the subject is frequently found to have undergone, when we return to it after a considerable interval; a development which cannot be reasonably explained in any other mode,

than by attributing it to an intermediate activity of the Cerebrum, which has automatically evolved the result without our consciousness. This was long since pointed out by Abraham Tucker, who says :—" With all our care to digest our materials, we cannot do it completely ; but after a night's rest, or some recreation, or the mind being turned into some different course of thinking, *she finds they have ranged themselves anew during her absence*, and in such manner as exhibits almost at one view all their mutual relations, dependences, and consequences ;—which shows that our organs do not stand idle the moment we cease to employ them, but continue the motions we put into them after they have gone out of sight, thereby working themselves to a glibness and smoothness, and falling into *a more regular and orderly posture than we could have placed them with all our skill and industry*."* This experience was thus recorded by Sir Benjamin Brodie :—

"It seems to me that on some occasions a still more remarkable process takes place in the Mind, which is even more independent of volition than that of which we are speaking; as if there were in the mind a principle of order, which operates without our being at the time conscious of it. It has often happened to me to have been occupied by a particular subject of inquiry; to have accumulated a store of facts connected with it ; but to have been able to proceed no further. Then, after an interval of time, without any addition to my stock of knowledge, I have found the obscurity and confusion in which the subject was originally enveloped to have cleared away; the facts have seemed all to settle themselves in their right places, and their mutual relations to have become apparent, although I have not been sensible of having made any distinct effort for that purpose."—(*Psychological Inquiries*, vol. i. p. 20.)

431. There is considerable ground to believe that the best *judgments* are often mentally delivered, in difficult cases, by the unconscious resolution of the difficulties in the way of arriving at a conclusion, when the question (after being *well considered* in the

* "*Light of Nature Pursued*," 2nd edition (1805), vol. i. p. 358.

first place) is left to *settle itself.* The following extract from the Report of a Lecture delivered by the Writer, not long since, expresses not merely his own experience, but what he has gathered from the experience of others :—

"It has on several occasions occurred to me to have to form a decision as to some important change, either in my own plans of life, or in those of members of my family, in which were involved a great many of what we are accustomed to call *pros* and *cons*,—that is, in which there was a great deal to be said on both sides. I heard the expression once used by a Naturalist, with regard to Classification, 'It is very easy to deal with the *white* and the *black;* but the difficulty is to deal with the *grey.*' And so it is in Life. It is perfectly easy to deal with the white and the black;—there are things which are clearly right, and things which are clearly wrong; there are things which are clearly prudent, and things which are clearly imprudent;—but a great many cases arise, in which even right and wrong may seem questionable, or opposing motives in themselves good may be so balanced that it is difficult to see where our duty lies; and again there are cases in which it is difficult to say *what* is prudent :—and I believe that in all such cases, where we are not hurried and pressed for a decision, our best plan is to let the question *settle itself* by Unconscious Cerebration; having first brought before our minds, as fully as possible, everything that can be fairly urged on both sides. We discuss the question in our own family circle; then we go to our friends, who very probably suggest considerations that did not occur to ourselves, but who cannot feel as *we* do the strength of some of the considerations involved; and then it will be found the best way to *put the matter altogether aside* for a month or two (if that time can be allowed), and to turn the current of thought and feeling into *some entirely different channel.* It is often wonderful, on returning to the subject after such an interval, to find how unhesitatingly the Mind then gravitates, how distinctly the balance of judgment then turns. I feel convinced that, in the habitually well-disciplined nature, this unconscious operation of the Brain, in balancing for itself all these considerations, in putting all in order (so to speak), and in working out the result, is far more likely to lead us to a good and true decision, than continual discussion and argumentation. For when one argument is pressed with what we feel to be undue force, this leads us to bring up something on the other side; so that we are driven into antagonism by what we think the undue pressure of the force which is being exerted. No true balance can be struck until this excitement has subsided; and when it *has*, the inclination of the balance will be determined by the whole previous training and discipline of our Minds, which will be the more likely to give to it the right direction, in proportion as we have *habitually* and *determinately* shaped our course of *conscious* action under the direction of the highest motives."

432. The Writer has received from a distinguished Prelate the following account of his own frequently-repeated experience of another form of the same kind of action:—

"I have for years been accustomed to act upon your principle of 'Unconscious Cerebration,' with very satisfactory results. I am frequently asked, as you may suppose, to preach *occasional* sermons; and when I have undertaken any such duty, I am in the habit of setting down and thinking over the topics I wish to introduce, without in the first instance endeavouring to frame them into any consistent scheme. I then put aside my sketch for a time, and give my mind to some *altogether different subject;* and when I come to write my sermon, perhaps a week or two afterwards, I very commonly find that the topics I set down have *arranged themselves*, so that I can at once apply myself to develope them on the plan in which they then present themselves before me."

433. The following example, furnished by O. Wendell Holmes, is interesting as one in which the individual was *conscious of the flow of an under-current* of mental action, although this did not rise to the level of distinct ideation:—

a. "I was told, within a week, of a business-man in Boston, who, having an important question under consideration, had given it up for the time as too much for him. But he was conscious of an action going on in his brain, which was so unusual and painful as to excite his apprehensions that he was threatened with palsy, or something of that sort. After some hours of this uneasiness, his perplexity was all at once cleared up by the natural solution of his doubts coming to him—worked out, as he believed, in that obscure and troubled interval."—(*Op. cit.*, p. 47.)

And it is well said by the same able author, who combines no small measure of the intuition of the Poet with the acquired knowledge of the Physiologist and the Physician :—

b. "I question whether persons who think most—that is, have most conscious thought pass through their minds—necessarily do most mental work. The tree you are sticking in 'will be growing when you are sleeping.' So with every new idea that is planted in a real thinker's mind: it will be growing when he is least conscious of it. An idea in the brain is not a legend carved on a marble slab: it is an impression made on a living tissue, which is the seat of active nutritive processes. Shall the initials I carved in bark increase from year to year with the tree? and shall not my recorded thought develope into new forms and relations with my growing brain?"—(*Op. cit.*, p. 68.)

434. The same mode of action seems to have a large share in the process of *invention*, whether Artistic or Poetical, Scientific or Mechanical. For it is a common experience of inventors (whether Artists, Poets, or Mechanicians), that when they have been brought to a stand by some difficulty, *the tangle will be more likely to unravel itself* (so to speak) *if the attention be completely withdrawn from it*, than by any amount of continued effort. The Writer has taken every opportunity that has presented itself to him, of asking *creators* in various departments of Art and Science, what their experience has been in regard to difficulties which they have felt, and which they have after a time overcome; and the experience has been almost always the same. They have kept the result which they have wished to obtain strongly before their attention in the first instance, just as we do when we "try to recollect" something we have forgotten, by thinking of everything likely to lead to it; but, if they do not succeed, they put it aside for a

time, and give their minds to something else, endeavouring to obtain either complete repose of mind, or refreshment by change of occupation ; and they find that either after sleep, or after some period of recreation by a variety of employment, just what they want "comes into their heads."—The following is told of Charlotte Bronte by her Biographer, Mrs. Gaskell :—

a. "She said that it was not every day that she could write. Sometimes weeks or even months elapsed before she felt that she had anything to add to that portion of her story which was already written. Then, some morning she would waken up, and the progress of her tale lay clear and bright before her in distinct vision, its incidents and consequent thoughts being at such times more present to her mind than her actual life itself."—(*Life*, p. 234.)

b. "Whenever she had to describe anything which had not fallen within her own experience, it was her habit 'to think of it intently many and many a night before falling to sleep, wondering what it was like, or how it would be;' till at length, sometimes after the progress of her story had been arrested at this one point for weeks, she wakened up in the morning with all clear before her, as if she had in reality gone through the experience, and then could describe it word for word as it had happened."—(*Life*, p. 425.)

So of the late Mr. Appold—the inventor of the centrifugal pump, which attracted much attention in the International Exhibition of 1851, as well as of many other ingenious applications of scientific principles to practical purposes—it is recorded that :—

c. "It was his habit, when a difficulty arose, carefully to consider the exact result he required ; and having satisfied himself upon that point, he would direct his attention to the simplest mode in which the end could be attained. With that view he would during the day bring together in his mind all the facts and principles relating to the case ; and the solution of the problem usually occurred to him in the early morning after sleep. If the matter was difficult, he would be restless and uneasy during the night; but after repose, when the brain had recovered from fatigue, and when in the quiet of the early morning no external influences distracted his attention, the resultant of all known scientific principles bearing upon the question presented itself to his mind."—(*Proceedings of the Royal Society*, vol. xv. p. 5.)

435. The following instances, communicated to the Writer by the gentleman (§ 422) in whose experience they have occurred, are of special interest, as showing, *first,* how *nearly complete* the "circuit of thought" (as in the subsequent illustration) may remain for a long time without being "closed;" and *secondly,* how readily its result may fade from the conscious memory, if it is not recorded at the moment of production. It is within the experience of most persons of active minds, that they can distinctly remember being struck by some particular "happy thought," which has afterwards entirely escaped them through not having been noted down at the time; and it is a prudent system, therefore, to have a memorandum-book always at hand, for the registration of all noteworthy ideas.

a. "When at school, I was fond of trying my hand at geometrical problems. One baffled me. I often returned to it, in fact kept by me an elaborate figure. Some years after, and when the problem had not been touched by me for some time, I had been sitting up till the small hours, deciphering a cryptograph for one of my pupils. Exulting in the successful solution, I turned into bed; and suddenly there flashed across my mind the secret of the solution of the problem I had so long vainly dealt with, this secret being a slight addition to my elaborate figure. The effect on me was strange. I trembled, as if in the presence of another being who had communicated the secret to me.

b. "Another time, an algebraical sum had plagued me for a day or two. I could not get the desired result. Some weeks after, on returning from a social gathering, I retired, thinking of the pleasant evening I had spent; when suddenly it flashed across me that there was an error in the sum as set. I leaped out of bed with the same mysterious feeling upon me, wrote down the involved expression with the suggested correction, worked the sum, and obtained the desired result. Strange to say, some weeks afterwards I took the sum from the book, but could not discover what change should be made; and it was not until I found the scrap of paper upon which I had worked it that night, that I could correct the sum in the book.

"I select the above from a large number of similar experiences. In fact, it is my habit to retire to bed, if I meet with any great difficulty in mathematics; and frequently, when my mind seems occupied with something else, and generally just as I seem falling asleep, the difficulty vanishes, the problem is solved. This has often caused me some anxiety; for I have regarded it as an *unhealthy* action of the brain."

436. We seem justified in citing as an example of the same process, one of the most admirable discoveries in modern Mathematics, —that of the method of "Quaternions." Its author, the late Sir W. Rowan Hamilton, was eminently distinguished for the possession of that Poetic faculty, which some have supposed to be incompatible in its very nature with the severe requirements of Mathematical study; but which shows itself in his exposition of his system, in "that exquisite charm of combined beauty, power, and originality, which made Hamilton himself compare Lagrange's great work to a scientific poem," and which flashes out in the following description he gave (in a letter to a friend) of the mode in which the first conception of his great discovery occurred to him:—

"To-morrow will be the fifteenth birthday of the Quaternions. They started into life, or light, full-grown, on the 16th of October, 1843, as I was walking with Lady Hamilton to Dublin, and came up to Brougham Bridge. That is to say, I then and there felt the galvanic circuit of thought *close;* and the sparks which fell from it were *the fundamental equations between* i, j, k; *exactly such* as I have used them ever since. I pulled out, on the spot, a pocket-book, which still exists, and made an entry, on which, *at the very moment,* I felt that it might be worth my while to expend the labour of at least ten (or it might be fifteen) years to come. But then it is fair to say that this was because I felt a *problem* to have been at that moment *solved,* —an intellectual *want relieved,*—which had *haunted* me for at least *fifteen years before.*"—(*North British Review*, Vol. xlv. p. 57.)

437. The following statement of the mode in which an important

scientific invention came into being, was furnished to the Writer by the inventor himself, Mr. F. H. Wenham; an amateur Optician of great ability, who has devoted much time and attention to the construction of the Microscope, and has devised many useful improvements, both in the instrument itself, and in the apparatus associated with it.

The first form of Binocular Microscope (designed to take advantage of the principle of Stereoscopic combination of two dissimilar perspectives, discovered by Wheatstone, § 168) was devised by M. Nachet, on a plan which might readily suggest itself to any well-informed Optician, who should give sufficient thought to the requirements of the case:—that of dividing the cone of rays proceeding from the object-glass, into its *right-hand* and its *left-hand* halves, by the interposition of an equiangular prism; and then subjecting each half to a second reflection, by which it should be brought into the required direction. This construction was perfect in theory, but had two practical defects; (1.) that both half-cones were subjected to two reflections, and to transmission through four surfaces, each of which changes involved a certain loss of light and a certain liability to error; and (2.) that the instrument could *only* be used as a *binocular* Microscope.—Now it occurred to Mr. Wenham, that it might be possible so to divide the cone of rays, that one-half of it should go straight on without any interruption, while the other half alone should be deflected by a single prism, passing through two surfaces only; whereby greater distinctness would be secured in the *direct* image, whilst, by the withdrawal of the prism, the instrument might be used in the ordinary way for purposes to which the Binocular microscope could not be applied. He thought of this a great deal, without being able to hit upon the form of prism which would do what was required; and as he was going into business as an Engineer, he put his microscopic studies entirely aside for more than a fortnight, attending only to his other affairs. One evening, after his day's work was done, and "while he was reading a stupid novel," thinking nothing whatever of his microscope, the form of the prism that should answer the purpose flashed into his mind. He fetched his mathematical instruments, drew a diagram of it, and worked out the angles which would be required; the next morning he made his prism, and found that it answered perfectly well; and it has been on this plan that all the "binoculars" hitherto in ordinary use in this country have been since constructed.

438. The more thoroughly, then, we examine into what may be termed the Mechanism of Thought, the more clear does it become that not only an *automatic*, but an *unconscious* action enters largely into all its processes. As O. Wendell Holmes has well remarked:—

a. "Our *definite ideas* are stepping-stones; how we get from one to the other, we do not know: something carries us; *we* [*i. e.* our conscious selves] do not take the step. A creating and informing spirit, which is *with* us, and not *of* us, is recognised every where in real and in storied life. It is the Zeus that kindled the rage of Achilles; it is the Muse of Homer; it is the Daimon of Socrates; it is the inspiration of the Seer; it is the mocking spirit that whispers to Margaret as she kneels at the altar; and the hobgoblin that cried "Sell him, sell him!" in the ear of John Bunyan; it shaped the forms that filled the soul of Michael Angelo, when he saw the figure of the great lawgiver in the yet unhewn marble, and the dome of the world's yet unbuilt Basilica against the black horizon; it comes to the least of us as a voice that will be heard; it tells us what we must believe; it frames our sentences; it lends a sudden gleam of sense or eloquence to the dullest of us all; we wonder at ourselves, or rather not at ourselves, but at this divine visitor, who chooses our brain as his dwelling-place, and invests our naked thought with the purple of the kings of speech or song."—(*Op. cit.* p. 59.)

439. But it is not *intellectual* work alone, that is done in this manner; for it seems equally clear that *emotional* states, or rather states which constitute Emotions when we become conscious of them, may be developed by the same process; so that our feelings towards persons and objects may undergo most important changes, without our being in the least degree aware, until we have our attention directed to our own mental state, of the alteration which has taken place in them.—A characteristic example of this kind of action, is afforded by an occurrence extremely common in real life, and con-

tinually reproduced in fiction; namely, the growing-up of a powerful attachment between individuals of opposite sexes, without either being aware of the fact; its full strength being only revealed to the consciousness of each, when circumstances threaten a separation, or when both are exposed to a common danger. An *éclaircissement* then takes place; the Love which each has come unconsciously to entertain for the other is first *self-revealed*, and then, that of each becoming apparent to the other, it suddenly bursts forth, like a smouldering fire, into full flame. The existence of a mutual attachment, indeed, is often recognised by a by-stander (especially if the perception be sharpened by jealousy, which leads to an intuitive interpretation of many minute occurrences that would be without significance to an ordinary observer), before either of the parties has made the discovery, whether as regards the individual *self*, or the beloved *object;* the Cerebral state manifesting itself in action, although no distinct consciousness of that state has been attained, chiefly because, the whole attention being attracted by the present enjoyment, there is little disposition to introspection.—The fact, indeed, is recognised in our ordinary language; for we continually speak of the "feelings" which we *unconsciously* entertain towards another, and of our not becoming aware of them until some circumstance calls them into activity.

440. Here again, it would seem as if the material organ of these Feelings tends to *form itself* in accordance with the impressions habitually made upon it; whilst we may be as completely unaware of the changes which have taken place in it, as we are of those by which passing events have been registered in our memory, until some circumstance calls-forth the conscious manifestation, which is the "reflex" of the new condition which the organ has acquired. And it is desirable, in this connection, to recall the fact that the Emotional state seems often to be determined by circumstances of which the individual has no Ideational consciousness, and especially by the emotional states of those by whom he is surrounded

(§189); a mode of influence which acts with peculiar potency on the minds of Children, and which is a most important element in their Moral education. As Dr. Bushnell (an American Divine) has well remarked :—

"Men are ever touching unconsciously the springs of motion in each other; one man, without thought or intention or even a consciousness of the fact, is ever leading some others after him. . . . There are two sorts of influence belonging to Man: that which is active and voluntary, and that which is unconscious; that which we exert purposely, or in the endeavour to sway another, as by teaching, by argument, by persuasion, by threatenings, by offers and promises, and that which flows out from us, unawares to ourselves. . . . The more stress needs to be laid on this subject of *insensible* influence, because it *is* insensible, because it is *out of mind*, and, when we seek to trace it, beyond a full discovery. If the doubt occur whether we are properly responsible for an influence which we exert insensibly, I reply that we are not, except so far as this influence flows directly from our character and conduct. And this it does even much more uniformly than our active conduct. In the latter we may fail in our end, though animated by the best motives, through a want of wisdom or skill; or, again, we may really succeed and do great good by our active endeavours, from motives altogether base and hypocritical. But the influences we exert unconsciously will hardly ever disagree with our real character. They are honest influences, following our character as the shadow follows the sun. And therefore we are much more certainly responsible for them, and for their effects on the world. They go streaming from us in all directions, though in channels that we do not see, poisoning or healing, around the roots of society, and among the hidden wells of character. If good ourselves, they are good; if bad, they are bad.

"Then, if we go over to others, that is to the subjects of influence, we find every man endowed with two inlets of impression: the ear and the understanding for the reception of speech; and the sympathetic powers, the sensibilities or affections, for tinder to those sparks of emotion revealed by looks, tones, manners, and general conduct. And these sympathetic powers, though not immediately rational, are yet inlets, open on all sides, to the understanding and character.

They have a certain wonderful capacity to receive impressions, and catch the meaning of signs, and propagate in us whatsoever falls into their passive moulds, from others. The impressions they receive do not come from verbal propositions, and are never received into verbal propositions, it may be, in the mind; and therefore many think nothing of them. But precisely on this account are they the more powerful.

"Influences of this kind are not insignificant because they are unnoticed and noiseless. How is it in the natural world? Behind the mere show, its outward noise and stir, Nature always conceals her hand of control. Who ever saw with the eye, for example, or heard with the ear, the exertions of that tremendous Astronomic force, which every moment holds the compact of the Physical universe together? The lightning is, in fact, but a mere firefly spark in comparison; but, because it glares on the clouds, and thunders so terribly in the ears, and rives the tree or the rock where it falls, many will be ready to think that it is a vastly more potent agent than gravity."—(*Unconscious Influence*, a Discourse, by Horace Bushnell, D.D.)

441. The unconscious influence of what may be called the Moral Atmosphere breathed during the earlier period of life, in forming the habits, and thereby determining the Mechanism of Thought and Feeling, is a subject of such great practical importance, as to have required separate treatment (§ 290). But it is not one of the least valuable teachings of that doctrine, that it should lead us (as Mr. Lecky has very justly pointed out) to a very large toleration for those, who, having been brought up in a different "school" from ourselves, entertain views entirely different from our own as to many questions of practical life; not being able to recognise what are to ourselves "self-evident" conclusions, and even regarding many things as "right" which seem to us to be plainly "wrong," or *vice versâ*. How completely such decisions are matters of *judgment*—depending not only upon the early direction of the Mind, but frequently also, it seems probable, upon *hereditary* tendencies,—has been already shown (§§ 210, 292). The

unconscious prejudices which we thus form, are often stronger than the *conscious;* and they are the more dangerous, because we cannot knowingly guard against them. And further (as Mr. Lecky has well remarked), though the reason, in her full strength, may pierce the clouds of prejudice, and may even rejoice and triumph in her liberty, yet "the conceptions of childhood will long remain latent in the mind, to reappear in every hour of weakness, when the tension of the reason is relaxed, and the power of old associations is supreme."

"This very painful recurrence, which occupies such an important place in all religious biographies, seems to be attached to an extremely remarkable and obscure department of Mental phenomena, which has only been investigated with earnestness within the last few years, and which is termed by Psychologists 'latent consciousness,' and by Physiologists 'unconscious cerebration,' or 'the reflex action of the brain.' That certain facts remain so hidden in the mind, that it is only by a strong act of volition they can be recalled to recollection, is a fact of daily experience; but it is now fully established that a multitude of events which are so completely forgotten that no effort of the will can revive them, and that the statement of them calls up no reminiscences, may nevertheless be, so to speak, imbedded in the memory, and may be reproduced with intense vividness under certain physical conditions. * * * It is in connection with these facts, that we should view that reappearance of opinions, modes of thought, and emotions, belonging to a former stage of our intellectual history, which is often the result of the Automatic action of the mind, when Volition is altogether suspended. * * * There can be little doubt that when we are actively reasoning, this automatic action of the mind still continues; but the ideas and trains of thought that are thus produced are so combined and transformed by the reason, that we are unconscious of their existence. They exist, nevertheless; and form (or greatly contribute to) our mental bias."—(*History of Rationalism*, vol. i. p. 101.)

CHAPTER XIV.

OF REVERIE AND ABSTRACTION ;—ELECTRO-BIOLOGY.

442. It has been shown (Chap. VI., Sect. 2) that the sequence of the Thoughts and Feelings, when left to follow their own course by the suspension of the controlling power of the Will, may be determined by suggestions either from *within* or from *without ;* that is, by the promptings of previous *ideational* states recorded in the Cerebrum, or by those of new *sensorial* impressions. In the former case, the attention is so engrossed by the objects which present themselves to the *internal* senses, that impressions made on the *external* are either not *felt* at all, or their *meaning* is not apprehended. But when the Mind is *not* following any definite direction of its own, one idea may be readily substituted for another by new *suggestions* from without ; and thus the whole state of the convictions, the feelings, and the impulses to action, may be altered from time to time, without the least perception of the strangeness of the transition.—Such are the characteristics of the states known as *Reverie* and *Abstraction ;* which are fundamentally the same in their character, though the *form* of their products differs with the temperament and previous habits of the individual, and with the degree in which his consciousness may remain open to external impressions,—*Reverie* being the automatic mental action of the Poet, *Abstraction* that of the Reasoner.

443. The Poet who is fond of communing with Nature in her various moods, and of resigning himself freely to her influences, is apt to give the reins to his Imagination, whilst gazing fixedly upon some picturesque cloud, or upon the ever-varying surface of a pebbly brook, or whilst listening to the breezy murmurs of a neighbouring

wood, or the gently-repeated ripple of the quiet waves; or he falls into a reverie as he sits before his winter fire, and contemplates the shapes and hues of its burning caverns, following with intent gaze every variation of light and shade produced by their ever-changing flames, and every alteration in form that results from the wasting combustion of their walls. In his attention to such monotonous series of impressions, his Will seems, as it were, to glide away; and his thoughts and feelings are thus left free to wander hither and thither, according as they are swayed by changes in the external impressions which prompt them, or by those seemingly erratic suggestions which proceed from that play of association to which we give the name of Fancy (§ 401).—On the other hand, it has been the constant habit of the Philosopher to reflect rather upon his own ideas, than upon the impressions he receives through his organs of sense: in fact, he purposely keeps these as much as possible outside his cognizance, in order that they may not exercise any distracting influence upon his thoughts; the promptings of fancy or imagination—if he should happen to possess any share of such endowments—are at once repressed, and the attention is kept steadily fixed upon the logical sequence of the ideas; and thus it happens that his Mind, even when he gives it up to its own automatic action, in the monotonous solitude of his study, works with more or less of logical consistency, and that the fabric which it rears possesses a unity and stability which is in striking contrast with the airy castle-building of the poetic day-dreamer.

444. In neither of these two states are we cognizant of the inconsistency of the notions which possess our minds, with the actual experience of realities. It is true that this inconsistency seldom rises to the same extravagant height, that it attains in true Dreaming. The incongruities of poetic Reverie are not often actually absurd; and the conclusions at which we arrive in a fit of intellectual Abstraction usually have a show of truth, even when

their correctness is altogether vitiated by some false step in the reasoning process. And this limitation seems to depend upon the fact, that when the train of suggestion is bringing some very extravagant notion before the consciousness, the shock to the lingering remnant of Common Sense which still survives, is enough to put an abrupt termination to the reverie.

445. But it is one of the most curious phenomena of the state of Abstraction, that external impressions, if received by the consciousness at all, are very often *wrongly* perceived; being interpreted in accordance with the ideas which happen to be dominant in the mind at the time (§ 186), instead of giving rise to those new ideas which ordinarily connect themselves with them, in virtue of the individual's habitual experience. The records of "absence of mind" are full of amusing instances of such misinterpretation. Nothing seems too strange for the individual to believe, nothing too absurd for him to do under the influence of that belief.

Thus of Dr. Robert Hamilton, a well-known Professor at Aberdeen, who was the author of many productions distinguished for their profound and accurate science, their beautiful arrangement, and their clear expression, we are informed that, "In public, the man was a shadow; pulled-off his hat to his own wife in the streets, and apologised for not having the pleasure of her acquaintance; went to his classes in the college on the dark mornings, with one of her white stockings on the one leg, and one of his own black ones on the other; often spent the whole time of the meeting in moving from the table the hats of the students, which they as constantly returned; sometimes invited them to call on him, and then fined them for coming to insult him. He would run against a cow in the road, turn round, beg her pardon, call her 'Madám,' and hope she was not hurt. At other times he would run against posts, and chide them for not getting out of his way." (See *New Monthly Magazine*, vol. xxvii. p. 510.)

446. The influence of some habitual form of thought often shows itself very curiously in the strange turn given to communica-

tions made to persons whose whole attention is engrossed by what is passing in their own minds. The well-known story of the philosopher, who, when interrupted in his meditations by the intelligence that his house was on fire, coolly replied to the servant who had burst in upon him with the terrible news, "Go and tell your mistress; you know that I never interfere about domestic matters," is not more incredible than the following circumstance, which, as the Writer has been informed on good authority, actually occurred in the case of the late celebrated German mathematician Gauss:—

Being engaged in one of his most profound investigations, at a time when his wife, to whom he was known to be deeply attached, was suffering from a severe illness, his study was one day broken into by a servant, who came to tell him that her mistress had suddenly become much worse. He seemed to *hear* what was said, but either he did not comprehend it, or immediately forgot it, and went on with his work. After some little time, the servant came again to say that her mistress was much worse, and to beg that he would come to her at once, to which he replied, "I will come presently." Again he relapsed into his previous train of thought, entirely forgetting the intention he had expressed, most probably without having distinctly realised to himself the import either of the communication itself or of his answer to it. For not long afterwards, when the servant came again, and assured him that her mistress was dying, and that if he did not come *immediately* he would probably not find her alive, he lifted up his head and calmly replied, "Tell her to wait till I come;"—a message he had doubtless often before sent, when pressed by his wife's requests for his presence, while he was himself similarly engaged.

447. There are many individuals, who, though not prone either to Reverie or to Abstraction as distinctly isolated states, are really the subjects of the same automatic mental action, either (1) when sleep is stealing over them, or (2) when they are passing out of sleep into the state of full wakeful activity. With some, it is true, the transition from the one state to the other is usually sudden and complete; the state of full activity giving place to one of entire

torpor, and *vice versâ.* But even these, if they will take the trouble to make observations upon their own consciousness, will find that on some occasions the process is much more gradual, and that it consists of several stages. In the first, the directing and controlling power of the Will is suspended, and the thoughts flow onwards automatically, as in Reverie or Abstraction, according to the direction which they may have previously received. Secondly, the ideas lose their ordinary coherence, so that the strangest and most inconsistent notions are brought into collocation; and in this state, as in reverie, it not unfrequently happens that we are recalled to full activity by the shock we receive from the absurdity of the images which thus rise before our view. But, thirdly, if this does not occur, the automatic activity seems gradually to subside; the succession of thoughts becomes less and less rapid, and they present themselves with diminishing vividness; and at last, as the ideational changes cease to make a definite impression on the consciousness, the state of complete repose supervenes.—On the other hand, on our first awaking, we frequently experience a considerable degree of mental confusion, especially if we find ourselves in an unaccustomed place. We do not know where we are, we do not recollect what has last occurred to us, we are almost destitute of the consciousness of personal identity. By degrees all these things come back to us, and the confusion of our ideas gives place to orderly arrangement. But it may be some little time before we can determinately perform any Mental operation which involves the Volitional direction of the thoughts; this being the power that is the first to leave us, and the last to be regained.

448. *Induced Reverie, or Electro-Biology.*—In the course of that important series of researches, carried on about thirty years ago by the late Mr. Braid, of Manchester, by which, in the Writer's opinion, more light has been thrown upon the reflex actions of the Cerebrum than by any other investigations, Mr. B. discovered

that there are many persons in whom a state may be artificially induced, resembling profound *reverie;* save that the subjects of it are yet more amenable than the subjects of natural reverie, to external suggestions conveyed to their minds through their Senses; whilst, in the absence of such suggestions, the Mind (having in itself no power of altering the current of its ideas) remains as completely "possessed" by its own internal train of thought, as it is in profound *abstraction.* This state may be superinduced in certain susceptible or "sensitive" individuals, upon the ordinary waking state, without a previous passage through the stage of insensibility; it being often sufficient for its induction, that the attention should be *fixed* for a few minutes, or even for a few seconds, upon any object whatever. By "leading" or suggestive questions addressed to his "sensitives," Mr. B. could not only bring them to feel any kind of sensation (pricking, burning, streaming, creeping, and the like) which he chose that they should describe, when magnets were drawn along their limbs, or elicit descriptions of small volcanoes of flame seen by them to issue from the poles of the magnet; but he could cause their hands to be powerfully attracted towards magnets or crystals, or to be repelled by them: in short, he could repeat all the phenomena adduced by Baron Von Reichenbach* as proofs of "odylic force," and this *as well without magnets as with them,* provided only that the "subjects" *believed* that some operation was being performed, and were led to expect some result (§§ 144—146).

a. The Writer himself witnessed a most remarkable series of such experiments, about the year 1847, upon a gentleman of high literary and scientific attainments, who possessed in an unusual degree the power of self-concentration. It was sufficient for him to place his hand upon the table, and to fix his attention upon it for half a minute, to be entirely unable to withdraw it, if assured in a determined tone

* *Researches in Magnetism, Electricity, Heat, Light, Crystallization, and Chemical Attraction, in their relations to the Vital Force.* Translated by Dr. Gregory. London, 1850.

that he *could not possibly* do so.—When his gaze had been steadily kept for a short time upon the poles of a Magnet, he could be brought to see flames issuing from them, of any form or colour that Mr. Braid chose to name. And when he had been desired to place his hand upon one of the poles, and to fix his attention for a brief period upon it, the peremptory assurance that he *could not* detach it was sufficient to hold it there with such tenacity, that Mr. Braid positively dragged him round the room, in a manner that most amusingly realised the German fairy story of the Golden Goose.—Some may, perhaps, think the Writer rather credulous in at once yielding his assent to the genuineness of such a strange performance. But the character and position of the "subject" of it were such as to place him beyond the suspicion of intentional deceit; and the Writer's previous inquiries had prepared him to find nothing too strange for his belief, that could be referred to the one simple and intelligible principle of "possession" by a "dominant idea" excited through *suggestion.*

449. Notwithstanding that Mr. Braid's investigations were thus carried on for several years, with every disposition on his part to enable the public to judge of their nature and results, they did not attract by any means the amount of general attention that might have been anticipated for them. But about the year 1850, "the world was turned upside down" by a couple of itinerant Americans, who styled themselves "professors" of a new art which they termed *Electro-Biology;* asserting that, by an influence of which the secret was known only to themselves, but which was partly derived from a little disk of zinc and copper held in the hand of the "subject" and steadily gazed on by him (whence the designation which they adopted), they could subjugate the most determined will, paralyse the strongest muscles, pervert the evidence of the senses, destroy the memory of even the most familar things or of the most recent occurrences, induce obedience to any command, or make the individual believe himself transformed into any one else,—all this, and much more, being done while he was still wide awake. They soon attracted large assemblages to witness their performances; and there was an appearance of good

faith about them which made a favourable impression, as they showed themselves ready to operate upon any who might offer, and seldom failed to elicit some of the most remarkable phenomena from individuals whose honesty could not be called in question, and who had previously been entire strangers to them. Ever on the watch, however, for any novelties in his favourite study, Mr. Braid set himself to inquire into the real nature of the so-called "electro-biological" process: and he soon proved that the little disk of copper and zinc may be replaced by any object which furnishes a *point d'appui* for the fixed gaze; the whole secret consisting in the induction—through the steady direction of the eyes to one point, at the ordinary reading distance, for a period usually varying from about five to twenty minutes—of a state of *reverie*, in every respect similar to that which had previously fallen under his notice. Thus, in place of a few peculiarly susceptible "subjects" difficult to be met with, and open to suspicion on various grounds, every member of the public came to be furnished with a ready means of experimenting for himself upon his own family and friends, the student upon his fellow-students, the officer on the members of his mess; everybody, in fact, upon somebody else on whom he felt that he could place reliance. "Electro-biology," or "Biology" (as it came to be very commonly designated), was not merely introduced at scientific *réunions*, but became a fashionable amusement, in some circles, at ordinary evening parties. And thus it happened that a very large proportion of the public became familiarised with its phenomena; though still labouring under the perplexing difficulty of not knowing what to believe as to their genuineness, or to what scientific principles to refer them if their genuineness were admitted.

450. Considered in their relation to other states in which the Mind is "possessed" by "dominant ideas," and acts in accordance with them, the Biological phenomena are so far from being absurd or incredible, that they are simply manifestations of a condition to

which we may frequently detect very close approximations within our ordinary experience; the most special peculiarity which attends them, consisting in the *method* by which the peculiar condition in question may be *artificially induced*—in such individuals, at least, as are constitutionally susceptible of its influence. In some "subjects," five, ten, or twenty minutes may be necessary to produce the effect; in others, a single minute, or even half a minute, is sufficient. It may be regarded as certain that the "biological" state may be thus induced in individuals who were previously quite incredulous in regard to its reality; so that it does not require any mental preparation on the part of the "subject." But it seems no less certain that the anticipation of the result tends to produce it in a shorter time than would otherwise be necessary; and it is for the most part among individuals who have repeatedly subjected themselves to the operation, that the greatest facility presents itself. The longer the steady gaze is sustained, the more is the Will of the individual withdrawn from the direction of his *thoughts*, and concentrated upon that of his *eyes*, so that at last it seems to become entirely transferred to the latter; and, in the mean time, the continued *monotony* is tending, as in the induction of Sleep or of Reverie, to produce a corresponding state of mind, which, like the body of a cataleptic subject, can be moulded into any position, and remains in that position until subjected to pressure from without.*

* It is curious that the artificial induction of similar states, varying in degree from simple Reverie to apparent Death, is practised among the Yogi, a set of Hindoo devotees, by whom it is connected with a system of religious philosophy very much akin to the "Spiritualism" of our own country. The subjects of the lower states are considered to be peculiarly susceptible of spiritual impressions; whilst those of the higher are supposed to be completely "possessed" by Brahma, the "supreme soul," and to be incapable of sin in thought, word, or deed.—It is a question of great interest whether the state of mind of those who were resorted to as "oracles" in ancient times was not very similar to this. See Mr. Plumptre's article on "Urim and Thummim" in Dr. William Smith's *Dictionary of the Bible.*

451. When this state is complete, the Mind of the Biologized "subject" seems to remain entirely dormant, until aroused to activity by some *suggestion* which it receives through the ordinary channels of sensation, and to which it responds as automatically as a ship obeys the movements of its rudder; the whole course of the individual's thought and action being completely under external direction. He is, indeed, for the time, a mere *thinking automaton.* His mind is entirely given up to the domination of any idea that may transiently possess it; and of that idea his conversation and actions are the exponents. He has no power of judging of the consistency of his idea with actual facts, because he cannot determinately bring it into comparison with them. He cannot of himself turn the current of his thoughts, because all his power of self-direction is in abeyance. And thus he may be played-on, like a musical instrument, by those around him; thinking, feeling, speaking, acting, just as *they will* that he should think, feel, speak, or act. But this is not, as has been represented, because *his* will has been brought into direct subjection to *theirs;* but because, his will being in abeyance, all his mental operations are directed by such suggestions as they may choose to impress on his consciousness.—This distinction may seem unimportant; but it is essential (in the Writer's opinion) to that comprehension of the true nature of this peculiar state, and of its relations with others, which gives to it its special place in Psychological Science.

452. In the public exhibitions of professed "Biologists," much assumption was made of a peculiar power possessed by the operator over his "subject;" his suggestions were conveyed in the form of commands, and the delusion was kept up by a frequent recourse to "passes" resembling those of the Mesmerists. There is no good reason to believe, however, that any such relation actually exists, save where it has been established by previous habit, or by a strong antecedent expectation on the part of the "subject." When an individual brings himself into this state of artificial Reverie or

Abstraction for the first time, and without any previous idea that he is to be controlled by one person rather than by another, he is amenable (as the Writer has frequently proved) to suggestions from *any* of the bystanders ; and the directing force of such suggestions depends in great degree upon the tone and manner in which they are given. But as previous expectation, or acquired habit, affect the facility with which this condition may be induced, so do they influence every part of its phemomena ; and if the "subject" be "possessed" with a previous conviction that a particular individual is destined to exert a special influence over him, the suggestions of that individual are obviously received with greater readiness, and are responded to with greater certainty, than are those of any other bystander.—This is the whole mystery of the relationship between the "Biologizer" and his "subject ;" a relationship which is quite conformable, on the one hand, to what we see in the daily experience of life, as to the influence acquired by certain individuals over the course of thought and action of others ; whilst on the other, it becomes, when still more concentrated and established, the source of that peculiar and exclusive *rapport*, which the Mesmerist claims to be able to establish between his "subject" and himself (§ 521). The assumption of the tone of command has simply the effect of strongly impressing the "subject" (whose condition in this respect resembles that of a Child) with the feeling of the *necessity* of the action enjoined ; and the earnest reiteration of the phrases "you must" or "you cannot," is found to be quite as efficacious as the vehement tone of mastery in which the directions are frequently given. So, again, the effect of the "passes" is merely to concentrate the attention of the "subject" upon the part to which the injunction refers ; for, as Professor Bennett pointed out, they are made over the part which is to move or to be fixed (as over the mouth when it is to be prevented from opening, or over the foot which is to be riveted to a certain spot of the floor), and not over the muscles by which the action is produced.

453. The Biologized "subject," like a person in an ordinary Reverie, must be considered as *awake*; that is, he has generally the use of all his senses, and preserves, in most cases, a distinct recollection of what takes place. There is every gradation, however, between this condition and that of true Somnambulism (§ 487); in which one or more of the inlets to sensation are closed, and no remembrance is afterwards preserved (save in a renewed condition of the same kind) of anything that may have been thought, felt, or acted. In fact, the two conditions are essentially the same in every respect, save their intensity; and the one graduates insensibly into the other. Different individuals preserve very different degrees of recollection of what may have passed in the Biological state; and this may be the case, too, with the same individual on different occasions. Sometimes everything can be retraced, sometimes only the general course of thought and action; sometimes the fact of Emotional excitement is more strongly remembered than that of the circumstances which produced it; and whilst, in other instances, not the slightest memorial trace remains of the most passionately expressed feelings, the particular incidents by which these were excited may have left their distinct impressions.

454. The same kind of variety shows itself in the psychical phenomena manifested during the persistence of the "Biological" state. Suggestions of different kinds are received and acted-on by different individuals with very varying degrees of readiness; and few are equally amenable to all. Thus we meet with one individual whose *muscular movements* may be entirely governed by the authoritative assurance "you *must* do this," or "you *cannot* do that;" his whole mind being, for the time, possessed with the fixed idea thus introduced of the absolute necessity of the action commanded, and of the impossibility of that which is forbidden. His hands being placed in contact with each other, he is assured that he cannot separate them; and they remain as if firmly glued together, in spite of all his apparent efforts to draw them apart. Or, the

hand of the operator being held up before him, he is assured that he cannot succeed in striking it ; and all his power seems, and actually is, inadequate to the performance of this simple action, so long as he remains convinced of its entire impossibility.

The Writer has seen a strong man thus chained down to his chair,—prevented from stepping over a stick on the floor,—or obliged to remain almost doubled upon himself in a stooping posture,—by the assurance that he *could* not move ; and when on the first occasion this assurance seemed not to have its full effect, its repetition, in a more vehement tone, was sufficient to retain him. So he has seen a very lively young lady struggling in vain for utterance, with a most ludicrous expression of distress, when assured that she could not open her mouth to speak a word; and he has been obliged to put forth all his strength to drag another lady across the threshold of the door, who had been thus convinced of the impossibility of her crossing it.

455. There is no end to the strange performances which may be thus called forth ; but they are all referable to the one simple principle already laid down as the characteristic of this state,—the "possession" of the mind by *a dominant idea*, which the individual himself has lost all power of testing by his previous or present experience, simply because he cannot himself direct his thoughts to any other object (§ 316). Of this "dominant idea," introduced by suggestions *ab extra*, all his actions are the direct expressions, so long as he remains possessed by it ; but as soon as his attention is directed into another channel, or his previous idea of the *necessity* or of the *impossibility* of an action (as the case may be) is dissipated by a word, a sign, or a look on the part of the individual who is thus directing his thoughts and actions, the potent spell by which he appeared to be enchained is at once dissolved, the effort to fulfil the supposed necessity immediately subsides, the most violent struggle with the assumed impossibility at once comes to an end, and the "subject" appears to be "himself again." Yet he is not so in

reality; for his volitional power is still withdrawn from the direction of his thoughts, so that the peremptory command of another exerts its former influence over him, even after a considerable interval may have elapsed. It is impossible to state how long this state may continue; the Writer has himself known it to last for several hours; and he is inclined to think that the Biologized "subject" does not usually regain his proper control over himself, until he has experienced the renovating influence of Sleep.

456. It will be frequently observed that the mandates of the operator are not immediately or implicitly obeyed. The "subject" makes attempts, which are often successful, to resist them; doing, though with difficulty, what has been asserted to be impossible; and refraining, though with obvious effort, from the performance of that which he has been assured that he *must* do. This is obviously due to the persistence of a certain degree of self-directing power, which preserves to the *imperfectly* Biologized "subject" some little capacity of judging and acting for himself: but such voluntary efforts may yet be defeated by the assumption of a more and more peremptory tone and manner on the part of the operator, who at last succeeds in impressing on the mind of his victim the assurance of the futility of further opposition. From henceforth (as we too often see in ordinary life, when a strong volition has brought a weak one into subjection to itself) the "subject" becomes the mere slavish tool of his arbitrary will; his actions being directly prompted by the ideas with which he is possessed, and thus falling into the category of *ideo-motor* (§ 235 *et seq.*) as distinguished from *volitional.*

457. In like manner, what has been described as a control of the *sensations* of the Biologized "subject," is really a control over his *belief* (§ 145). A glass of water is presented to him, and he is directed to drink it, with the assurance at the same time that it is milk, coffee, porter, wine, or any other liquid the operator may

choose to name. The liquid is tasted, and all the indications of approval may be given by the "subject," who obviously believes most firmly that he is actually partaking of the liquor in question; his Ideational consciousness being so fully "possessed" by the strong assurance which has been conveyed to his mind through his sense of Hearing, that the impressions made by the liquid itself upon his sight and taste are not sufficient to correct the erroneous notion. Here, as in regard to control over the Muscular movements, a very curious result often presents itself, in consequence of the imperfect degree in which the mind of the "subject" is possessed by the notion which the operator has endeavoured to impress upon him. He often, after tasting or looking at the liquid, expresses hesitation, or downright disbelief in the asserted metamorphosis; and reiterated and very forcible assurances may be required to convince him that it is anything else than what it really is. Convinced, however, he usually is at last; although it is a very curious fact that some Biologized "subjects," whose Muscular movements are entirely amenable to the control of the operator, never give up their Senses to his direction; whilst, on the other hand, some of those who may be most successfully played-on as regards their Sensations, altogether resist the influence of suggestion with respect to their Muscular movements, continuing to keep them under their own exclusive control. Nay, further, the Writer has seen instances in which the "subject" would believe himself to be *tasting* anything which the operator might please to assure him that he ought to taste, but was instantly disabused by *looking at* the liquid, if its appearance was inconsistent with that representation; whilst, on the other hand, another would *see* milk or porter, wine or coffee, as he was directed to see it, but would instantly set himself right when directed to *taste* the liquid.

458. Nothing can be more amusing, however, than to experiment upon a "subject" who has no misgivings of this kind, but whose Perceptive Consciousness is entirely given up to the direc-

tion of external suggestions. He may be made to exhibit all the manifestations of delight, which would be called forth by an unlimited supply of the viands or liquors of which he may happen to be most fond; and these may be turned in a moment into expressions of the strongest disgust, by simply giving the word that the liquid which he is imbibing so eagerly is something which he holds in utter abomination. Or, when he believes himself to be drinking a cup of tea or coffee, let him be assured that it is so hot that he cannot take more than a sip at a time, neither persuasion nor bribery will induce him to swallow a mouthful at once; yet, a moment afterwards, if assured that he can do so without inconvenience, he will be ready to swallow the whole at a draught. Tell him that his seat is growing hot under him, and that he cannot remain upon it, however strongly he may endeavour to do so; and he will fidget uneasily for some time, and at last start up with all the indications of having really found his place no longer bearable. Whilst he is firmly grasping a stick in his hand, let him be assured that it will burn him if he continue to hold it, or that it is becoming so heavy that he can no longer sustain it, and he will presently drop it, with gestures conformable to the impression with which his mind is occupied.

459. To those who would say "there is no proof that all this is not acted," or, "nothing can be easier than to pretend such obedience," it can only be replied, that as the proof rests upon the double basis of the conformity of the phenomena to known principles, and the character of the "subjects,"—these phenomena having been frequently presented by individuals entirely beyond suspicion, who had never previously witnessed such experiments, and who were altogether incredulous with regard to the reality of this peculiar state,—their genuineness cannot be fairly called in question, even were they far more strange than they prove to be when carefully investigated. Those, on the other hand, who dispose of them by turning them over to the limbo of "imagina-

tion," have more reason on their side; for it is quite true that the Biologized "subject" *imagines* himself to be something different from the reality, and that it is from his perverted conceptions that all his strange performances proceed. But it is clearly from the suspension of Volitional control over the direction of the thoughts, that the possibility of this perversion arises.

460. It has been already shown (§§ 142-7, 186) that Sense-perceptions of various kinds may be excited in the mind, not merely by impressions made upon the corresponding *organs of sense*, but also by *ideas* with which the mind becomes possessed through other channels. And applying this principle (fully recognised by every Scientific Psychologist) to the case before us, we shall see that it affords the key which unlocks the whole of this part of the "biological" mystery. For when the "subject" is assured, whilst drinking a glass of water, that it is coffee or porter, that assurance, taking firm possession of his consciousness, produces the very same effect upon it, as would be induced by the actual contact of the liquid in question with his tongue and palate. He tastes it (so to speak) with his mind, though he does not taste it with his tongue; and it is the mental, not the bodily impression, that constitutes the actual Perception. This false perception is not contradicted by the inconsistent impression transmitted from the Organ of Sense; because it is the characteristic of the Biological state, that the mind of the "subject," being *entirely* possessed by the idea which may chance to be before it at the time, can entertain no other, and is incapable therefore of bringing it to the test of experience. And it thus becomes a mere question of the relative strength of the two suggestions,—that conveyed by the assurances of the bystanders, and that derived from the sensory impression. The latter may prevail in the first instance, and may yet be overcome by the augmented force which the former will derive from repetition with added earnestness or vehemence.

461. It is only necessary to glance at some of the most familiar features of Insanity, to be assured that the strangest perversions of Sense-perceptions exhibited by the Biologized "subject," have their counterparts in those morbid states in which the controlling power of the will is altogether suspended, and the mind is possessed, not transiently but enduringly, by some "dominant idea" (§ 559).

462. Passing now to the higher Psychical phenomena of the Biological condition, we find that even such of these as are most extraordinary, or even incredible, in the apprehension of the uninstructed observer, are readily explained on the same general principle. The operator assumes the power of controlling the Memory of his "subject;" and proves that he possesses it, by assuring him that he cannot remember his own name, the first letter of the alphabet, or something else equally familiar. The "subject" exhibits a puzzled and somewhat vacant aspect, and confesses that he cannot recall the desiderated idea. Nothing is more simple than the explanation of this phenomenon, when we call to mind that the very simplest act of determinate Recollection involves a Volitional change in the direction of our thought, *from* the idea which may occupy the consciousness at the moment, *towards* that which it is desired to bring before it (§ 372). It must be within the experience of every one—if not to have forgotten his own name (as many "absent" gentlemen have done)—at any rate to have lost the recollection of some name, fact, or date, usually most familiar to the consciousness, and not at once to be able to find the clue for its recovery (§ 419). The state of the Biologized "subject" precisely resembles this in *kind*, being simply more intense in *degree*. Entirely "possessed" as he is with the one idea which may at the time be present to his mind, and unable to escape from it by any determinate act of his own, the "subject," peremptorily assured of his inability to remember the most familiar thing, surrenders himself to the conviction thus enforced upon him; even his own

name, or the first letter of the alphabet, being as much beyond the reach of his Mental apprehension, whilst his Volitional power remains thus paralysed, as a bunch of grapes lying on a plate at his side would be beyond the grasp of his hands, if his arms had been smitten with a complete palsy. In fact, there is a complete parallelism between his bodily and mental state, whilst in this condition; the Will being temporarily withdrawn from control over both alike (§ 312). He is unable to lay hold of a bank-note of a hundred pounds, though offered him as a reward for his successful effort, if he has been completely possessed with the conviction that he *cannot* stretch out his arm towards it. And he is unable to lay his Mental grasp upon any idea, however familiar, if he has been completely possessed by the assurance that he *cannot* succeed in bringing it to his recollection.

463. So, again, the loss of the sense of personal Identity, or the actual change of personality, which the Biological operator asserts that he is able to induce, and of which the Writer has seen some very amusing and (he is satisfied) genuine cases, is readily explicable upon the same principle of suggestion acting upon a mind entirely amenable to it, and not able by any effort of its own to escape from the idea thus forced upon it. Mr. A. is repeatedly assured that he is Mrs. B., or Mrs. C. is brought by reiterated assertion to the belief that she is Dr. D.; their own Common Sense does not correct this absurd perversion, because the sense of personal Identity is dependent upon memory (§ 364), and they can *recollect* nothing when forbidden to do so; and, when once under its domination, all their language and actions are conformable to their metamorphosed personality, or, at least, to their conception of it. (This state has its parallel in that of some of our greatest Actors—especially of the female sex—who become so completely engrossed in the "parts" they play, as to lose altogether, for the time, the sense of their own personality, and to *be* rather than to *act* the characters they have assumed.) It is not by any means in all Biologized

"subjects" that we meet with a capability of being thus affected; for there are many whose muscular movements, and whose ordinary course of thought and feeling, can be entirely directed by external suggestion, who yet obstinately cling to their own personality. But when the metamorphosis *is* made, it is usually complete; and nothing can be more remarkable than the assumption of the tone, manner, habits of thought, forms of expression, and other characteristic peculiarities of the individual whose personality the "subject" has been made to assume.

a. The Writer can never forget the intensity of the lackadaisical tone in which a Lady thus metamorphosed into the worthy Clergyman on whose ministry she attended, and with whom she was personally intimate, replied to the matrimonial counsels of the Physician to whom he (she) had been led to give a long detail of his (her) hypochondriacal symptoms—"A wife for a dying man, doctor!" No *intentional* simulation could have approached the exactness of the imitation, alike in tone, manner, and language, which spontaneously proceeded from the idea with which the fair "subject" was possessed, that she herself experienced all the discomforts, whose detail she had doubtless frequently heard from the real sufferer.

464. The precise counterpart of this condition is one of the commonest forms of Insanity. Every large asylum contains patients who imagine themselves to be kings, queens, princes, lords, bishops, and the like; nay, the metamorphosis may proceed to yet greater extremes, the lunatic persisting that he is the Holy Ghost, Jesus Christ, or even the Eternal Father. No reasoning can dispossess him of this conviction: because, whilst his mind remains possessed with this "dominant idea," all the arguments that can be employed are to his apprehension entirely irrelevant. And even in our ordinary experience of life, we meet with individuals who are possessed by notions scarcely less absurd, from which they cannot be driven by any appeals to their Common Sense, simply because the "dominant idea" presents itself to their consciousness with greater force than does any other that can be brought before it.

465. So, again, by a judicious use of the principle of suggestion, the thoughts of the Biologized "subject" may be readily directed into any channel whatever by appropriate hints; and descriptions may be called forth, by leading questions, of any scene which the operator may choose to name. This "mental travelling," as it has been called, is not accomplished with equal readiness on the part of any "subject;" and the manifestations of it, as given in the replies elicited, are obviously determined by *the previous knowledge and habits of thought* of the individual "subject," where they are not directly suggested by the words or tone of the questioner.

The same Lady who underwent the metamorphosis into a hypochondriacal Clergyman, was kind enough to ascend in a balloon at the Writer's request, and to proceed to the North Pole in search of Sir John Franklin, whom she found alive; and her description of his appearance and that of his companions was given with an inimitable expression of sorrow and pity.

466. It has thus been shown by the analysis of the principal phenomena of the "Biological" state, how easily they may be all reduced to the one simple principle of *suggestion*, acting on a mind which has lost for a time the power of volitional direction; and how much this state of mind, anomalous as it appears at first view, has in common with mental conditions with which every one is more or less familiar. Such being the case, there would seem no reason to doubt the genuineness of these phenomena, notwithstanding that, in particular instances, they may have been simulated for the purpose of satisfying the spectators at a public exhibition, or for the gratification of a love of fun or mischief on the part of the performer; and the chief marvel lies in the discovery that a continued steady gaze at a fixed object will *induce* this peculiar state in certain individuals,—chiefly such as are constitutionally predisposed to Abstraction or Reverie, or who possess that kind of imaginative power, which transports them without

effort into scenes and circumstances altogether different from those which really surround them. The proportion of such individuals is stated, by those who have extensively experimented upon this subject, to be from one in twelve to one in twenty; so that in a company of fifty or sixty persons, there are pretty sure to be two or three who will prove to be good Biological "subjects," if they take the appropriate means. The undue repetition of such experiments, however, and especially their frequent repetition upon the same individuals, are to be strongly deprecated; for the state of Mind thus induced is essentially a morbid one; and the reiterated suspension of that volitional power over the direction of the thoughts, which is the highest attribute of the Human mind, can scarcely do otherwise than tend to its permanent impairment.

467. One of the most remarkable of all the phenomena of this condition, however, yet remains to be considered; namely, the superinduction of genuine *Sleep*, which may often be accomplished in a few minutes, or even seconds, by the expressed determination of the operator that the "subject" *shall* sleep, or even, in some cases, by the simple prediction that he *will*. This has been repeatedly witnessed by the Writer; who has assured himself of the genuineness of the condition by all the tests he could venture to apply, as well as by his reliance upon the good faith of the parties operated on. Here, again, however, we find that the greatest apparent marvel disappears under an intelligent consideration of the case; for the first great step in the induction of sleep—the reduction of the spontaneous activity of the mind—has been already gained by the antecedent process, which in many individuals will of itself produce the whole effect. And when the Biologized "subject" is left in a state of perfect inactivity, and the whole attention is concentrated upon the idea of sleep, it seems quite consistent with our knowledge of the conditions which most favour the supervention of ordinary sleep, that the undisturbed and imperturbable

monotony of impression, though continued but for a short time, should be adequate to produce the result.

468. The duration of this Sleep, however, and the mode of its termination, may be decided in a most remarkable manner by the impression made upon the minds of the "subject" before passing into it. If he be directed to go to sleep for a short time only, and to awake spontaneously, he will do so; and the same result will ensue upon a like suggestion conveyed in other ways.

a. Thus, the Writer has seen a lady sent off to sleep, by the conviction that a handkerchief held beneath her nose was charged with chloroform; the same symptoms were observable as if she had actually inhaled the narcotic vapour (which she had really done on two or three previous occasions), and she gradually passed into a state of profound insensibility, from which, however, she awoke spontaneously in the course of a few minutes, as she would have done had she been really "chloroformed." But this same lady, having been put to sleep by the assurance of the operator that she could not remain awake for two minutes, and having also received from him the injunction not to awaken until called upon by him to do so, resisted all the Writer's attempts to awaken her by any ordinary means he could employ; showing no sign of consciousness when a large hand-bell was rung close to her ear, when she was somewhat roughly shaken, or when a feather was passed fully two inches up her nostril. Her slumber appeared likely to be of indefinite duration; but it was instantly terminated by the operator's voice, calling the lady by her name in a gentle tone. The Writer was assured by Sir James Simpson that in one instance a patient of his thus slept for thirty-five hours, with only two short intervals of permitted awakening.

469. The influence thus exerted over the duration of the Sleep and the mode of its termination, and the susceptibility of the "subject" to certain sensory impressions whilst utterly insensible to all others, are points of extreme interest; and have a very important bearing on those phenomena of Mesmerism, which have been supposed to indicate a peculiar relation between the Mesmerizer

and his "subject." That they are entirely conformable to certain well-known phenomena of Sleep and of natural Somnambulism, will be hereafter made apparent (§§ 480, 488). And whatever may be considered as the most feasible explanation of those facts, that same explanation will be found equally applicable to the phenomena now under consideration. Thus, when B goes to sleep at the suggestion or bidding of A, and is also told by A that she will awake spontaneously at a certain hour, which really happens, the case differs in no essential respect from that of the sleeper who spontaneously awakens in accordance with a pre-formed determination; the requisite state of mind being produced by the assurance of another, instead of by the intention of the individual herself. Or, again, when B is told, on going to sleep, that she is to awake at the sound of A's voice, and that no other sounds are to recall her to consciousness, all of which really occurs; the case does not differ from those to be presently cited, except in the production of the peculiar susceptibility to the one kind of sound by a mental impression forced (as it were) upon the individual, instead of by the habit of attention to it. In the one instance, as in the other, the effect is obviously dependent upon the *previous mental state* of the "subject" of it; and there is no need to refer it to any new or special force, so long as we have evidence, not merely of the *spontaneous* occurrence of such impressible states —exceptional though they be,—but of the possibility of *inducing* similar states in many to whom they are not habitual, by the adoption of a method which gives to the "dominant idea" a complete mastery over the mind into which it has been introduced.

CHAPTER XV.

OF SLEEP, DREAMING, AND SOMNAMBULISM.

470. A large portion of the Life of every Human being is passed in a state of more or less complete suspension of the Animal powers of sense and motion; the continued maintenance of those powers requiring periodic intervals of repose, which seem to be employed in the removal of the products of the "waste" of the Nervous and Muscular tissues that is produced by the exercise of them, and in the repair of the deteriorated mechanism by Nutritive regeneration. The degree of this suspension, however, varies so greatly, as to render it inappropriate to include under one category all the states that are intermediate between ordinary profound Sleep and complete Wakefulness. And as, in the preceding Chapter, we have considered the most characteristic of those which may be regarded as modifications of the *waking* state, so we shall now treat of those of which *sleep* may be taken as the type. In so doing, it will be convenient to commence with the state of *ordinary profound* Sleep, which may be defined as one of complete *suspension of sensorial activity:* the consciousness of the Ego being neither excited by impressions made on the nerves of his *external* senses and transmitted *upwards* to his Sensorium, nor by the *downward* transmission through the nerves of his *internal* senses of the results of changes taking place in his Cerebrum (§ 100). The activity of the entire Axial Cord *below* the Sensorium, however, is not in the least diminished; and thus not merely do the movements of Respiration continue without interruption, while those of Swallowing may be excited by the appropriate stimulus (§ 48); but *other* reflex movements, not (like the foregoing) directly related

to the maintenance of the Organic functions, may be called-forth, with such a semblance of *adaptiveness*, that their performance is commonly regarded as indicative of a partial, though momentary, awakening. There is, however, no evidence that the consciousness of the sleeper is aroused, merely because he withdraws a limb from a source of irritation, puts down his hand to rub any part of his body on which such an irritation is acting, or even turns round in his bed after lying long in the same position. For we find ample evidence in the results of experiments on Animals, and in pathological observations on Man, that *secondarily*-automatic actions involving a combination or sequence of movements adapted to a definite purpose, may come to be performed without the least Consciousness (§§ 67—71).

It is said that the Dacoits or professional thieves of India have been known to steal a mattress from beneath a sleeper, by taking advantage of this tendency. They begin with intensifying his sleep, by gently fanning his face; and then, when they judge him to be in a state of profound insensibility, they gently tickle whatever part of his body may lie most conveniently for their purpose. The sleeper withdrawing himself from this irritation towards the edge of the mattress, the thief again fans his face for a while, and repeats the tickling, which causes a further movement. And at last the sleeper edges himself off the mattress, with which the thief makes away.

471. It is characteristic, however, of this state, that while *ordinary* sense-impressions do not awaken the consciousness, the sleeper can be aroused to activity by impressions which are either in themselves of *extraordinary* strength, or which exert a special effect on the nerve-centres in virtue of a peculiar *receptivity* of the latter (§ 480). And it is by this that natural *sleep* is distinguished from the morbid state of insensibility termed *coma;* which, so long as the suspended activity affects the Sensorium alone, resembles profound sleep in every particular, save that the patient cannot be aroused from it; but in which, if the cause of it be

sufficiently potent (as happens in apoplexy, narcotic poisoning, &c.), the suspension of activity extends downwards to the Respiratory nerve-centres (§ 62), so that death ensues from the stoppage of the movements of breathing. Between these two conditions, however, there is every gradation. For the effect of an overdose of Opium or any other powerful narcotic, shows itself at first in ordinary sleep, from which the patient may be aroused by calling him by name; but he gradually becomes more and more insensible to sense-impressions of any kind, and at last no stimulus will draw from him the least manifestation of consciousness. Even natural sleep, when following upon extreme fatigue, may be so intensified as almost to resemble coma, as in the following examples:—

a. It is on record, that, during the heat of the battle of the Nile, some of the over-fatigued boys fell asleep upon the deck: and during the attack upon Rangoon, in the Burmese War, the Captain of one of the steam-frigates most actively engaged, worn-out by the excess of continued mental tension, fell asleep, and remained perfectly unconscious for two hours, within a yard of one of his largest guns, which was being worked energetically during the whole period.

b. So even the severest bodily pain yields before the imperative demand occasioned by the continued exhaustion of the powers of the sensorial centres: thus Damiens slept upon the rack during the intervals of his cruel sufferings; the North American Indian at the stake of torture will go to sleep on the least remission of agony, and will slumber until the fire is applied to awaken him; and the Medical practitioner has frequent illustrations of the same fact.

c. Previously to the shortening of the hours of work, Factory-children frequently fell asleep whilst attending to their machines, although well aware that they should incur severe punishment by doing so.

On the other hand, a series of gradational states between ordinary profound sleep and the condition of full activity of the animal and psychical powers, is often exhibited during the transition from one to the other (§ 447); as well in the process of going to sleep, as in that of awaking. Any attempt, therefore, to give a Physiological *rationale* of the difference between the sleeping and the waking states, must take account of these intermediate phases.

472. There is strong reason to believe that in profound Sleep there is a greatly diminished activity of the Blood-circulation through the Brain; but whether that diminution is the *cause*, or the *effect*, of the diminished functional activity of the organ, is a question on which there is more ground for difference of opinion.

In the experimental inquiries of Mr. A. Durham, made by removing (under chloroform) a portion of the skull of a dog, so as to expose the cortical layer of the Cerebrum, it was observed that as the effects of the chloroform passed off, and the animal sank into a natural sleep, the surface of the brain, which had previously been turgid with blood and inclined to rise into the opening through the bone, became pale, and sank below its level. On the animal being roused after a time, a blush seemed to start over the surface of the brain, which again rose into the opening through the bone. And as the animal was more and more excited, the brain-substance became more and more turgid with blood, numerous vessels which were invisible during the sleep being now conspicuous, and those before visible being greatly distended. After a short time the animal was fed; and when it again sank into repose, these vessels contracted again, and the surface of the brain became pale as before.—(*Guy's Hospital Reports*, 1860, p. 153).

b. Similar experiments, with the like results, have been made by Dr. W. A. Hammond, of New York.

c. Dr. J. Hughlings Jackson having examined, by means of the ophthalmoscope, the condition of the retina during profound sleep, found it paler, and its arteries more contracted, than in the waking state.—(*Royal Lond. Ophthalm. Hosp. Reports*).

The value of this last observation depends, first, upon the fact that the retina may developmentally be regarded as a kind of off-shoot from the Optic Ganglion, with which it corresponds in structure (§ 38); and secondly, upon the circumstance that no disturbance of the circulation was produced (as in Mr. Durham's experiments) by operative interference.—Now that the contraction of the vessels is the *cause*, not the effect, of the reduction of the Cerebral activity, may be fairly inferred from the fact already stated, that the entire stoppage of the arterial flow produces immediate and complete insensibility (§ 41). And it has been found by the experiments of Dr. A. Fleming, that a state closely resembling profound sleep may be induced with great certainty, by simple compression of the carotid arteries in the neck :—

"The best mode of operating is to place the thumb of each hand under the angle of the lower jaw, and, feeling the artery, to press backwards and obstruct the circulation through it. . . . It is sometimes difficult to catch the vessel accurately; but once it is fairly under the thumb, the effect is immediate and decided. There is felt a soft humming in the ears; a sense of tingling steals over the body; and, in a few seconds, complete unconsciousness and insensibility supervene, and continue so long as the pressure is maintained. On its removal, there is confusion of thought, with return of the tingling sensation, and in a few seconds consciousness is restored. *The mind dreams with much activity, and a few seconds appear as hours, from the number and rapid succession of thoughts passing through the brain.* The period of profound sleep, in my experiments, has seldom exceeded fifteen seconds, and never half a minute."—That the effect is not due to the obstruction of the internal jugular vein (which must be more or less compressed at the same time with the carotid artery), appears from the fact that it is most decided and rapid when the arterial pulsation is distinctly controlled by the finger, and some pallor shows itself in the face; whilst it is manifestly postponed and rendered imperfect when an impediment to the return of the blood through the veins is indicated by any approach to lividity of the face.—(*Brit. and For. Med.-Chirurg. Review*, vol. xv., p. 530.)

473. The Writer has thus come to agree with Mr. Charles H. Moore (*On going to Sleep*), that the state of Sleep is essentially dependent on a reduction of the enormous blood-supply which is essential to the functional activity of the Brain (§ 42); and that

this reduction is effected by the control which the *vaso-motor* system of Nerves (§ 113) has over the calibre of the Arteries. That such a reduction may be *suddenly* effected, and may as suddenly give place to enlargement (as when a person turns pale and blushes by turns), is in accordance with the frequently sudden passage from waking to sleep or from sleep to waking: which is yet more remarkable in the state known as *hysteric coma,*—the supervention of which will sometimes interrupt a patient while speaking; whilst its departure, after a longer or shorter period of complete insensibility, will show itself in her completion of the broken sentence, without any consciousness of the suspension. On the other hand, there is ample evidence that the passage from contraction to dilatation of the vessels, and from dilatation to contraction, may be *gradual;* and this corresponds with the *gradational* passage between the sleeping and the waking state (§ 447). Moreover, if there should be a contracted state of the vessels of the part of the Sensorium which receives the nerves of the *external* senses, whilst there should be an only partial reduction in the blood-supply of the portion which serves as the centre to the nerves of the *internal* senses, we have the *rationale* of the state of Dreaming; in which there is consciousness of Cerebral changes, whilst there is none of sense-impressions. Finally, a strong Physiological probability in favour of this view* is afforded

* It does not appear to the Writer to be in any way opposed to it, that a state of Coma may be induced either by pressure within the cranium, or by an obstruction to the free return of venous blood. For in the first of these cases, the pressure will produce a direct reduction of the arterial blood-supply, in consequence of the unyielding nature of the brain-case: while in the second, the obstruction will extend backwards, so as to *retard* the flow of blood through the capillaries, even though they may be themselves distended. And in the case of *Asphyxia*, it was long since shown by Dr. Kay (now Sir James Kay Shuttleworth), that the obstruction to the passage of blood through the lungs gradually diminishes the stream which should pass back to the heart, and thence into the general circulation; so that the left side of the heart at last becomes empty, and there is no blood at all to be sent into the arteries of the Brain.

by the fact now clearly established, that the *occasional* increase of activity in many Glandular organs, is distinctly regulated by the "vaso-motor" system of Nerves (§ 565).

474. When the ordinary waking activity has continued during a considerable proportion of the twenty-four hours, a sense of fatigue is usually experienced, which indicates that the Brain requires repose; and it is only under some very strong physical or moral stimulus, that the Mental energy can be sustained through the whole cycle. In fact, unless some decidedly abnormal condition of the Cerebrum be induced by the protraction of its functional activity, Sleep will at last supervene, from the absolute inability of the organ to sustain any further demands upon its energy, even in the midst of opposing influences of the most powerful nature. That the strongest Volitional determination to remain awake, is forced to give-way to sleep, when this is required by the exhaustion of nervous power, must be within the experience of every one; and the only way in which the will can even retard its access, is by determinately fixing the attention upon some definite object (§ 118), and resisting every tendency in the thoughts to wander from this. It does not appear to be of any consequence, whether this exhaustion be produced by the active exercise of volition, reflection, emotion, or simple sensation; still we find that the *volitional* direction of the thoughts, in a course different from that in which they tend spontaneously to flow, is productive of far more exhaustion than the automatic activity of the mind (§ 228); whilst, on the other hand, an excess of *automatic* activity, whether as regards the intellectual operations or emotional excitement, tends to prevent sleep. This is particularly the case when the feelings are strongly interested: thus, the strong desire to work-out a result, or to complete the survey of a subject, is often sufficient to keep-up the intellectual activity as long as may be requisite (a state of restlessness, however, being often induced, which prevents the access of sleep for some time longer); so,

again, anxiety or distress is a most frequent cause of wakefulness. It is generally to be observed that the state of *suspense* is more opposed to the access of sleep, than the greatest joy or the direst calamity when certainty has been attained: thus it is a common observation, that criminals under sentence of death sleep badly, so long as they entertain any hopes of a reprieve; but when once they are satisfied that their death is inevitable, they usually sleep more soundly, and this even on the very last night of their lives.—But although an excess of automatic activity is opposed, so long as it continues, to the coming-on of Sleep, yet it cannot be long protracted without occasioning an extreme exhaustion of nervous power, which necessitates a long period of tranquillity for its complete restoration.

475. Whilst, however, the necessity for Sleep arises out of the state of the Nerve-centres, there are certain external conditions which favour its access; and these, in common parlance, are termed its "predisposing causes." Among the most powerful of these, is the *absence* of sensorial impressions: thus, darkness and silence usually promote repose; and the cessation of the sense of muscular effort, which takes place when we assume a position that is sustained without it, is no less conducive to slumber. There are cases, however, in which the *continuance* of an accustomed sound is necessary, instead of positive silence, the cessation of the sound being a complete preventive of sleep: thus it happens that persons living in the neighbourhood of the noisiest mills or forges, cannot readily sleep elsewhere; for their Nerve-centres, having *grown to* a particular set of constantly-recurring impressions (§ 138), are as much affected by the *want* of them, as those of ordinary persons are by their incidence. Again, the *monotonous repetition* of sensorial impressions is often more favourable to sleep than their complete absence. Thus it is within the experience of every one, that the droning voice of a heavy reader on a dull subject, is often a most

effectual hypnotic; in like manner, the ripple of the calm ocean on the shore, the sound of a distant waterfall, the rustling of foliage, the hum of bees, and similar impressions upon the auditory sense, are usually favourable to sleep; and the muscular and tactile senses may be in like manner affected by an uniform succession of gentle movements, as we see in the mode in which nurses "hush off" infants, or in the practice of gently rubbing some part of the body, which has been successfully employed by many who could not otherwise compose themselves to sleep. The reading of a dull book acts in the same mode through the visual sense; for the eyes wander-on from line to line and from page to page, receiving a series of sensorial impressions which are themselves of a very monotonous kind, and which only tend to keep the attention alive, in proportion as they excite interesting ideas.

476. In these and similar cases, the influence of external impressions would seem to be exerted in withdrawing the Mind from the distinct consciousness of its own operations (the loss of which is the transition-state towards that of complete unconsciousness), and in suspending the directing power of the Will. And this is the case, even where the attention is in the first instance *volitionally* directed to them; as in some of the plans which have been recommended for the induction of sleep, when there exists no spontaneous disposition to it. Thus it has been recommended that the attention should be determinately fixed upon the respiratory acts; and that the entering air should be mentally followed in its course through the air-passages, down into the lungs, and then out again. In other methods, the attention is fixed upon some internal train of thought, which, when once set-going, may be carried-on automatically; such as counting numbers, or repeating a French, Latin, or Greek verb. In either case, when the sensorial consciousness has been once steadily fixed, the monotony of the impression (whether received from the organs of Sense, or from the Cerebrum) tends to retain it there; so that the Will abandons,

as it were, all control over the operations of the mind, and allows it to yield itself up to the soporific influence. This last method is peculiarly effectual, when the restlessness is dependent upon some mental agitation; provided that the will has power to withdraw the thoughts from the exciting subject, and to reduce them to the tranquillizing state of a mere mechanical repetition.

477. Though the access of Sleep is sometimes quite sudden, the individual passing at once from a state of complete mental activity to one of entire torpor, it is more generally gradual; and various intermediate phases may be detected, some of which bear a close resemblance to the state of reverie (§ 447). When we *try* to compose ourselves to sleep, we "drop the reins" of our thoughts, and let them wander as they will; and it sometimes happens that we find ourselves suddenly "pulled-up," as it were, by the strange incongruity between some idea which has been brought before our consciousness by suggestive association, and our Common Sense, which has not, as in dreaming, altogether abandoned its post of guardianship. So, the transition from the state of sleep to that of wakeful activity, *may* be sudden and complete, although it usually consists of a succession of stages; the complete consciousness of the Ego's relation to the external world, and the power of directing his thoughts and actions to any subject about which he may be required to exert himself, being the last to return to him. There may be a rapid alternation of these different states; the loss and recovery of the waking consciousness being many times repeated in the course of a few minutes, when the circumstances are such as to prevent the access of profound sleep by the recurrence of sensory impressions: as when a man on horseback, wearied from want of rest, lapses at every moment into a dozing state, from which the loss of the balance of his body as frequently and suddenly arouses him. So, when a man going to sleep in a sitting posture gradually loses the support of the muscles which keep his head erect, his head droops by degrees and at last falls forwards

on his chest; the slight shock thence ensuing partially arouses and restores his voluntary power, which again raises the head; and we see this *partial* awakening recur over and over again, without any *complete* awakening. Similar fluctuations occur in regard to sense-perceptions; and these may be often artificially induced by very simple means.

"We find, for example, one condition of sleep so light, that a question asked restores consciousness enough for momentary understanding and reply; and it is an old trick to bring sleepers into this state, by putting the hand into cold water, or producing some other sensation, not so active as to awaken, but sufficient to draw the mind from a more profound to a lighter slumber. This may be often repeated, sleep still going on; but make the sound louder and more sudden, and complete waking at once ensues. The same with other sensations. Let the sleeper be gently touched, and he shows sensibility, if at all, by some slight muscular movement. A ruder touch excites more disturbance and motion, and probably changes the current of dreaming; yet sleep will go on; and it often requires a rough shaking, particularly in young persons, before full wakefulness can be obtained." * * * "It is certain that the faculties of sensibility and volition are often unequally awakened from sleep. The case may be stated, familiar to many, of a person sleeping in an upright posture, with the head falling over the breast; in whom sensibility is suddenly aroused by some external impression, but who is unable, for a certain time, to raise his head, though the sensation produced by this delay of voluntary action is singularly distressing."

Thus, as Sir Henry Holland has justly remarked,* what we call Sleep is not a single state capable of being distinctly differentiated from that of waking activity; but is a gradational series of states, intermediate between that of complete possession of the mental faculties, and that of complete suspension of all psychical

* See the excellent Chapter on "Sleep," from which the preceding extracts are taken, in his *Medical Notes and Reflections*, and his *Chapters on Mental Physiology*.

action. And among those intermediate states between sleep and waking, which either occur spontaneously, or can be induced in numerous individuals by very simple processes (§§ 449, 493), there are several which exhibit peculiarities that are not in themselves at all less remarkable, than are those which are regarded with so much wonder by the uninformed observer, when induced by the asserted Mesmeric influence, and paraded as specimens of its power. (See § 469.)

478. It is unquestionable that the supervention of Sleep may be promoted by the strong *previous expectation* of it; and this is true, not merely of ordinary natural sleep, but of the states of artificial Reverie and Somnambulism. Every one knows the influence of habit, not only in regard to "time," but also as to "place and circumstance," in predisposing to Sleep. Thus, the celebrated pedestrian Capt. Barclay, when accomplishing his extraordinary feat of walking 1000 miles in as many successive hours, obtained at last such a mastery over himself, that he fell asleep the instant he lay down. And the sleep of soldiers, sailors, and others, who are prevented by "duty" from obtaining regular periods of repose, but are obliged to take their rest at short intervals, may be almost said to come at command; nothing more being necessary to induce it, than the placing the body in an easy position, and the closure of the eyes. It is related that the Abbé Faria, who acquired notoriety through his power of inducing Somnambulism, was accustomed merely to place his patient in an arm-chair, and then, after telling him to shut his eyes and collect himself, to pronounce in a strong voice and imperative tone the word "dormez," which was usually successful. The Writer had frequent opportunities of satisfying himself, that the greater success which attended the Hypnotic mode of inducing Somnambulism (§ 493), in the hands of Mr. Braid, its discoverer, than in that of others, chiefly lay in the mental condition of his subjects, who came to him for the most part under the confident expectation of its production,

and were further assured by a man of very determined will, that it *could not* be resisted. And it is one of the most curious phenomena of the "Biological" state (§ 467), that, in many subjects at least, Sleep may be induced in a minute or less, by the positive assurance, with which the mind of the individual becomes possessed, that it *will* and *must* supervene. (See also § 518.)

479. The influence of previous Mental states is yet more remarkable, in determining the effects produced upon the sleeper by different Sensory impressions. The general rule is, that *habitual* impressions of any kind have much less effect in arousing the slumberer, than those of a new or unaccustomed character. An amusing instance of this kind has been related, which, even if not literally true, serves extremely well as an illustration of what is unquestionably the ordinary fact:—

A gentleman, who had obtained a passage on board a ship of war, was aroused on the first morning by the report of the morning gun, which chanced to be fired just above his berth; the shock was so violent as to cause him to jump out of bed. On the second morning, he was again awoke, but this time he merely started and sat-up in bed; on the third morning, the report had simply the effect of causing him to open his eyes for a moment, and turn in his bed; on the fourth morning, it ceased to affect him at all; and his slumbers continued to be undisturbed by the report, so long as he remained on board.

It often happens that sleep is terminated by the *cessation* of an accustomed sound, especially if this be one whose monotony or continuous repetition had been the original inducement to repose. Thus, a person who has been read or preached to sleep, will awake, if his slumber be not very profound, on the cessation of the voice; and a naval officer, sleeping beneath the measured tread of the watch on deck, will awake if that tread be suspended.—In this latter case, the influence of the simple cessation of the impression will be augmented by the circumstance next to be alluded-to, which is of peculiar interest both in a Physiological and Psychological point of view, and is practically familiar to almost every one.

480. The awakening power of Sensory impressions is greatly modified by our *habitual state of mind* in regard to them. Thus, if we are accustomed to attend to these impressions, and our perception of them is thus increased in acuteness, we are much more easily aroused by them, than we are by others which are in themselves much stronger, but which we have been accustomed to disregard.

a. Thus, most sleepers are awoke by the sound of *their own names* uttered in a low tone; when it requires a much louder sound of a different description to produce any manifestation of consciousness. The same thing is seen in comatose states; a patient being often found capable of being momentarily aroused by shouting his name into his ear, when no other sound produces the least effect.—The Medical practitioner, in his first profound sleep after a laborious day, is awoke by the first stroke of the clapper of his night-bell, or even by the movement of the bell-wire which precedes it.—The Telegraph-clerk, however deep the repose in which he has lost the remembrance of his previous vigils, is recalled to activity by the faintest sound produced by the vibration of the signalling needle, to whose indications he is required to give diligent heed.—The Mother, whose anxiety for her offspring is for a time the dominant feeling in her mind, is aroused from the refreshing slumber in which all her cares have been forgot, by the slightest wail of uneasiness proceeding from her Infant charge.

These familiar facts cannot be explained upon the supposition that the sleep, prevented from becoming profound by the persistence of the previous excitement, is consequently interrupted by trifling disturbances: for in all these instances the sleeper may remain unaffected by much louder sounds, which have not the same relation to his (or her) previous mental state:—

b. Thus the Doctor's wife shall not be aroused by the full peal of the night-bell, whose first tingle awakes her snoring husband; and he may go forth upon his errand and return to his couch, without disturbing the slumbers of his partner. But her turn next comes; the cries of her child arouse her maternal vigilance; and she may spend hours in attempting to soothe it to repose, which are passed by her husband in a state of blissful unconsciousness.—This is no imaginary picture, but one of daily, or rather nightly, occurrence. It is the very familiarity of these facts, which prevents their import from being duly apprehended.

The following remarkable example of this class of phenomena was mentioned to the Writer by the late Sir Edward Codrington:—

c. When a young man, he was serving as Signal-lieutenant under Lord Hood at the time of the investment of Toulon; and being desirous of obtaining the favourable notice of his commander, he applied himself to his duty—that of watching for signals made by the "look-out" frigates—with such energy and perseverance, that he often remained on deck eighteen or nineteen hours out of the twenty-four, going below only to sleep. During the few hours which he spent in repose, his slumber was so profound that no noise of an ordinary kind, however loud, would awake him; and it used to be a favourite amusement with his comrades, to try various experiments devised to test the soundness of his sleep. But if the word "signal" was even *whispered* in his ear, he was instantly aroused, and was fit for immediate duty; the constant direction of his mind towards this one object, having given to the impression produced by the softest mention of its name, a power over his Brain which no other could exert.

It seems impossible to account for these facts in any other way, than by attributing to the Nerve-centres a peculiar Physical *receptivity* for impressions of some particular class, which they have *acquired* in virtue of the previous direction of the mind to them. To affirm that it is the persistent *consciousness* of the Ego which makes the brain receptive to these impressions, is equivalent to saying that consciousness is never entirely suspended even during the most profound sleep. But although we have no right to affirm with certainty that the consciousness of *cerebral* changes *is* so suspended, yet the complete suspension of *sensorial* consciousness in profound sleep seems to be as clearly indicated as such a condition can possibly be (§ 125).

481. But it is not requisite that the Sensory impression should be one habitually attended to during the waking hours; for it is generally sufficient to produce the effect, that the attention should have been strongly fixed upon it, previously to the access of the sleep, as one at which the slumberer is to be aroused. Thus, the traveller who requires to set forth upon his journey at an early hour in the morning, and has given directions to be called accordingly, is awakened by a gentle tap at the door of his chamber; although he may have previously slept through a succession of far louder noises with which he had no concern. And the student who has set his heart upon rising at a particular hour, in order to continue some literary task, is aroused by *that* recurrence of the stroke of the clock, although no other may have affected him throughout the night, and although he may have habitually slept to a later hour without being disturbed by it. Nay, more, there are many individuals who have the power of determining, at the time of going to rest, the hour at which they shall awake; and who arouse themselves at the precise time fixed upon,—not from the restless sleep which such a determination would ordinarily induce (the Writer, for example, would be prevented by it from obtaining an hour of continuous repose through the whole night), but from a slumber that remains unbroken until the appointed time arrives.—This fact is even more significant than the preceding; and seems to point to a kind of *unconscious chronometry*, which is in some way connected with the sequence of the Organic functions (§ 82 *d*). The whole series of such phenomena has a peculiar interest, in connection with the pretensions advanced by Mesmerizers to exercise a special control over the "subjects" of their manipulations.

482. *Dreaming.*—We have hitherto spoken of Sleep in its most complete or profound form; that is, the state of complete unconsciousness. But with the absence of consciousness of external things, there may be a state of Mental activity, of which we are more or less distinctly cognizant at the time, and of which our subsequent remembrance in the waking state varies greatly in completeness. The chief peculiarity of the state of *dreaming* appears to be, that there is an *entire* suspension of volitional control over the current of thought, which flows-on automatically, sometimes in a uniform, coherent order, but more commonly in a strangely incongruous sequence. The former is most likely to occur, when the mind simply takes-up the train of thought on which it had been engaged during the waking-hours, not long previously; and it may even happen that, in consequence of the freedom from distraction resulting from the suspension of external influences, the reasoning processes may thus be carried-on during sleep with unusual vigour and success, and the imagination may develope new and harmonious forms of beauty.

a. Thus, Condorcet saw in his dreams the final steps of a difficult calculation which had puzzled him during the day; and Condillac tells us that, when engaged in his "Cours d'Etude," he frequently developed and finished a subject in his dreams, which he had broken off before retiring to rest. Coleridge's dream-poem of Kubla Khan has been already noticed (§ 231*d*); a similar occurrence happened in the Writer's own family; and two cases of the like kind are given by Miss Cobbe (*Macmillan's Magazine*, April, 1871). So the Sonata which Tartini thought in his dream that he heard the arch-fiend play at his request, and which he afterwards endeavoured to note down, was of course the production of his own brain. And the Writer has been assured by a distinguished artist, that he once had revealed to him in a Dream the solution of a difficulty which had been for some time puzzling him, as to the mode of working-out a part of a picture he was painting; the finished work presenting itself to his mental vision with such vividness, that, on awaking, he at once went to his picture, and was able to record enough of the impression to guide him in its completion.

The more general fact is, however, that there is an entire want of any ostensible coherence between the ideas which successively present themselves to the consciousness; and yet, like the "Biologized" subject (§ 455), we are completely unaware of the incongruity of the combinations which are thus formed. It has been well remarked that *nothing surprises us in dreams.* All probabilities of "time, place, and circumstance" are violated; the dead pass before us as if alive and well; even the sages of antiquity hold personal converse with us; our friends upon the antipodes are brought upon the scene, or we ourselves are conveyed thither, without the least perception of the intervening distance; and occurrences, such as in our waking state would excite the strongest emotions, may be contemplated without the slightest feeling of a painful or pleasurable nature.

b. We have in dreams, as Miss Cobbe remarks (*Macmillan's Magazine*, Nov. 1870), a manifestation of that "myth-making" tendency of the human mind, which is continually "transmuting sentiments into ideas." Even during the waking state, our minds are ever at work of this sort, "giving to airy nothing" (or at least to what is merely a subjective feeling) "a local habitation and a name." The automatic action of the Brain during sleep proceeds on the same track. Our sentiments of love, hate, fear, anxiety, are each one of them the fertile source of a whole series of illustrative dreams; which have their parallel in the delusions that spring out of any fixed emotional perversion in Insanity (§ 559). The difference between the two states, indeed, is only that the one is transient, the other persistent. "Dreams," as the Laureate says, "are true while they last," that is, they are true *to us*, because it is our nature to believe in our own states of consciousness. But we awake, and "behold it was a dream." Our Common Sense at once repudiates its incongruities, how vivid soever our recollection of them may be; and it is only when we feel that the occurrences *might* have taken place, that we are tempted to put the question to ourselves, "Did this really happen, or did I dream it?"—Of this tendency to compose ingenious fables explanatory of the phenomena around us, which has given rise to the Mythology of Greece and Rome, India and Scandinavia, &c., the effect of sense-impressions in shaping the course of Dreams gives a most remarkable series of examples (§ 485). "Have we not here, then," (says Miss Cobbe), "evidence that there is a real law of the human mind causing us constantly to compose ingenious fables explanatory of the phenomena around us,—a law which only sinks into abeyance in the waking hours of persons in whom the reason has been highly cultivated, and which resumes its sway even over *their* well-tutored brains when they sleep?"

As the suspension of the power of forming Common Sense judgments prevents us from being struck with the improbabilities or impossibilities of our dreams, so the suspension of the power of forming Moral judgments usually prevents any check of conscience from being felt, even by persons whose waking hours are profoundly imbued with moral feeling.

"We commit in dreams," says Miss Cobbe (*loc. cit.*), "acts for which we should weep tears of blood if they were real, and yet never feel the slightest remorse. The familiar check of waking hours, 'I must not do it because it would be unjust or unkind,' never once seems to arrest us in the satisfaction of any whim which may blow about our wayward fancies in sleep. A distinguished philanthropist, exercising for many years high judicial functions, continually commits forgery; and only regrets the act when he learns that he is to be hanged. A woman whose life was devoted to the instruction of pauper children, seeing one of them make a face at her, doubled him up in the smallest compass, and poked him through the bars of the lion's cage. One of the most benevolent of men (the late Mr. Richard Napier), who shared not at all in the enthusiasm of his warlike brothers, ran his best friend through the body, and ever after recalled the extreme gratification he had experienced on seeing the point of his sword come out through the shoulders of his beloved companion."

483. Thus it may be said that a great part of our dreams

consists in the exercise of our constructive Imagination (§ 396), working *automatically* without guidance or restraint. And the *creative* faculty, in those who possess it, occasionally evolves conceptions which seem to pass all experience (§ 408); though the subsequent memory of them is usually too vague to allow of their being turned to good account.—There can be no doubt that the materials of our dreams are often furnished by the "traces" left upon the brain by occurrences long since past, which have completely faded-out of the *conscious* memory. And there is similar reason for believing that the course of dreams is sometimes determined by the "traces" of impressions, which, if they ever really affected the consciousness of the Ego, did so in such a slight and transient manner as not to be at all remembered.

Many curious instances are on record, in which particulars that the memory has been repeatedly and vainly called-on to retrace during the waking state, have presented themselves in dreams with great vividness; and thus lost documents have been recovered, and explanations have been furnished of perplexing difficulties. In some of these cases (of which an interesting collection will be found in Dr. Abercrombie's "Inquiries concerning the Intellectual Powers,") it would appear that the mind of the dreamer had worked upon suggestions which would not have given his waking thoughts the same direction. And it seems probable that the limitation of the Cerebral action to one set of impressions, gives to these, as in the state of attention, an unusual potency; so that from this action ideas are evolved, which would not have suggested themselves through the consciousness. It is probably in this mode that we are to explain (where the case is not one of mere *coincidence*) the fact of *discoveries* of crimes or of intended crimes being made through dreams; some impression, or succession of impressions, having been left upon the Ego, which suggested nothing to his *waking* consciousness, but which took full possession of the limited capacity that remained to him when dreaming.—The following circumstance recently mentioned to the Writer by an eminent Judge, one of whose mental experiences has already been cited (§ 362), affords a characteristic illustration of this kind of

Cerebral action. Having been retained, before his elevation to the Bench, in a case which was to be tried in the North of England, he slept at the house of one of the parties in it; and dreamed through the night that lizards were crawling over him. He could not imagine what had suggested such an idea to his mind, until, on going into the apartment in which he had passed the evening, he noticed a mantel-piece clock on the base of which were figures of crawling lizards. This he must have *seen* without *noticing* it; and the sight must have left a "trace" in his brain, though it left no record in his conscious memory.

484. One of the most remarkable of all the peculiarities in the state of Dreaming, is the *rapidity* with which trains of thought pass through the mind; for a dream in which a long series of events has seemed to occur, and a multitude of images has been successively raised-up, has been often certainly known to have occupied only a few minutes, or even seconds, although whole years may seem to the dreamer to have elapsed. There would not appear, in truth, to be any limit to the amount of thought which may thus pass through the mind of the dreamer, in an interval so brief as to be scarcely capable of measurement; as is obvious from the fact, that a dream involving a long succession of supposed events, has often distinctly originated in a sound which has also awoke the sleeper, so that the whole must have passed during the almost inappreciable period of transition between the previous state of sleep and the full waking consciousness.* Hence it has been argued by some, that *all* our dreams really take place in the momentary passage between the states of sleeping and waking; but such an idea is not consistent with the fact, that the

* The only phase of the waking state, in which any such intensely-rapid succession of Thoughts presents itself, is that which is now well attested as a frequent occurrence, under circumstances in which there is imminent danger of Death, especially by Drowning; the whole previous life of the individual seeming to be presented instantaneously to his view, with its every important incident vividly impressed on his consciousness, just as if all were combined in a picture, the whole of which could be taken-in at a glance.

course of a dream may often be traced, by observing the successive changes of expression in the countenance of the dreamer, or by listening to the words he utters from time to time. It seems, however, that those dreams are most distinctly remembered in the waking state, which have passed through the mind during the transitional phase just alluded-to; whilst those which occur in a state more allied to Somnambulism, are more completely isolated from the ordinary consciousness.—There is a phase of the dreaming state, which is worthy of notice as marking another gradation between this and the waking state; that, namely, in which the dreamer has a consciousness that he is dreaming, being aware of the unreality of the images which present themselves before his mind. He may even make a voluntary and successful effort to prolong them if agreeable, or to dissipate them if unpleasing; thus evincing the possession of a certain degree of that directing power, the entire want of which is the characteristic of the true state of dreaming.

485. But the sensibility to external impressions may not be entirely suspended in Dreaming; and it is curious that even where sensations are not perceived by the mind of the dreamer as proceeding from external objects, they may affect the course of its own thoughts; so that the character of the dreams may be in some degree predetermined by such an arrangement of sensory impressions as is likely to modify them. This is especially the case in regard to the dreamy state induced by certain narcotics, such as Opium, or Hachisch (§ 537); and as sense-impressions are here intensified in a most extraordinary degree, so does it seem likely that in ordinary dreaming the course of thought may be affected by sense-impressions too faint to be perceived at all in the waking state.

Thus General Sleeman mentions that while charged to put down Thuggee in India, being in pursuit of Thugs up the country, his wife one morning urgently entreated him to move their tents from the spot—a lovely opening in a jungle—where they had been pitched the previous evening. She said she had been haunted all night by the sight of dead men. Information received during the day induced the General to order digging under the ground whereon they had camped; and beneath Mrs. Sleeman's tent were found fourteen corpses, victims of the Thugs. "It is easily conceivable," says Miss Cobbe (*loc. cit.*), "that the foul odour of death suggested to the lady, in the unconscious cerebration of her dream, her horrible vision. Had she been in a state of Mesmeric trance, the same occurrence would have formed a splendid instance of supernatural revelation."

Again, Dr. Reid tells us that having had his head blistered on account of a fall, and a plaster having been put on it which pained him excessively during the night, on falling asleep towards morning he dreamed very distinctly that he had fallen into the hands of a party of Indians, and was scalped. Now here, as Miss Cobbe says, "the number of mental operations needful for the transmutation of a blistered head into a dream of Red Indians, is very worthy of remark. First, Perception of pain, and allotment of it to its true place in the body. Secondly, Reason seeking the true cause of the phenomenon. Thirdly, Memory suppressing the real cause, and supplying from its stores of knowledge an hypothesis of a cause suited to produce the phenomenon. Lastly, Imagination stepping in precisely at this juncture, fastening on this suggestion of memory, and instantly presenting it as a *tableau vivant*, with proper decorations and 'local colour.' The only Intellectual faculty which remains dormant seems to be the Judgment, which has allowed Memory and Imagination to work regardless of those limits of probability which would have been set to them awake. If, when awake, we feel a pain which we do not wholly understand, say a twinge of the foot, we speculate upon its cause only within the very narrow series of actual probabilities. It may be a nail in our boot, a chilblain, a wasp, or so on. It does not even cross our minds that it may be a sworn tormentor with red-hot pincers; but the very same sensation experienced asleep will very probably be explained by a Dream of the sworn tormentor or some other cause which the relations of time and space render equally inapplicable."—(*Macmillan's Magazine*, April, 1871.)

486. In ordinary Dreaming, then, there may be intense Cerebral

activity; but this is entirely of the *automatic* kind. And it is quite possible that this activity may go on without affecting the consciousness of the sleeper, so as to evolve important ideational results, without any recollection on his part of having dreamed of them (§ 434). But, again, with the suspension of Sensorial activity, which seems more complete in regard to special than to common sensation, there is also an entire suspension of Muscular activity, save such as is purely reflex (§ 69); the only movements which express what is going on within, being slight gestures or changes of countenance. If the dreamer *acts* his dreams, that is, if the ideational or emotional state calls forth Muscular movements corresponding to it, we term the state not Sleep, but *Somnambulism*, or sleep-walking; this being the most common form of movement executed by the dreamer. Between the two states there is a gradational transition. There are many, for instance, who *talk* much in their sleep, yet never attempt to leave their beds and *walk*. And among sleep-*talkers* there are some who merely utter meaningless sequences of words, or strangely jumbled phrases, and are utterly incapable of being influenced by *suggested* ideas; whilst there are others who give utterance to a coherent train of thought, still without any receptivity of external suggestion; and others, again, obviously hear what is said to them, and attend to it or not according to the impression it makes upon them (§ 488).

487. *Somnambulism.*—It seems common to every phase of this condition, that there is the same want of Volitional control over the current of thought, and the same complete subjection of the consciousness to the idea which may for a time possess it, as in dreaming: but the Somnambulist differs from the ordinary dreamer in possessing such a control over his nervo-muscular apparatus, as to be enabled to execute, or at any rate to attempt, whatever it may be in his mind to do; while some of the inlets to sensation ordinarily remain open, so that the Somnambulist may *hear*, though he does not *see* or *feel*, or may *feel*, while

he does not *see* or *hear*. The Muscular Sense, indeed, seems always active; and many of the most remarkable performances both of *natural* and of *induced* Somnambulism, seem referable to the extraordinary intensity with which impressions on it are perceived, in consequence of the exclusive fixation of the attention on its guidance (§ 128).—The phenomena of Somnambulism present a very curious diversity, which in some respects corresponds to the difference between Abstraction and Reverie (§§ 442, 443). Sometimes, as in the former of these states, the Somnambulist's attention is so completely fixed upon *his own* trains of thought, that he is only conscious of such external impressions as are in harmony with them; and a definite and connected sequence of ideas is not unfrequently followed-out, with a steadiness and consistency which contrasts very strikingly with the strange incongruities and abrupt transitions of an ordinary dream. When this is the case, we may usually trace the operation of some one dominant idea or feeling—the key-note (so to speak) of the entire piece,—to which all the thoughts which pass through the mind are related, and of which everything that is done by the body is an expression. A Mathematician will work out a difficult problem; an Orator will make a most effective speech; a Preacher will address an imaginary congregation with such earnestness and pathos as deeply to move his real auditors; a Musician will draw forth most enchanting harmonies from his accustomed instrument; a Poet will improvise a torrent of verses; a Mimic will keep the spectators in a roar of laughter at the drollness of his imitations. The Reasoning processes may be carried on with remarkable accuracy and clearness, so that the conclusion may be quite sound, if the data have been correct and adequate; and it is a very remarkable fact that their purely Automatic action in this state will frequently evolve conclusions which Volitional exertion has vainly striven to attain. The following are well-attested examples of this singular phenomenon; in which not merely did solutions of the difficulties present themselves to the mind (as in dreaming, § 483), but these expressed themselves in appropriate bodily actions.—The first case is given by Dr. Abercrombie, on the authority of the family of a distinguished Scottish lawyer of the last age :—

a. "This eminent person had been consulted respecting a case of great importance and much difficulty; and he had been studying it with intense anxiety and attention. After several days had been occupied in this manner, he was observed by his wife to rise from his bed in the night, and go to a writing-desk which stood in the bedroom. He then sat down, and wrote a long paper which he carefully put by in his desk, and returned to bed. The following morning he told his wife that he had had a most interesting dream;—that he had dreamt of delivering a clear and luminous opinion respecting a case which had exceedingly perplexed him; and that he would give anything to recover the train of thought which had passed before him in his dream. She then directed him to the writing-desk, where he found the opinion clearly and fully written out; and this was afterwards found to be perfectly correct." (*Intellectual Powers*, 5th Edit., p. 306.)

The following was narrated by the Rev. John de Liefde, as the experience of a brother Clergyman, on whose veracity he could fully rely :—

b. "I was a Student of the Mennonite Seminary at Amsterdam, and attended the Mathematical lectures of Professor Van Swinden. Now, it happened that once a Banking-house had given the Professor a question to resolve, which required a difficult and prolix calculation; and often already had the Mathematician tried to find out the problem; but as, to effect this, some sheets of paper had to be covered with figures, the learned man at each trial had made a mistake. Thus not to fatigue himself, he communicated the puzzle to ten of his students—me amongst the number; and begged us to attempt its unravelling at home. My ambition did not allow me any delay. I set to work the same evening, but without success.

Another evening was sacrificed to my undertaking, but fruitlessly. At last I bent myself over my figures for a third evening. It was winter, and I calculated till half-past one in the morning—all to no purpose! The product was erroneous. Low at heart, I threw down my pencil, which already that time had beciphered three slates. I hesitated whether I would toil the night through, and begin my calculation anew; as I knew that the Professor wanted an answer the very same morning. But lo! my candle was already burning in the socket; and, alas, the persons with whom I lived had long gone to rest. Then I also went to bed; my head filled with ciphers; and tired in mind I fell asleep. In the morning, I awoke just early enough to dress and prepare myself to go to the Lecture; vexed at heart at not having been able to solve the question, and at having to disappoint my teacher. But, O wonder! as I approach my writing table, I find on it a paper, with figures in my own hand, and (think of my astonishment!) the whole problem on it solved quite aright, and without a single blunder. I wanted to ask my *hospita* whether any one had been in my room; but was stopped by my own writing. Afterwards I told her what had occurred, and she herself wondered at the event, for she assured me no one had entered my apartment. Thus I must have calculated the problem in my sleep, and in the dark to boot; and what is most remarkable, the computation was so succinct, that what I saw now before me on a single folio sheet, had required three slatefuls closely beciphered on both sides during my waking state. Professor Von Swinden was quite amazed at the event, and declared to me that whilst calculating the problem himself, he had never once thought of a solution so simple and concise." (*Notes and Queries*, Jan. 14, 1860.)

Another case of a similar kind has been lately communicated to the Writer, by the gentleman to whom he is indebted for his own experiences of unconscious cerebration (§§ 422, 435) :—

c. "My father, when a student of Divinity at Basle, was required in due course to compose a discourse for public delivery on a given text of scripture. All power to grapple with the subject seemed gone from him; and he was for days in a state of nervous agitation, unable to deal with the matter in any way satisfactory to himself. The evening before the day of ordeal, he composed something, and lay down utterly disgusted with his performance. He fell asleep; dreamed of a novel method of handling and illustrating the subject; awoke; leaped out of bed to commit the ideas to paper; and, on opening his desk, found that they were so committed already in his own writing, the ink being hardly dry."

It is a frequent defect, however, of the Intellectual operations carried on in this condition, that, owing to their very intensity and exclusiveness, the attention is drawn off from the considerations which ought to modify them; and thus it happens that the result is often palpably inconsistent with the teachings of ordinary experience; which, if they present themselves to the consciousness at all, are not perceived by it with sufficient vividness for the exercise of their due corrective influence.

488. Now in this form of Somnambulism, there is usually as complete an insensibility as in ordinary sleep to all sensory impressions, excepting to *such as fall-in with the existing current of ideas.* No ordinary sights or sounds, odours or tastes, pricks, pinches, or blows, make themselves felt; and yet, if anything be addressed to the sleep-talker through either of his senses, which is in harmony with the notion that occupies his mind at the time, he may take cognizance of it, and interweave it (as it were) with his web of thought, which may receive a new colour or design therefrom.

Thus a young lady was formerly known to the Writer, who, when at school, frequently began to talk, after having been asleep an hour or two; her ideas almost always ran upon the events of the previous day; and, if encouraged by leading questions addressed to her, she would give a very distinct and coherent account of them, frequently disclosing her own peccadilloes and those of her school-fellows, and expressing great penitence for the former, whilst she seemed to hesitate about making known the latter. To all ordinary sounds, however, she seemed perfectly insensible. A loud noise would awake her, but was never perceived in the sleep-talking state; and if the interlocutor addressed to her any questions or observations that did not fall in with her train of thought, they were completely disregarded. By a little adroitness, however, she might be led to talk upon almost any subject,—a transition being *gradually* made from one to another, by means of leading questions.

489. It is an important and distinctive feature of the Somnambulistic state, that neither the trains of thought which have passed through the mind, nor the actions which have resulted from them, are usually remembered in the waking state; and that, if any recollection of them be preserved, they are retraced only as passages of an ordinary dream. Both the trains of thought and the occurrences of the somnambulistic state, however, are frequently remembered with the utmost vividness *on the recurrence of that state*, even at a very distant interval; and of this interval, however long it may have been, there seems to be no consciousness whatever. The same thing happens, but more rarely, in ordinary dreaming; the dreamer sometimes recollecting a previous dream, and even taking it up and continuing its thread;—a circumstance which marks the close affinity of this state to that of somnambulism, since it is only when the dream-idea possesses the fixity and congruity characteristic of the latter, that it shows this tendency to recurrence. The following case, which happened in the Writer's own family, affords a good exemplification of the "acted dream," and of the continuity of the impression from one such state to another, whilst it was altogether lost to the waking consciousness:—

A Servant-maid, rather given to sleep-walking, missed one of her combs; and being unable to discover it, on making the most diligent search, charged the fellow-servant who slept in her room with having taken it. One morning, however, she awoke *with the comb in her hand;* so that there can be no doubt that she had put it away on a previous night, without preserving any waking remembrance of the occurrence; and that she had recovered it when the remembrance of its hiding-place was brought to her, by the recurrence of the state in which it had been secreted.

490. Many of the most characteristic features of this form of Somnambulism are presented by the following case, which occurred within the Writer's own experience :—

The subject of it was a young lady of highly nervous temperament; and the affection occurred in the course of a long and trying illness, in which all the severest forms of hysterical disorder had successively presented themselves. The state of Somnambulism usually supervened in this case upon the waking state; instead of arising, as it more commonly does, out of the condition of ordinary sleep.—In this condition, her ideas were at first entirely fixed upon one subject, the death of her only brother, which had occurred some years previously. To this brother she had been very strongly attached; she had nursed him in his last illness; and it was perhaps the return of the anniversary of his death, about the time when the somnambulism first occurred, that gave to her thoughts that particular direction. She talked constantly of him, retraced all the circumstances of his illness, and was unconscious of anything that was said to her which had not reference to this subject. On one occasion she mistook her sister's husband for her lost brother; imagined that he was come from heaven to visit her; and kept up a long conversation with him under this impression. This conversation was perfectly rational on her side, allowance being made for the fundamental error of her data. Thus she begged her supposed brother to pray with her; and on his repeating the Lord's Prayer, she interrupted him after the sentence "forgive us our trespasses," with the remark, "But *you* need not pray thus; *your* sins are already forgiven." Although her eyes were open, she recognized no one in this state,—not even her own sister, who, it should be mentioned, had not been at home at the time of her brother's last illness.

On another occasion it happened that, when she passed into this condition, her sister, who was present, was wearing a locket containing some of their deceased brother's hair. As soon as she perceived this locket, she made a violent snatch at it, and would not be satisfied until she had got it into her own possession, when she began to talk to it in the most endearing and even extravagant terms. Her feelings were so strongly excited on this subject, that it was judged prudent to check them; and as she was inaccessible to all entreaties for the relinquishment of the locket, force was employed to obtain it from her. She was so determined, however, not to give it up, and was so angry at the gentle violence used, that it was found necessary to abandon the attempt; and having become calmer, after a time, she passed off into ordinary sleep. Before going to sleep, however, she placed the locket under her pillow, remarking, "Now I have hid it safely, and they shall not take it from me." On awaking in the morning, she had not the slightest consciousness of what had passed; but the impression of the excited feelings still remained; for she remarked to her sister, "I cannot tell what it is that makes me feel so; but every time that S—— comes near me I have a kind of shuddering sensation," the individual named being a servant, whose constant attention to her had given rise to a feeling of strong attachment on the side of the invalid, but who had been the chief actor in the scene of the previous evening. This feeling wore off in the course of a day or two.

A few days afterwards, the somnambulism again recurred; and the patient, being upon her bed at the time, immediately began to search for the locket under her pillow. In consequence of its having been removed in the interval (in order that she might not, by accidentally finding it there, be led to inquire into the cause of its presence, of which it was thought better to keep her in ignorance) she was unable to find it; at which she expressed great disappointment, and continued searching for it, with the remark, "It *must* be there; I put it there myself a few minutes ago, and no one can have taken it away."—In this state, the presence of S—— renewed her previous feelings of anger; and it was only by sending S—— out of the room, that she could be calmed and induced to sleep.

This patient was the subject of many subsequent attacks, in every one of which the anger against S—— revived; until the current of thought changed, no longer running exclusively upon what related to her brother, but becoming capable of direction by *suggestions* of various kinds presented to her mind, either in conversation, or, more directly, through the several organs of sense.

491. Here, then, we perceive the complete limitation of the consciousness to the one train of ideas which was immediately connected with the object of strong affection, and the want of

receptivity for all impressions which did not call forth a responsive association in the Somnambulist's mind. Her recognition of the locket which her sister wore, when she did not recognize the wearer, was extremely curious; and may be explained in two modes, each of them in accordance with the known facts of somnambulism. Either the concentration of her thoughts on this one subject caused her to remember only that which was *immediately* connected with her brother; while her want of recognition of her sister might be due to the absence of the latter at the time of his death, which caused her to be less connected with him in the thoughts of the patient. Or it may have happened that she was directed to this locket by the sense of Smell, which is frequently exalted in the *hypnotic* state to a very remarkable degree (§ 498). The continuity of the train of thought from one fit to the next, was extremely well marked in this instance; and the prolongation of the emotional disturbance throughout the interval, without any idea as to the *cause* of that disturbance, is a feature of peculiar interest, as showing that some organic impression must have been left by the mental operations of the Somnambulist, of which her waking consciousness could take no ideational cognizance.—The personal experience of most persons will furnish them with facts of the same order: a sense of undefined uneasiness often remaining as a consequence of a troubled dream, of whose course there is no definite remembrance; and this uneasiness sometimes manifesting itself especially in regard to certain persons or objects, the sight of which calls forth a vague recollection that they have been recently before the mind in some peculiarly disagreeable association. When the entire engrossment of the thoughts by some one subject, and the intensity of their occupation upon it, which constitute the characteristic features of Somnambulism, are kept in mind, it is not surprising that the impressions which they leave behind should possess a remarkable degree of strength, and should re-act with unwonted potency.

492. But there is another and very different phase of the Somnambulistic state, in which the mind, though not less possessed *for the time* by its own idea than it is in the preceding form, is yet capable of having the direction of its thoughts, and consequently of the bodily actions which they prompt, readily altered by external impressions. Between these two forms, again, there is every shade of transitional gradation; the facility with which the mind of the somnambulist is amenable to the guidance of external Suggestions, being always inversely proportional to the degree in which it is possessed by some one dominant idea.

a. Of the form of natural Somnambulism in which the influence of external impressions is complete, so that all the actions of the subject of it are performed in respondence to them, the case of the Officer who served in the expedition to Louisburgh in 1758 (given by Dr. Abercrombie, on the authority of Dr. James Gregory), is an apt illustration. This is frequently spoken of as a case of Dreaming; but as the dream was *acted*, it most legitimately falls under the present head. The course of this individual's dreams could be completely directed by whispering into his ear, especially if this was done by a friend with whose voice he was familiar (another illustration of the fact that the consciousness of sensory impressions in this condition is in great degree governed by the degree of attention habitually paid to them in the waking state); and his companions in the transport were in the constant habit of thus amusing themselves at his expense. "At one time they conducted him through the whole progress of a quarrel, which ended in a duel; and when the parties were supposed to be met, a pistol was put in his hand, which he fired, and was awakened by the report. On another occasion they found him asleep on the top of a locker or bunker in the cabin, when they made him believe he had fallen overboard, and exhorted him to save himself by swimming. He immediately imitated all the motions of swimming. They then told him that a shark was pursuing him, and entreated him to dive for his life. He instantly did so, with such force as to throw himself entirely from the locker upon the cabin floor, by which he was much bruised, and awakened of course. After the landing of the army at Louisburgh, his friends found him one day asleep in his tent, and evidently much annoyed by the cannonading. They then made him believe that he was engaged, when he expressed great fear, and showed an evident disposition to run away. Against this they remonstrated, but at the same time increased his fears, by imitating the groans of the wounded and the dying; and when he asked, as he often did, who was down, they named his particular friends. At last they told him that the man next to himself in the line had fallen, when he instantly sprang from his bed, rushed out of the tent, and was aroused from his danger and his dream together by falling over the tent-ropes. After these experiments he had no distinct recollection of his dreams, but only a confused feeling of oppression and fatigue, and used to tell his friends that he was sure they had been playing some trick upon him."—(*Intellectual Powers*, 5th Edit., p. 278).

493. *Induced Somnambulism, or Hypnotism.*—We have now to consider the phenomena of a state which bears the same relation to the preceding, that the "Electro-Biological" bears to Reverie and Abstraction; its peculiarity consisting in its being *artificially induced.* The method, discovered by Mr. Braid, of producing this state of artificial Somnambulism, which was appropriately designated by him as *Hypnotism,* consists in the maintenance of a fixed gaze, for several minutes consecutively, on a bright object placed somewhat above and in front of the eyes, at so short a distance that the convergence of their axes upon it is accompanied with a sense of effort, even amounting to pain. This process, it will be at once perceived, is of the same kind as that employed for the induction of the "Biological" state (§ 449); the only difference lying in the *greater intensity* of the gaze, and in the more complete concentration of Will upon the direction of the eyes, which the nearer approximation of the object requires for the maintenance of the convergence.—In Hypnotism, as in ordinary Somnambulism, no remembrance whatever is preserved in the waking state, of anything that may have occurred during its continuance; although

the previous train of thought may be taken up and continued uninterruptedly, on the next occasion that the hypnotism is induced. And when the mind is not excited to activity by the stimulus of external impressions, the hypnotized subject appears to be profoundly asleep; a state of complete torpor, in fact, being usually the first result of the process just described, and any subsequent manifestation of activity being procurable only by the prompting of the operator. The hypnotized subject, too, rarely opens his eyes; his bodily movements are usually slow; his mental operations require a considerable time for their performance; and there is altogether an appearance of heaviness about him, which contrasts strongly with the comparatively wide-awake air of him who has not passed beyond the ordinary "Biological" state.

494. In regard to the influence of external Suggestion in directing the current of thought and action, the two states are essentially the same; and it is unnecessary, therefore, to repeat with regard to Hypnotism what has been already described so fully. There seems, however, to be a state of greater *concentration* about the Hypnotized somnambule, than exists in the Biologized subject. The *whole man* seems given to each perception. No doubts or difficulties present themselves to distract the attention; and, in consequence, there is a greater apprehensiveness for suggestions, and their results are more vividly displayed. This is the case especially in regard to *emotional* states, which are aroused with the greatest facility, and which can be governed by a word, or even by a tone, or—as Mr. Braid discovered—by the subject's own Muscular sense, which suggests to his mind ideas or feelings corresponding to the attitude or gesture into which he may be brought by the operator. Thus, if the hand be placed upon the top of the head, the Somnambulist will frequently, of his own accord, draw his body up to its fullest height, and throw his head slightly back; his countenance then assumes an expression of the most lofty pride, and his whole mind is obviously possessed

by that feeling. Where the first action does not of itself call forth the rest, it is sufficient for the operator to straighten the legs and spine, and to throw the head somewhat back, to arouse that feeling and the corresponding expression to its fullest intensity. During the most complete domination of this emotion, let the head be bent forward, and the body and limbs gently flexed; and the most profound humility then instantaneously takes its place. Of the reality and suddenness of these changes the Writer is fully assured, not only from having been an eye-witness of them on various occasions, but also from the reliance he places on the testimony of an intelligent friend, who submitted himself to Mr. Braid's manipulations, but retained enough self-consciousness and voluntary power to endeavour to exercise some resistance to their influence at the time, and subsequently to retrace his course of thought and feeling. This gentleman declares that, although accustomed to the study of character and to self-observation, he could not have conceived that the whole mental state should have undergone so instantaneous and complete a metamorphosis, as he remembers it to have done, when his head and body were bent forward in the attitude of humility, after having been drawn to their full height in that of self-esteem. These phenomena are most graphically described by Dr. Garth Wilkinson, in the following extract:—

a. "The preliminary state is that of Abstraction produced by fixed gaze upon some unexciting and empty thing (for poverty of object engenders abstraction), and this abstraction is the logical premise of what follows. Abstraction tends to become more and more abstract, narrower and narrower; it tends to unity, and afterwards to nullity. There, then, the patient is, at the summit of attention, with no object left, a mere statue of attention, a listening, expectant life; a perfectly undistracted faculty, dreaming of a lessening and lessening mathematical point; the end of his mind sharpened away to nothing. What happens? Any sensation that appeals is met by this brilliant attention, and receives its diamond glare; being perceived with a force of leisure of which our distracted life affords only the rudiments. External influences are sensated, sympathized with, to an extraordinary degree; harmonious music sways the body into graces the most affecting; discords jar it, as though they would tear it limb from limb. Cold and heat are perceived with similar exaltation; so also smells and touches. In short, *the whole man appears to be given to each perception.* The body trembles like down with the wafts of the atmosphere; the world plays upon it as upon a spiritual instrument finely attuned.

b. "This is the *natural* Hypnotic state, but it may be modified artificially. The power of suggestions over the patient is excessive. If you say, 'What animal is it?' the patient will tell you it is a lamb, or a rabbit, or any other. 'Does he see it?' 'Yes.' 'What animal is it *now?*' putting depth and gloom into the tone of *now*, and thereby suggesting a difference. 'Oh!' with a shudder, 'it is a wolf!' 'What colour is it?' still glooming the phrase. 'Black.' 'What colour is it *now?*' giving the *now* a cheerful air. 'Oh! a beautiful blue!' spoken with the utmost delight. And so you lead the subject through any dreams you please, by variation of questions and of inflections of the voice; and *he sees and feels all as real.*

c. "Another curious study is the influence of the patient's *postures* on his mind in this state. Double his fist, and pull up his arm, if you dare, for you will have the strength of your ribs rudely tested. Put him on his knees, and clasp his hands; and the saints and devotees of the artists will pale before the trueness of his devout actings. Raise his head while in prayer, and his lips pour forth exulting glorifications, as he sees heaven opened and the majesty of God raising him to his place; then in a moment depress the head, and he is dust and ashes, an unworthy sinner, with the pit of hell yawning at his feet. Or compress the forehead, so as to wrinkle it vertically, and thorny-toothed clouds contract in from the very horizon; and, what is remarkable, the smallest pinch and wrinkle, such as will lie between your nipping nails, is sufficient nucleus to crystallize the man into that shape, and to make him all foreboding; as, again, the smallest expansion in a moment brings the opposite state, with a full breathing of delight. Raise the head next, and ask (if it be a young lady) whether she or some other is the prettier; and observe the inexpressible hauteur, and the puff sneers let off from the lips, which indicate a conclusion too certain to need utterance. Depress the head, and repeat the question, and mark the self-abasement with which she now says, '*She* is,' as hardly worthy to make the comparison. In this state, whatever posture of any passion is induced, the passion comes into it at once, and dramatizes the body accordingly."—("*The Human Body and its Connection with Man*," p. 473.)

495. The suggestion of the ideas connected with particular actions, through the same channel, is not less curious. Thus, if the hand be raised above the head, and the fingers be bent upon the palm, the notion of climbing, swinging, or pulling at a rope, is called up; if, on the other hand, the fingers are bent when the arm is hanging at the side, the idea excited is that of lifting some object from the ground; and if the same be done when the arm is advanced forwards in the position of striking a blow, the idea of fighting is at once aroused, and the Somnambulist is very apt to put it into immediate execution. On one occasion on which the Writer witnessed this result, a violent blow was struck, which chanced to alight upon a second somnambulist within reach; *his* combativeness being thereby excited, the two closed, and began to belabour one another with such energy that they were with difficulty separated. Although their passions were at the moment so strongly excited, that, even when separated, they continued to utter furious denunciations against each other, yet a little discreet manipulation of their muscles soon calmed them, and put them into perfect good humour.

496. Not only may the Mind be thus played upon through impressions communicated to it from the body, but it can re-act upon the body in a way which at first sight appears almost incredible, but which is in perfect conformity with the principles already laid down (§ 267). Thus, an extraordinary degree of power may be thrown into any set of Muscles, by assuring the Somnambulist

that the action which he is called upon to perform is one which he can accomplish with the greatest facility. The Writer saw one of Mr. Braid's hypnotized subjects—a man so remarkable for the poverty of his physical development, that he had not for many years ventured to lift a weight of twenty pounds in his ordinary state—take up a quarter of a hundred-weight upon his little finger, and swing it round his head with the greatest apparent facility, upon being assured that it was as light as a feather. On another occasion he lifted a half-hundred weight on the last joint of his fore-finger, as high as his knee. The personal character of this individual placed him, in the opinion of those to whom he was well known, above all suspicion of deceit; and the impossibility of any trickery in such a case would be evident to the educated eye, since, if he had practised such feats (which very few, even of the strongest men, could accomplish without practice), the effect would have made itself visible in his muscular development. Consequently, when the same individual afterwards declared himself unable, with the greatest effort, to lift a handkerchief from the table, after having been assured that he could not possibly move it, there was no reason for questioning the truth of his conviction, based as this was upon the same kind of suggestion as that by which he had been just before prompted to what seemed an otherwise impossible action (§ 454). It is well known to Physiologists, that in our ordinary Volitional contraction of any muscle, we do not employ more than a small part of it at any one time; whilst, on the other hand, every experienced Medical practitioner knows that in Convulsive contraction far more force is often put forth, than the strongest exertion of the will could bring into action.

497. In like manner may various other Muscular movements be induced, by the exclusive direction of the Somnambule's attention to their performance, of which the same individual would not be capable in the natural state. One of the most remarkable of these

phenomena was the exact imitation of Mademoiselle Jenny Lind's vocal performances, which was given by a factory girl, whose musical powers had received scarcely any cultivation, and who could not speak her own language grammatically. The Writer was assured by most competent witnesses, that this girl, in the hypnotized state, followed the Swedish Nightingale's songs in different languages so instantaneously and correctly, as to both words and music, that it was difficult to distinguish the two voices. In order to test the powers of this Somnambule to the utmost, Mademoiselle Lind extemporised a long and elaborate chromatic exercise, which the girl imitated with no less precision, though in her waking state she durst not even attempt anything of the sort.

498. So, again, the Writer has seen abundant evidence that the sensibility of a Hypnotized subject may be exalted to an extraordinary degree, in regard to some particular class of impressions; this being due, as before, to the concentration of the whole attention upon the objects which excited them. Thus he has known a youth in the hypnotized state find out, by the sense of Smell, the owner of a glove which was placed in his hand, from amongst a party of more than sixty persons; scenting at each of them one after the other, until he came to the right individual. In another case, the owner of a ring was unhesitatingly found out from amongst a company of twelve; the ring having been withdrawn from the finger before the Somnambule was introduced. The Writer has seen other cases, again, in which the sense of Temperature was extraordinarily exalted,—very slight differences, inappreciable to ordinary touch, being at once detected; and any considerable change, such as the admission of a current of cold air by the opening of a door, producing the greatest distress.—Some of the most remarkable examples of this kind, however, are afforded by that exaltation of the *muscular* Sense, which seems to be an almost

constant character of the Somnambulistic state, replacing the sense of sight in the direction of the movements (§ 128). That Sleep-walkers can clamber walls and roofs, traverse narrow planks, step firmly along high parapets, and perform other feats which they would shrink from attempting in their waking state, is simply because they are *not distracted* by the sense of danger which their Vision would call-up, from concentrating their exclusive attention on the guidance afforded by their Muscular sense (§ 192).

499. Many pages might be filled with the record of such marvels, which present themselves alike in natural, and in artificial or induced Somnambulism: but all such phenomena are easily reducible to the general principles we have already laid down as characteristic of this state;—namely (1), the *entire engrossment* of the Mind with whatever may be for a time the object of its attention; so that sensory impressions are perceived with extreme vividness, long-forgotten ideas retraced with the most remarkable distinctness, and muscular movements performed with extraordinary energy and the most precise adaptiveness;—and (2) the *passive receptivity* of the Mind (when not previously engrossed by some dominant idea of its own) for any notion that may be suggested to it; the particular course which such suggested train of ideas will take, being much influenced by the temperament of the "subject," and by the previous habits of thought and feeling.

500. But there is one point that was brought into prominent relief by Mr. Braid's experiments, which is too important, on account of its bearing on the supposed *curative* powers of Mesmerism and Spiritualism, to be passed by in this general sketch. The influence of the state of "expectant attention" upon the Organic functions of the body (Chap. XIX.) being fully admitted among scientific Physiologists, there need be no difficulty in making the further admission, that the peculiar *concentration* of the attention which can be obtained in the Hypnotic state, should produce still more striking results. And there is nothing in the least degree in-

credible, therefore, in the phenomena which Mr. Braid recorded, many of which the Writer himself witnessed. The pulsations of the heart and the respiratory movements may be accelerated or retarded; and various secretions may be altered both in quantity and quality, of which the following is a striking example :—

a. A lady, who was leaving off nursing from defect of milk, the baby being thirteen months old, was hypnotized by Mr. Braid ; and whilst she was in this state he made passes over the right breast to call her attention to it. In a few moments her gestures showed that the baby was sucking, and in two minutes the breast was distended with milk, at which, when subsequently awakened, she expressed the greatest surprise. The flow of milk from that side continued most abundant ; and, in order to restore symmetry to her figure, Mr. Braid subsequently produced the same change on the other ; after which she had a copious supply of milk for nine months.

The removal of morbid deposits under the same influence, seems quite as well attested as the charming-away of warts (§ 570), and the Physiologist who holds with the illustrious Müller, that "an idea that a structural defect will certainly be removed by a certain act, increases the organic action in the part," will see no inherent improbability in the following statement :—

b. A female relative of Mr. Braid was the subject of a severe rheumatic fever, during the course of which the left eye became seriously implicated, so that after the inflammatory action had passed away, there was an opacity over more than one half of the cornea, which not only prevented distinct vision, but occasioned an annoying disfigurement. Having placed herself under Mr. Braid's hypnotic treatment for the relief of violent pain in her arm and shoulder, she found, to the surprise alike of herself and Mr. B., that her sight began to improve very perceptibly. The operation was therefore continued daily ; and in a very short time the cornea became so transparent that close inspection was required to discover any remains of the opacity.—(*Neurhypnology*, p. 175.)

The Writer has known other cases, in which secretions that had been morbidly suspended, have been re-induced by this process; and is satisfied that, if applied with skill and discrimination, it would take rank as one of the most potent methods of treatment which the Physician has at his command. The channel of influence is obviously the *vaso-motor* system of Nerves; which, though not directly under subjection to the Will, is peculiarly affected by Emotional states (§§ 113, 565).

CHAPTER XVI.

OF MESMERISM AND SPIRITUALISM.

"When the Mind is once pleased with certain things, it draws all others to consent, and go along with them; and though the power and number of instances that make for the contrary, are greater, yet it either attends not to them, or despises them, or else removes them by a distinction, with a strong and pernicious prejudice to maintain the authority of the first choice unviolated. And hence in most cases of Superstition, as of Astrology, Dreams, Omens, Judgments, &c., those who find pleasure in such kind of vanities, *always observe where the event answers, but slight and pass by the instances where it fails, which are much the more numerous.*"—Bacon, *in Novum Organon.*

501. "What to Believe?" as to that diversified series of phenomena termed "Mesmeric" and "Spiritualistic," is a question which most persons have at times asked themselves during the last few years; and to which the responses have varied with the amount of information possessed by each questioner, with his previous habits of thought, and with his tendency to credulity or to scepticism,—his love of the marvellous and occult, or his desire to bring everything to the test of Science and Common Sense.

502. Some there are, who persist in the determination to disbelieve in the genuineness of *all* the asserted facts; designating them as "all humbug," and maintaining that none but fools or knaves could uphold such nonsense. Such persons, however, must now find themselves in the unenviable predicament of being obliged to place some of their best friends in one or other of these two categories; since it is impossible to go into any kind of society, literary or scientific, professional or lay, gentle or simple, without finding a large proportion of intelligent and truthful persons, such as would be regarded as trustworthy on all other subjects,

who affirm that they have been themselves the actors in some or other of the performances in question, and that, however strange the phenomena may seem, they are nevertheless genuine.

503. Others, again, admit such of the facts as seem to them least repugnant to Common Sense; but, without attempting to give any rational explanation of these, consider that they have sufficiently disposed of them by characterising them as "all Imagination:" not informing us, however, whether it be the *actors* or the *spectators*, whose imaginative faculty has been worked upon; or in what way the mysterious performances of a *clairvoyante* or a *medium* are related to those glorious creations, which have sprung from the legitimate exercise of the imaginative faculty in a Shakspere or a Milton, a Mozart or a Beethoven, a Raffaelle or a Turner.

504. The members of the Medical profession, accustomed to the vagaries of Hysteria, and recognizing the hysterical constitution in a large proportion of the subjects of Mesmeric and Spiritualistic agency, have too generally satisfied themselves with the phrase "all hysterical:" a reply which affords no real information to those inquirers who think that their doctors ought to help them to a solution of such difficulties, and which has now been fully proved to be incorrect by the fact, that steady, sensible, middle-aged men, having all their wits about them, are sometimes found to be as good subjects of certain of these operations, as the susceptible young females who are deservedly regarded with so much suspicion.

505. Then there is a class of *partial* believers, who admit that there is "something in it"—they cannot exactly tell what; and who are sorely puzzled between the dictates of their own Common Sense, and the assurances pressed on them, that what *they* find rather too strong for their belief, is just as well authenticated as what they profess themselves disposed to receive.

506. And the ascending series is terminated by that assemblage of *horough-going* believers, who find nothing too hard for "spiritual"

agency, nothing improbable (much less impossible) in any of its reputed performances; and who recognise in the wondrous revelations of a *clairvoyante* or a *medium*, and in the dispersion of a tumour,—in the communications of departed spirits with their surviving friends, and in the rotation of a table,—in the induction of profound insensibility during the performance of a severe operation, and in the oscillations of a suspended button,—in the subjugation of the actions of one individual to the will of another, and in the flexure of a hazel twig,—in everything, in short, great and small, which they cannot otherwise explain,—the manifestations of some "occult" power, to be ranked among the Cosmical forces, but not to be identified with any one of those previously admitted; which is capable, not only of raising heavy tables from the ground, and keeping them suspended in the air, but of making musical instruments play without being touched by visible hands, and even of transporting living men and women through the air, and bringing them into apartments of which all the entrances had been securely closed.

507. It is a phenomenon of no small interest to the student of Human Nature, that from the *first* of these classes the transition should often be immediate and abrupt to the *last*. Every one has heard of determined scoffers, who, having been teased into "assisting" at a Mesmeric or a Spiritualistic *séance*, have left it metamorphosed into true believers; and the conversion of these individuals is triumphantly cited by the partisans of these systems, as a bright example of the progress of truth. *Magna est veritas, et prevalebit*, is their continual cry; and we are assured that we have only to witness the facts as these have done, to be ourselves convinced.—But it requires no great discrimination to see that little value is to be attached to opinions thus embraced. There is a class of persons who begin by straining at gnats, and end by swallowing camels. At first they take exception to everything, declare there *must* be fraud somewhere, and consequently

refuse their confidence to any one of the parties concerned; but then, having allowed themselves to be put in communication with a *clairvoyant* or a *medium*, are so staggered by the wonderful revelations they receive, that they pass at once to the opposite extreme, and seem to yield their belief the most readily to that which makes the largest demands upon their credulity. It is an old observation that some of the greatest sceptics in Religion are the most credulous in other matters; and of one such person it was happily said by a distinguished wit, that "She believes anything that is not in the Bible." It is, in fact, from the very same disposition to jump at important conclusions, without due examination,—to hazard a decision, where a sound judgment would still hold the balance free to oscillate—to give up the entire mind to any idea which makes a strong impression upon it, without opening it to the reception of those considerations which might modify, if not entirely metamorphose, its condition,—that a large proportion of Mankind become utter sceptics on the one hand, or thorough-going believers on the other. A feather's weight will often turn the scale, when it is vibrating between these two states.

508. The antithesis to this condition is presented by those sincere and earnest seekers after Truth, who see in the class of facts under consideration a group of natural phenomena strongly calling for scientific and painstaking investigation; who enter upon the inquiry with all the assistance that a knowledge of Physiology and Pathology, mental as well as bodily, can afford them; and with minds trained in those habits of philosophical scrutiny, which are far more valuable than any amount of *mere* knowledge;—who commence with the systematic study of the phenomena which are least removed from the previous circle of admitted principles, and, if satisfied of *their* reality by careful and testing investigations, set themselves to consider whether they can be legitimately brought within that circle, or whether its radius may be fairly lengthened so

as to comprehend them, or whether it is necessary to have recourse to the hypothesis of a new and hitherto unknown agency for their explanation ;—who are not ashamed to halt and hesitate, when they cannot obtain the definite basis which they require for a sound decision ;—and who thus proceed, step by step, clearing the ground as they advance ; so that the way they have already made is rendered plain to all who may come after them, and the most promising paths are opened for further researches.

509. Now, to this class of inquirers, whom the true Philosopher, whatever be his special object of pursuit, welcomes as his most valuable coadjutors, Mesmerists and Spiritualists have ever shown a decided repugnance. "All or nothing" seems to be the motto of the latter, who act as if a rational explanation of any one of their marvels were a thing to be deprecated rather than welcomed. In order to reconcile this discouraging treatment with their loud professions of readiness to court investigation, they have been obliged to have recourse to the hypothesis, that, just as a damp atmosphere around an electrical machine prevents a high state of electric tension from being maintained, the presence of even a candid sceptic weakens the Mesmeric or the Spiritual force ; and this, not merely when he manifests his incredulity by his language, his tones, his looks, or his gestures, but even when he keeps it concealed beneath the semblance of indifference.

The following are the rules laid down, by M. Deleuze (*Histoire Critique du Magnetisme Animal*, 1813), for the attainment of curative success :—" Forget for a while all your knowledge of Physics and Metaphysics.—Dismiss from your mind all objections that may occur. —Imagine that it is in your power to take the malady in hand, and throw it on one side.—Never reason for six weeks after you have commenced to study.—Remove from the patient all persons who might be troublesome to you.—Have an active desire to do good, a firm belief in the power of Magnetism, and an entire confidence in employing it. In short, repel all doubts, desire success, and act with simplicity and attention."—This is tantamount to saying, " Be very credulous; be very persevering; reject all past experience; and do not listen to reason."

510. It is attributable, then, to the difficulties which honest truth-seeking investigators have encountered, through finding themselves disqualified by their very scepticism from evoking the phenomena of which they were in search, and through being treated as enemies, or (at best) as suspicious allies, by those to whom they might naturally look for what they require, that they are not in a position to elucidate *all* the marvels which are continually being brought before the public, by witnesses whose truthful testimony as to *what they themselves believe* must be accepted as valid, if Human testimony is to be accepted on any subject whatever. For it is the experience of the Writer, as it is of many of his sceptical friends, that none of the marvellous phenomena which are related to them as of indubitable occurrence, can be brought to recur in their presence; so that the evidence on which they are called upon to believe in their *actual* occurrence, entirely consists in the *belief of others*, as to matters on which it has been amply proved, (§§ 142–147) that there is an extraordinary tendency to self-deception and inaccuracy of memory (§ 365) under the influence of "dominant ideas."

511. The most important contribution to the elucidation of what is really true in Mesmeric, and (by anticipation) in Spiritualistic phenomena, was undoubtedly that made by the researches of Mr. Braid, on that state of *artificial* or *induced* Somnambulism to which he gave the appropriate designation of Hypnotism, and of which the principal features were described in the last chapter. For while this condition may be induced by the individual himself (or herself), without any external agency whatever, its phenomena are so essentially the same with those of the (so-called) Mesmeric Somnambulism, as to afford the most valuable assistance in the analysis of the real nature of the latter. Many grave sources of error have thus been eliminated; for not merely do we at once get rid—so far as the phenomena in question are concerned—

of any pretext for the assumption of a new and special force passing forth from the eyes or finger-ends of the operator into the body of his subject, but we are also able to select our own subjects from among a large body of individuals, of all ranks, ages, and temperaments, who are found to present the requisite susceptibility; we can vary the conditions of our experiments at our pleasure, so as to educe the greatest variety of results for comparison with each other; and we can apply any tests we may be able to devise, for discriminating between the fictitious and the real, the acting of a part, and the natural expression of the subject's genuine state of mind.

512. It has been shown that the Hypnotized subject, like the natural or the mesmeric Somnambulist (§ 494), is entirely destitute of the power of *self*-direction over either his ideas, his feelings, or his actions; and seems entirely amenable to the will of *another*, who may govern the course of his subject's thoughts at his own pleasure, and may thus oblige him to perform any actions which he may choose to determine. The clue to the real nature of this condition, however, having been found by Mr. Braid, in the undisturbed and concentrated operation of that principle of *suggestion*, which has long been well known to Psychologists, he subsequently, under the guidance of this idea, followed-up the investigation of its varied manifestations with great zeal and intelligence. The general result of these investigations was to give a definite basis for the study of the reputed phenomena of both Mesmerism and Spiritualism; enabling the inquirer who is armed with the knowledge of them to distinguish what is *probable* from what is *incredible*,—what may be readily admitted as Scientific truth, from what must be unhesitatingly rejected as depending either on fraud or on self-deception.

513. In the first place, then, it may be freely admitted that Mesmerized subjects have exhibited all the phenomena which are identical with, or analogous to, those which are presented in the

states of Electro-biology and Hypnotism; the artificial reverie and somnambulism which are produced by the Mesmerist, being in all essential particulars the same as those which are self-induced by the fixation of the vision. A state resembling the Biological reverie can be induced by Mesmerism, as well as the condition of true Somnambulism; and it may be witnessed in somnambules who have not been completely restored to the natural command of their faculties,—any direction given to them being automatically obeyed, although they had been awakened in the ordinary mode. It is unquestionable, moreover, that the mode in which these conditions are usually produced by the Mesmerizer, is such as to produce a *monotony of impression* and a *fixation of the attention.* Some, for instance, content themselves with directing the subject to gaze fixedly at their eyes; the effect of which will be presumably the same as that of looking at a shilling in the hand, or at Mr. Braid's lancet-case. In fact, the "lively young lady" formerly referred to (§ 454) was "biologized" either by staring at her own fingers or at the eyes of the operator; and the effect was precisely the same in both cases, her *rapport* with the operator being as complete in the one case as in the other. Other Mesmerizers employ certain strokings and waftings of the hand, termed "passes;" and these have a two-fold effect, serving to produce that monotony of impression which is favourable to the access of the sleep, and also to direct the attention of the patient towards any part upon which it may be intended specially to act.

514. All the ordinary methods of the Mesmerist, then, may be legitimately considered to act in the manner in which they operate when practised by those who employ them merely as means of fixing the attention of the "subject;" and their efficacy may be fairly explained on the principles already so fully dwelt on. The question of any mysterious Magnetic or other dynamical agency, which is the fundamental article of faith in the Mesmeric creed,

must, therefore, be decided by quite a different kind of evidence;—that, namely, which should demonstrate the possibility of the induction either of the somnambulistic state, or of some other characteristic phenomenon, *without any consciousness on the part of the "subject" that any agency was being exerted.*

515. The Writer does not hesitate to express the conviction, based on long, protracted, and careful examination of the evidence adduced to prove the existence of a Mesmeric force acting independently of the consciousness of the "subject," that there is none which possesses the least claim to acceptance as scientific truth. It is far more difficult to guard against sources of fallacy,—arising out of the guesses or anticipations at which the "sensitives" are marvellously ready, and the unconscious intimations of what is expected, of which they are wonderfully alert in taking advantage (not intentionally but suggestively),—than most persons who have not carefully studied these phenomena are at all aware: and those who go into the inquiry with a prepossession that the result will be affirmative, are certainly not those who are to be trusted as to their exclusion of all possibilities of error. It has been repeatedly found that Mesmerizers who had no hesitation in asserting that they could send particular individuals to sleep, or affect them in other ways by an effort of "silent will," have altogether failed to do so *when the subjects were carefully kept from any suspicion that such will was being exercised;* whilst, on the other hand, sensitive subjects have repeatedly gone to sleep *under the impression that they were being mesmerized from a distance,* when the supposed Mesmerizer was not even thinking of them.

516. The following experiment, one of several tried by Dr. Noble of Manchester, is worthy of being cited in detail; as affording a good illustration of the precautions which are needed to obtain trustworthy results:—

"An intelligent and well-educated friend had a female servant, whom he had repeatedly thrown into a sleep-waking state, and on whom he had tried a variety of experiments, many of which we ourselves witnessed. We were at length informed that he had succeeded in magnetizing her from another room, and without her knowledge; that he had paralysed particular limbs by a fixed gaze, unseen by the patient; and we hardly know what besides. These things were circumstantially related to us by many eye-witnesses; amongst others, by the medical attendant of the family, a most respectable and intelligent friend of our own. We were yet unsatisfied: we considered that these experiments were so constantly going on, that the presence of a visitor, or the occurrence of anything unusual, was sure to excite expectation of some Mesmeric process. We were invited to come and judge for ourselves, and to propose whatever test we pleased. Now, had we visited the house, we should have felt dissatisfied with any result; we therefore proposed that the experiment should be carried out at our own residence; and it was made under the following circumstances:—The gentleman, early one evening, wrote a note, as if on business, directing it to ourselves. He thereupon summoned the female servant (the mesmeric subject), requesting her to convey the note to its destination, and to wait for an answer. The gentleman himself, in her hearing, ordered a cab, stating that if any one called he was going to a place named, but was expected to return by a certain hour. Whilst the female servant was dressing for her errand, the master placed himself in the vehicle, and rapidly arrived at our dwelling. In about ten minutes afterwards the note arrived, the gentleman in the meantime being secreted in an adjoining apartment. We requested the young woman, who had been shown into our study, to take a seat whilst we wrote the answer; at the same time placing the chair with its back to the door leading into the next room, which was left ajar. It had been agreed that after the admission of the girl into the place where we were, the Magnetizer, approaching the door in silence on the other side, should commence operations. There, then, was the patient, or 'subject,' placed within two feet of her magnetizer,—a door only intervening, and that but partially closed,—but she, all the while, perfectly free from all idea of what was going on. We were careful to avoid any unnecessary conversation with the girl, or even to look towards her, lest we should raise some suspicion in her own mind. We wrote our letter (as if in answer) for nearly a quarter of an hour, once or twice only making an indifferent remark; and on leaving the room for a light to seal the supposed letter, we beckoned the operator away. No effect whatever had been produced, although we had been told that two or three minutes were sufficient, even when mesmerizing from the drawing-room, through walls and apartments, into the kitchen. In our own experiment the intervening distance had been very much less, and only one solid substance interposed, and that not completely; but here, we suspect, was the difference—*the 'subject' was unconscious of the magnetism and expected nothing.*" —(*British and Foreign Medical Review*, 1845, vol. xix. p. 478.)

517. The following is a converse experiment performed by the same acute investigator:—

"We were one evening in company with a young lady, who, we had been informed, had evinced high Mesmeric susceptibility. We requested permission to test this ourselves, and were obligingly permitted to do so. Accordingly we commenced to *magnetize* the lady, by keeping our thumbs in apposition with those of our subject, and fixing the gaze at the same time upon her eyes, with all the intensity our will could command; in a few minutes a sort of hysteric somnolency ensued. Having satisfied ourselves thus far, we demagnetized. We next proceeded to *hypnotize* the same lady, adopting Mr. Braid's mode of directing the stare at a fixed point. The result varied in no respect from that which had taken place in the foregoing experiment; the duration of the process was the same, and its intensity of effect neither greater nor less. De-hypnotization again placed us where we were. And now we requested our patient to rest quietly at the fire-place, to think of just what she liked, and to look where she pleased, excepting at ourselves, who retreated behind her chair, saying that a new mode was about to be tried, and that her turning round would disturb the process. We very composedly took up a volume which lay upon a table, and amused ourselves with it for about five minutes; when, on raising our eyes, we could see by the excited features of other members of the party, that the young lady was once more *magnetized.* We were informed by those who had attentively watched her during the progress of our little stratagem, that all had been in every respect just as before. The lady herself, before she was undeceived, expressed a distinct consciousness of having *felt our unseen passes streaming down the neck.*" —(*Op. cit.* p. 477.)

518. That the *expectation* of the result affords an adequate explanation of its occurrence, was also the conclusion arrived at by M. Bertrand, who was the first to undertake a really scientific investigation of the phenomena of Mesmerism. The following is one out of several cases of this kind, which occurred within his own experience :—

"I had, amongst others, a female Somnambule who exhibited very curious phenomena, of a character not to be doubted. Being after some time compelled to be absent, I left her in the hands of one of my friends, who was very anxious to continue the treatment. The perusal of a great number of works, and my conversations with Magnetizers who doubted of nothing, suggested to me to see if I could not influence my Somnambulist, in spite of the distance of a hundred leagues which separated me from her. I wrote in consequence to my friend, and sent to him a little magnetized note, which I prayed him to place upon the stomach of the patient; I indicated the epigastrium, because I had always heard this locality mentioned in these experiments. The experiment was made; it succeeded, and the patient had a sleep accompanied with all the customary phenomena. However, I did not conceal from myself that as the patient had been apprized of the experiment which we were anxious to try, it might be that the sleep, although quite real, had been produced by the imagination alone. I therefore made another trial, to know what to think of it. I wrote a second letter, which I did not magnetize, and sent it as if it had been magnetized, warning the patient that it would cause her to fall into somnambulism; in fact, she fell into this state, which presented all the characters which had been usual. I communicated this result of my experiment to the magnetizers whom I frequented; they appeared greatly surprised thereat; and, not being able to recognize the power of Imagination in a manner so marked, they pretended that if the last letter had produced the effect which I stated, it was only because, in writing it, I had (even unintentionally) impregnated it with my fluid. I set about an experiment which should teach me what was the real state of the case. I asked one of my friends to write a few lines in my place, and to strive to imitate my writing, so that those who should read the letter should mistake it for mine (I knew he could do so). He did this; our stratagem succeeded; and the sleep was produced just as it would have been by one of my own letters."—(*Du Magnétisme Animale en France*, Paris, 1826.)

519. That the Imagination would supply the place of what he maintained to be the real Mesmeric influence, was admitted even by Dr. Elliotson :—

"Mere imagination was at length sufficient; for I one day told her and two others that I would retire into the next room, and mesmerize them through the door. I retired, shut the door, performed *no* mesmeric passes, but tried to forget her, walked away from the door, and busied myself with something else,—even walked through into a third room; and, on returning in less than ten minutes from the first, found her soundly asleep, and she answered me just as was usual in her sleep-waking condition."—*Zoist*, 1846, vol. xxxvi. p. 47.

520. It is further to be borne in mind, that (putting aside the hypothesis of intentional deceit) the "sensitives" are often affected by impressions so slight as to be imperceptible to others. In this mode they may become aware, through slight differences in temperature or in odour, whether certain objects presented to them have, or have not, been mesmerized by contact with the hand of the operator. There would be nothing, in such an exaltation of ordinary Sensibility, at all transcending the accepted verities of science. (See §§ 128, 498.)

521. But, it is asserted, the existence of a special Mesmeric force is proved by the existence of the peculiar *rapport* between the Mesmerizer and his "subject," which is not manifested towards any other individuals, save such as may be placed *en rapport* with the subject by the mesmeriser. Nothing is more easy, however, than to explain this on the principle of "dominant

ideas." It will be readily understood from what has preceded, that if the mind of the subject be so yielded up to that of the mesmerizer, as to receive and act upon any impression which the latter forces upon, or even suggests to it, the idea of such a peculiar relation is as easily communicable as any other, and may exert a complete domination over the subject through the whole of the sleep-waking state. Hence the commands or suggestions of the mesmerizer meet with a response which those of no other individual may produce; in fact, the latter usually seem to be unheard by the somnambule, simply because they are not related to the dominant impression,—a phenomenon of which the experience of *natural* somnambulism is continually presenting examples (§ 488). And further, it being a fact that individuals, of what may be termed the "susceptible" constitution, have brought themselves, by the habit of obedience, into complete subjection to the expressed or understood will of some other person, even in the waking state, without any mesmeric influence whatever (§ 452), it is easy to understand how such a habit of attending to the operator, and to him alone, should be peculiarly developed in the state of Somnambulism, in which the mind has lost its self-acting power, and is the passive recipient of external impressions.—The same explanation applies to the other phenomena of this *rapport*, such as its establishment with any bystander by his joining hands with the mesmerizer and the somnambule; for it is quite sufficient that the somnambule should be previously possessed with the idea that this new voice will thus be audible to her, and that she must obey its behests, for it to produce all the same effects upon her as that of the mesmerizer had previously done.

522. The History of Mesmerism, indeed, when candidly and philosophically examined (as it was by M. Bertrand), affords abundant evidence in support of the foregoing position: for the *rapport* was not discovered until long after the practice of Mesmerism had come into vogue, having been unknown alike to Mesmer himself and to his

immediate disciples; and its phenomena only acquired constancy and fixity, in proportion as their (supposed) "laws" were announced and received. It is an important fact that Mesmerizers who began to experiment for themselves without any knowledge of what they were to expect, and who succeeded in producing a great variety of remarkable phenomena, never discovered this *rapport;* though they obtained immediate evidence of it, when once the idea of it was put into their own minds, and thence transferred into those of their subjects.—In all the experiments witnessed by the Writer which seemed to indicate its existence, the previous idea had either been distinctly present, or it had obviously been suggested by the methods by which the mesmeric somnambulism had been induced: whilst in a large number of other cases which have fallen under his notice, but of which the subjects had not been among the *habitués* of mesmeric *séances,* the phenomena of this class could not be made to show themselves; the consciousness of the somnambulist *not* being limited to the mesmerizer, or to those placed by him *en rapport* with them, but being either equally limited, or as equally extended, to all around.

523. It is not within the scope of the present treatise, to enter into a detailed examination of the multitudinous phenomena of Mesmerism. All that can be now attempted is to give the clue by which those who desire to use it, may explore this mysterious labyrinth for themselves. And the same may be said of the phenomena of like kind, to which public attention has more recently been called under the designation of Spiritualism. To those who, in common with the Writer, have gone through a course of study of Mesmeric phenomena, under the guidance of a scientific knowledge of those parts of Psychology which relate to the subject, it is perfectly obvious that the same general principles afford the clue to the explanation of what is *genuine,* in the latter as in the former class of phenomena; and that the same tendencies to *self-deception* conspire to produce, in minds predisposed to it, an acceptance of

beliefs which are altogether repugnant to educated Common Sense (§ 385).

524. The asserted phenomena of *Mesmerism* and *Spiritualism* may be fairly grouped under the following Classes :—

I. Those whose genuineness may be readily admitted, *without any extraordinary weight of evidence* in their support; since they are quite conformable to our previous knowledge, and are readily assignable to principles determined by the Scientific study of the class of facts to which they belong.

II. Those which, *not* being conformable to our previous knowledge, or explicable upon principles already admitted, cannot be accepted as genuine, *without a greater body of satisfactory evidence in their favour;* but which, not being in absolute opposition to what we deem the best established Laws of Nature, we may receive upon adequate evidence, without doing violence to our Common Sense,—holding ourselves ready to seek their explanation in a more extended acquaintance with the powers of Mind and of Matter.

III. Those which not only lie *beyond* our existing knowledge, but are in direct *contrariety* to it. Here, even though the *external* evidence in their favour should be the same with that on which the facts of the preceding groups find a secure support, the *internal* evidence is altogether antagonistic to their reception (§ 321); and its force in the well-ordered mind must remain conclusive against the validity of all statements, save those which shall have been carefully, sagaciously, and perseveringly investigated, by observers fully qualified for the task by habits of philosophical discrimination, by *entire freedom from prejudice*, and by *a full acquaintance with the numerous and varied sources of fallacy which attend this particular department of inquiry.*

These being the rules of all other branches of Scientific research, there is no reason why they should be departed from in one which so pre-eminently needs a constant reference to the canons of sound philosophy. Entertaining, as the Writer has been forced to do, an extremely low opinion of the logical powers of the great bulk of the upholders of the Mesmeric and Spiritualistic systems,*—their belief being founded on certain foregone conclusions which they have adopted without due examination, and which they sustain by nothing better than insensate reiteration,—it has given him the greatest astonishment to find any men of high attainments in particular departments of Science committing themselves to the extraordinary proposition, that if we admit the reality of the *lower* phenomena (Class I.), the testimony which we accept as good for them ought to convince us of the *higher* (Classes II. and III.).† Such men seem totally oblivious of the difference between *external* and *internal* evidence,—the testimony of our *senses* (or of those of other individuals), and that of our *sense.*

525. Under the *first* of these categories may be ranked the occurrence of Mental states showing every gradation between mere *expectant attention* and the most profound *coma;* the characteristic which is common to all these states being the more or less complete surrender of the guiding and controlling action of the Will, while there is a greater or less degree of receptivity for external impressions. The current of mental activity thus comes to be essentially *automatic*; and the corrective action of Common Sense—which is the *general resultant* of antecedent experience—being thus suspended (as in dreaming), the Ego comes to believe implicitly in the ideas which may possess him at the time,

* Thus, nothing is more common than the assertion that Faraday's *experimental demonstration* that "table-turning" is produced by unconscious muscular action (§ 245) *does not touch the question;* because the operators *know* that they do *not* themselves move the tables. When, by the use of Faraday's or any other properly constructed testing apparatus, they can *prove* that the tables on which hands are placed will turn or tilt *without receiving any pressure from those hands,* they will have made out a case for further enquiry; but until they have done so, their assertions are of no scientific value.

† This is the position which has been repeatedly taken by Mr. Alfred R. Wallace; a gentleman whose admirable researches as a Naturalist have justly gained for him a reputation of the highest order.

whether these have been directly suggested to him by external prompting, or have been evolved by the operations of his own mind. Not only may all the phenomena which have been described under the heads of Electro-Biology (§§ 448-469) and Somnambulism (§§ 487-500), but a great variety of others too numerous to particularize—such as the induction of cataleptic rigidity or of convulsive movements in particular groups of Muscles, or of unconsciousness to Sensory impressions of a particular class or affecting a particular part of the body,—which are adduced as evidence of a special Mesmeric power or Spiritual agency, be thus accounted-for in accordance with definite Physiological principles. For to the very same "possession," voluntarily permitted by persons who yield themselves up to the domination of a certain set of ideas, we are fully justified in attributing an amount of *self-deception* as to matters of fact, which—extraordinary as it may seem—is perfectly intelligible on Scientific principles. Nothing is more common at the present time, than for the advocates of Spiritualism to appeal to "the evidence of their own senses" as conclusive in regard to any thing done by "the spirits"; and to claim that their testimony and that of others should be received as that of honest and truthful witnesses to what Common Sense rejects as altogether preposterous and incredible: such persons being altogether ignorant of the fact well known to the Physiologist and Psychologist, that, when the Mind has been previously possessed by a "dominant idea," *nothing is more fallacious* than the "evidence of the senses" (§§ 139-148, 186, 187).

526. To this category, again, may be referred a large variety of Movements by which those mental states are expressed in action; as Table-talking, Planchette-writing, Spirit-drawing, and the like. Putting aside the large amount of intentional deception practised by those who trade upon public credulity in regard to such matters, there can be no reasonable doubt that all these thing

are done in good faith, by persons who honestly believe themselves to be "mediums" of communication with the spiritual world (§ 252 *b*). Some of these communications come to them spontaneously in a state resembling profound Reverie; while others are made in answer to questions having more or less of a suggestive character. Now where the answers given are such as the "medium" himself (or herself) *may* have given, there is obviously no ground for affirming that they indicate any occult agency; any more than the replies of Biologized subjects, or of Hypnotic or Mesmeric somnambules. In some cases it seems likely that the supposed "revelations" are merely the reproductions of impressions long since recorded, which have so completely passed out of the *conscious* memory, that the "medium" may honestly believe that they never either *had* or *could have* been made. Such cases, being precisely paralleled by the recovery of the lost "traces" in Dreaming and Somnambulism (§§ 483, 487*a*), obviously in no respect lie outside the ordinary course of Psychical action.

527. But *secondly*, it is affirmed of the Spiritual "mediums," as of Mesmeric "clairvoyants," that they occasionally give information as to matters of fact, of which they cannot conceivably have become aware through any ordinary channel; so that they must be credited either with the possession of some "psychic force" at present unknown to Science, or with the reception of communications from another sphere of existence. Now in regard to a large proportion of these cases, it may be unhesitatingly asserted that they would break down altogether, if submitted to the same searching enquiry that has been bestowed upon others of their kind, by men who were specially armed with a knowledge of the probable sources of fallacy. It need not be imputed to the narrators of them, that they have intentionally stated what they know to be untrue; for experience shows that the memory of the most truthful persons is very treacherous in regard to matters as to which they have a preformed bias; so that round a nucleus of

truth an accretion of error will often form, without the least intention to deceive on the part of the contributors to it (§§ 365, 366). In the course of his own enquiries on this subject, the Writer has had numerous opportunities of observing the readiness with which occurrences have been caught-at by sympathizing witnesses, and worked-up into marvels; some of which were obviously the results of suggestions,— sometimes designedly made by himself; while others were mere guesses, often very wide of the mark, which were *made* to fit the facts, by progressive though unintentional modification,—the myth-making process just referred to.

528. But in addition to the cases in which *no intentional* deception has been practised, the Scientific enquirer has to deal with those in which the results have been obtained by a system of cheating and trickery, devised to play upon the credulity of those who are predisposed to fall into the trap laid for them, and veiled by ingenious artifices from the detection of such as are desirous and able to expose it. For the performers in such cases are sufficiently keen-sighted to perceive the existence of a sceptical disposition on the part of any of the "circle," and take their measures accordingly; assigning as a pretext that the Mesmeric agency cannot manifest itself, or that "the spirits" will not be propitious, unless certain conditions are complied with, which are tantamount to the exclusion of all thorough scrutiny. Or, if they submit to these tests, it is with the distinct warning that no manifestation is likely to be vouchsafed to such suspicious sceptics,— which, so far as the Writer's experience has extended, has been the invariable result. Scientific men have been continually taunted with their unwillingness to investigate phenomena of this class; but those who have had to encounter these *negative* results over and over again, are not to be blamed if they not only refuse to accept the testimony of those who have *not* been trained in habits of scientific investigation, as to matters on which they

know that there are peculiar liabilities to error, but decline to waste valuable time in repetitions of similar futile attempts to obtain manifestations that will bear being submitted to such tests as would be required in any other department of enquiry.

529. That a large number of the so-called Physical manifestations have no other existence than in the belief of those who report them, the Writer has come to feel a complete assurance, alike from what he has himself witnessed, and from the testimony of others. But it may be said,—"Why do you believe the "evidence of *your* senses, and tell us in the same breath that we "are *not* to believe that of *ours?* And why do you accept the "testimony of the witnesses on *your* side, and refuse to credit that "of the witnesses who confirm *our* statements?"—The answer is simple. When either our own senses, or the testimony of others, inform us of something that is either accordant with inherent probability, or is not discordant with inherent possibility, we receive that evidence as valid, until it is rebutted by some counter-proof. But if either our own senses, or the testimony of others, inform us of something that is *entirely inconsistent* with inherent possibility, we *refuse to accept the information*, feeling assured that a fallacy must lurk somewhere.

Thus when we witness the deceptions of a clever Conjuror, we know perfectly well that we are *not* to trust the evidence of our own senses; and we set our ingenuity to work to discover how the trick is done. Thus in some cases it is managed by pure sleight of hand; one thing being substituted for another with such dexterity and rapidity, that—the attention of the observer being purposely distracted—his eye does not follow the movement. In others, it depends on certain optical pre-arrangements, which make us believe that we see what we really do not see; as in the case of the "talking head," which seems to rest on a table that appears to be supported only on four legs, so that we suppose ourselves to be looking beneath it at the drapery *behind*,—instead of which we are really looking at a pair of mirrors meeting at a right angle in front, so as to reflect to us the drapery at the *sides*, the body of the "head" being concealed behind them.

530. Now it has happened over and over again in the Writer's experience, that what *he* considered as simple facts admitting a perfectly natural explanation, were interpreted as the results of some occult agency, Mesmeric or Spiritual, as the case might be. And from these cases the transition is easy to others, in which subjective sensations are referred to objective realities. Thus when two spiritualistic performers, in perfectly good faith, asserted that a table rose from the floor beneath their hands, whilst a third person, who was carefully watching the feet of the table, declared that one of them had never left the ground, it turned out that the first assertion entirely rested upon their mental conviction that they had "felt it pressing upwards against their hands,"—a tactile sensation obviously producible by their expectation of such an occurrence (§ 146). And so, when Mr. Varley assures us that he has seen, in broad daylight, a large dining-table lifted bodily off the floor, and moved in the direction which he mentally requested it to take, we have to consider whether it is more consistent with inherent probability that Mr. Varley interpreted subjective visual perceptions produced by his mental expectation (§ 186), as objective realities, or that the table was actually raised, either by his own "psychic force," or by the agency of disembodied spirits (*Quarterly Review*, Oct. 1871, pp. 330, 348). The process by which the mind of a person given up to the "possession" of dominant ideas, is first led to *misinterpret actual occurrences*, and then (as in dreams, § 482 *b*) to *invent objective explanations* of his own sensations, is perfectly familiar to all who have carefully studied the phenomena of Insanity (§ 559). And every one who accepts as facts, merely on the evidence of his own senses, or on the testimony of others who partake of his own beliefs, what Common Sense tells him to be much more probably the fiction of his own imagination—even though confirmed by the testimony of hundreds

affected with the same epidemic delusion,—must be regarded as the subject of "a diluted insanity."

531. At the same time, every one who admits that "there are more things in heaven and earth than are dreamt of in our philosophy," will be wise in maintaining a "reserve of possibility" as to phenomena which are not altogether *opposed* to the Laws of Physics or Physiology, but rather *transcend* them. Some of the Writer's own experiences have led him to suspect that a power of intuitively perceiving what is passing in the mind of another, which has been designated as "thought-reading," may, like certain forms of sense-perception (§§ 128, 498), be extraordinarily exalted by that entire concentration of the attention, which is characteristic of the states we have been considering. There can be no question that this divining power is naturally possessed in a very remarkable degree by certain individuals, and that it may be greatly improved by cultivation, So far, however, as we are acquainted with the conditions of its exercise, it seems to depend upon the unconscious interpretation of indications (many of them indefinable) furnished by the expression of the countenance, by style of conversation, and by various involuntary movements; that interpretation, however, going in many instances far beyond what can have been learned by experience as to the *meaning* of such indications. Some very curious examples of this kind are related in the Autobiography of Heinrich Zschokke, who, according to his own statement (p. 170), possessed this power in a very remarkable degree, frequently being able to describe not only the general course, but even many particulars, of the past life of a person whom he saw for the first time, and of whose history he knew nothing whatever.—Looking at Nerve-force as a special form of Physical energy, it may be deemed not altogether incredible that it should exert itself from a distance, so as to bring the Brain of one person into direct dynamical communication with that of another, without the intermediation either of verbal language or of movements of expression. A

large amount of evidence, sifted with the utmost care, would be needed to establish even a *probability* of such communication. But would any Man of Science have a right to say that it is *impossible?*

532. The case is altogether different, however, in regard to the *third* order of asserted facts; which every one whose mind has been trained in a conviction of the universality of the Law of Gravitation, *must* regard as incredible. That a living woman should be caught up by "the spirits," conveyed two miles through the air over the streets of London, and then introduced into a room of which the doors and windows were fastened, and to which there was no other access than the chimney, can only be believed even as a possibility by such as have entirely surrendered their Common Sense *quoad* this particular subject. And the explanation of the accordance of testimony as to the asserted fact, lies in the previous condition of the witnesses, whose minds were possessed with the expectation of its occurrence. Those who have studied the history of Epidemic Delusions, and especially that of the Witch-persecutions which took place in Great Britain and New England not two centuries ago, will at once see the parallel between the two cases.

Thus in 1657, Richard Jones, a sprightly lad of twelve years old, living at Shepton Mallet, was bewitched by one Jane Brooks; he was seen to rise in the air, and pass over a garden wall some thirty yards; and at other times was found in a room with his hands flat against a beam at the top of the room, and his body two or three feet from the ground, *nine people at a time seeing him in this position.* Jane Brooks was accordingly *condemned and executed* at Chard Assizes, in March, 1658.

The fact that such beliefs not only have been, but even now are, entertained by "educated" men and women, is a most curious manifestation of the myth-making tendency which seems inherent in Human nature, and which ever and anon breaks out in some new form. It is not a little curious, however, that, as Mr. Edward Tylor has pointed out, the various asserted Physical manifestations of modern Spiritualism are but repetitions of those which constitute, even at the present day, the means by which the Sorcerers of various uncultivated races maintain an influence over their dupes.

> "The received Spiritualistic theory belongs to the philosophy of savages. As to such matters as apparitions or possessions, this is obvious; and it holds in more extreme cases. Suppose a wild North American Indian looking on at a spirit-séance in London. As to the presence of disembodied spirits, manifesting themselves by raps, noises, voices, and other physical actions, the savage would be perfectly at home in the proceedings; for such things are part and parcel of his recognized system of Nature. The part of the affair really strange to him would be the introduction of such arts as spelling and writing, which do belong to a different state of civilization from his. The issue raised by the comparison of savage, barbaric, and civilized Spiritualism, is this:—Do the Red Indian medicine-man, the Tatar-necromancer, the Highland ghost-seer, and the Boston medium, share the possession of belief and knowledge of the highest truth and import, which, nevertheless, the great intellectual movement of the last two centuries has simply thrown aside as worthless? Is what we are habitually boasting of, and calling new enlightenment, then, in fact, a decay of knowledge? If so, this is a truly remarkable case of degeneration; and the savages whom some Ethnographers look on as degenerate from a higher civilization, may turn on their accusers, and charge them with having fallen from the high level of savage knowledge."—(*Primitive Culture*, vol. i. p. 141.)

Those who yield their ready assent to the claims set up by pretenders to occult powers of any kind, are really placing themselves on the level of the poor Greenlander who buys a fair wind from his Angekok, or of the credulous servant-girl who is cheated out of her savings by the cunning old woman who promises so to "rule the planets" as to bring her love-affair to a favourable issue.

CHAPTER XVII.

OF INTOXICATION AND DELIRIUM.

533. There is no class of aberrant Mental phenomena which is more deserving of careful scientific study, than that which is produced by the introduction into the Blood of substances which are foreign to its composition, and which have the special property of perverting its normal action on the Brain. For, in the first place, these phenomena bring into strong relief the contrast between that augmented *automatic* activity of the Cerebrum, which manifests itself in the rapid succession of thoughts, the vividness of images, and the strong excitement of feelings,—and the diminished *volitional* control, of which we have the evidence in the incoherence of thought, the incongruity of the imaginary creations, and the extravagance of the feelings. And in the second place, it is perfectly clear that this disturbance of purely *psychical* action, affecting not merely what may be regarded as the functions of the Brain, but the exercise of that attribute of Man's nature which seems most strongly indicative of a Power beyond and above it (§ 26), is produced by agencies purely *physical.* For it is not only that the balance between the automatic activity of the Brain, and the directing and controlling power of the Will, is disturbed by the *exaltation* of the former, so as to give it a predominance over the latter. On the contrary, the absolute *weakening* of Volitional control is clearly a primary effect of these agencies; being as strongly manifested when the automatic activity (as often happens) is reduced, as when it is augmented. And this weakening is still more obvious, when not merely the quality of the Blood, but the nutrition of the Brain, has been deteriorated by the prolonged

action of "nervine stimulants;" the Will becoming, as it were, paralysed, so that the mental powers are not under its command for any exertion whatever, while even its controlling power over bodily movement may be greatly diminished.

534. The states of Mind temporarily produced by *intoxicating agents*,—Alcohol, Opium, Hachisch, and the like,—are closely akin to one another in this fundamental character; as they are also to the *delirium* of fevers or other diseases, which is due to the introduction of a morbid matter into the Blood, whereby a *zymosis* or fermentation of its own materials is produced, which gives it a poisonous action on the Brain. In the second case, as in the first, the effect is transient; the poison being gradually eliminated from the circulation by the excretory apparatus (including the respiratory organs), so that the blood regains its original purity. And it is this *temporary* character alone, which differentiates the mental perversion of Intoxication and Delirium, from that which is *persistent* in Insanity (§ 550). Now although Alcoholic intoxication usually differs in some of its phenomena from the dreamy reverie produced by Opium, as this again does from the *fantasia* of the Hachisch, yet the differences in the states produced by any one of them—especially alcohol—in different individuals, are not less remarkable than those which ordinarily characterize the action of these different intoxicants. And as the one last named has been made the subject of special study (by experiment on his friends as well as on himself) on the part of a French Physician thoroughly conversant with the parallel phenomena of Insanity, it will be convenient to take his account of its action as furnishing the type of what may be called artificial Delirium.*

535. The Hachisch is a peculiar preparation of the *Cannabis Indica* or Indian Hemp, which has been used in the Levant as an

* Du Hachisch et d'Aliénation Mentale, Études Psychologiques; par Dr. T. Moreau (de Tours); Paris, 1845.

Intoxicating agent from a very remote period; the Assassins—a peculiar military and religious order of Mussulmans, founded in Persia in the eleventh century, whose representatives are still to be found in Bombay—deriving their name (originally Hachischin) from the use made of it by their chief to bring his followers into blind devotion to his service. It is a curious feature in the action of the Hachisch, that, except when under the complete influence of a very powerful dose, the person who has taken it does not altogether lose his power of introspection, and is subsequently able to retrace most of what he has felt and acted during the state of excitement. Its effects vary extremely, not only according to the dose that is taken, but also according to the susceptibility of the individual; and there are some persons on whom it seems to produce no impression whatever. A small dose seems usually to produce no other effect than a moderate exhilaration of the spirits, or, at most, a tendency to unseasonable laughter; and the first result of a dose sufficient to produce what is termed in the Levant the *fantasia*, is usually an intense sentiment of *happiness*, which attends all the operations of the mind.

"It is really *happiness*," says M. Moreau, "which is produced by the Hachisch; and by this I imply an enjoyment entirely moral, and by no means sensual, as we might be induced to suppose. This is surely a very curious circumstance; and some remarkable inferences might be drawn from it; this for instance among others,—that every feeling of joy and gladness, even when the cause of it is exclusively Moral,—that those enjoyments which are least connected with material objects, the most spiritual, the most ideal,—may be nothing else than sensations purely physical, developed in the interior of the system, as are those procured by the Hachisch. At least, so far as relates to that of which we are internally conscious, there is no distinction between these two orders of sensations, in spite of the diversity in the causes to which they are due; for the Hachisch eater is happy, not like the gourmand or the famished man when satisfying his appetite, or the voluptuary in gratifying his amative desires, but like him who hears tidings which fill him with joy, like the miser counting his treasures, the gambler who is successful at play, or the ambitious man who is intoxicated with success."—(*Op. cit.*, p. 54.)

536. Most persons will be able to recall analogous states of exhilaration, and the reverse condition of depression, in themselves; the former being characterized by a feeling of general well-being, a sentiment of pleasure in the use of all the bodily and mental powers, and a disposition to look with enjoyment upon the present, and with hope to the future; whilst in the latter state there is a feeling of general but indefinable discomfort. Every exertion, whether mental or bodily, is felt as a burden; the present is wearisome, and the future is gloomy (§ 156). These, like all other phases of Human nature, are faithfully delineated by Shakspere. Thus Romeo gives expression to the feelings inspired by the first state:—

> "My bosom's lord sits lightly in his throne,
> And all this day an unaccustomed spirit
> Lifts me above the ground with cheerful thought."
> (*Romeo and Juliet*, v., 1.)

While the reverse state is delineated by Hamlet in his well-known soliloquy—

> "I have of late (but, wherefore, I know not) lost all my mirth, foregone all customs of exercises: and, indeed, it goes so heavily with my disposition, that this goodly frame, the earth, seems to me a sterile promontory; this most excellent canopy, the air, look you, this brave o'erhanging firmament, this majestical roof fretted with golden fire, why, it appears no other thing to me, than a foul and pestilent congregation of vapours." (*Hamlet*, ii., 1.)

In the conditions here referred to, the same feelings of pleasure and of discomfort attend *all* the operations of the Mind,—the merely sensational, and the ideational. In the state of exhilaration, we feel a gratification from Sensations which at other times pass unnoticed, whilst those which are usually pleasurable are remarkably enhanced; while in like manner, the trains of Ideas which are started being generally attended with similar agreeable feelings, we are said to be under the influence of the pleasurable or elevating Emotions. On the other hand, in the state of depression we feel an indescribable discomfort from the very sensations which before produced the liveliest gratification; and the thoughts of the past, the present, and the future, which we before dwelt on with delight, now excite no feelings but those of pain, or at best of indifference. These conditions are essentially Physical (§ 552).

537. One of the first appreciable effects of the Hachisch, is the gradual weakening of that power of *volitionally* controlling and directing the thoughts, which is so characteristic of the vigorous mind. The individual feels himself incapable of fixing his attention upon any subject; the continuity of his thoughts being continually drawn off by a succession of disconnected ideas, which force themselves (as it were) into his mind, without his being able in the least to trace their origin. These speedily engross his attention, and present themselves in strange combinations, so as to produce the most impossible and fantastic creations. By a strong effort of the Will, however, the original thread of the ideas may still be recovered, and the interlopers may be driven away; their remembrance, however, being preserved, like that of a dream recalling events long since past. These lucid intervals become progressively of shorter duration, and can be less frequently procured by a voluntary effort; for the internal tempest becomes more violent, the torrents of disconnected ideas are so powerful as completely to arrest the attention, and the mind is gradually withdrawn altogether from the contemplation of external realities, being conscious only of its own internal workings. There is always preserved, however, a much greater amount of "self-consciousness" than exists in ordinary Dreaming; the condition rather corresponding with that in which the sleeper knows that he dreams, and, if his

dream be agreeable, makes an effort to prolong it, being conscious of a fear lest he should by awaking cause the dissipation of the pleasant illusion. It is another characteristic of the action of the Hachisch, that the succession of ideas has *at first* less of incoherence than in ordinary Dreaming, and the ideal events do not so far depart from possible realities; the disorder of the mind being primarily manifested in errors of sense, in false convictions, or in the predominance of one or more extravagant ideas. These ideas and convictions are generally not altogether of an imaginary character, but are rather *suggested* by external impressions; these impressions being erroneously interpreted by the perceptive faculties, and giving origin, therefore, to fallacious notions of the objects which excited them. It is in that more advanced stage of the "fantasia," which immediately precedes the complete withdrawal of the mind from external things, and in which the self-consciousness and Volitional power are weakened, that this perverted impressibility becomes most remarkable; more especially as the general excitement of the feelings causes the erroneous notions to have a powerful effect in arousing them:—

"We become the sport of impressions of the most opposite kind; the continuity of our ideas may be broken by the slightest cause. We are turned, to use a common expression, by every wind. By a word or a gesture our thoughts may be successively directed to a multitude of different subjects, with a rapidity and a lucidity which are truly marvellous. The mind becomes possessed with a feeling of pride, corresponding with the exaltation of its faculties, of whose increase in energy and power it becomes conscious. It will be entirely dependent on the circumstances in which we are placed, the objects which strike our eyes, the words which fall on our ears, whether the most lively sentiments of gaiety or of sadness shall be produced, or passions of the most opposite character shall be excited, sometimes with extraordinary violence; for irritation shall rapidly pass into rage, dislike to hatred and desire of vengeance, and the calmest affection to the most transporting passion. Fear becomes terror; courage is developed into rashness, which nothing checks, and which seems not to be conscious of danger; and the most unfounded doubt or suspicion becomes a certainty. *The mind has a tendency to exaggerate everything* ; and the slightest impulse carries it along. Those who make use of the Hachisch in the East, when they wish to give themselves up to the intoxication of the *fantasia*, take care to withdraw themselves from everything which could give to their delirium a tendency to melancholy, or excite in them anything else than feelings of pleasurable enjoyment; and they profit by all the means which the dissolute manners of the East place at their disposal."—(*Op. cit.*, p. 67.)

538. The disturbance of the perceptive faculties is remarkably shown in regard to *time* and *space*. Minutes seem hours, and hours are prolonged into years; and at last all idea of time seems obliterated, and the past and present are confounded together. M. Moreau mentions as an illustration, that on one evening he was traversing the passage of the Opera when under the influence of a moderate dose of Hachisch: he had made but a few steps, when it seemed to him as if he had been there two or three hours; and, as he advanced, the passage appeared to him interminable, its extremity receding as he pressed forwards. But he gives another more remarkable instance. In walking along the Boulevards, he has frequently seen persons and things at a certain distance presenting the same aspect as if he had viewed them through the large end of an opera-glass; that is, diminished in apparent size, and therefore suggesting the idea of increased distance.—This erroneous perception of space is one of the effects of the *Amanita muscaria*, an intoxicating fungus used by the Tartars; a person under its influence being said to take a jump or a stride sufficient to clear the trunk of a tree, when he wishes only to step over a straw or a small stick. Such erroneous perceptions are common enough among Lunatics, and become the foundations of fixed illusions; whilst in the person intoxicated by Hachisch there is still a certain consciousness of their deceptive character.

539. Though all the Senses appear to be peculiarly impressible in

this condition, yet that of hearing seems the one through which the greatest influence may be exerted upon the mind, especially through the medium of musical sounds. The celebrated artist, M. Theodore Gaultier, describes himself as hearing sounds from colours, which produced undulations that were perfectly distinct to him. But he goes on to say that the slightest deep sound produced the effect of rolling thunder; his own voice seemed so tremendous to him, that he did not dare to speak out, for fear of throwing down the walls, or of himself bursting like a bomb; more than five hundred clocks seemed to be striking the hour with a variety of tones, &c., &c. Of course those individuals who have a natural or an acquired "musical ear," are the most likely to be influenced by the concord or succession of sweet sounds; and in such, the simplest music of the commonest instrument, or even an air sung by a voice in a mediocre style, will excite the strongest emotions of joy or melancholy, according as the air is cheerful or plaintive; the mental excitement being communicated to the body, and being accompanied with muscular movements of a semi-convulsive nature. This influence of music is not merely sensual, but depends, like that of other external impressions, upon the associations which it excites, and upon the habitual disposition to connect with it the play of the Imaginative faculties.

540. A somewhat similar experience from another intoxicant, is recorded of himself by Dr. Laycock:—

"On a certain night, when a sufferer from severe pain and great weakness, he took one drop of Fleming's tincture of aconite, and slept. About midnight he became sensible of a novel state of perception, obscure at first, but shaped at last into strains of grand aërial music in cadences of exquisite harmony, now dying away round mountains in infinite perspective, now pealing along ocean-like valleys. Knowing by previous studies that it was a hallucination of perception, he at last listened to ascertain the cause, and found it was the rattle of a midnight train entering an adjoining railway station. Thus, under the changes induced in the Brain by a

drop of tincture of aconite, the harsh rattle of the iron vibrating on the air in the silence of a summer midnight was changed into harp-like aërial music, such not only as 'ear had not heard,' but no conceivable art of man could realise. Associated therewith was also a suggested terrestrial vision of space of infinite extent and grandeur."
—(*Mind and Brain*, 2nd Edit., vol. i. p. 422.)

Such phenomena, as Dr. Laycock justly remarks, indicate the possibility that even the highest Intuitions of Genius are the expressions of appropriate changes in the Brain-tissue.

541. It is seldom that the excitement produced by the Hachisch fixes itself upon any particular train of ideas, and gives rise to a settled delusion; for in general one set of ideas chases another so rapidly, that there is not time for either of them to enchain the attention and settle itself in the intellect; more especially since (as already remarked) there is usually such a degree of self-consciousness preserved throughout, as prevents the Ego from entirely yielding himself up to the suggestions of his ideational activity. M. Moreau mentions, however, that on one occasion, having taken an overdose, and being sensible of unusual effects, he thought himself poisoned by the friend who had administered it, and persisted in this idea in spite of every proof to the contrary,—until it gave way to another, namely, that he was dead, and was about to be buried; his self-consciousness however, being yet so far preserved, that he believed his body only to be defunct, his soul having quitted it. But when this is altogether suspended, as it seems to be by a larger dose, the erroneous ideas become transformed into convictions, taking full possession of the mind; although sudden gleams of Common Sense still burst through the mists of the imagination, and show the illusive nature of the pictures which the "internal senses" have impressed on the Sensorium. All this—as every one knows, who has made the phenomena of Insanity his study—has its exact parallel in the different stages of mental derangement: the illusive ideas

and erroneous convictions being in the first instance capable of being dissipated by a strong effort of the Will, gradually exerting a stronger and stronger influence on the general current of thought, and at last acquiring such complete mastery over it, that the Reason cannot be called into effectual operation to antagonize them (§ 562).

542. In Opium-dreams and reveries, it would seem from the description given by De Quincey (*Confessions of an English Opium eater*) that the mind is less susceptible of the suggestive influence of *present* sense-impressions, the course of thought and feeling being rather determined by the recurrence of *past* ideas. And it is curious that here again the multiplication or intensification of the images, so as to give rise to ideal conceptions of which the range seemed to be *infinite*, either in Number, in Time, or in Space, should be one of the most constant phenomena. How far this was due to the imaginative temperament of De Quincey himself, may be a matter of question; but the fact that such Mental conceptions, transcending all actual experience, could be called into existence by Physical agencies, has no slight significance. He tells us the four following facts in regard to one particular period of his Opium-dreams, as specially noticeable:—

"1. Whatsoever I happened to call-up and to trace by a voluntary act upon the darkness, was very apt to transfer itself to my dreams; so that I feared to exercise this faculty: for, as Midas turned all things to gold, that yet baffled his hopes and defrauded his human desires, so whatsoever things capable of being visually represented I did but think of in the darkness, immediately shaped themselves into phantoms of the eye; and, by a process apparently no less inevitable, when thus once traced in faint and visionary colours, like writings in sympathetic ink, they were drawn out by the fierce chemistry of my dreams, into insufferable splendour that fretted my heart.

"2. For this and all other changes in my dreams, were accompanied by deep-seated anxiety and gloomy melancholy, such as are wholly incommunicable by words. I seemed every night to descend, not metaphorically but literally, into chasms and sunless abysses, depths below depths, from which it seemed hopeless that I could ever re-ascend. Nor did I, by waking, feel that I had re-ascended. This I do not dwell upon ; because the state of gloom which attended these gorgeous spectacles, amounting at least to utter darkness, as of some suicidal despondency, cannot be approached in words.

"3. The sense of Space, and in the end the sense of Time, were both powerfully affected. Buildings, landscapes, &c., were exhibited in proportions so vast as the bodily eye is not fitted to receive. Space swelled, and was amplified to an extent of unutterable infinity. This, however, did not disturb me so much as the vast expansion of Time ; I sometimes seemed to have lived for 70 or 100 years in one night ; nay, sometimes had feelings representative of a millennium passed in that time, or, however, of a duration far beyond the limits of human experience.

"4. The minutest incidents of childhood, or forgotten scenes of later years, were often revived : for I could not be said to recollect them ; for if I had been told of them when waking, I should not have been able to acknowledge them as parts of my past experience. But placed as they were before me, in dreams, like intuitions, and clothed in all their evanescent circumstances and accompanying feelings, I *recognized* them instantaneously."—(*Op. cit.*, Ed. 1853, pp. 139-142.)

A very curious example of the suggestive influence of a past impression, and the magnification of that impression by the peculiar susceptibility produced by the previous mental life, is presented by De Quincey's account of the results of the chance visit of a Malay beggar :—

"The Malay has been a fearful enemy for months. I have been every night, through his means, transported into Asiatic scenes, which always filled me with such amazement at the monstrous scenery, that horror seemed absorbed, for a while, in sheer astonishment. Sooner or later came a reflux of feeling that swallowed up the astonishment, and left me, not so much in terror as in hatred and abomination of what I saw. Over every form, and threat, and punishment, and dim sightless incarceration, brooded a sense of eternity and infinity that drove me into an oppression as of madness."—(*Op. cit.*, p. 152.)

543. The almost complete paralysis of Will produced by the prolonged abuse of Opium, has been graphically described by the same powerful writer. From the studies which he had formerly pursued with the greatest interest, he shrank with a sense of powerless and infantine feebleness, that gave him an anguish the greater from remembering the time when he grappled with them to his own hourly delight; and an unfinished work to which he had dedicated the blossoms and fruits of his powerful intellect, seemed nothing better than a memorial of hopes defeated, of baffled efforts, of materials uselessly accumulated, of foundations laid that were never to support a superstructure. In this state of volitional but not intellectual debility, he had for amusement turned his attention to Political Economy, for the study of which his previous training had eminently fitted him; and after detecting the fallacies of many of the doctrines then current, he found in the treatise of Mr. Ricardo that which satisfied his intellectual hunger, and gave him a pleasure and activity he had not known for years. Thinking that some important truths had escaped even "the inevitable eye" of Mr. Ricardo, he made great progress in what he designed to be an "Introduction to all future systems of Political Economy;" arrangements were made for printing and publishing the work, and it was even twice advertised. But he had a preface to write, and a dedication, which he wished to make a splendid one, to Mr. Ricardo; and he found himself quite unable to accomplish this, so that the arrangements were countermanded, and the work laid on the shelf.

"I have thus," he continues, "described and illustrated my intellectual torpor, in terms that apply, more or less, to every part of the four years during which I was under the Circean spells of opium.

But for misery and suffering, I might, indeed, be said to have existed in a dormant state. I seldom could prevail on myself to write a letter; an answer of a few words, to any that I received, was the utmost that I could accomplish; and often *that* not until the letter had lain weeks, or even months, on my writing-table. Without the aid of M. all records of bills paid, or to *be* paid, must have perished: and my whole domestic economy, whatever became of Political Economy, must have gone into irretrievable confusion. I shall not afterwards allude to this part of the case: it is one, however, which the opium-eater will find, in the end, as oppressive and tormenting as any other, from the sense of incapacity and feebleness, from the direct embarrassments incident to the neglect or procrastination of each day's appropriate duties, and from the remorse which must often exasperate the stings of these evils to a reflective and conscientious mind. The opium-eater loses none of his moral sensibilities or aspirations: he wishes and longs, as earnestly as ever, to realize what he believes possible, and feels to be exacted by duty; but his intellectual apprehension of what is possible infinitely outruns his power, not of execution only, but of power to attempt. He lies under the weight of incubus and nightmare: he lies in sight of all that he would fain perform, just as a man forcibly confined to his bed by the mortal languor of a relaxing disease, who is compelled to witness injury or outrage offered to some object of his tenderest love:—he curses the spells which chain him down from motion:—he would lay down his life if he might but get up and walk; but he is powerless as an infant, and cannot even attempt to rise."—(*Op. cit.*, pp. 136-138.)

It is quite obvious that it was not from *intellectual* but from *volitional* torpor, that De Quincey suffered. That he could master such a work as Ricardo's, still more, that he could not only detect but supplement its deficiencies, shows that his Intellect was unimpaired. But he was, in regard to the Volitional use of his mental faculties, exactly in the condition of the patients formerly mentioned, who were prevented by paralysis of Will from performing the most simple bodily movements, though the Nervo-muscular apparatus was uninjured (§ 312).

544. It would seem that in whatever way the exertion of Volitional power is related to the condition of the Brain, this exertion is interfered-with by the use of Intoxicating agents, *before* there is any serious perversion of the automatic activity. And this may be especially noticed in *Alcoholic* intoxication; the usual tendency of which is to produce a greater change in the *actions* of the unhappy subject of it, than is ordinarily induced either by Opium or by Hachisch. For whilst the tendency of these is to act upon the moral feelings and sentiments, the action of alcohol more commonly manifests itself in the excitement of the lower propensities. As soon as the liquor begins to exert *any* effect upon the Brain, its operation shows itself in quickening either the Ideational or the Emotional activity, or both combined, and, at the same time, in weakening the Volitional control. It was in this condition that Theodore Hook's powers of Improvisation displayed themselves most remarkably (§ 399); and that Hartley Coleridge could hold a rustic audience enchained by the succession of stories that flowed from his exhaustless fountain of invention. Many men under this influence are more generous and conceding than in their perfectly sober condition, so that they are ready to grant favours and make agreements which their better judgment disapproves,—a circumstance of which those who have a point to gain from them are not slow to take advantage. Those, on the other hand, in whose constitutions the lower animal propensities habitually predominate, are subject to an exaltation of these from a very slight alcoholic stimulus; and their power of self-control being at the same time weakened, they become the slaves of any brutal passion that the slightest provocation may arouse. It is in this primary stage of Alcoholic excitement, that a large number of "crimes of violence," as well as of minor offences, are committed; as is shown by the remarkable reduction in these which took place in the Navy, immediately that the "evening grog" was stopped. The following very characteristic instance of this kind was related to the

Admiralty Committee on whose recommendation this change was made :—

"I had a Marine," said Capt. Drew, "who was constantly complained against for quarrelling and fighting, and disobedience to the orders of his sergeant. At length I began with flogging him, and told him that I would increase his punishment every time that I had a complaint against him. This I had to do twice; and as the man was constantly excited, it appeared to me that the man's reason must be affected. I therefore applied to the Surgeon, and asked him to examine the man, to see whether he was not a fit subject for invaliding; but the surgeon reported that he was as fine and healthy a young man as there was in the ship. I then did not think myself justified in flogging him again, but took upon myself to do an illegal act with a good intention; and when we came into harbour (in the West Indies) I hired a cell in the gaol, and kept him there three days upon bread and water. When the man came out of gaol, I told him that whenever I had a complaint against him, as sure as we came into harbour I would send him to gaol; but that if he would choose to alter his conduct, I would start afresh with him and forget everything that had happened. He said that he was very much obliged to me; and he came to me the next day, and asked me if I would stop his allowance of grog, and let him be paid for it. I did so, and never had another complaint against the man while I was in the ship."

How purely *physical* is this agency, is strikingly shown by the experiments of Dr. Huss, of Stockholm, upon dogs; for when these animals, having been dosed with brandy during several months, were in the advanced stage of the disease (which he was studying for the benefit of Humanity), designated by Dr. H. *Alcoholismus chronicus*, although scarcely able to stand, they were always aroused from their apathetic condition by the sight of other dogs, endeavouring even in their weakened state, to attack and bite them; and this irritability showed itself to the very last.

545. There is, in fact, no abrupt transition between the "sober" and the "drunken" state; but a gradual weakening of Volitional

control, a gradually increasing confusion of the thoughts, and a gradual augmentation of the turbulence of the passions, in proportion as the alcoholized blood takes more and more hold of the Brain. When the government of the Will is completely overthrown, and the excited passions rage uncontrolled, the drunkard may be most truly said to be a madman, and is, like him, *at the time* completely irresponsible for his actions; since, even if some glimmering consciousness of their criminality should still remain, he has lost all power either of restraining his vehement impulses, or of withdrawing himself from their influence. His responsibility arises from his having knowingly and voluntarily given up the reins of Reason and Conscience, and subjected himself to the domination of his evil passions; so that his better nature loses its due supremacy, and he becomes the mere instrument of his insane impulses. It has been argued with considerable plausibility, that a man ought not to be punished for any crime he may commit in a state of Intoxication, since he is then in a state of "temporary insanity;" but that he should be punished as severely for having brought himself into that state. This would doubtless be the most *logical* mode of dealing with the criminal; but as it would require that *every* drunkard should be held guilty of a crime equal in gravity to murder, such punishment could obviously not be enforced. The time may perhaps come, when the man who voluntarily resigns that self-directing power which is the noblest gift of his Creator, and gives himself over to the domination of rage, lust, jealousy, or any other bad passion which may be excited by the action of alcohol on his brain, may be regarded as not less criminal than an engine-driver who should raise the fire of his locomotive to an extra heat, and bring up its steam to its highest pressure, and then abandon it, after starting it on a career of destruction.

546. The closeness of the affinity between the states of Insanity and alcoholic Intoxication is further made apparent by the

extreme readiness with which the balance of reason is disturbed by a small quantity of liquor, in those unfortunate individuals in whom there exists a predisposition to mental derangement. The power of Volitional control being already feeble, it is easily overthrown; and the propensities or passions which are always unduly excitable, are readily aroused into morbid activity by this provocation; so that a very few glasses of wine, or a small quantity of spirits, are sufficient to induce what may be regarded either as a fit of Drunkenness or a paroxysm of Insanity,—the two influences concurring to produce the mental disturbance, which neither of them would have alone sufficed to bring about. Not unfrequently the state thus induced is one of temporary *Monomania* (§ 559 *a*); the mind becoming possessed by a particular emotional state, which governs the conduct, and leads to the perpetration of atrocious crimes. Thus at least two instances of this kind have occurred within the recollection of the Writer, in which the Captain of a ship, having been thus seized with the belief that his crew was in a state of mutiny, has killed one of them after another, in (as he believed) rightful self-defence.—Such a predisposition may arise from previous injury or disease affecting the Brain (tropical sun-stroke being often alleged as the cause of it), or it may be inherited; and it exists in peculiar force in those who have an hereditary tendency to insanity derived from drunkenness on the part of the parents (§ 299 *a*). Cases are continually occurring, in which drunken outrages are committed by individuals thus circumstanced, in whose excuse it is alleged that a very small quantity of liquor is sufficient to inflame their passions and destroy their self-control. But this does not constitute any real apology, except in the case of the *first* outbreak; since their consciousness of their peculiar liability ought to lead them most rigidly to abstain from that indulgence which they know to destroy their power of self-government.

547. The debasing influence of continued Alcoholic excess is

unfortunately but too apparent. Cases like that of Hartley Coleridge, in which it seems only to excite the *higher* part of the Intellectual and Moral nature to an irregular activity, are extremely rare. Far more generally, while weakening the Will and exciting the lower propensities, it blunts theMoral sense also; and the wretched victim becomes so completly the slave of his tyrannical appetite for drink, that he is ready to gratify it at any sacrifice. This Moral degradation is perhaps even more marked in Women than in Men; for the drunkenness of the former (especially in the upper ranks of society) being usually *secret*—at least in the first instance,—whilst in the latter it is generally open, it can only be practised by deceit and fraud; and when the habit has obtained such a dominance that the customary restraints are thrown aside, there is a more complete abandonment of self-respect. In either sex, it is the *physical craving* produced by the continued action of the stimulant upon the nutrition of the Nervous system (§ 155), which renders the condition of the habitual drunkard one with which it is peculiarly difficult to deal by purely *moral* means. Vain is it to recall the motives for a better course of conduct, to one who is already familiar with them all, but is destitute of the Will to act upon them; the seclusion of such persons from the reach of alcoholic liquors, for a sufficient length of time to free the blood from its contamination, to restore the healthful nutrition of the brain, and to enable the recovered mental vigour to be wisely directed, seems to afford the only prospect of reformation; and this cannot be expected to be permanent, unless the patient determinately adopts and steadily acts on the resolution to *abstain entirely* from that, which, if again indulged in, will be poison alike to his body and to his mind, and will transmit its pernicious influence to his offspring.

548. The ordinary *Delirium* of disease corresponds in all its essential characters with that which is induced by the introduction of intoxicating agents into the blood. "In its highest

degree," says Dr. Todd,* "it is a complete disturbance of the Intellectual actions; the thoughts are not inactive, but rather far more active than in health; they are uncontrolled, and wander from one subject to another with extraordinary rapidity; or, taking up one single subject, they twist and turn it in every way and shape, with endless and innumerable repetitions. The thinking faculty seems to have escaped from all control and restraint, and thought after thought is engendered without any power of the patient to direct and regulate them. Sometimes they succeed each other with such velocity, that all power of perception is destroyed, and the mind, wholly engrossed with this rapid development of thoughts, is unable to perceive impressions made upon the senses; the patient goes-on unceasingly raving, apparently unconscious of what is taking-place around him; or it may be, that his senses have become more acute, and that every word from a bystander, or every object presented to his vision, will become the nucleus of a new train of thought; and, moreover, such may be the exaltation of his sensual perception, that subjective phenomena will arise in connection with each sense, and the patient fancies he hears voices or other sounds, whilst ocular spectra in various forms and shapes appear before his eyes and excite further rhapsodies of thought."

The following circumstance, mentioned to the Writer whilst he was a student at Edinburgh, remarkably illustrates the influence of suggestions derived from external sources, in determining the current of thought.—During an epidemic of Fever which had occurred some time previously, and in which an active delirium had been a common symptom, it was observed that many of the patients of one particular Physician were possessed by a strong tendency to throw themselves out of the window, whilst no such tendency presented itself in unusual frequency in the practice of others. The Author's informant,

* Lumleian Lectures on the Pathology and Treatment of Delirium and Coma; 1850.

Dr. C., himself a distinguished Professor in the University, explained this tendency by what had occurred within his own knowledge; he having been himself attacked by the fever, and having been under the care of this physician, his friend and colleague Dr. A. Another of Dr. A's patients, whom we shall call Mr. B, seems to have been the first to make the attempt in question; and, impressed with the necessity of taking due precautions, Dr. A. then visited Dr. C., *in whose hearing* he gave directions to have the windows properly secured, as Mr. B. had attempted to throw himself out. Now Dr. C. distinctly remembers, that although he had not previously experienced any such desire, it came upon him with great urgency as soon as ever the idea was thus suggested to him; his mind being just in that state of incipient delirium, which is marked by the temporary dominance of some one idea, and by the want of volitional power to withdraw the attention from it. And he deemed it probable that, as Dr. A. went on to Mr. D., Dr. E., &c., and gave similar directions, a like desire would be excited in the minds of all those who might happen to be in the same impressible condition.

549. It must be remarked that there is usually a greater disorder of the perceptive faculty in Delirium, than in ordinary dreaming; for in the former condition, the erroneous images are more vividly conceived-of as having an existence *external to the mind*, than they are in the latter; the illusory visual and auditory perceptions, which are often excited by *real* sense-impressions (§ 186), having all the force of reality, and being the original *source* of ideas, instead of (as seems to be rather the case in dreaming) their *products*.—For whilst, in true Dreaming, all the images which we believe ourselves to see, or the sounds that we fancy ourselves to hear, seem to result from changes in the Sensorium excited by Cerebral influence, there is evidently in Delirium a disordered action of the Sensorium itself, of which spectral illusions and other false perceptions are the manifestation. This peculiarity probably depends upon a primary affection of the Sensorial centres by the morbid poison. The two affections seem combined in *Delirium tremens.* This state, which constitutes a

connecting link between Intoxication and Insanity, seems rather to arise from perverted and imperfect nutrition of the Brain, than from poisoning of the blood; for it may be produced by other agencies which depress the Nervous power, such as great loss of blood, the shock of severe injuries, or extreme cold. It is characterized by a low restless activity of the Cerebrum, manifesting itself in muttering delirium, with occasional paroxysms of greater violence; and the nature of this delirium almost always shows the mind of the subject of it to be possessed with the apprehension of some direful calamity. He imagines his bed to be covered with loathsome reptiles; he sees the walls of his apartment covered with foul or terrific spectres; and he supposes the friends or attendants who stand around, to be fiends come to drag him down into a fiery abyss beneath. Here we have, as in the case of false perceptions (§§ 186,187), a misinterpretation of actual Sense-impressions, under the influence of a dominant Emotional state.

CHAPTER XVIII.

OF INSANITY.

550. FROM the condition of *temporary* derangement of the functional action of the Brain, which results from the presence of poisons in the Blood, we pass to that in which the derangement is *persistent*. Between the state of the well-balanced Mind, in which the habit of Self-control has been thoroughly established, so that its whole activity is directed by the Moral Will of the Ego,—and that of the raving madman, whose reasoning power is utterly gone, who is the sport of uncontrollable passion, and is lost to every feeling of affection, of right, and even of decency,—vast as the interval may seem, there is an insensible gradation. For, as has been heretofore more than once remarked, there are many individuals abroad in the world, who are so much more governed by Impulse than by Reason, that they can scarcely be accounted as altogether sane; whilst there are many others, who knowingly surrender the control which they originally possessed over their course of thought and action, to the domination of a fixed idea, which gradually acquires a complete mastery over them (§ 561). It is not the purpose of this Treatise, however, either to discuss the general subject of Insanity, or to attempt to draw the line which separates it from Sanity,—which is no more possible *scientifically* (though usually not difficult *in practice*), than to draw a definite line between bodily health and disease. All that it is here desired to do, is to show, on the one hand, the relation between the phenomena of Insanity and those of healthful activity of the Mind; and, on the other, between

its disordered Psychical manifestations and morbid conditions of the Brain or the Blood.

551. In the first place, it may be unhesitatingly affirmed that there is nothing in the Psychical phenomena of Insanity which distinguishes this condition from states that may be temporarily induced in minds otherwise healthy; for they are all referable either to *excess* or to *deficiency* of normal modes of mental action. That which is common to every form of Insanity, which is frequently its first manifestation, and which, in so far as it exists, renders the Lunatic irresponsible for his actions, is *deficiency of volitional control* over the current of thought and feeling, and consequently a want of self-direction and self-restraining power over the conduct. With this, there may be a *general* disturbance either of *intellectual* or of *emotional* activity, or of both combined, constituting *Mania;* or there may be a *partial* or *limited* disorder, arising from excess or deficiency of some particular tendency, constituting *Monomania.* Not unfrequently an attack which begins with violent Mania, will subside into a chronic and comparatively harmless Monomania; but, on the other hand, Monomaniacal patients are often subject to paroxysms of Mania; and, even when there is no such general disturbance, the smallest touch on the "sore place" (§ 559 *a*) may induce a dangerous outbreak of passion, which the subject of it has no power to control.

552. It is unquestionable that in a large proportion of cases of settled Insanity, there is an impairment of the due *nutrition* of the Cerebrum; and this, which is often an hereditary defect, may arise *de novo,* like abnormal changes in the nutrition of other parts, from deficiency or perversion in the formative power of the Nervous tissue, or from an imperfect supply or an altered character of the Blood. Of the influence of deficient or perverted formative power in the Tissue, we have examples in the insanity resulting from mechanical injuries of the Brain, and

from excessive "wear" of the organ by forced activity. Of the effects of deterioration in the character of the Blood, we have illustrations in the insanity that is often linked-on with constitutional diseases of which such deterioration is a marked feature, as well as in that which is so frequent a consequence of habitual alcoholic excess. These conditions may exist in combination; and it is probably by such a combination that many of the so-called "moral causes" of insanity operate. For there can be little doubt that Emotional excitement, from its immediate relation to Nerve-force (§ 265), has a direct influence on the formative capacity of the Cerebrum; whilst, on the other hand, we know that it has so great an influence over the Organic functions, that it can produce very decided alterations in the condition of the Blood (Chap. XIX.). But without any serious perversion of the *nutrition* of the Cerebrum, its *action* may be disturbed, either by the presence of some poisonous agent in the Blood, or by functional disturbance in other parts of the Nervous system. We have seen that the Delirium of intoxication, or of fever, is, whilst it lasts, a true Insanity; and it ceases because the poison is eliminated from the circulation. But there are many diseases in which there is a continual production of a poison within the system, whereby the normal train of mental action is deranged so long as the blood is tainted by it: the indication of treatment is here obviously to check this production, and to depurate the blood; and when this has been effectually accomplished, the healthy action of the brain is immediately restored, which would not have been the case if its nutrition had been seriously impaired. Most persons have experienced the extreme emotional depression and incapacity for intellectual exertion, which are consequent upon certain derangements of the digestive function, and especially upon disordered action of the liver,—a cloud passing away (as it were) from the mental vision, a weight being lifted off "the spirits," by a dose of blue-pill; and it is unquestionable that many forms of Insanity, in which extreme dejection is a prominent symptom, but which may also include intellectual delusions, are solely dependent upon this cause. So, a functional disturbance of the Cerebrum is often induced by the irregular action of other parts of the Nervous system, especially those connected with the reproductive apparatus. Of this we have examples in certain peculiar forms of disordered mental action, which are connected with "hysterical" states of the female system,—in particular, mutability and irritability of temper, and disposition to cunning deceit; and it is a singular fact, well known to Medical Jurists, that girls about the age of puberty, and suffering under functional irregularities, are sometimes "possessed" by a propensity to set fire to their dwellings. It frequently happens that agencies of both classes jointly contribute to the result: some long-continued defect of nutrition (very often arising from hereditary constitution) serving as the "predisposing cause;"—whilst violent mental emotion, or depravation of the blood by noxious matter of some kind, acts as the "exciting cause,"—the two together producing that effect, which neither would singly have brought-about.

553. The state of *Mania* is usually characterized by the combination of complete derangement of the Intellectual powers, with passionate excitement upon every point which in the least degree affects the Feelings. There is, however, a considerable amount of variety in the symptoms of Mania, depending upon differences in the relative degree of *intellectual* and of *emotional* disturbance. For there may be such a derangement of the former, as gives-rise to complete incoherence in the succession of ideas, so that the reasoning power is altogether suspended; and yet there may be at the same time an entire absence of emotional excitement, so that the condition of the mind is closely allied to that of Dreaming or of rambling Delirium. On the other hand, the intellectual powers may be themselves but little disturbed,

the trains of thought being coherent, and the reasoning processes correctly performed; but there may be such a state of general emotional excitability, that nothing is *felt* as it should be, and the most violent passion may be aroused and sustained by the most trivial incidents, or by the wrong ideas which are formed by the mind as a consequence of their misinterpretation (§ 264). Between these two opposite states, and that in which the disturbance affects at the same time the intellectual and the emotional part of the mental nature, there is a complete succession of transitional links; but, underlying all phases of this condition (these often passing into each other in the same individual), there is one constant element, namely, the deficiency of Volitional control over the succession of thought and feeling. This deficiency appears to be a primary element in those forms which essentially consist in Intellectual disturbance; whilst in those of which Emotional excitement is the prominent feature, it results apparently from the overpowering mastery that is exercised over the Will, by the states of uncontrollable passion which succeed each other with little or no interval. It seems probable, however, from the phenomena of Intoxication (§§ 537, 544), that the very same agency which is the cause of the undue emotional excitability, also tends to produce an absolute diminution in the power of volitional control.

554. It is chiefly (but not solely) in those cases in which the Cerebral power has been weakened by a succession of attacks of Mania, Epilepsy, or some other disorder which consists in a perverted action of the whole organ, that we find the *intellectual* powers specially and permanently disordered; the succession of thought becoming incoherent, and the perception of those relations of ideas on which all reasoning processes depend, being more or less completely obscured. The failure usually shows itself *first* in the power of volitional direction, and especially in the faculty of recollection. In proportion as the mind is unable

to bring the results of past experience to bear on its present operations, do these lose their connectedness and consistency; and at last all the ordinary links of association appear to be severed, and the succession of ideas seems altogether disconnected, as in the most incoherent kinds of Dreaming. All this may take place with or without emotional excitement; not unfrequently the latter occurs in paroxysms, which interrupt the otherwise tranquil life of the subjects of this form of Insanity; and it is not at all incompatible with this condition, that there should be a special excitabilility upon some one point, which, owing to the annihilation of the Volitional controlling power, acquires a temporary predominance whenever it is called into play. It is the general characteristic, however, of this type of Insanity, that there are no *settled* delusions; the mind not being disposed to dwell long upon any one topic, but wandering-off in a rambling manner, so as speedily to lose all trace of the starting-point. Such patients are unable to recollect what passed through their thoughts but a few minutes previously; if any object of desire be placed before them, which it requires a consistent reasoning process to attain, they are utterly unable to carry this through; and the direction of their desires is perpetually varying, and may be readily altered by external suggestion. Cases of Intellectual insanity, depending (as this form of the disease usually does) upon structural disorder of the Cerebrum, are less amenable to treatment than are those of the other forms presently to be described; and their tendency is usually towards complete fatuity.

555. There may, however, be no primary disorder of the Intellectual faculties; and the Insanity may essentially consist in a tendency to disordered *emotional* excitement; which affects the course of thought, and consequently of action, without disturbing the reasoning processes in any other way than by supplying wrong materials to them (§ 264). Now the Emotional disturbance may be either *general* or *special:* that is, there may

be a derangement of feeling upon almost every subject, matters previously indifferent becoming invested with strong pleasurable or painful interest, things which were previously repulsive being greedily sought, and those which were previously the most attractive being in like manner repelled; or, on the other hand, there may be a peculiar intensification of some one class of feelings or impulses, which thus acquire a settled domination over the whole character, and cause every idea with which they connect themselves to be presented to the mind under an erroneous aspect.—The first of these forms, now generally termed *Moral Insanity*, *may* and frequently *does* exist without any disorder of the intellectual powers, or any delusion whatever; it being (as we shall presently see) a result of the generality of the affection of the emotional tendencies, that no one of them maintains any constant hold upon the mind, one excitement being (as it were) driven-out by another. Such patients are among those whose treatment requires the nicest care, but who may be most benefited by judicious influences. Nothing else is requisite, than that they should exercise an adequate amount of self-control; but the best-directed moral treatment cannot enforce this, if the patient do not himself (or herself) co-operate. Much may be effected, however, as in the education of children, by presenting adequate *motives* to self-control; and the more frequently this is exerted, the more easy does the exertion become.—This form of Insanity is particularly common among females of naturally "quick temper," who, by not placing an habitual restraint upon them selves, gradually cease to retain any command over it. The Writer well remembers that when going with Dr. Conolly through one of the wards on the female side of the Lunatic Asylum at Hanwell, Dr. C. remarked to him,—"It is my belief that two-thirds of the women here have come to require restraint, through the habitual indulgence of an originally bad temper."

556. The more limited and settled disorder of any one portion

of the Emotional nature, however, gives an entirely different aspect to the character, and produces an altogether dissimilar effect upon the conduct. It is the essential feature of this state, that some one particular tendency acquires a dominance over the rest; and this may happen, it would seem, either from an extraordinary exaggeration of the tendency, whereby it comes to overmaster even a strongly-exercised volitional control; or, on the other hand, from a primary weakening of the volitional control, which leaves the predominant bias of the individual free to exercise itself. Again, the exaggerated tendency may operate (like an ordinary emotion), either in directly prompting to some kind of action which is the expression of it, or in modifying the course of thought, by habitually presenting erroneous notions upon the subjects to which the disordered feeling relates, as the basis of intellectual operations.

557. The first of these forms of *Monomania* is that which is known as *impulsive* insanity; and the recognition of its existence is of peculiar importance in a juridical point of view. For whilst the Law of England only recognizes as *irresponsible*, on the ground of Insanity, those who are incapable of distinguishing right from wrong, or of recognizing the consequences of their acts, it is unquestionable that many criminal actions are committed under the irresistible dominance of some insane impulse, the individual being at the time perfectly aware of the evil nature of those actions, and of his amenableness to punishment for them.

The following very characteristic example of the Homicidal form of impulsive insanity, was given in the Report of the Morningside (Edinburgh) Lunatic Asylum for the year 1850.—The case was that of a female, who was not affected with any disorder of her Intellectual powers, and who laboured under no delusions or hallucinations, but who was tormented by "a simple abstract desire to kill, or rather, for it took a specific form, to strangle. She made repeated attempts to effect her purpose, attacking all and sundry, even her own nieces and other relatives; indeed, it seemed to be a matter of indifference to her *whom* she strangled, so that she succeeded in killing *some one.* She recovered, under strict discipline, so much self-control as to be permitted to work in the washing-house and laundry; but she still continued to assert that she 'must do it,' that she was 'certain she would do it some day,'—that she could not help it, that 'surely no one had ever suffered as she had done,'—was not hers 'an awful case;' and, approaching any one, she would gently bring her hand near their throat, and say mildly and persuasively, 'I would just like to do it.' She frequently expressed a wish that all the men and women in the world had only one neck, that she might strangle it. Yet this female had kind and amiable dispositions, was beloved by her fellow-patients, so much so that one of them insisted on sleeping with her, although she herself declared that she was afraid she would not be able to resist the impulse to get up during the night and strangle her. She had been a very pious woman, exemplary in her conduct, very fond of attending prayer-meetings, and of visiting the sick, praying with them and reading the Scriptures, or repeating to them the sermons she had heard. It was the second attack of Insanity. During the former she had attempted suicide. The disease was hereditary, and it may be believed that she was strongly predisposed to morbid impulses of this character, when it was stated that her sister and mother both committed suicide. There could be no doubt as to the sincerity of her morbid desires. She was brought to the Institution under very severe restraint, and the parties who brought her were under great alarm upon the restraint being removed. After its removal, she made repeated and very determined attacks upon the other patients, the attendants, and the officers of the Asylum, and was only brought to exercise sufficient self-control by a system of rigid discipline. This female was perfectly aware that her impulses were wrong, and that if she had committed any crime of violence under their influence, she would have been exposed to punishment. She deplored, in piteous terms, the horrible propensity under which she laboured."—In the Report of the same institution for 1853, it is mentioned that this female had been re-admitted, after nearly succeeding in strangling her sister's child under the prompting of her homicidal impulse. "She displays no delusion or perversion of ideas, but is urged-on by an abstract and uncontrollable impulse to do what she knows to be wrong, and deeply deplores."

Such impulses may drive the subjects of them to kill, to commit a rape, to steal, to burn, and so on, without any malicious feeling towards the persons injured; and many instances have occurred, in which the individuals thus affected have voluntarily withdrawn themselves from the circumstances of whose exciting influence they were conscious, and have even begged to be put under restraint.

558. It is a remarkable fact, moreover, and one that strikingly confirms the view of the nature of Emotional states which has been previously advocated (§ 260), that the insane impulse appears to be not unfrequently the expression of a dominant *idea*, with which there is no such association of pleasurable feeling as makes the action prompted by it an object of *desire*, but which operates by taking full possession of the mind, and by forcing the body (so to speak) into the movements which express it. The individual thus affected regards himself as the victim of a *necessity* which he cannot resist, and may be perfectly conscious (as when the impulse proceeds from a strong desire) that what he is doing will be injurious to others or to himself. This state bears a close resemblance to that of the Biologized "subject," who is peremptorily told, "You *must* do this," and does it accordingly (§ 454); and it is one that is particularly liable to be induced in persons who habitually exercise but little Volitional control over the direction of their thoughts, by the influence of suggestions from without, and especially by occurrences which take a strong hold of their attention.

a. To this condition are to be referred many of the Insane actions which are commonly set down to the account of *imitation.* This term would be best restricted to that state of mind, in which there is an *intention* to imitate; for what is called "involuntary imitation" is merely the expression of the fact, that the consciousness of the performance of a certain act by one individual gives-rise to a tendency to its performance by the other (§ 259 *b*, *c*), as in the case of the act of yawning. So, the commission of suicide or homicide, after an occurrence of the same kind which has previously fixed itself strongly upon the attention, is an *ideo-motor* action, prompted by a *suggesting idea*. Thus, it is well known that after the suicide of Lord Castlereagh, a large number of persons destroyed themselves in a similar manner. Within a week after the "Pentonville Tragedy," in which a man cut the throats of his four children, and then his own, there were two similar occurrences elsewhere. After the trial of Henriette Cornier for child-murder, which excited a considerable amount of public discussion on the question of homicidal insanity, Esquirol was consulted by numerous mothers, who were haunted by a propensity to destroy their offspring.

b. The following is a remarkable example of the *sudden* domination of a morbid impulse, to which no tendency seems to have been previously experienced, and which appears to have been altogether devoid of any emotional character. Dr. Oppenheim, of Hamburg, having received for dissection the body of a man who had committed suicide by cutting his throat, but who had done this in such a manner that his death did not take place until after an interval of great suffering, jokingly remarked to his attendant,—"If you have any fancy to cut your throat, don't do it in such a bungling way as this; a little more to the left here, and you will cut the carotid artery." The individual to whom this dangerous advice was addressed, was a sober, steady man, with a family and a comfortable subsistence; he had never manifested the slightest tendency to suicide, and had no motive to commit it. Yet, strange to say, the sight of the corpse, and the observation made by Dr. O., suggested to his mind the idea of self-destruction; and this took such firm hold of him that he carried it into execution, fortunately, however, without duly profiting by the anatomical instructions he had received; for he did not cut the carotid, and recovered.

559. In most forms of Monomania, however, there is more or less of disorder in the *ideational* process, leading to the formation of positive *delusions* or *hallucinations*, that is to say, of fixed beliefs or dominant ideas which are palpably inconsistent with

reality. These delusions, however, are not attributable to original perversions of the Reasoning process, but arise out of the perverted Emotional state. This gives rise, in the first place, to a *mis-interpretation of actual facts or occurrences,* in accordance with the prevalent state of the feelings (§ 264). Thus, a lunatic who is possessed with an exaggerated feeling of his own importance, may suppose himself to be a sovereign prince; and under the influence of this dominant idea, looks upon the place of his confinement as his palace, believes his keepers to be his obsequious officers, and his fellow-patients to be his obedient subjects; the plainest fare is converted into a banquet of the choicest dainties, and the most homely dress into royal apparel. His condition, therefore, closely corresponds with that of a Biologized subject, whose mind may become possessed *for a time* by similar ideas through the influence of external suggestion (§ 451), and who is not undeceived by their discordance with objective realities, because the force with which the consciousness is impressed by the latter, is less than that with which it is acted-on by the former. Now and then, perhaps, the Lunatic, like the Biologized subject, is visited by a gleam of common-sense, which enables him to view certain objects in their true light, so that he becomes sensible of some inconsistency between his real and his imaginary condition; as when a patient in a Scotch pauper-lunatic asylum, after dilating upon the imaginary splendours of his regal state, confessed that there was one thing which he could not quite comprehend, namely, that all his food tasted of oatmeal! In a more advanced state of the disorder, however, ideas which have had their origin in the *imagination* alone, and which it has at first presented faintly and transiently, are habitually dwelt on in consequence of the interest with which they are invested; and at last become *realities* to the consciousness of the Ego, simply because he does not bring them to the test of actual experience.

a. The Writer remembers to have heard the following case from Dr. A. T. Thomson:—He was requested to see a gentleman whose friends were desirous of placing him under restraint, being well assured of his Insanity from the supervention of uncontrollable outbreaks of temper (to which he had never previously given way), though they could find no ostensible ground in his conversation or actions, which would legally justify the use of coercive measures. Several medical men had been consulted, who had failed to obtain any such justification, notwithstanding that they had employed all the means which their experience dictated for gaining an insight into the nature of his disorder. Dr. Thomson having been introduced to him as a scientific man in whose conversation he would feel interested, was struck, on entering the room, with the evidence of paroxysms of violent passion afforded by the shivering of a large pier-glass, the fracture of the arms and legs of chairs, and other damages to the handsome furniture of the apartments; and he felt convinced that there was some perversion of this gentleman's feelings or intellect, which it was his business to discover. For this purpose he directed the conversation into a great variety of channels; and being himself a man of very comprehensive information and fluent speech, and finding a ready response on the other side, he ran through a great variety of topics in the course of a couple of hours. He said that he had never enjoyed a more agreeable or instructive conversation; his patient being evidently a gentleman of great attainments in literature, science, and art, and having a most original as well as pleasing manner of expressing himself upon every subject that came before him. Dr. Thomson was beginning to despair of finding out the mystery of his disorder, when it chanced that Animal Magnetism was adverted to on which the patient began to speak of an influence which some of his relatives had acquired over him by this agency, described in the most vehement language the sufferings he endured through their means, and vowed vengeance against his persecutors with such terrible excitement, that it was obviously necessary, alike for their security and his own welfare, that he should be placed under restraint.

Here, it is obvious, the Emotional excitement was the essence of the disorder, and the Intellectual delusion was merely the expression of it.

560. This view of the Emotional source of most, if not all, of

the *delusions* of the Insane, occurred to the Writer in early life, through having had his attention strongly drawn to a case in which he had the opportunity of observing from its commencement the progressive formation of such delusions, and in which the varying tenacity of their hold over the intellectual *belief* (which sometimes appeared disposed to get rid of them) corresponded exactly with the varying degrees of intensity of the dominant emotion. His subsequent experience of other forms of Monomania, and the results of his inquiries among those who have made Insanity their special study, have fully confirmed this view.

a. Thus Dr. Skae remarks in the "Morningside Report" for 1853, that "nothing can be further from the truth, than to believe that in every case of Insanity there must be some delusion, or some perturbation of the Intellect. Of all the features of Insanity, *morbid impulses, emotions*, and *feelings*, and the *loss of control over them*, are the most essential and constant. Delusions, illusions, and hallucinations are, comparatively speaking, the accidental concomitants of the disease. The former, perhaps, invariably accompany the invasion of disease; the latter are frequently only developed during its progress, and are sometimes never present at all."

b. It is not a little interesting, in this connexion, as well as in the additional relation which it indicates between Insanity and the various phases of Delirium, Dreaming, &c., that the *particular delusion* seems often to be suggested by accidental circumstances, the mind being previously under the influence of some morbid tendency which has given the *general direction* to the thoughts. Thus we find it mentioned in the "Morningside Report" for 1850, that the Queen's public visit to Scotland seemed to give a special direction to the ideas of several individuals who became insane at that period, the attack of insanity being itself in some instances traceable to the excitement produced by that event. One of the patients, who was affected with puerperal mania, believed that, in consequence of her confinement having taken place on such a remarkable occasion, she must have given birth to a person of royal or divine dignity.—During the religious excitement which prevailed at the time of the "disruption" of the Scottish Church, an unusually-large number of patients were admitted into the various asylums of Scotland, labouring under delusions connected with religion; the disorder having here also doubtless commenced in an exaggeration of this class of *feelings*, and the erroneous *beliefs* having been formed under their influence.—Again, in the Report of the same Institution for 1851, it is stated that, as in former instances "the current topics of the day gave colouring and form to the delusions of the disordered fancy. We have thus had no less than five individuals admitted during the year, who believe themselves the victims of Mesmeric agency (a sort of "Mesmeric mania" having been prevalent in Edinburgh during that period); "three of the inmates talked much of California, and of the bags full of gold which they had obtained from the diggings; and one of them arrived at the persuasion that his body was transmuted into gold."

561. Every one who observes the ordinary working of his own mind, must be aware how differently he looks at the *very same occurrences*, according to the state of Feeling he is in at the time; and no judicious man will allow himself to act upon any conclusion he may have formed under the influence of emotional excitement. It is, in fact, in the *persistence* and *exaggeration* of some emotional tendency, leading to an erroneous interpretation of everything that may be in any way related to it, that Insanity very frequently commences; and it is in this stage that a strong effort at self-control may be exerted with effect, not merely in keeping down the exaggerated emotion, but in determinately directing the thoughts into another channel. For there can be no doubt that while the tendency to *brood upon* a particular class of ideas and on the feelings connected with them, gives them, if this tendency be habitually yielded to, an increasing dominance,—so that they at last take full possession of the mind, overmaster the will, and consequently direct the conduct,—there is a stage in which the will *has* a great power of preserving the right balance, by steadily resisting the "brooding" tendency, calling-off the attention from the contemplation of ideas which *ought not* to be entertained (§ 271), and directing it into some entirely different channel. The records of Crime abound in cases in which murder or attempt

to murder has been committed under the dominance of an idea or feeling, that has taken such complete possession of the mind, as to render the Ego no longer morally responsible for his act *at the time* of its commission; but for which act he is nevertheless remotely responsible (like the drunkard, § 545), because he has allowed himself to become thus possessed, when the means of escape lay in his own power. And in the infliction of punishment, the same principle ought to be applied to both cases,—that of bringing the strongest possible deterrent motives to bear upon the minds of those who are meditating such criminalities.

An extremely good example of the deterrent influence of a judiciously-devised punishment, was afforded by the stop which was put to the repeated alarms to which the Queen was subjected, after the real attempt upon her life made by Oxford. The motive in his case seemed to be nothing else than morbid vanity; which was gratified by his being tried for high-treason, and made an object of public notoriety. Being found "not guilty" on the ground of Insanity (to which it was proved that he had an hereditary predisposition), and being placed in Bethlehem Hospital as a lunatic, no corrective impression as to *punishment* was made upon the class from which he sprung; and the like morbid love of notoriety led one young fellow after another to threaten the life of the Queen, by presenting pistols or other weapons when she appeared in public. In order to protect her from the repetition of this outrage, a bill was carried through the Legislature in the shortest possible time, making the offence of presenting any fire-arm at the Queen (even if unloaded), a *disgraceful one*, to be punished with *whipping;* and no more was heard of such attempts for many years, the next attempt—that of Lieutenant Pate—being the result of "brooding" over some fancied injuries.

562. It is singular how closely the ordinary history of the access of *Monomania* corresponds with that of intoxication by Hachisch. A man who has been for some time under the strain of severe mental labour, perhaps with the addition of emotional excitement, breaks down in mental and bodily health; and becomes subject to morbid ideas, of whose abnormal character he is in the

first instance quite aware. He may see spectral illusions, but he knows that they are illusive. He may hear imaginary conversations, but is conscious that they are empty words. He feels an extreme depression of spirits, but is willing to attribute this to some physical cause. He exhibits an excessive irritability of temper, but is conscious of his irascibility and endeavours to restrain it. He has strange thoughts respecting those who are most dear to him, suspects his wife of infidelity, his children of wilful disobedience, his most intimate friends of injurious designs; but he has still intelligence enough to question the validity of these suspicions, and shrinks from giving them permanent lodgment in his breast. Dark visions of future ruin and disgrace flit before him; but he may refuse to contemplate them, may be reasoned into the admission of their utter baselessness, and may second the efforts of his friends to direct his thoughts and feelings into a different channel. It is in this stage that change of scene, the withdrawal from painful associations, the invigoration of the bodily health, and the direction of the Mental activity towards any subject that has a healthful attraction for it, exert a most beneficial influence (§ 271 *a*); and there can be no doubt that many a man has been saved from an attack of Insanity, by the resolute determination of his Will *not* to yield to his morbid tendencies.—But if he should give way to these tendencies, and should dwell upon his morbid ideas instead of endeavouring to escape from them, they come at last to acquire a complete mastery over him; and his Will, his Common sense, and his Moral sense, at last succumb to their domination. The visual appearances which he at first dismissed as unreal, become to his mind objects of actual sight; the airy words are conversations which he distinctly hears, and to which he gives full credence, however repugnant their import may be to his sober sense; his suspicions of wife, children, and friends acquire the force of certainties, although they may not have the slightest basis of reality; the conviction of impending ruin is ever before him, and he makes no effort to escape from it; no reasoning can now dispel his delusions; no proof, however clear to the sane mind, can demonstrate the groundlessness of his notions. His temper, now entirely uncontrolled, becomes more and more irritable; the slightest provocations occasion the most violent outbreaks; and these are excited, not merely by the exaggeration or misinterpretation of actual occurrences, but by the fictions of his own imagination. No conception can be too obviously fallacious or absurd, as judged by the sound intellect, to command his assent and govern his actions; for when the directing power of the Will is altogether lost, he is as incapable as a Biologized or Hypnotized subject, of testing his ideas by their conformity to the general result of his previous experience (§ 451), or of keeping his emotions under due control.

563. But, it may be said, if Insanity be the expression of disordered *physical* action of the Cerebrum, it is inconsistent to expect that a man can control this by any effort of his own; or that *moral* treatment can have any efficacy in the restoration of mental health. Those, however, who have followed the course of the argument expounded in this Treatise, will have no difficulty in reconciling the two orders of facts. For whilst the disordered physical action of the Cerebrum, *when once established*, puts the automatic action of his mind altogether beyond the control of the Ego, there is frequently a stage in which he has the power of so directing and controlling that action, as *to prevent the establishment of the disorder*; just as, in the state of perfect health, he has the power of forming habits of Mental action, to which the nutrition of the Brain responds, so as ultimately to render them automatic (§§ 287, 288). And so, the judicious Physician, in the treatment of an insane patient, whilst doing everything he can to invigorate the bodily health, to ward off sources of mental disturbance, and to divert the current of thought and feeling from a morbid into a healthful channel, will sedulously watch for every opportunity of

fostering the power of self-control, will seek out the motives most likely to act upon the individual, will bring these into play upon every suitable occasion, will approve and reward its successful exercise, will sympathize with failure even when having recourse to the restraint which it has rendered necessary, will encourage every renewed exertion, and will thus give every aid he can to the re-acquirement of that Volitional direction, which, as the bodily malady abates, is alone needed to prevent the recurrence of the disordered mental action. It is when the patient has so far recovered, as to be capable of being made to feel that he *can* do what he *ought*, if he will only *try*, that moral treatment becomes efficacious. And thus the judicious Physician, when endeavouring either to ward-off or to cure Mental disorder, brings to bear upon his patient exactly the same power as that which is exerted by an Educator of the highest type (§ 290, III). Each has the high prerogative of calling into exercise that element in Man's nature which is the noblest gift of his Creator, enabling him to turn to the best account whatever mental endowments he may possess, "for the glory of God, and the good of Man's estate."

CHAPTER XIX.

INFLUENCE OF MENTAL STATES ON THE ORGANIC FUNCTIONS.

564. It has been shown in the preceding Chapters how close is the dependence of the normal action of the Brain upon an adequate supply of pure Blood: serious *reduction* in its quantity at once producing *deficient* mental activity; whilst a *depravation* of its quality occasions a *perversion* of that activity. And thus it comes to pass that very slight departures from the health of the Body exert a most powerful influence upon our *intellectual*, and still more upon our *emotional* condition, through the deterioration they produce in the circulating fluid (§ 552). The functional activity of the Brain is also affected, through its nervous connections, by the physical condition of remote parts of the body; various aberrant phenomena being traceable to such "morbid sympathies."* But what we have now to consider is the converse power exerted by Mental states over the functions of Nutrition and Secretion, so as to modify not merely the movements, but the molecular actions, of various parts of the body. This power (it has been already shown, §§ 112-115) is for the most part exercised through the Sympathetic system of Nerves; and whilst the regulation of the calibre of the arteries, which determines the *quantity* of blood supplied to each part, seems to be effected through the motor fibres which that system receives from the Cerebro-spinal, its influence over the condition of the Blood itself, and the use that is made of it, appears to be exerted through its own proper fibres and ganglia.

* See especially Dr. Laycock's Treatise "*On the Nervous Diseases of Women.*"

565. Much of the action exerted by the *Vaso-motor* system of Nerves (§ 113) has obvious reference to the *harmonization* of the Organic functions with each other. Thus, to take a very simple and familiar case, when a particle of dust lodges between the eye and the eyelid, an increased flow of tears is produced by the dilatation of the artery that supplies the lachrymal gland; so as, if the particle be not too large, to wash it down into the inner corner of the eye, from which it may be easily removed. So, again, the introduction of food into the mouth produces an immediate flow of saliva for its mastication; while at the same time there is an outpouring of gastric juice into the stomach, in preparation for its digestion; the production of both these secretions being due to the increase of the supply of blood proceeding in the one case to the salivary glands, and in the other to the gastric follicles contained in the coats of the stomach. But a flow of saliva may be occasioned in a hungry man, by the sight, the smell, or even the thought, of savoury food; and it has been ascertained by experiments on dogs, that a flow of gastric juice takes place into their stomachs, when, after long fasting, attractive food is placed before them. So, the free secretion of milk, excited by suction applied to the nipple, is also producible in the nursing mother by the sight, by the cry, or even by the thought of her infant (§ 500 *a*), which occasions the dilatation of the mammary artery (analogous to the act of blushing) that permits the rush of blood to the Mammary gland known as "the draught."—Now in none of these cases has the *will* any influence whatever; the mental state which determines the result being an *emotional* one, which may be linked-on either to a perception or to an idea, according as the object that calls it forth is actually or only "subjectively" present. And the direct influence of the Emotions upon the quantity of these Secretions is shown by numerous other facts.

a. Thus, the secretion of Tears, which is continually being formed to an extent sufficient to lubricate the surface of the eyes, is poured out in great abundance under the moderate excitement of the emotions, either of joy, tenderness, or grief. It is checked, however, by violent grief; and it is a well-known indication of moderated sorrow, when tears "come to the relief" of the sufferer.

b. So, the Salivary secretion may be suspended by strong emotion; a fact of which advantage is taken in India for the discovery of a thief among the servants of a family,—each of them being required to hold a certain quantity of rice in his mouth during a few minutes, and the offender being generally distinguished by the dryness of his mouthful.

c. That the Gastric secretion may be entirely suspended by powerful emotion, clearly appears as well from the results of experiments on animals, as from the well-known influence exerted by a sudden Mental shock (whether painful or pleasurable), in dissipating the appetite for food, and in suspending the digestive process when in active operation. Several other secretions are affected in a similar manner by emotional excitement: thus the special odoriferous secretions of many animals are poured forth under alarm with such potency as to constitute their special means of defence; and in some human beings the cutaneous secretion becomes strongly ammoniacal, when either fear or bashfulness is strongly excited.

566. There is no Secretion, however, on the *quality* as well as the *quantity* of which Emotional states have so obvious an influence, as they have upon that of milk; and this point, being one of great practical importance, as well as of scientific interest, will be here dwelt-on in some detail,—conclusive evidence of such alterations being afforded by the disorder produced by the altered secretion in the digestive system of the infant, which is a more delicate apparatus for testing its quality, than any that the chemist could devise. The following general statements on this subject were made by Sir Astley Cooper, as the result of extended and careful enquiries:—

a. "The secretion of Milk proceeds best in a *tranquil state of mind*, and with a cheerful temper; then the milk is regularly abundant, and agrees well with the child. On the contrary, a *fretful temper* lessens the quantity of milk, makes it thin and serous, and causes it to disturb the child's bowels, producing intestinal fever and much griping. *Fits of anger* produce a very irritating milk, followed by griping in the infant, with green stools. *Grief* has a great influence on lactation, and consequently upon the child. The loss of a near and dear relation, or a change of fortune, will often so much diminish the secretion of milk, as to render adventitious aid necessary for the support of the child. *Anxiety of mind* diminishes the quantity, and alters the quality, of the milk. The reception of a letter which leaves the mind in anxious suspense, lessens the draught, and the breast becomes empty. If the child be ill, and the mother is anxious respecting it, she complains to her medical attendant that she has little milk, and that her infant is griped and has frequent green and frothy motions. *Fear* has a powerful influence on the secretion of milk. I am informed by a medical man who practises much among the poor, that the apprehension of the brutal conduct of a drunken husband will put a stop for a time to the secretion of milk. When this happens, the breast feels knotted and hard, flaccid from the absence of milk, and that which is secreted is highly irritating; and some time elapses before a healthy secretion returns. *Terror*, which is sudden and great fear, instantly stops this secretion." Of this, two striking instances, in which the secretion, although previously abundant, was completely arrested by this emotion, are detailed by Sir A. Cooper.

There is even evidence that the Mammary secretion may acquire an actually *poisonous* character, under the influence of violent mental excitement; for certain phenomena which might otherwise be regarded in no other light than as simple coincidences, appear to justify this inference, when interpreted by the less striking but equally decisive facts already mentioned.

b. "A carpenter fell into a quarrel with a soldier billeted in his house, and was set-upon by the latter with his drawn sword. The wife of the carpenter at first trembled from fear and terror, and then suddenly threw herself furiously between the combatants, wrested the sword from the soldier's hand, broke it in pieces, and threw it away. During the tumult, some neighbours came in and separated the men. While in this state of strong excitement, the mother took up her child from the cradle, where it lay playing and in the most perfect health, never having had a moment's illness; she gave it the breast, and in so doing sealed its fate. In a few minutes the infant left-off sucking, became restless, panted, and sank dead upon its mother's bosom. The physician who was instantly called-in, found the child lying in the cradle, as if asleep, and with its features undisturbed; but all his resources were fruitless. It was irrecoverably gone."*

In this interesting case, the milk seems to have undergone a change which gave it a powerful sedative action upon the susceptible nervous system of the infant.

c. Similar facts are recorded by other writers.—Mr. Wardrop mentions ("*Lancet*," No. 516), that having removed a small tumour from behind the ear of a mother, all went well until she fell into a violent passion; and the child, being suckled soon afterwards, died in convulsions. He was sent-for hastily to see another child in convulsions, after taking the breast of a nurse who had just been severely reprimanded; and he was informed by Sir Richard Croft, that he had seen many similar instances.—Three others are recorded by Burdach ("*Physiologie*," § 522). In one of them, the infant was seized with convulsions on the right side and hemiplegia on the left, on sucking immediately after its mother had met with some distressing occurrence. Another case was that of a puppy, which was seized with epileptic convulsions, on sucking its mother after a fit of rage.

The following, which occurred within the Writer's own knowledge, is perhaps equally valuable to the Physiologist, as an example of the similarly-fatal influence of undue emotion of a different character; and should serve, with the preceding, as a salutary warning to Mothers, to prevent themselves from brooding over depressing ideas, as they would from indulging in passionate excitement.

d. A lady having several children, of which none had manifested any particular tendency to cerebral disease, and of which the youngest was a healthy infant a few months old, heard of the death (from acute hydrocephalus) of the infant child of a friend residing at a distance, with whom she had been on terms of close intimacy, and whose family had increased almost simultaneously with her own. The circumstance naturally made a strong impression on her mind; and she

* Dr. Von Ammon, quoted in Dr. A. Combe's excellent little work on "The Management of Infancy."—See also Dr. Kellog's case, quoted in Dr. Tuke's "Illustrations of the Influence of the Mind upon the Body."

seems to have dwelt upon it the more, as she happened at that period to be separated from the rest of her family, and to be much alone with her babe. One morning, shortly after having nursed it, she laid the infant in its cradle, asleep and apparently in perfect health; her attention was shortly attracted to it by a noise; and on going to the cradle, she found her infant in a convulsion, which lasted a few moments and then left it dead.

Now, although the influence of the Emotion is less unequivocally displayed in this case than in the preceding, it can scarcely be a matter of doubt; since it is natural that no feeling should be stronger in the Mother's mind under such circumstances, than the fear that her own beloved child should be taken from her, as that of her friend had been; and it is probable that she had been particularly dwelling on it, at the time of nursing the infant on that morning.

567. There is abundant evidence that a *sudden* and *violent* excitement of some depressing emotion, especially terror, may produce a severe and even a fatal disturbance of the Organic functions; with general symptoms (as Guislain has remarked) so strongly resembling those of sedative poisoning, as to make it highly probable that the *blood* is *directly* affected by the emotional state, through nervous agency; and, in fact, the emotional alteration of the secretions seems much more probably attributable to some such affection of the Blood, than to a primary disturbance of the secreting process itself. Although there can be no doubt that the *habitual* state of the emotional sensibility has an important influence upon the general activity and perfection of the nutritive processes,—as is shown by the well-nourished appearance usually exhibited by those who are free from mental anxiety as well as from bodily ailment, contrasted with the "lean and hungry look" of those who are a prey to continual disquietude,—yet it is not often that we have the opportunity of observing the production of disorder in the Nutrition of any specific part, by such influence. The two following cases, however, in which *local* disorder of nutrition followed upon power-

ful emotion, determined as to their seat by the intense direction of the attention to a particular part of the body, rest upon excellent authority.

a. "A lady, who was watching her little child at play, saw a heavy window-sash fall upon its hand, cutting off three of the fingers; and she was so much overcome by fright and distress, as to be unable to render it any assistance. A surgeon was speedily obtained, who, having dressed the wounds, turned himself to the mother, whom he found seated, moaning, and complaining of pain in her hand. On examination, three fingers, corresponding to those injured in the child, were discovered to be swollen and inflamed, although they had ailed nothing prior to the accident. In four-and-twenty hours, incisions were made into them, and pus was evacuated; sloughs were afterwards discharged, and the wounds ultimately healed."—(*Carter on the Pathology and Treatment of Hysteria*, p. 24.)

b. "A highly intelligent lady known to Dr. Tuke related to him that one day she was walking past a public institution, and observed a child, in whom she was particularly interested, coming out through an iron gate. She saw that he let go the gate after opening it, and that it seemed likely to close upon him, and concluded that it would do so with such force as to crush his ankle; however, this did not happen. 'It was impossible,' she says, 'by word or act to be quick enough to meet the supposed emergency; and, in fact, I found I could not move, for such intense pain came on in the ankle, corresponding to the one which I thought the boy would have injured, that I could only put my hand on it to lessen its extreme painfulness. *I am sure I did not move so as to strain or sprain it.* The walk home—a distance about a quarter of a mile—was very laborious, and in taking off my stocking I found *a circle round the ankle, as if it had been painted with red currant juice, with a large spot of the same on the outer part.* By morning the whole foot was inflamed, and I was a prisoner to my bed many days.' "—(*Influence of the Mind upon the Body*, p. 260.)

568. The influence of the state of *expectant attention*, in modifying the processes of Nutrition and Secretion, is not less remarkable than we have seen it to be in the production of muscular movements (§ 238 *et seq.*) The Volitional *direction of the consciousness* to a part, independently of emotional excitement, suffices

to call forth sensations in it, which seem to depend upon a change in its circulation (§ 129); and if this state be kept up automatically by the *attraction* of the attention, the change may become a source of modification, not only in the functional action, but in the Nutrition of the part. Thus, there can be no doubt that real disease often supervenes upon fancied ailment, especially through the indulgence of what is known as the *hypochondriacal* tendency to dwell-upon uneasy sensations; these sensations being themselves, in many instances, purely "subjective" (§ 143). In many individuals (especially females) whose sympathies are strong, a pain in any part of the body may be produced by witnessing in another, or, even by hearing described, the sufferings occasioned by disease or injury of that part; and if this pain be attended-to and believed-in as an indication of serious mischief, injurious consequences are very likely to follow. So, again, the self-tormenting hypochondriac will imagine himself the victim of any malady that he may "fancy;" and if this fancy should be sufficiently persistent and engrossing, it is not unlikely to lead to real disease of the organ to which it relates. This persistent direction of the attention has a much greater potency, when combined with the *expectation* of a particular result; and thus it happens that the spells of pretenders to occult powers, in all ages and nations, often produce the predicted maladies in the subjects who are credulous enough to believe in their efficacy. Such was formerly the case among the Negroes of the British West Indies, to such a degree that it was found necessary to repress what were known as "Obeah practices" by penal legislation; a slow pining-away, ending in death, being the not uncommon result of the fixed belief on the part of the victim, that "Obi" had been put upon him by some old man or woman reputed to possess the injurious power. So great, indeed, was the dread of these spells, that the mere threat of one party to a quarrel to "put Obi" upon the other, was often sufficient to terrify the latter into submission.

And there is adequate ground for the assertion, that even amongst the better instructed classes of our own country, a fixed belief that a mortal disease had seized upon the frame, or that a particular operation or system of treatment would prove unsuccessful, has been in numerous instances the real occasion of a fatal result.

569. But, on the other hand, the same Mental state may operate beneficially, in checking a morbid action and restoring the healthy state. That the *confident expectation of a cure* is the most potent means of bringing it about, doing that which no Medical treatment can accomplish, may be affirmed as the generalized result of experiences of the most varied kind, extending through a long series of ages. For it is this which is common to methods of the most diverse character; some of them,—as the Metallic Tractors, Mesmerism, and Homœopathy,—pretending to some physical power; whilst to others, as the invocations of Prince Hohenlohe, and the commands of Dr. Vernon or the Zouave Jacob, some miraculous influence was attributed. It has been customary, on the part of those who do not accept either the "physical" or the "miraculous" hypothesis as the interpretation of these facts, to refer the effects either to "imagination" or to "faith;"—two mental states apparently incongruous, and neither of them rightly expressing the condition on which they depend. For although there can be no doubt that in a great number of cases the patients have *believed themselves* to be cured, when *no real amelioration* of their condition had taken place, yet there is a large body of trustworthy evidence, that permanent amendment of a kind perfectly obvious to others, has shown itself in a great variety of local maladies, when the patients have been sufficiently possessed by the *expectation* of benefit, and by *faith* in the efficacy of the means employed.* "Any system of treatment," it has been recently remarked, "however absurd, that can be 'puffed' into public notoriety for efficacy,—any individual who, by accident

* A valuable collection of such evidence is contained in Dr. Tuke's "Illustrations of the Influence of the Mind upon the Body in Health and Disease," Chap. XVI.

or design obtains a reputation for the possession of a special gift of healing,—is certain to attract a multitude of sufferers, among whom will be several who are capable of being *really* benefited by a strong assurance of relief, whilst others, for a time, *believe* themselves to have experienced it. And there is, for the same reason, no religious system that has attained a powerful hold on the minds of its votaries, which cannot boast its 'miracles' of this order."

"*a.* Nothing, for example, can be more complete than the attestation of a very remarkable cure which took place in the Nunnery of Port Royal, in the person of one of the young scholars, a niece of Pascal, who was affected with an aggravated *fistula lachrymalis*, at a time when the hostility of the Jesuits and the Jansenists was at its height. The poor girl had been threatened with the 'actual cautery' by the eminent surgeon under whose care she was, as the only way of getting rid of the disease of the bones of the nose, which manifested itself in intolerable fœtor; and the day was fixed for its application. Two days previously, however, the patient walked in procession before a 'Holy Thorn,' which was being exhibited with great ceremony in the chapel of the convent; and was recommended by the nuns, as she passed before the altar, to apply the precious relic to her eye, and implore relief from the dreaded infliction. This she did, no doubt, with the most childlike confidence and heartfelt sincerity; and her faith was rewarded by the favourable change which took place within a few hours, and which had so far advanced by the time of the surgeon's next visit, that he wisely did not interfere, the cure in a short time becoming complete. Of course, this 'miracle' was vaunted by the Jansenist party as indicating the special favour of the Virgin, whilst the Jesuits could scarcely bring themselves to believe in its reality. A most careful enquiry was made by direction of the Court; the testimony of the surgeons and others, who knew the exact condition of the patient both before and after the 'miracle' (that condition being patent to their observation), was conclusive; and the reality of the cure could no longer be denied, though it remained inconceivable to the Jesuits that a miracle should have been worked in favour of their opponents.—Full details of this remarkable incident are given in Mrs. Schimmelpenninck's 'History of the Port Royalists.'

"*b.* No fact of this kind rests on a wider basis of testimony, than the efficacy of the royal touch in the 'king's evil.' The readers of Macaulay's 'History' will remember that when the honest good sense of William the Third made him refuse to exercise the power with which he was undoubtedly credited by the great mass of his subjects, an overwhelming mass of evidence was brought together as to the 'balsamic virtues of the royal hand.' Not only theologians of eminent learning, ability, and virtue, gave the sanction of their authority to this belief; but some of the principal surgeons of the day certified that the cures were so numerous and rapid that they could not be attributed to any natural cause, and that the failures were to be ascribed to want of faith on the part of the patients. Charles the Second, in the course of his reign, had 'touched' near a hundred thousand persons; and James, in one of his progresses, 'touched' eight hundred persons in Chester Cathedral. William's refusal to continue the practice brought upon him the outcries of the parents of scrofulous children against his cruelty; whilst bigots lifted up their hands and eyes in horror at his impiety. Jacobites sarcastically praised him for not presuming to arrogate to himself a power which belonged only to legitimate sovereigns; and even some Whigs thought that he acted unwisely in treating with such marked contempt a superstition which had so strong a hold on the vulgar mind.

"*c.* There are, probably, persons yet living who remember the reputed efficacy of 'Perkins's Metallic Tractors,' which was made the subject of a very careful investigation by Dr. Haygarth, an eminent physician of Bath, and Mr. Richard Smith, a distinguished surgeon of Bristol, in the early part of the present century. These gentlemen satisfied themselves that real benefit was often derived from the use of the Tractors, which were supposed to exert the 'galvanic agency' then newly discovered; but that the same benefit was obtainable from the similar manipulation of two pieces of wood painted to resemble them, the *faith* of the patient being the condition required.

"*d.* Within our own recollection, the 'miracles' of Prince Hohenlohe were as well attested as any of the kind that have been worked before or since; these were succeeded by the therapeutic marvels of Mesmerism, which can all be accounted for by the like agency; and within the last few years we have seen the 'spiritual' cures of Dr. Newton at least equalled by those worked by the Zouave Jacob."

"Each reputation of this kind has its period of growth, maturity, decline, and death; and we should confidently anticipate that before the lapse of many years, the 'Spiritual' cures will, in like manner, have passed into the limbo of forgotten wonders of the same description, if it were not that the belief in them is only one of the manifestations of a morbid condition of the popular mind, the origin of which unfortunately lies very deep in its constitution."—(*Quarterly Review*, Oct. 1871, pp. 323-5).

570. There is no more satisfactory example of the influence of Expectant Attention "pure and simple," than is afforded by the charming-away of warts: for the disappearance of these excrescences has so frequently occurred within the experience of trustworthy observers, in close connection with this *psychical* treatment, that we must disbelieve in the efficacy of *any* remedies, if we do not accept this.

a. "In one case," says Dr. Tuke (*op. cit.*, p. 365), "a relative of mine had a troublesome wart on the hand, for which I made use of the usual local remedies, but without effect. After they were discontinued, it remained *in statu quo* for some time, when a gentleman 'charmed' it away in a few days."

b. The same author continues:—"A surgeon informs me that some years ago his daughter had about a dozen warts on her hands. They had been there about eighteen months, and her father had applied caustic and other remedies without success. One day a gentleman called, and, in shaking hands with Miss C., remarked upon her disfigured hand. He asked her how many she had; she replied that she did not know, but thought about a dozen. 'Count them, will you,' said the caller; and, taking out a piece of paper, he solemnly took down her counting, remarking, 'You will not be troubled with your warts after next Sunday.' By the day named the warts disappeared, and did not return."

c. Two similar cases have occurred within the Writer's personal knowledge. In one, the warts were disposed of by "counting;" in the other, by touching each singly with coloured water;—the assured conviction of its success being (of course) the condition of the efficacy of the "spell."

571. It is not a little remarkable that the influence of Mental states should be unmistakeably manifested, not only in maladies in which Nervous disorder has a large share, but also in some—as Scurvy and Gout—which seem to depend upon the existence of a definite perversion in the condition of the Blood.

a. Thus, during the Siege of Breda in 1625, the garrison having been reduced to a state of extreme distress by Scurvy in its severest form, attended with a great mortality, so that the city was on the point of capitulating, the Prince of Orange managed to send word that the sufferers should soon be provided with medicines of the greatest efficacy. Three small phials, containing a decoction of camomile, wormwood, and camphor, were put in the hands of each physician; and it was publicly given out that three or four drops were sufficient to impart a healing virtue to a gallon of liquor,—not even the commanders being let into the secret. The effect of the soldiers' faith in the efficacy of the "prince's remedy" was most marvellous; for not only was the further spread of the disease checked, but a large proportion of those who were then suffering under it, including many who had been for some time completely invalided, recovered very rapidly. (See *Lind on Scurvy*, p. 352.)—It is well known, on the other hand, that mental depression is one of the most potent of the "predisposing causes" of Scurvy.

b. There are numerous well-authenticated cases, in which a severe fit of Gout has been suddenly dissipated by violent emotion. And Dr. Rush recorded one in which an old farmer, languishing under severe infirmity caused by repeated attacks of this disease, was not only cured of the particular fit, but was restored to perfect health, by the fright and anger brought on by the careless driving of one of his sons, which caused the window-sash near which he was lying to be broken-in.—(See Tuke, *op. cit.*, pp. 368.)

572. In all ages, the possession of men's minds by "dominant ideas" has been most complete, when these ideas have been *religious* aberrations. And hence it is only to be expected that the effects of such "possession" should exert an unusually powerful influence on the Organic functions, as we have seen it to do on Muscular actions (§§ 258, 259). There is to the Writer's mind,

therefore, nothing either incredible or miraculous in the numerous recorded cases of "stigmatization," *i.e.*, the appearance of wounds upon the hands and feet, on the forehead, and on the side,—corresponding with those of the crucified Jesus,—from which blood has periodically flowed. The subjects of these cases were mostly "Ecstaticas;" *i.e.*, females of strongly Emotional temperament, who fell into a state of profound Reverie, in which their minds were entirely engrossed by the contemplation of their Saviour's sufferings, with an intense direction of their sympathetic attention to his several wounds. And the power which this state of Mind would have on the local action of the corresponding parts of their own bodies (§ 567, *a*, *b*), gives a definite Physiological *rationale* for what some persons accept as genuine miracles, and others repudiate as the tricks of imposture.

a. The most recent case of this kind, that of Louise Lateau, has undergone a scrutiny so careful, on the part of Medical men determined to find out the deceit, if such should exist, that there seems no adequate reason for doubting its genuineness. This young Belgian peasant had been the subject of an exhausting illness, from which she recovered rapidly after receiving the Sacrament; a circumstance which obviously made a strong impression on her mind. Soon afterwards, blood began to issue every Friday from a spot in her left side; in the course of a few months, similar bleeding spots established themselves on the front and back of each hand, and on the upper surface of each foot, while a circle of small spots formed on the forehead; and the hæmorrhage from these recurred every Friday, sometimes to a considerable amount. About the same time, fits of "ecstasy" began to occur, commencing every Friday between 8 and 9 A.M., and ending at about 6 P.M.; interrupting her in conversation, in prayer, or in manual occupations. This state appears to have been intermediate between that of the Biologized and that of the Hypnotized subject; for whilst as unconscious as the latter of all Sense-impressions, she retained, like the former, a recollection of all that had passed through her mind during the "ecstasy." She described herself as suddenly plunged into a vast flood of bright light, from which more or less distinct forms soon began to evolve themselves;

and she then witnessed the several scenes of the Passion successively passing before her. She minutely described the cross and the vestments, the wounds, the crown of thorns about the head of the Saviour; and gave various details regarding the persons about the cross,—the disciples, holy women, Jews, and Roman soldiers. And the progress of her vision might be traced by the succession of actions she performed at different stages of it; most of these being movements expressive of her own emotions; whilst regularly about 3 P.M. she extended her limbs in the form of a cross. The fit terminated with a state of extreme physical prostration: the pulse being scarcely perceptible, the breathing slow and feeble, and the whole surface bedewed with a cold perspiration. After this state had continued for about ten minutes, a return to the normal condition rapidly took place.—These last phenomena, which were paralleled to a certain degree in Mr. Braid's experiments, seem quite beyond the power of intentional simulation; while the tests applied to determine the possibility of the artificial production of the stigmata and of the issue of blood from them, appear no less conclusive as to their non-simulation. (*Macmillan's Magazine*, April, 1871.)

As the transudation of Blood from the skin through the perspiratory ducts (apparently through the rupture of the walls of the cutaneous capillaries) under strong Emotional excitement, is a well-authenticated fact (see Tuke, *op. cit.*, p. 267), there is nothing in the foregoing narration that the Physiologist need find any difficulty in accepting.

CHAPTER XX.

OF MIND AND WILL IN NATURE.

573. The views expressed in the preceding pages as to the constitution of the Mind of Man, and its relation to his Bodily Organism, appear to the Writer to be capable of legitimate extension to the notion which we form of the Mind of the Deity in its relation to that Universe, whose phenomena, so interpreted, are but a continual revelation of His universal presence and ceaseless agency. And he deems it desirable here to advert to this subject (foreign though it may seem to the proper object of this Treatise), for the sake of showing not merely that the doctrine herein propounded is strictly conformable to the highest teachings of Religion, but that it affords some guidance towards the solution of difficulties which have perplexed many deep-thinking men, and which have (especially of late) tended to keep Science and Religion in mutual antagonism, instead of in that harmonious co-operation which should spring from the intimacy of their relationship. For, as Mr. Martineau has well said, "Science discloses the Method of the world, but not its Cause; Religion its Cause, but not its Method; and there is no conflict between them, except when either forgets its ignorance of what the other alone can know." (*National Review*, Vol. xv., p. 398.) If, then, a means can be found for their complete reconcilement, it is obvious that each will gain by their accord :—Science, by being led to regard all the phenomena of Nature as manifestations of the constant and all-pervading energy of a Mind of infinite perfection ;—Religion, by obtaining that expansion and definition of its ideas as to the unlimited range

and predetermined order of the Divine operations, which the Scientific conception of them can alone afford (§§ 586-588).

574. In the first place, it should be clearly understood that Science is nothing else than Man's Intellectual representation of the phenomena of Nature,—his conception of the Order of the Universe in the midst of which he is placed. That conception is formulated in what he terms *Laws of Nature;* which, in their primary sense, are simply expressions of *phenomenal uniformities*, having no *coercive* power whatever. The whole problem of the Scientific investigation of Nature from this point of view has been thus stated by Mr. J. S. Mill in his *System of Logic*:—" What are the fewest assumptions, which being granted, the Order of Nature as it exists would be the result? What are the fewest general propositions, from which all the Uniformities existing in Nature could be deduced?"—Of such propositions we have a characteristic type in Kepler's three Laws of Planetary Motion; which simply express the systemátized results he obtained by his comparison of the observed places of Mars, and the definite proportion he discovered between the times and distances of the six Planets known to him; without affording any valid ground for the assurance that the same Laws would hold good elsewhere. And the motions of other Planets could no more be said to be "explained" or "accounted for," by their conformity to those of Mars, than the fall of a stone to the ground would be "explained" or "accounted-for" by the statement that "*all* stones (unsupported) fall to the ground,"—the multiplication of similar phenomena only leading to the conclusion that there is a common Cause for the whole, without giving any indication of what that cause is.—The utmost hold that purely phenomenal Laws can have upon our minds, is derived from that Belief in the Uniformity of Nature, which has become one of our primary Tendencies of Thought (§ 201). As Mr. Herbert Spencer has well remarked (*First Principles*, p. 142):—

" All minds have been advancing towards a belief in the constancy of surrounding co-existences and sequences. Familiarity with special uniformities has generated the abstract conception of Uniformity—the idea of *Law*; and this idea has been in successive generations slowly gaining fixity and clearness. . . . Wherever there exist phenomena of which the dependence is not yet ascertained, these most cultivated intellects, impelled by the conviction that here, too, there is some invariable connection, proceed to observe, compare, and experiment; and when they discover the law to which the phenomena conform, as they eventually do, their general belief in the universality of Law is further strengthened. . . . . This habitual recognition of Law distinguishes modern thought from ancient thought."

To speak of such *phenomenal* Laws, however, as *governing* phenomena, is altogether unscientific; such laws being nothing else than comprehensive expressions of aggregates of particular facts, and giving no *rationale* of them whatever.

575. When, however, we not only look at bodies in motion, but try to resist their motion by an exertion of our own, or use a similar exertion in giving motion to a body at rest, we are led by our own sense of effort in making it to an entirely new conception, that of *Force*; and no advance in the Philosophy of Science has been greater, than that which has of late years extended the notion of Force, from the agency which produces or resists the Motion of masses, to the agencies which are concerned in producing the molecular changes which we refer to Heat, Light, Electricity, Magnetism, &c. The Man of Science of the present day is thus enabled to attach a distinct idea to that *efficient causation*, which Logicians have continually denied, but which the Common Sense of Mankind has universally recognized.—When the Cause of any event is spoken of, in common parlance, we certainly attach to the term the idea of *power*, at the same time that we include the notion of the *conditions* under which that power operates; and this view of the case can be shown to be scientifically correct. For though the Logician may define the "cause" of any event to be "the antecedent, or the con-"currence of antecedents, on which it is invariably and uncondition-

ally consequent" (Mill, *Op. cit.*), it is uniformly found, when this assemblage of antecedents is analyzed, that they may be resolved into two categories, which may be distinguished as the *dynamical* and the *material;* the former supplying a *force* or *power* to which the change must be attributed, whilst the latter afford the *conditions* under which that power is exerted.

Thus in a Steam-Engine, we see the dynamical agency of Heat made to produce Mechanical power, by the mode in which it is applied—first, to impart a mutual repulsion to the particles of water,—and then, by means of that mutual repulsion, to give motion to the various solid parts of which the machine is composed. And thus, if asked what is the cause of the movement of the Steam-Engine, we distinguish in our reply between the *dynamical* condition supplied by the Heat, and the assemblage of material conditions afforded by the collocation of the boiler, cylinder, piston, valves, &c.—So, again, if we are asked what is the cause of the movement of a Spinning-mule, we refer it to its connection by bands or wheels with some shaft, which itself derives its power to move from a Steam-engine or a Water-wheel; these material collocations here again serving to supply the conditions under which the Force (which, in each case, is ultimately Heat) becomes operative.—In like manner, if we inquire into the cause of the germination of a Seed, which has been brought to the surface of the earth, after remaining dormant, through having been buried deep beneath the soil, for (it may be) thousands of years, we are told that the phenomenon depends upon warmth, moisture, and oxygen: but out of these we single Heat as the *dynamical* condition; whilst the oxygen and the water, with the organized structure of the seed itself, and the organic compounds which are stored up in its substance, constitute the *material conditions.*

The Material conditions, in fact, merely furnish the *fuel* and the *mechanism;* it is the Force or Power that *does the work.*

576. The strictest Scientific inquiry, then, *must* recognize Dynamical agency as fundamentally distinct from Material conditions; and by this recognition we bring our Scientific conception into harmony with the universal Consciousness of Mankind, which, as Sir John Herschel has truly said, is as completely in

accord in regard to the existence of a real and intimate connection between Cause and Effect, as it is in regard to the existence of an External World. Now as it is universally admitted that our notion of that External World would be very incomplete, if our Visual perceptions were not supplemented by our Tactile, so our interpretation of the phenomena of the Universe would be very inadequate, if we did not mentally co-ordinate the idea of Force with that of Motion; and it has been by such co-ordination that all those higher conceptions of the Order of Nature have been arrived at, which it is the greatest glory of Science to have attained, and which are the chief sources of Man's power.

577. The grandest and most comprehensive of these, the Law of Gravitation, is an expression of the fact that everywhere and under all circumstances, two masses of matter *attract* one another in certain definite ratios; and the term "attract" implies that they are drawn together by a *force*, similar to that which directly impresses itself upon our consciousness by the *sense of effort* we experience when we lift a pound-weight from the ground, and which furnishes the *unit* by which the weights of the Earth, the Sun, and the other bodies of the Solar System are estimated. From this *dynamical assumption*, combined with those other assumptions that are embodied in the Newtonian "Laws of Motion," not only have all the observed Uniformities both of Celestial and Terrestrial Motion (including those embodied in Kepler's Laws) been derived deductively, but predictions have been drawn in advance of observation, of which the verification has never yet been wanting. And thus a Law which expresses the invariable conditions (so far as known to us) of the action of a Force of whose existence we are directly cognizant, impresses our minds with an assurance of *its universality and its constancy*, which no mere generalized expression of "co-existences and sequences" could carry with it.

578. In like manner, the Undulatory Theory of Sound, Light, Heat, and Actinism, and the Doctrine of Electric and Magnetic

Polarity, are *dynamical* conceptions, whose gradual approximation in completeness and simplicity to the Law of Gravitation, constitutes the strongest assurance of their truth. Another hypothesis recently propounded, that of Molecular Motion, is regarded by some of our most advanced Physicists as likely to bring under one and the same dynamical expression the Newtonian Law of Universal Gravitation, the Laws of Diffusion of Liquids and Gases, and many other subordinate Laws of which no *rationale* has yet been given. And the intimate relations which have been proved to exist between Chemical Affinity, on the one hand, and Heat, Electricity, and even Mechanical Force, on the other, leave it scarcely open to question that the Laws of Chemical Combination, which are at present only in their *phenomenal* stage, will ere long be brought within a definite *dynamical* expression. This is also the tendency of modern Physiological inquiry; which, abandoning the old doctrine of a "Vital Principle" as a mere refuge for ignorance, fixes its attention on the dynamical conception of a Vital Force, differing in its manifestations from Heat, Electricity, Mechanical Force, &c., as the several manifestations of those Forces differ from each other; but definitely "correlated" with them, as they are correlated to each other. It has been one object of this Treatise to show that the like "correlation" exists between Nerve-force (one of the *modes* of Vital Force) and Mind-force; and that the latter is thus mediately brought into relation with the Physical Forces of Nature.

579. The culminating point of Man's Intellectual interpretation of Nature, may be said to be his recognition of the Unity of the Power, of which her phenomena are the diversified manifestations. Towards this point all Scientific inquiry now tends. For the Convertibility of the Physical Forces, the Correlation of these with the Vital, and the intimacy of that *nexus* between Mental and Bodily activity, which, explain it as we may, cannot be denied, all lead upward towards one and the same conclusion,—the

source of all Power in Mind; and that philosophical conclusion is the apex of a pyramid, which has its foundation in the primitive instincts of Humanity. By our own remote progenitors, as by the untutored savage of the present day, every change in which Human agency was not apparent, was referred to a particular Animating Intelligence.* And thus they attributed not only the movements of the Heavenly bodies, but all the phenomena of Nature, each to its own Deity. These deities were invested with more than human power; but they were also supposed capable of human passions, and subject to human capriciousness. As the Uniformities of Nature came to be more distinctly recognized, some of these deities were invested with a dominant control, while others were supposed to be their subordinate ministers. A serene majesty was attributed to the greater Gods who sit above the clouds; whilst their inferiors might "come down to earth in the likeness of men." With the growth of the Scientific study of Nature, the conception of its harmony and unity gained ever-increasing strength. Among the most enlightened of the Greek and Roman philosophers, we find a distinct recognition of the idea of the Unity of the directing Mind from which the Order of Nature proceeds; for they obviously believed that, as our modern poet has expressed it,—

> "All are but parts of one stupendous whole,
> Whose body Nature is, and God the Soul."

And thus, whilst the deep-seated instincts of Humanity, and the profoundest researches of Philosophy, alike point to Mind as the one and only source of Power, it is the high prerogative of Science to demonstrate the *unity* of the Power which is operating through the limitless extent and variety of the Universe, and to trace its *continuity* through the vast series of ages that have been occupied in its evolution.

* See Tylor's *Primitive Culture* (London, 1871), Chaps. xi.-xvii.

580. But if such be the legitimate tendency of Scientific inquiry, the question arises why—especially in these days—so many of its votaries should place themselves in an attitude of direct antagonism to Religion. The answer to this question seems threefold; and each point needs a separate consideration.

581. In the *first* place, there has been, for several centuries past, a constant endeavour on the part of the upholders of Theological creeds and Ecclesiastical systems, either to repress Scientific inquiry altogether, or to limit its range. While accepting, with the rest of the world, those results of Scientific labour which contribute to their own comfort or enjoyment,—making no objection to Science, so long as it confines itself to giving them Steam-engines and Railroads, Gas-lighting and Electric Telegraphs,—such Theologians maintain that the minds of men who devote the best powers of their lives to the search for the Truth as it is in Nature, are to be "cribb'd, cabin'd, and confined" by narrow interpretations of the Bible; and now think to put down the great Scientific hypothesis which is engaging much of the best thought of our time, by citing the text, "God made man in His own image," just as, three centuries ago, they declared the Copernican system to be a pernicious error, because Joshua commanded the Sun and Moon to stand still, and even yet denounce Geologists as sceptics or even infidels, because they refuse to accept as revealed truth that God made heaven and earth in six days, and rested the seventh day.—It is not strange, then, that Men of Science should not only rebel against such self-constituted domination, but should repudiate the whole system of Belief of which it is the expression. For all History shows that nothing drives men to the extreme of license, so surely as tyrannical restriction. The *juste milieu* can only be found by those who are free to seek for it.

582. In the *second* place, there is in what claims to be the "orthodox" systems of Theology, so much that runs counter to

the strongest and best instincts of Humanity, that those who have been led by Scientific study to build up their "fabric of thought" on the basis of their own Intellectual and Moral Intuitions, find it impossible to fit into this (§ 321) a set of doctrines which are altogether conformable to it. They cannot reconcile, for example, the everlasting damnation of all such as are unable to accept a body of unintelligible Dogma, with any conception they can form either of a Righteous Creator or of a Loving Father. Nor can they conceive that either the performance of the Baptismal rite, or the Atoning sacrifice of a Divinity, can be the condition on which depends the rescue of an innocent child from eternal torment. Nor, again, can they regard as acceptable worship of a Beneficent God, the utterance of imprecations conceived in a spirit of hatred and vengeance worthy only of the votaries of a Moloch, and altogether inconsistent with those later teachings of Christianity, which have revealed to us the highest and holiest elements of our own nature, elements which most distinctly bear the impress of the Divine.—So long as this is the haven to which "orthodox" Theology invites Men of Science, the great mass of them will most assuredly avoid it altogether; and, unless they can find an anchorage elsewhere, will drift away into either vague *un*belief or absolute *dis*belief.

583. While Theologlical Systems are thus answerable for two sources of Scientific antagonism to Religion, a *third* arises out of the tendencies of Scientific research itself. For the more constant and invariable the great Agencies of Nature are found to be, and the more what at first seemed exceptional phenomena are brought within the domain of Law,—the more, on a superficial view, does it appear as if the Order of Nature were simply mechanical, going on *of itself*, as it has done through all the past, and will continue to do through the future. But a deeper scrutiny has shown us that the Man of Science cannot dispense with the notion of a Power always working throughout the Mechanism of the Universe;

and that, on Scientific grounds alone, this Power may be regarded as the expression of Mind. And anything else than unvarying Uniformity in the mode of operation of that Mind, would be an indication of its defect rather than of its perfection. For if all the agencies of Nature are the unconscious ministers of an All-wise and All-powerful Ruler, they will work-out His bidding like the disciplined members of a large and well-ordered household, in which everyone knows his work and does it. Surely it would be strange if any who should watch these servants in the performance of their several duties,—should study the successions of every hour, should find each doing at a certain prefixed time and place exactly that which proves most suitable to the occasion, and should thus finally arrive at a conception of the harmony and completeness of the whole scheme of domestic economy,—were to be led by this very harmony and completeness to regard that as a mere mechanical routine, which is really the silent invisible action of the directing Will, and were to see the operation of that Will only in such departures from the system as may be required to meet contingencies for which no human foresight can provide.*

584. Turning now to the other side of the inquiry, we have to consider briefly what is the real basis of Religion, what is its essential nature, and whether there is anything necessarily inconsistent in its teachings with the results of Scientific inquiry.—The only secure basis for Religion consists in Man's own *religious consciousness;* since it is as impossible that any Revelation should make a man religious, whose inner nature does not respond to its teachings, as that any instruction should make a man a Musician, who has not got a "musical ear." This Religious Consciousness may be considered to arise, in the first place, from that capacity for the *ideal* which seems to be the distinctive attribute of Man (§97); whilst it is progressively developed, refined, and elevated, in accordance

* See Dr. Chalmers's Sermon on *The Constancy of Nature and the Faithfulness of God*, in Vol. vii. of his collected Works.

with the general expansion of his Intellect, the exaltation of his Moral Sense, and the purification of his Aspirations, so as to shape out for itself, more and more distinctly, a Divine Ideal (§ 213). And in proportion as every one determinately endeavours to bring his own character and conduct into accordance with that ideal (§ 339), is he (in the truest sense of the term) a religious man.

585. Now whilst the conception which each individual forms of the Divine Nature will depend in great degree upon his own habits of thought, there are two extremes towards one or other of which most of the current notions on this subject may be said to tend, and between which they have oscillated in all periods of the history of Monotheism. These are *Pantheism* and *Anthropomorphism.* —Towards the Pantheistic aspect of Deity, we are especially led by the philosophic contemplation of His agency in external Nature; for in proportion as we fix our attention exclusively upon the "laws" which express the orderly sequence of its phenomena, and upon the "forces" whose agency we recognize as their efficient causes, do we come to think of the Divine Being as the mere *first principle* of the Universe,—as an all-comprehensive "law" to which all other laws are subordinate, as that most general "cause" of which all the Physical forces are but manifestations. This conception embodies a great truth, and a fundamental error. Its truth is the recognition of the universal and all-controlling agency of the Deity, and of His presence *in* Creation rather than on the outside of it. Its error lies in the absence of any distinct recognition of that *conscious volitional* agency, which is the essential attribute of Personality; for without this there can be no Moral Government, and Man's worthiest aspirations after the Divine Ideal would have no real object.—The Anthropomorphic conception of Deity, on the other hand, arises from the too exclusive contemplation of *our own* nature as the type of the Divine: and although, in the highest form in which it may be held, it represents the Deity as a Being in whom all Man's

noblest attributes are expanded to Infinity, yet it is practically limited and degraded by the impossibility of *fully* realizing such an existence to our minds; the failings and imperfections incident to our Human nature being attributed to the Divine, in proportion as the standard of Intellectual and Moral development attained by each individual limits his idea of possible excellence. Even the lowest form of any such conception, however, embodies (like the Pantheistic) a great truth, though mingled with a large amount of error. It represents the Deity as a *person;* that is, as possessed of that Intelligent Volition, which we recognize in ourselves as the source of the power we determinately exert, through our bodily organism, upon the world around; and it invests Him also with those Moral attributes, which place Him in sympathetic relation with His sentient creatures. But this conception is erroneous, in so far as it represents the Divine Nature as restrained in its operations by any of those limitations which are inherent in the very constitution of Man; and, in particular, because it leads those who accept it, to think of the Creator as "a remote and retired mechanician, inspecting from without the engine of creation to see how it performs," and as either leaving it entirely to itself when once it has been brought into full activity, or as only interfering at intervals to change the mode of its operation.

586. Now the Truths which these views separately contain, are in perfect harmony with each other; and the very act of bringing them into combination effects the elimination of the errors with which they were previously associated. For the idea of the universal and all-controlling agency of the Deity, and of His immediate presence throughout Creation, is not found to be in the least degree inconsistent with the idea of His personality; when that idea is freed from the limitations which cling to it in the minds of those who have not expanded their Anthropomorphic conception by the Scientific contemplation of Nature. And the

Man of Science who studies not only the Mechanism of Nature, but the Forces which give Life and Motion to that Mechanism, and who fixes his thought on that conception of *force* as an expression of *will*, which we derive from *our own experience* of its production, is thus led to recognize the universal and constantly sustaining agency of the Deity in every phenomenon of the Universe; and to feel that in the Material Creation itself, he has the same distinct evidence of His personal existence and ceaseless activity, as he has of the agency of Intelligent Mind in the creations of Artistic Genius, or in the elaborate products of Mechanical skill, or in those written records of Thought and Feeling which arouse our own Psychical nature into kindred activity.

587. That any *antagonism* should be supposed to exist between those "Laws" which express the Uniformities of Nature discovered by Science, and the Will of the Author of Nature as manifested in those uniformities,—so as for the acceptance of the former to exclude the notion of the latter,—can only arise either from an unworthy conception of the Deity as an arbitrary and capricious ruler, or from an unphilosophical conception of the real meaning of Science as the intellectual interpretation of Nature. It is on the *highest*, not on the *lowest*, form of Human Will, that we should base our ideas of the Divine;—upon such a Will as sets before it a great and good object, steadily perseveres in the course that leads towards its accomplishment, shapes its mode of operation to the best of its limited foreknowledge, is not discouraged by temporary failures, and finally succeeds because the means employed were *on the whole* adapted to bring about the result. Now, if the Foreknowledge be infinite, there will be no failures, because all fruitless efforts will be prevented by the prevision of the inadequacy of the means. And if the Power be infinite, there will be no limitation of choice, except as to the means which will *best* conduce to the end in view.

588. Hence there is a perfect conformity between the Scientific idea of "Law," as expressive of Uniformity of action, and the Theological idea of "Will" exerting itself with a fixed purpose according to a predetermined plan; and of the existence of such a plan, the Revelations of Science furnish Theology with its best evidence. For the Immutability of the Divine Nature is nowhere more clearly manifested, than in that *continuance of the same mode of action*,—not merely through the limited period of Human experience, but, as we have now strong reason to believe (on Scientific grounds alone), from the very commencement of the present system of the Universe,—which enables us to discern somewhat of the Plan on which the Creator has acted, and is still acting. If His every action were immediately prompted by present contingencies, instead of being the result of predetermination based on perfect knowledge of the future, there could be no Law. If that knowledge were, like Man's, imperfect, though we might trace a *general* method when the arrangements were viewed in their totality, the *details* would have much of that unsteadiness and occasional want of consistency, which we perceive in the actions of even the best-regulated Human mind. The *laws* would be made to bend to the necessities of the time; and new interpositions would be continually necessary, to correct the errors that would occasionally arise in the working of the machine. So far, however, is this from being the case in the Divine operations, that, in the only department of Science (Astronomy) in which the Philosopher has been able, from the simplicity of the phenomena, to attain to anything like a complete generalization of them, he has every reason to believe that the same Laws have been in operation from the beginning, or, in other words, that the work of Creation was commenced upon a plan so perfect, that no subsequent change in this plan has been required.

589. The Scientific sense of the term "Law," therefore,—considered simply as Man's expression of Uniformity of Sequence

within the range of his limited experience,—so far from being in antagonism with the motion of Will, is only in antagonism with that idea of inconstancy in its mode of exercise, which belongs to a Theology now disowned by the best thinkers of our time. Not even Mr. Herbert Spencer could express himself on the constancy of Nature (§ 574) more explicitly than did Dr. Chalmers more than forty years ago :—

"It is no longer doubted by men of science, that every remaining semblance of irregularity in the Universe is due, not to the fickleness of Nature, but to the ignorance of Man,—that her most hidden movements are conducted with a uniformity as rigorous as Fate,—that even the fitful agitations of the weather have their law and their principle,—that the intensity of every breeze, and the number of drops in every shower, and the formation of every cloud, and all the recurring alternations of storm and sunshine, and the endless shiftings of temperature, and those tremulous vibrations of the air which our instruments have enabled us to discover, but have not enabled us to explain,—that still, they follow each other by a method of succession, which, though greatly more intricate, is yet as absolute in itself as the order of the seasons, or the mathematical courses of astronomy. This is the impression of every philosophical mind with regard to Nature; and it is strengthened by each new accession that is made to Science. The more we are acquainted with her, the more are we led to recognise her constancy, and to view her as a mighty though complicated machine, all whose results are sure, and all whose workings are invariable."—(*Chalmers's Works*, vol. vii., p. 204.)

This constancy, on which every man counts in the plans and intentions he forms for his future action, is, to him who has found the reconcilement between Science and Religion, nothing else than an abiding testimony to the Infinity of the Divine Perfections.

590. There seems to be another source, however, for the supposed antagonism between the notion of Law and that of Will, as the governing and sustaining power of the Universe; namely, the idea that when God is said to "govern by law," it is implied that some agency exists *between* Himself and Nature. This idea seems to

have its origin in the imperfect analogy supplied by Human legislation,—an analogy so misleading that it might be wished that the term "law" could be altogether banished from Science, if it were not that, when carefully examined, any Law of Man's devising is found to be nothing else than an expression of certain *predetermined uniformities of action of the Governing Power.* Whether that Power be wielded by a single individual who rules by his personal supremacy, or be vested in him as the impersonation of the Will of the community, or be directly exerted by the community itself, the action of the Law upon those who are subject to it, is simply the constant, though silent, operation of such Power; for the law loses all its coercive efficacy, the moment that the power which enforces it is withdrawn by the overthrow or the paralysis of the Government which exercised it. Now if the Law, as first laid down by a Human legislator, prove inadequate, though backed by adequate Power, to produce the desired effect, he modifies or changes it; such alteration being required simply on account of his limited foreknowledge. But supposing his Foreknowledge to be Infinite, all the results of any exertion of his Will that he might embody in a Law, would be so completely foreseen in the first instance, that (supposing him possessed also of adequate Power) he could adapt his Law to the purpose it is to serve, with such perfection as to render any subsequent alteration unnecessary.—In regard to the Physical Universe, then, it might be better to substitute for the phrase "Government *by* laws," "Government *according to* laws;"—meaning thereby, the direct exertion of the Divine Will, or operation of the First Cause, in the Forces of Nature, according to certain constant Uniformities, which are simply nnchangeable, because—having been originally the expressions of Infinite Wisdom—any change would be for the worse.

591. Looking at the Deity, finally, in His relation to his Human offspring, we draw a like conclusion from the best results of our own

limited experience. For if a loving Father had foreknowledge enough to form at the outset all his future plans for the education of his children, and wisdom enough to adapt these plans in the best possible manner to their respective characters as progressively developed, and to all the conditions in which they may hereafter find themselves, and power enough to carry these plans into operation, so that the course of events would not require the alteration of one tittle in their fulfilment,—surely this would be a far more perfect manifestation of a Paternal character, than the continual change in his schemes which the Human parent is usually obliged to make, in order to adapt them to the purpose he has in view. The perpetual recurrence of *obvious design*, in the latter case, may be, to an ordinary bystander, more suggestive of the intentions of the parent; but the more profound observer will take another view, and will have reason to doubt, from the necessity of the perpetual change, the absolute wisdom of the controlling power. The notion of constancy and invariability in the Creator's plan, therefore,—by referring *all* those provisions for Man's benefit which He has placed before him, either in possession or in prospect, to the period when this present system of things had a beginning,—simply *antedates* the exercise of His discerning Love; and so far from our idea of its nature losing any of its force on this account, it ought to be strengthened and enlarged, in precisely the same ratio as our ideas of His Power and Wisdom are extended by the elevation of the point from which we view His operations, and the consequent range of the survey we take.

592. Thus, then, if Theologians will once bring themselves to look upon Nature, or the Material Universe, as the embodiment of the Divine Thought, and at the Scientific study of Nature as the endeavour to discover and apprehend that Thought (to have "thought the thoughts of God" was the privilege most highly esteemed by Kepler), they will see that it is their duty, instead of holding themselves altogether aloof from the pursuit of Science,

or stopping short in the search for Scientific Truth wherever it points towards a result that seems in discordance with their preformed conceptions, to apply themselves honestly to the study of it, as a Revelation of the Mind and Will of the Deity, which is certainly not less authoritative than that which He has made to us through the recorded thoughts of religiously-inspired Men, and which is fitted, in many cases, to afford its true interpretation. And they cannot more powerfully attract the Scientific student to Religion, than by taking up his highest and grandest Thought, and placing it in that Religious light which imparts to it a yet greater glory. They will then perceive that although, if God be *outside* the Physical Universe, those extended ideas of its vastness which modern Science opens to us, remove Him further and further from us, yet, if he be embodied *in* it, every such extension enlarges our notion of His being. As Mr. Martineau has nobly said:—"What, indeed, have we found by moving out along all radii into the Infinite? That the whole is woven together in one sublime tissue of intellectual relations, geometric and physical,—the realized original, of which all our Science is but the partial copy. That Science is the crowning product and supreme expression of Human reason. . . . . Unless, therefore, it takes more mental faculty to construe a Universe than to cause it, to read the Book of Nature than to write it, we must more than ever look upon its sublime face as the living appeal of Thought to Thought." But the Theologian cannot rise to the height of this conception, unless he is ready to abandon the worship of every idol that is "graven by art and man's device;"—to accept as a fellow-worker with himself every truth-seeker who uses the understanding given him by "the inspiration of the Almighty" in tracing out the Divine Order of the Universe;—and to admit into Christian communion every one who desires to be accounted a disciple of Christ, and humbly endeavours to follow in the steps of his Divine Master (§ 339).

APPENDIX.

THE whole of the foregoing Treatise was in type, when the Writer became acquainted with the remarkable results of the Experimental Researches which have been recently prosecuted by Dr. Ferrier into the Functions of different parts of the Brain.* Of these results, some afford full confirmation of certain general doctrines set forth in the preceding pages, giving to those doctrines an unexpected definiteness and extension; others are clear additions to our previous knowledge, which are not only important in themselves, but valuable as affording a clue to further inquiry; whilst of others, again, the precise meaning seems at present obscure. As none of them, however, appear to the Writer in any way opposed to his previous teachings, he has thought it preferable, in the present stage of the inquiry, to leave these as they stand; and to give a separate account of Dr. Ferrier's results, so far as they have yet been made public, with the conclusions which they at present seem to himself to justify.

Dr. Ferrier's researches were made by the localized application of an Electric current to different parts of the Cortical substance of the Cerebrum, and to other ganglionic centres forming part of the Brain; the animal having been previously rendered insensible by Chloroform, so that the movements excited by this stimulation may be regarded as the direct products of the *physical* changes it induced.

The method of applying Electricity found most effective, was that which is known as *faradization;* namely, the use of the interrupted current of an induction-coil, which could be increased or diminished at pleasure to meet the requirements of the case. As a rule, the current was not stronger than could be borne without great discomfort on the

* Medical Reports of the West Riding Lunatic Asylum, vol. iii., 1873.

tip of the tongue; but very considerable variations in its tension were found necessary, in order to produce the same effects in different animals, and at different times in the same animal. These variations were obviously related in part to the condition of the Blood-circulation: the excitability of the Nerve-centres being lowered alike by the effect of the "shock" of the severe operation necessary to expose the brain to the requisite extent, which enfeebles the action of the heart (§ 41); by the inevitable loss of Blood, which both directly diminishes the supply, and reduces the heart's action; and also by the depressing influence of the continued administration of chloroform upon the heart's action. It thus came about, that after great hæmorrhage, and when the brain no longer showed the pulsation which marks the influx of successive waves of blood, the strongest faradization failed to stimulate. But, again, the excitability of the nerve-centres was further depressed by the direct influence of the chloroform; though this seems to be much less strongly exerted upon the Motor, than it is upon the Sensorial centres. And, lastly, the excitability was progressively reduced by repeated stimulation; so that at last no movement could be called forth, though the current was much increased in strength. Yet even then, after an intermission of the experiments, during which the use of the chloroform was suspended, and the powers of the animal were refreshed by food and by the respiration of pure air, the excitability was found to have been restored. This is obviously in full accordance with the principle stated in (§§ 404–2); for while the *nutritive components* of the blood supply the materials whereby the nerve-substance generates and repairs itself, and its "potential energy" accumulates, it is by the combination of the *oxygen* of the blood with those materials, that the *potential* is converted into *actual* energy (§ 308).

The faradization of the cortical substance of the Cerebrum immediately produces an intense *hyperæmia*; that is, a greatly-increased flow of blood through its closely-compacted plexus of minute vessels. This is made apparent in two ways; first, in the increase in the size of the vessels, which is obvious to the eye; and second, in the profuse flow of venous blood that takes place from the great sinuses, which, in the quiescent state of the brain, had ceased to bleed. Several applications of the electrodes to the cortical substance of the Cerebrum in rabbits, convert it into the likeness of that morbid growth, mainly consisting of blood-vessels and connective tissue, which is known as "fungus hæmatodes."

The first set of Dr. Ferrier's experiments had reference to the pathology of Epilepsy; and were made by applying the two electrodes to points of the Cerebral surface at some distance from each other. This application excited either partial or general convulsion; the most severe fits being induced when the electrodes were applied at the greatest distance.

"In all cases," says Dr. Ferrier, "whether the fits were partial or more general, the immediate antecedent was an excited hyperæmic condition of the cortical matter of the hemispheres. The irritation was entirely confined to the surface of the hemispheres; the electrodes being simply applied, without causing mechanical or deep-seated lesion." And not only was there in every case a distinct interval between the application of the electrodes and the first convulsive movement; but there was occasionally "a distinct interval of time *after the withdrawal of the stimulation*, before the condition of the grey matter had reached the pitch of tension requisite for an explosive discharge. This of itself is sufficient to show that the effects were not due to conducted currents or direct stimulation of the motor nerves of the muscles, but to an abnormal excitability or irritability of parts, whose function, it might be inferred, was to initiate those changes which would result in normal contraction of the muscles affected."—(*Op. cit.*, p. 39.)

Nothing, it must be obvious, could more strikingly illustrate the principle, that while the ordinary Circulation suffices to keep up the nervous tension of the ganglionic centres to the point required for motorial discharge by automatic or volitional "closure of the circuit," the higher state of tension induced by *hyperæmia* is itself sufficient to produce *spontaneous* motorial discharges (§§ 308–311).

It seems to the Writer that the above principles may be fairly extended to *all* the cases in which local faradization produced special motor results; for although the *immediateness* of these results in the state of normal excitability might seem to indicate they proceeded from *direct* stimulation of the ganglionic nerve-substance itself, yet when the excitability is depressed by exhaustion, a distinct interval is perceptible. And that the "expressive" movements excited by the localized faradization really depend (like the epileptic convulsion) upon the hyperæmic state which it induces, seems to be further indicated by their frequent persistence after the discontinuance of the stimulation;—a dog, for example, continuing to hold up his head and wag his tail, or to utter cries as of pain (though completely stupefied by chloroform), after the withdrawal of the electrodes from the respective centres of these actions. For this could scarcely be the case, if the stimulus acted *directly* on the nerve-substance; whilst it is quite consistent with the fact that the hyperæmic state does not immediately subside. And further, it is a consideration of no small importance, that no motor effect is produced by the faradization of the fibres which make-up the medullary substance of the Cerebrum, which might be expected to respond to this stimulus if it acted directly upon the elements of the nerve-tissue, as in the galvanization of an ordinary nerve.

The general characteristic of the Movements called forth by the local stimulation of the cortical substance of the Cerebrum, is that they are such as involve the co-ordination of several distinct muscular actions; and resemble those which, in an animal in possession of its senses, we should regard as expressive of Ideas and Emotions.

Thus, in a Cat, the application of the electrodes at point 2 (Fig. 14) caused "elevation of the shoulder and adduction of the limb, exactly as when a cat strikes a ball with its paw;" at point 4, "immediate corrugation of the left eye-brow, and drawing downwards and inwards of the left ear;" at point 5, "the

Fig. 14.

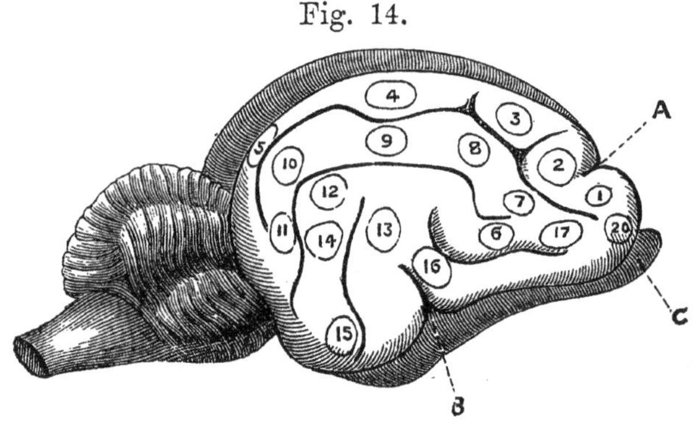

SIDE VIEW OF BRAIN OF CAT:—A, crucial sulcus dividing anterior convolutions; B, fissure of Sylvius; C, olfactory bulb.

animal exhibits signs of pain, screams and kicks with both hind legs, especially the left, at the same time turning its head round and looking behind in an astonished manner;" at point 6, "clutching movement of the left paw, with protrusion of the claws;" at point 13, "twitching backwards of the left ear, and rotation of the head to the left and slightly upwards, as if the animal were listening;" at point 17, "restlessness, opening of the mouth, and long-continued cries as if of rage or pain;" at point 18 (on the under side of the hemisphere, not shown in the figure), "the animal suddenly starts up, throws back its head, opens its eyes widely, lashes its tail, pants, screams and spits as if in furious rage;" and at point 20, "sudden contraction of the muscles of the front of the chest and neck, and of the depressors of the lower jaw, with panting movement."—Similar results were so constantly obtained, with variations obviously depending upon the degree of excitability and the strength of the stimulus, that the localization of the centres of these and other actions was placed beyond doubt; the movements of the paws being centralized in the region between points 1, 2, and 6; those of the eyelids and face between 7 and 8; the lateral movements of the head and ear in the region of points 9 to 14; and the movements of the mouth, tongue, and jaws, with certain associated movements of the neck, being localized in the convolutions bordering on the fissure of Sylvius (B), which marks the division between the anterior and middle lobes of the Cerebrum,—the centre for opening the mouth being in front of the under part of the fissure, while that which acts in closure of the jaws is more *in* the fissure.

A similar series of experiments on Dogs gave results that closely accorded with the foregoing; allowance being made for the somewhat different disposition of the convolutions, in accordance with the different habits of the animals, showing itself in the higher development of the centres for the paw in cats, and for the tail in dogs.

Thus when the electrodes were applied at point 9 (Fig. 15), "the tail was moved from side to side, and ultimately became rigidly erect;" within the circle 10, the application "elicited only cries, as if of pain;" at point 14 a continued application gave rise to the following remarkable series of actions:—"It began with wagging of the tail and spasmodic twitching of the left ear. After the cessation

Fig. 15.

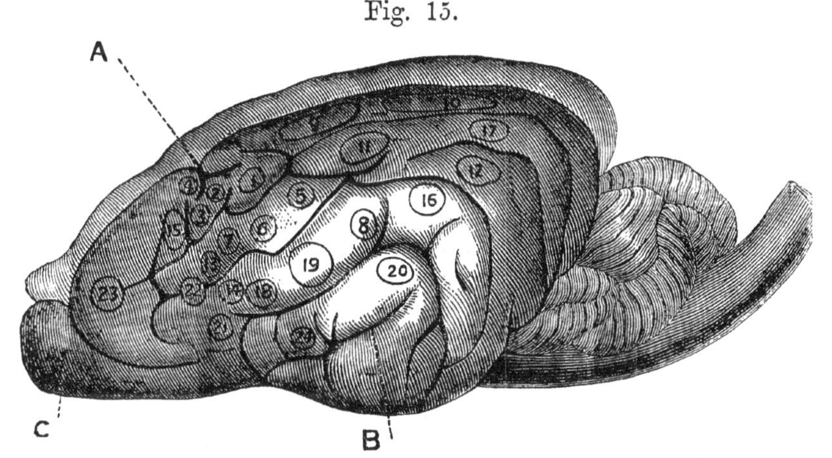

SIDE VIEW OF BRAIN OF DOG:—A, crucial sulcus; B, fissure of Sylvius; C, olfactory bulb.

of the more violent spasms, the animal held up its head, opened its eyes wide with the most animated expression, and wagged its tail in a fawning manner. The change was so striking, that I and those about me at first thought that the animal had completely recovered from its stupor. But notwithstanding all attempts to call its attention by patting it and addressing it in soothing terms, it looked steadfastly in the distance with the same expression, and continued to wag its tail for a minute or two, after which it suddenly relapsed into its previous state of narcotic stupor." The application of the electrodes to point 21 produced "drawing back of the head and opening of the mouth, with a feeble attempt at a cry or growl (the animal very much exhausted). Repeated applications of the electrodes to this point and its neighbourhood caused whining and growling noises," like those which a dog makes in its sleep, and which are supposed to indicate that it is dreaming.

Similar experiments having been made upon Rabbits, the results were again as accordant as it would be fair to expect; especially considering the difficulty in exactly localizing the different centres, which arises from the absence of the landmarks afforded by the convolutions.

Fig. 16.

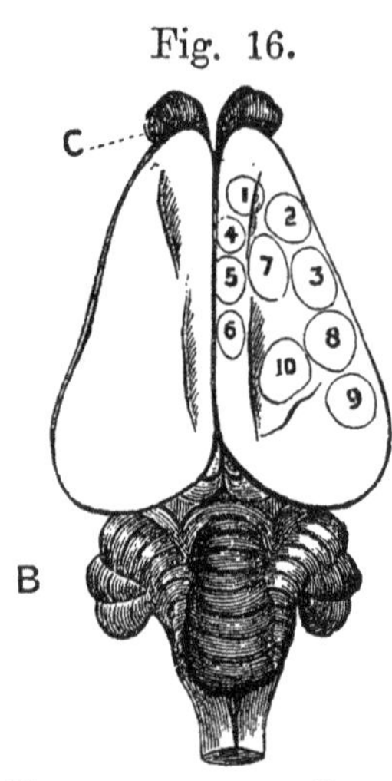

UPPER SURFACE OF BRAIN OF RABBIT :—A, Cerebrum ; B, Cerebellum ; C, Olfactory bulb.

It is curious that in this animal the centres of the mouth-movements seem to be the most highly developed; when these (2, 7, Fig. 16) are faradized, "there are munching movements of the upper lip, and grinding of the jaws, as if the animal were eating vigorously."

Dr. Ferrier has since made a series of experiments on the Monkey; of which the details are as yet unpublished, but which seem to be yet more remarkable than the preceding, in the far greater variety of the movements called forth from different centres, and their more distinctly expressive character. These results will also be of peculiar interest, on account of the close conformity which the simple arrangement of the convolutions in the Monkey bears to their more complex disposition in the Human Cerebrum (§ 106). They correspond with those of the previous experiments in this important particular—that those centres of movement which may be regarded as giving expression to attributes that Man shares with the higher Mammalia, are all located in the *anterior* lobes and the *anterior portion* of the *middle* lobes, the part of Man's Cerebrum which corresponds with the entire Cerebrum of the lower Mammalia (§ 106). In the Cat and the Dog, which have the middle lobes fully developed, stimulation of their posterior portion produces no respondent movement. And not only is this the case in the Monkey also; but *the whole of the posterior lobe is similarly irresponsive*, as is also that *front* portion of the *anterior* lobes, which, in all the higher Mammalia, as in Man, is produced by that forward as well as lateral development, which markedly distinguishes it from the corresponding part of the Cerebrum of the Rabbit. What may be the special functions of these parts, we can scarcely do more than guess at; but the negative fact just stated may be considered as a decided confirmation of the conclusion arrived at by the Writer twenty-seven years ago, on the basis of Comparative Anatomy and Embryology *,—that the *posterior* lobes of the Cerebrum are the

* See his Review of "Noble on the Brain and its Physiology," in the "British and Foreign Medical Review," for October, 1846.

instruments, *not* (as maintained by Phrenologists) of those passions and propensities which Man shares with the lower Animals, but of attributes peculiar to Man, which we fairly may suppose to consist in such Mental operations of a purely Intellectual character as do not express themselves in bodily action.

Before considering the further conclusions to be drawn from the results of Dr. Ferrier's experiments on the Cerebrum, it will be well to take into account those which he obtained from the application of the like stimulation to other ganglionic centres forming part of the Brain. When either of the *Corpora Striata* (§ 89), was thus excited to activity, an immediate and rigid *pleurosthotonos*, or bending of the body to one side, was excited in the opposite half of the body; the head being made to approximate the tail, the muscles of the face and neck being thrown into rigid tonic spasm, and the fore and hind limbs fixed and rigidly flexed. Apparently every muscle or group of muscles represented in the convolutions, along with the lateral muscles of the body, were stimulated to contraction from the corpus striatum; the predominance of the flexors over the extensors, however, being very marked. Similar excitation of the *Thalami Optici*, on the other hand, gave no motor result whatever; from which it may be concluded that they have no direct connection with movement. That the irritation did not call forth cries or other signs of pain, might be supposed equally conclusive against the idea that these ganglionic centres are instrumentally connected with sensation; but when it is borne in mind, not only that the animal was under the influence of chloroform, but that the connection of the irritated Thalamus with the Cerebral centres of the movements which express pain had been destroyed in order to expose this ganglion, the absence of any such expression seems adequately accounted for.—Experiments on the *Corpora Quadrigemina* (or Optic Ganglia) were chiefly made on Rabbits, in which these centres are relatively very large and are easily exposed (§ 85). The application of the electrodes to the anterior tubercles immediately calls forth a violent *opisthotonos*, or backward flexure of the body; so that, if the animal be not tied down, it executes a backward summersault which throws it off the table. The jaws are always violently clenched, and the pupils are dilated. These results do not militate against the idea of the connection of these ganglionic centres with the sense of Vision, which seems to be well established by other evidences; but they show that they are also motor centres, especially for the extensor muscles. Stimulation of the posterior tubercles occasioned noises of various kinds.

Dr. Ferrier's experiments on the *Cerebellum* have led him to the unexpected conclusion that it is the ganglionic centre of the motor nerves of the Eye; every kind of movement of the eye-balls,—even rotation on their antero-posterior axes,—being capable of excitation by stimulating some particular portion of the organ. The localization of the centres of combined movements of the two eyeballs in particular lobules of the Cerebellum in the Rabbit was extremely curious.

Thus, when the electrodes were applied to the median lobe at point 1 (Fig. 17), "the right eye moved outwards, and the left inwards, in a horizontal plane;" at points 2 and 3, the "right eye moved inwards, and the left outwards, on the same horizontal plane." Thus it appears that the middle lobe regulates those horizontal movements of the eyes which are *harmonious* but not *symmetrical;* and that the upper part of the median lobe, and its middle and lower parts, are in functional antagonism. When the electrodes were applied to point 4 on the *left* lateral lobe, "the right eye moved downwards and outwards, the left eye upwards and inwards;" and when the corresponding point of the *right* lateral lobe was stimulated, "the right eye moved upwards and inwards, and the left eye downwards and outwards;" while conjoint irritation of both lateral points neutralizes both effects. When the middle division of the left lateral lobe was irritated at point 5, a downward movement of the right eye, and an upward movement of the left eye, were combined with a rotation of each globe on its antero-posterior axis, the left in the direction of the hands of a clock, the right in the contrary direction. But when the irritation was applied at point 6 to the lowest division of the left lateral lobe, the two eyes rotated on their antero-posterior axes in the same direction, contrary to that of the hands of the clock, so that their vertical meridians retained their parallelism.—This last action is what takes place automatically when we fix our gaze at any object, and incline our head to the right side; the rotation of the eyeballs in the opposite direction serving to keep the image of the object on the same spot of the retina, just as do the automatic movements of the eyeballs in the vertical or horizontal plane, when the head is moved upwards and downwards, or from side to side (§ 21).

Fig. 17.

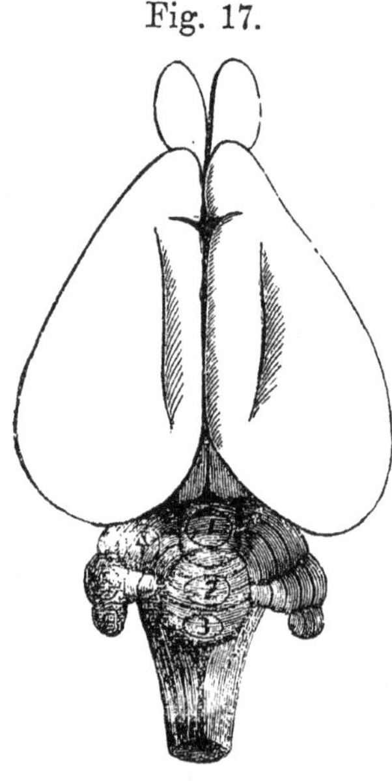

Upper surface of Brain of Rabbit.

These results throw great light upon the obscurity which previously enveloped the precise function of the Cerebellum. That it was in some way concerned in the regulation and co-ordination of the Muscular movements, especially those concerned in the maintenance of the equilibrium of the body, has long been a general opinion among Physiologists, based in part on the results of experiments, and in part upon Pathological observation; the doctrine of the Phrenologists, who regarded it as the organ of the generative instinct, having been long abandoned as untenable. But of the manner in which this power was exerted, nothing could be said to be precisely known. Now there can be no question as to the intimate relation between the guiding sensations we derive from Vision, and the co-ordination of our ordinary movements of Locomotion (§ 192). In the affection termed "nystagmus," which consists in a restless motion of the eye-balls from side to side, there is a difficulty in maintaining the equilibrium; and in the state called "locomotor ataxy," in which disease of the posterior columns cuts off the Cerebellum from its normal relation with the Spinal cord, it is impossible for the patient to maintain his equilibrium with his eyes shut (§ 80). So in the giddiness which most persons experience when they have rapidly turned round-and-round several times, it can scarcely be doubted that part at least of the result is occasioned by confusion of those visible perceptions, which would come through the oculo-motorial centres.—That this regulation of the movements of the Eyes, and the harmonization with them of the general movements of the body, constitute the entire function of the Cerebellum, it would be premature yet to assert; but Dr. Ferrier's experiments seem clearly to establish the first of these, and strongly to indicate the second, as essential parts of its action.

We now return to the inquiry as to the *import* of the experimental results previously detailed, in regard to the localization of Cerebral action in the production of Movement, and its relation to Mental states.

In the *first* place, they unmistakably prove the correctness of the doctrine, that the Cerebrum, like the nerve-centres on which it is superposed, has a *reflex action* of its own; which manifests itself in the production of co-ordinated movements, such as, in the normal condition of the animal, would be the expressions of Ideas and Emotions called forth by Sensations. The Cortical ganglion is ordinarily excited to activity by the nerve-force transmitted upwards along the ascending fibres from the Sensorium; this calls forth respondent physical changes in its substance, which changes excite the states of consciousness that we designate as Ideas and Emotions; and respondent Movements are involuntarily called forth, which we regard as expressions of those states. The same movements are called forth (as in other instances) by stimulation applied to their motor centres; which are now proved to be definitely localized in the Cerebral convolutions.

But, *secondly*, it seems equally clear that these movements are called forth, not by the Mental states themselves, but by the Cerebral changes which are their *physical antecedents*. For we can scarcely believe that Ideas and Emotions can be called up by faradization of the cortical substance, in animals completely stupefied by chloroform. And if we attribute any of those "expressive" actions which are called forth by such localized stimulation, to the states of consciousness they would ordinarily represent,* we cannot refuse the like character to the Epileptic convulsion called forth by the more general stimulation; a supposition at once disproved by the fact, that in the typical forms of Epilepsy, convulsive movements, such as have now been traced to "discharging lesions" of the Cerebrum,† take place *without any consciousness whatever*. Viewed in this aspect, Dr. Ferrier's results obviously afford additional support to the doctrine of "Unconscious Cerebration;" by showing that important Cerebral modifications of which only the *results* make themselves known, may take place outside the "sphere of consciousness" (§§ 416–418).

In the *third* place, we seem able to draw from these experimental results a more definite *rationale* than we previously possessed, as to the automatic performance in Man, of movements which originally proceeded from intentional direction. For it is clear that in Dogs, Cats, Rabbits, &c., the co-ordinated actions which result from localized stimulation of the Cerebral convolutions,—expressing, by the Nervous mechanism proper to each species, the Mental states naturally called up by their sensational experiences,—are as truly the "reflex actions of the Cerebrum," as the simpler forms of movement are of the Axial Cord. Now the Nervous mechanism of Man, as has been pointed out over and over again in the preceding pages, *forms itself* in accordance with the modes in which it is habitually called into action; and thus, it may well be believed, any special modes of co-ordinated movement to which an individual has been trained, or has trained himself, come to be so completely the reflex actions of particular

* Dr. Ferrier was himself so much impressed in one case by the *intelligent* character of the succession of actions thus called-forth, as to speak of it as "evidently an acted dream." But if *this* was, then *every other* must be regarded in the same light; and the Writer fails to see in what the evidence of consciousness consists. It seems to him that it might just as well be said that the headless body of a Frog is animated by a directing Will, when one leg wipes off an irritant applied to the other (§ 67).

† This view of the origin of those forms of Epilepsy which commence with convulsive spasm of the muscles ordinarily put in action voluntarily, as distinguished from those which primarily affect the muscles of Respiration whose centre of action is the Medulla Oblongata, is due to the clinical sagacity of Dr. Hughlings Jackson.

centres of his Cerebrum, that, if we could stimulate those centres by Electricity, respondent movements of the kind acquired by such special training would be the result. And since we now seem justified in asserting that such movements may be *executed* unconsciously, we may further regard it as at any rate conceivable that they may be *excited* unconsciously, even though such excitement comes through one of the organs of Special Sense.

The following statement recently made to the Writer by a gentleman of high intelligence, the Editor of a most important Provincial Newspaper, would be almost incredible, if cases somewhat similar were not already familiar to us :— "I was formerly," he said, "a Reporter in the House of Commons ; and it several times happened to me, that having fallen asleep from sheer fatigue towards the end of a debate, I found, on awaking after a short interval of entire unconsciousness, that I had continued to note down correctly the speaker's words. —I believe," he added, "that this is not an uncommon experience among Parliamentary Reporters." (Compare §§ 71, 194.)—The reading aloud with correct emphasis and intonation, or the performance of a piece of music, or (as in the case of Albert Smith) the recitation of a frequently-repeated composition, whilst the conscious mind is *entirely engrossed* in its own thoughts and feelings, may thus be accounted for without the supposition that the Mind is actively engaged in two different operations at the same moment ; which would seem tantamount to saying that there are two Egos in the same organism.

But, *fourthly*, these results entirely harmonize with the view formerly expressed (§§ 23, 89), that the Cerebrum does not act immediately on the motor nerves, but that it plays downwards on the motor centres contained within the Axial Cord ; from which, and not from the Cerebral convolutions, the motor nerves take their real departure. For although either mechanical or electric stimulation of a motor nerve in any part of its course, of the motor columns of the Spinal Cord, or of the fibrous strands which constitute the upward prolongations of these into the *corpora striata*, calls forth muscular contractions, no such contractions were excited by the faradization of the fibres of which the medullary substance of the Cerebrum is composed, any more than by mechanical stimulation. And the fact that all the muscles concerned in the ordinary movements of the body can be thrown into contraction by stimulation of these lower centres,—the extensors through the *corpora quadrigemina*, while the flexors predominated when the *corpora striata* were stimulated,—seems to show that the office of the Cerebrum is not immediately to evoke, but to co-ordinate and direct the muscular contractions excited through these antagonistic primary centres ; just as it controls the Respiratory movements whose centre is in the *medulla oblongata*.

In the *fifth* place, these experiments throw great light on the "crossed" action of the several ganglionic centres contained within the skull, which had previously been a matter of considerable obscurity; some phenomena of disease appearing to show that the motor centres of one side act on the nerves of the opposite side exclusively, whilst others seem to indicate that those of one side may affect the muscles on both sides. Anatomical investigation favoured the latter view, by showing that whilst some of the motor strands (*corpora pyramidalia*) which connect the Brain with the Spinal Cord, decussate, or cross to the opposite side, others pass continuously downwards without decussation. Now Dr. Ferrier found that the motor action of the *corpora striata* is strictly limited to the muscles of the opposite side of the body; being probably exerted solely through the decussating strands. On the other hand, the motor action of the *corpora quadrigemina* is not thus limited; the extensors of both sides being called into contraction by the application of the stimulus to either lateral half of the anterior pair; so that they would seem to act through both the decussating and the non-decussating strands. In his experiments on the *Cerebral Hemispheres*, again, Dr. Ferrier found the motor action to be generally limited to the opposite side of the body; though in some movements, particularly those of the mouth, it was obvious that muscles of both sides were put in action. Now it is well known that extensive destruction of the substance of either hemisphere, if resulting from the *gradual* action of disease, may occur without any obvious loss of voluntary movement; though sudden injuries of a certain severity occasion paralysis of the opposite half of the body, which, however, is usually incomplete and of transitory duration. And Dr. F. inclines to accept the conclusion drawn by Dr. Broadbent from clinical observation, that the movements which are most independent on the two sides, are those which are most completely paralyzed by injury to one side of the Cerebrum; whilst those in which the co-operation of the muscles on both sides is required, may be sustained by the action of either hemisphere. Not improbably the great transverse commissure (*corpus callosum*) here comes into action, enabling either hemisphere singly to do the work—to a certain extent—of both; while there seems some ground for the belief that the *left* hemisphere, which chiefly directs the movements of the *right* half of the body, is the "driving" side. For in all save "left-handed" persons, any movement which *may* be initiated by either limb, is almost sure to be initiated by the right: thus in beginning to walk, we almost invariably put the right foot foremost; and a person desired to hold up his hand, will as probably hold up his right hand.*

* In the well-known case of the murder of Mr. Blight by Patch, in which the

But, *sixthly*, we have to inquire how far these experimental results justify the belief that there is any such localization of strictly *mental* states, as there is of the centres of the expression of those states in movement. And as to this it must be confessed that we are still very much in the dark,—the only fact that seems to afford any clue to the solution of the mystery, being the apparent coincidence between the motor centre of the lips and tongue in the lower animals, and that region in the human Cerebrum of which disease is so often found to be associated with *Aphasia* (§ 355). This association, however, seems by no means so constant as to establish a *causative* relation between the Physical and the Psychical state; and a careful examination of the phenomena of Aphasia would probably lead to the conclusion, that several distinct forms of disorder have been grouped under one designation. The typical Aphasia consists (as stated in § 354), in the loss of the memory of words, or rather of the power of recalling them; the patient understanding what is said to him, but not being able to reply verbally, because he is unable to call to mind the words which would express his thoughts. Such patients are exactly in the condition of the "Biologized" subject (§ 462), who, being assured that he cannot recollect his own name, finds himself absolutely unable to do so. But in other instances, it would seem as if the defect were not so much in the want of the *memory* of words, as in the want of power to *express* them vocally; and this, not from paralysis of the nerves of speech, but from an interruption to the action of the Will on the motor centres (§ 312). And although there would seem strong ground for the belief that the memory of particular classes of ideas *may* be thus localized, yet it would be certainly premature to affirm that either the phenomena of disease, or the results of experiments, at present justify the belief that the region in question is the seat of the memory of words.—The analogy afforded by the specialization of *downward* (motor) action, would lead us to anticipate that a like contralization may exist for *upward* (sensory) action; and that particular parts of the convolutions may be the special centres of the classes of perceptional Ideas that are automatically called up by sense-impressions; and anatomical investigation, particularly in the lower animals—in which such ideas may be supposed to prevail almost to the exclusion of the intellectual ideas—may not improbably throw light on this relation. But in regard to those Mental processes which mainly consist in the selection, classification,

sagacity of Sir Astley Cooper enabled him to infer from an examination of the local circumstances, that the pistol must have been fired by a left-handed man, the prisoner, when called upon to plead and hold up his right hand, held up his left.

and comparison of distinct Ideas (§ 227), whether these processes be carried on volitionally or automatically, it still seems to the Writer just as absurd as it formerly did *, to suppose that there can be special "organs" for their performance, such as those named Comparison and Causality in the Phrenological system.

* British and Foreign Medical Review, October, 1846.

INDEX.

Printed in the United States
By Bookmasters